SEAPOWER AND SPACE

From the Dawn of the Missile Age to Net-Centric Warfare

NORMAN FRIEDMAN

NAVAL INSTITUTE PRESS
Annapolis, Maryland

© Norman Friedman 2000

All rights reserved. No part of this book may be reproduced without written permission of the publishers.

First published in Great Britain by Chatham Publishing, 61 Frith Street, London W1V 5TA

Published and distributed in the United States and Canada by the Naval Institute Press, Beach Hall, 291 Wood Road, Annapolis, Maryland 21402-5034

Library of Congress Catalog Card No. 99-69320

ISBN 1-55750-897-6

This edition is authorized for sale only in the United States, its territories and dependencies, and Canada.

Manufactured in Great Britain.

Contents

List of Plates		6
1.	Introduction	7
2.	Satellites and their Mechanics	13
3.	Getting into Space: Boosters	20
4.	Polaris and Precise Navigation	46
5.	Passing the Word: Reliable Communications	54
6.	Finding Targets: Reconnaissance	87
7.	A New Kind of Naval Warfare	129
8.	Dealing with an Emerging Soviet Threat	173
9.	Enter Tomahawk: OTH Targeting	209
10.	Defending the Fleet: The Outer Air Battle	230
11.	Copernicus	249
12.	The Global Positioning System	266
13.	The Navy and the Battle Ashore	282
14.	A New Kind of War?	301
Notes		318
Bibliography		367
Glossary		371
Index		373

List of Plates

Between pages 192 and 193

USS *Shiloh* firing a Tomahawk missile. (U.S. Department of Defense)
HMS *Marlborough*. (Crown Copyright)
HMS *Ark Royal*. (MoD (UK))
The first Delta II booster launches a GPS satelllite. (U.S. Air Force)
The GRAB and Transit 2A satellites. (U.S. Navy)
DSCS II satellite. (TRW Systems Group)
The U.S. Fleet Satellite. (U.S. Department of Defense)
NATO III satellite. (Aeronutronic Ford Corp.)
DSCS III satellite. (U.S. Air Force)
Hughes UHF Follow-On satellite. (Hughes Space and Communications Co.)
Skynet 4 satellite. (Matra Marconi)
Skynet 5 satellite. (Lockheed Martin)
The new Asia Cellular Satellite. (ACeS)
Matra Marconi concept for Skynet 5. (Matra Marconi)
Milstar satellite. (Lockheed Martin)
Helios imaging satellite. (Matra Marconi)
'Shaddock' cruise missile. (Author)
'Bear-D' (Tu-95Ts) radar aircraft. (U.S. Navy)
Russian synthetic-aperture radar satellite. (Author)
'Echo II' submarine with 'Punch Bowl' antenna. (MoD(UK))
'Oscar' class cruise missile submarine. (Crown Copyright)
Tu-22M 'Backfire' bombers. (Author)
Kh-15S missiles aboard a 'Backfire'. (Author)
U.S. Defense Support Program IR early warning satellite. (TRW)
GPS satellite.
The U.S. view of the new style of warfare. (TRW)
Over-the-horizon warfare. (Lockheed Martin)
Littoral warfare. (Raytheon)
Tomahawk launched by HMS *Splendid*. (MoD(UK))
Tomahawk approaching target. (General Dynamics)

ONE

Introduction

The advent of space systems has transformed naval warfare, at least for those navies with access to such resources. It has, for example, brought U.S. surface warships back to the category of capital ships from that of secondary fleet escort they had occupied since the middle of the Second World War. In a larger sense, access to space systems makes possible a new style of warfare, in which ships fire their weapons at distant targets on the basis of information taken from distant sensors. It is closely related to other changes which are sometimes called the 'Revolution in Military Affairs' or which conform to the U.S. Joint Chiefs of Staff Vision 2010. The very visible part of this transformation, for the U.S. Navy, has been the increasing importance of Tomahawk missiles fired by surface ships and by submarines.

The underpinnings of this transformation are, however, far less visible. Some of the key systems, developed during the Cold War, are rarely discussed, because they are usually seen more as sources of intelligence than as operational sensors. Their connection to the ongoing revolution in naval warfare is therefore largely unappreciated. That may have real consequences. Although the U.S. Navy is the largest user of space systems within the U.S. defence establishment, the U.S. Air Force is seeking control of U.S. military space activities. The Air Force is unlikely to share the Navy's views of appropriate priorities in space systems. Because few naval officers have been involved directly in the use of space systems, particularly the reconnaissance systems which have made so great a difference, they may not be inclined to insist on naval priorities. The issue is critical because the Cold War systems, on which the Navy was able to draw at very low cost, must now be replaced. In an era of very tight budgets, the Navy may well find it tempting to allow the Air Force to shoulder the budgetary burden – and to take control of the character of next-generation systems.

As a token of the transformation of U.S. naval warfare, in the autumn of 1998 two American surface warships in the Indian Ocean fired Tomahawk cruise missiles at targets in Afghanistan and in the Sudan, far too distant for them to detect using their own sensors. Later in 1998 other U.S. warships fired hundreds of Tomahawks at targets in Iraq, largely neutralising Iraqi air defences so that follow-on aircraft could attack in safety. These strikes typified a new kind of naval warfare – which was possible only thanks to the combination of missiles and the array of space assets the United States had developed over the past four decades. The targets were detected and evaluated by sensors and intelligence systems very remote from the shooters. Many of those sensors were space-based. Target and other data reached the shooters via space-based communications nets; no other kind of communications system had the requisite capacity. The attacks were planned on board the firing ships, based on data they obtained via space links, while the accuracy of those attacks was largely due to navigational systems which depend on space satellites. Moreover, the effects of the attacks were evaluated by space-based sensors.

War had been moving in this direction for a long time. For example, in 1986 U.S. carrier aircraft attacked Libya. Their targets were certainly detected by satellites, which could obtain details of the target areas without alerting the Libyans. However, communication capacity was limited. U.S. satellites kept taking pictures of the target areas, but there was no instant way to relay those pictures to the strike force. Clearly the targets might have changed between the time when the force went to sea (carrying satellite photos the carrier planners used for targeting) and when the strike aircraft arrived over Tripoli. Using carrier-based reconnaissance aircraft for updates would have ruined the surprise attack. The pilots made up some of the difference, applying their judgement to the somewhat outdated information with which they took off.

At that time the U.S. Navy was in the throes of a revolution which the space assets had made possible. For about a decade its watchword had been Over the Horizon Targeting (OTH-T) for anti-ship missiles, particularly Tomahawk (which was then a long-range anti-ship missile). Now there was an anti-air equivalent, the Outer Air Battle (OAB). Both were attempts to deal with Soviet attackers (ships and aircraft, respectively) before they could get within shooting range. To some extent the new American tactics mirrored a transformation in the Soviet Navy, which had been prompted by the emergence of long-range anti-ship missiles, delivered both by bombers and by submarines and surface ships, in the 1950s. By the 1970s Soviet anti-ship operations were increasingly dependent on space-based sensors.

Both the U.S. and the Soviet transformations might be considered parts of a longer-term shift towards information-oriented naval warfare, what is now called 'net-centric warfare'. This in turn depends heavily on the availability of space-based systems to detect and track targets, to provide tactical information to shooters, and even to make that information useable by providing precise navigation.[1] An information-centric approach to naval warfare shifts primary attention from the enemy's warships to the information system which makes them effective. Using long-range weapons, the Soviet Navy could disperse its ships very widely, concentrating their fire against major U.S. units. The U.S. Navy would have found it very difficult to deal with all of the dispersed units; the alternative was a direct attack against the Soviet information system which located the targets and which provided that information to the shooters. Advocates of a new kind of U.S. naval warfare often call for the use of dispersed platforms which would concentrate their fire at a distant target. In their case the advantage claimed is not so much invulnerability to counterattack as the ability to deal very quickly with widely separated targets. Again, the efficacy of the strategy depends on the information system which backs it up.

This book is the story of the linked revolution of long-range missiles and their space-based supporting systems, which transformed the U.S. and Soviet navies. Particularly for the United States, it was bound up with an earlier revolution, the advent of nuclear weapons, delivered first by bombers and then by missiles. Many of the space systems developed to support this other revolution fed into the later one, radically changing the capabilities of the U.S. fleet. The changes were obscured because so little seemed to happen to the ships themselves. Yet each ship now has new access to remote radars, or to new kinds of passive sensors, or to telescopes capable of seeing hundreds of miles. To make use of information from external sensors the ship has to know exactly where she is in relation to those sensors. Precision navigation – precise gridlock – becomes extremely important.

The external sensors can collect far more information than their shipboard predecessors; they cover a much larger area, throughout which the ship's long-range weapons can be effective. It takes powerful computers to make sense of what the external sensors can provide. The revolution, then, can be described as information-centred, its success depending on the way in which space systems and computers can be made to interact. The term 'net-centric warfare' is used by some in the U.S. Navy to emphasise the importance of the net carrying information between sensors (many of them space-based), decision-makers, and shooters. The idea that the new kind of warfare is information-centred has become particularly fashionable in a world of computers connected to the Internet. However, in the end what matters is how well the shooters can fire their weapons to destroy their targets.

Even without the connection to long-range missiles, satellites in space are natural complements to navies moving freely over the world's oceans. Only they can be in place (however fleetingly) wherever the fleet is – or is likely to go. As artificial astronomical bodies they have revolutionised navigation. For the first time they have given ships far out to sea reliable global communication. For the few navies which posses them, satellite sensors have drastically reduced the stealth which navies – and aircraft – used to enjoy virtually from the time they left port or flew beyond the radar horizon of the coast.

During the Cold War, it was widely imagined that war had been reduced to strategic nuclear conflict. Thus civilians understood military space operations usually to mean efforts to assure one side or the other that it could be sure of warning if the other struck first; or, if war came, that its own ballistic missiles could destroy the other side. Space reconnaissance seemed to mean mainly taking the photographs on which strategic targeting was based, or listening in on the enemy's most vital strategic communications. Yet in reality the main impact of nuclear weapons was to make a big strategic war extremely unlikely. Non-nuclear war was much likelier. The evolving space systems were probably most important for the way they affected naval operations. By the 1980s, the U.S. Navy, not the U.S. Air Force, was the single heaviest user of U.S. military space systems. They had already drastically changed the threat the U.S. Navy faced. Perhaps it was appropriate that one of the first major leaked satellite photographs was an image of the first true Soviet carrier, *Kuznetzov*, taking shape at Nikolaev in 1984 (another showed the prototype of the Tu-160 'Blackjack' strategic bomber).

Because the naval use of space is generally overshadowed in the public mind by other military functions, such as strategic reconnaissance ('spy satellites') and early warning against missile attack, U.S. – and Soviet – interest in space warfare during the 1970s and 1980s was largely misunderstood. It seemed obvious, for example, that Soviet anti-satellite weapons were intended to blind the U.S. leadership as a first stage in a nuclear first strike. Later the Soviets would use similar charges to try to stop a U.S. anti-satellite programme. However, the early-warning and communications spacecraft were in very high orbits, and therefore probably entirely safe from such weapons. But the naval targeting satellites, which operated in low orbits, were a very different proposition. During the Cold War they were vital to each side's plans to destroy the other's fleets, and they were therefore well worth destroying. With the Cold War over, they and other low-altitude satellites are still the key to the transformed U.S. Navy. Without them, the Tomahawks and other missiles which do the damage would be far less effective.

At first the exploitation of space was overwhelmingly for military ends. Now the balance has turned to commercial applications – mainly to communications and navigation. Within the past two or three years, the balance of investment has apparently shifted from military to civilian space. One major question for the future must be whether the U.S. Navy, which now uses most of the militarised space systems (the Russians have largely dropped out), will find that other navies can gain very similar advantages by exploiting the new commercial systems. For example, will the sanctuary of the open seas be lost? This sort of question can be approached only by examining – as this book does – just how the naval applications have worked, and just how they compare with emerging commercial operations. The conclusion seems to be that other navies can gain some, but by no means all, the advantages currently enjoyed by the U.S. Navy; that some key applications are unlikely to have civilian counterparts.

It is also worth reflecting that, at least for the United States, the kind of mobility associated with space systems is increasingly important to all the services. The fixed foreign base structure is both increasingly vulnerable (to political veto) and increasingly irrelevant (since crises are likely to erupt in unexpected places); it is also far too expensive to maintain. As a consequence, all the services have adopted more or less expeditionary outlooks. None of them can expect to rely on much infrastructure in or near the operational area. As a consequence, all of them will increasingly demand satellite services. Some of those services cannot easily be expanded, so competition may become intense. The story of the UHF satellites, which are now heavily used by the Army and the Air Force in addition to the Navy, is a case in point. It may become increasingly difficult for the Navy to demand the sort of systems it needs to operate effectively far from home. For example, commercial systems like Iridium are designed mainly to serve land areas (they recharge their batteries over the 'dead' sea areas). That is quite sufficient for the Army, but hardly so for the Navy. What happens if Iridium becomes a *de facto* Defense Department standard?

Given the increasing importance of military space for the United States, and the very high cost of space systems, there is a real possibility that space applications will be entirely centralised under a Space Force allied to (or part of) the U.S. Air Force. There is already a unified Space Command to handle operational matters. Unfortunately each service uses space in a very distinctive way. To unify the use of space under one service would almost inevitably be to deny the others what they need. Given this divergence of requirements, unification would probably have much the devastating effect that unification of air services had on the British after 1918, or that unification of army and air force services had on the U.S. Army's close air support capabilities after 1947. Conversely, the success of U.S. Navy and Marine aviation would suggest a strong argument against such unification. The problem, it should be stressed, is not that air forces or space forces are *per se* malignant, but that the different service users of space systems really do have distinctive requirements. Sharing (under joint systems) is not the same as asking a single super-agency for services.

Acknowledgements

This book could not have been written without considerable assistance from many people. I am particularly grateful to the public relations staff of the U.S. Naval Research Laboratory, to the staff of the U.S. Navy's Operational Archives (particularly to John Walker), and to Patrick Reidy of the U.S. Navy's Space Command. David Isby and Steve Zaloga were extremely helpful. Others helped catch errors in my accounts of current communications satellite systems. However, I am responsible for the text and for any errors it may contain. I would particularly emphasise that all accounts of reconnaissance satellites in this book were taken from unclassified and unofficial sources. I would like to thank my wife, Rhea, without whose love and understanding – and support – this project could not have been completed.

TWO

Satellites and their Mechanics

A satellite stays in orbit above the Earth due to a combination of its speed (about 18,000 miles per hour at low altitude) and the Earth's gravity. Like any other object, it is attracted towards the Earth by gravity. It is also moving forward at high speed – so fast that instead of hitting the Earth, it falls *around* it – in orbit. If it moves fast enough, of course, an object escapes the Earth altogether. If it is not fast enough, it falls to Earth. It has been a long-standing joke in the missile industry that satellites and ballistic missile warheads both follow orbits, but that only the latter intersect the surface of the Earth.

Placing a satellite in orbit, then, requires both a vertical boost (which determines its altitude) and a horizontal one to provide sufficient speed along the satellite's orbit. Because the Earth rotates, a satellite fired due east enjoys a significant boost from the Earth's motion. That is one advantage enjoyed by Cape Canaveral, in Florida: its natural launch direction is to the east, over water. On the other hand, the U.S. West Coast launch site, Vandenberg, must fire its weapons either west (against the direction of the Earth's spin), or more or less due south (and slightly to the east).

As a satellite orbits, the Earth rotates beneath it. From the point of view of naval forces on the surface, what matters is the satellite's path over the surface, its ground track, which combines the motions of satellite and Earth to form a curving path. As the Earth rotates, the path moves to the west from orbit to orbit. How much depends on the satellite's period. The Earth turns 15 degrees of longitude each hour. Thus a satellite with a 120-minute period traces out ground paths 30 degrees apart (the path also moves slightly due to the irregular shape of the Earth) at the equator. Apart from this slight change, the satellite would make twelve complete orbits each day, returning to its original orbit after 24 hours.

The laws of physics dictate that all satellite orbits be elliptical (a circle is a special type of ellipse), and that their planes always pass through the centre of the Earth (because the attraction of gravity is directed to the centre of the Earth). For example, a satellite cannot orbit the Earth at a fixed latitude (other than at the equator), because the force of gravity, directed toward the centre of the Earth, pulls the satellite south or north. Alternatively, one might say that such an orbit is prohibited because its plane would not pass through the centre of the Earth. Due to the north-south pull of gravity, a satellite launched due east from a point at a given latitude orbits the Earth with an inclination (the angle between its orbit and the north-south axis of the Earth) equal to its latitude. For example, because Cape Canaveral is at 28.5 degrees North, satellites launched from it have inclinations of at least 28.5 degrees. If the satellite is fired at an angle to due east, it can take up a higher inclination. However, in such cases more fuel is needed, because the satellite does not take full advantage of the Earth's spin. The main Soviet satellite launch site was at Tyuratam (46.5 degrees North). However, China lies due east of Tyuratam, so to avoid dropping boosters onto Chinese territory the Soviets often had to launch on a more northerly course, with a minimum inclination of 51 degrees (they also launched south, over Afghanistan).[1]

As an example of the cost of using anything other than an easterly orbit, the U.S. Space Shuttle was designed to place 30 tonnes in a low Earth orbit (altitude about 300km) using an easterly launch direction. However, the same Shuttle could place only 19 tonnes in a polar orbit (90 degrees inclination), and only 14 tonnes in a somewhat westerly orbit (98 degrees inclination).[2]

In each case, satellites can be inserted into other orbits, such as equatorial ones (inclination 0 degrees), but only at a cost. They are initially fired into minimum-inclination parking orbits. When they cross the equator, they are boosted into the desired orbits. That takes an extra motor, and less satellite can be placed in the desired orbit. This consideration explains why the French placed their national launch site (now used by the European Space Agency) at Kourou, in French Guiana (latitude about 5 degrees North). It also explains the attraction of sea launch platforms (one of which is currently in operation, launching Russian-made rockets), which can be placed at the equator – and of schemes in which a large aeroplane substitutes for the initial rocket stage.[3] In each case, the satellite can be launched in a much more advantageous way. Note that because ballistic missiles follow portions of satellite-like orbits, they too can follow only certain paths.

The natural laws governing satellite motion limit the extent to which a satellite can manoeuvre in space; whatever thrust may be applied to it, the satellite is also subject to the very strong attraction of the Earth. Because a very short part of any orbit approximates a straight line, over short distances a satellite in orbit can steer directly towards a given point in space. However, once the manoeuvre has begun, the satellite is simply in a new orbit, subject to the same laws as before. It cannot, for example, afford to lose so much forward velocity that it falls out of space and back into the atmosphere. A satellite attempting to change the plane of its orbit (see below) can do so only when its own orbit intersects the desired plane. If the two orbits do not intersect, the satellite must be thrusted into an intermediate (transfer) orbit which intersects both, or into a succession of such orbits.[4] Such manoeuvres require both calculation and timing. Thus it is not a trivial matter, for example, to bring a reconnaissance satellite into position to observe a given point on the Earth at a desired time – or, for that matter, to intercept an enemy satellite.

Satellite orbits are generally described in terms of perigee (closest point of approach to the Earth) and apogee (farthest distance from the Earth); for a circular orbit, the two are the same. The laws of physics (as expressed in Kepler's Second Law) dictate that, as the satellite moves, an imaginary line connecting it to the centre of the Earth sweeps out equal areas in equal times. The further the satellite is from the Earth, the larger the area associated with any particular segment of its orbit. Thus a satellite in a highly elliptical orbit moves much more rapidly when it is closest to Earth (near perigee) than when it is furthest away (near apogee). That is, near apogee the satellite may seem to hang in the sky. One consequence is that a reconnaissance satellite with a highly elliptical orbit (to bring it closest to its target) cannot spend very much time over its target area.

The higher the satellite, the slower it runs in its orbit, and the longer it takes to circle the Earth. According to Kepler's Third Law, the square of the period is proportional to the cube of the long axis of the orbit (the semi-major axis).[5] The time to make the lowest practicable orbit (an altitude of about 240km) is approximately 89 minutes. On the other hand, a satellite at an average altitude of 19,360 nautical miles (about 23,000 land miles) takes 24 hours to complete an orbit. In a circular orbit along the equator, then, the Earth rotates under the satellite just as quickly as the satellite moves over it. The satellite remains apparently fixed over one point on the Earth. Such an orbit is called *geostationary* or *geosynchronous*.

Note that only satellites orbiting over the equator can be geostationary; they seem more or less to hang in space over points on the equator. There is no way to make a satellite hang in space above any other point on the Earth. To achieve the equivalent of this means maintaining a constellation of satellites, so that one is always near the desired point. Thus, maintaining sustained coverage, for example for radar warning, is inevitably expensive. The lower the satellites, the less time they spend over any one area, and thus the more such satellites are needed for quasi-continuous coverage. There is no way to launch a single satellite so that, for example, it is always over a battle group. Conversely, a constellation that always provides coverage over one battle group will provide the same coverage over many other battle groups (limited, of course, by the north-south area of the Earth the satellites overfly).

Because of the gravitational effects of the Sun and the Moon, no satellite is truly geostationary. Uncorrected, the inclination of the satellite would drift between the desired 0 and 15 degrees in a 55-year cycle. Because orbital inclination is never quite zero, the satellite traces out a figure-8 pattern over the ground. This motion can have real significance for ships using such satellites for communication. At the same time, the irregularly distributed mass of the Earth tries to pull a geosynchronous satellite into one of two 'geosynchronous graveyards', at 75 degrees East and 105 degrees West respectively. Thus the satellite needs onboard propulsion to maintain its position and attitude. Otherwise it drifts.

Satellites are generally not fired directly into a geosynchronous orbit. Instead, they are placed in highly elliptical orbits whose apogees reach geosynchronous altitude, called geosynchronous transfer orbits (GTOs). When the satellite reaches apogee, a relatively small additional boost in the right direction can place it in geosynchronous orbit. A variation on this theme is to launch the satellite into a low circular orbit, then fire a booster to kick it into a transfer orbit (the perigee of which is the altitude of the low orbit). That is the technique used, for example, when the U.S. Space Shuttle launches geosynchronous satellites.[6]

A geosynchronous satellite's footprint is a circle on the Earth's surface. It cannot cover the whole hemisphere, because the cone connecting the satellite with the Earth is only 162 degrees wide. That is, at a latitude of 81 degrees North or South of the equator a line from satellite to Earth just grazes the surface. Nothing north of 81 degrees North (or south of 81 degrees South) is visible to the satellite. Effective radio operation requires that the satellite be 5 degrees above the horizon, so the geostationary satellite is effective only between latitudes 76 degrees North and South. The limits are often set somewhat closer to the equator. For the high Arctic, for example, something else is needed.

The satellite's orbit lies in a plane, which cuts through the Earth (and through the centre of the Earth). Apart from the effects of other heavenly bodies, the orientation of the plane is fixed in space as the Earth rotates beneath it and moves around the Sun. This is a striking consequence of the nature of inertia (in this case, the angular inertia associated with the satellite's motion in orbit). The orbit itself can move around on the plane in which it is set. For example, imagine a satellite in a highly eccentric orbit, which is set in a plane cutting the Earth at an angle. At one time the satellite might reach its farthest point (apogee) above the Earth, to the north. As its orbit itself rotated around the plane, the farthest point might be near the equator, and eventually it might point below the Earth, to the south, before continuing to rotate back up to the equator and then to the north.

Most satellite orbits are inclined to the Earth's north-south axis. The portion of the Earth the satellite's orbits cover depends on its inclination, the angle to true north at which it orbits. For example, a satellite with an inclination of 45 degrees and a circular orbit will cover the Earth between 45 degrees North and 45 degrees South latitude. The steeper the inclination, the further north coverage will extend. From the point of view of the ground, a satellite in circular orbit spends relatively little of its time near the northern or southern end of its orbit. To get more time there, a satellite must follow a highly elliptical orbit, called a 'Molniya' orbit after the Soviet communications spacecraft which first used it. These orbits are inclined at about 63 degrees, with their apogees above the northern hemisphere.

Any 24-hour orbit will pass over the same ground path again and again, because the Earth makes a complete rotation between revisits by the satellite. For example, a satellite in a circular 24-hour north-south (0 inclination) orbit will make a figure-eight path over the Earth, always crossing the same spot at the equator. Such a satellite will fly over every point on the Earth.

Because the Earth is not spherical, the plane in which a satellite orbits gradually rotates (precesses) in a direction opposite to that in which the satellite orbits (*ie*, to the west if a satellite orbits east). The amount of rotation depends on orbital height (it is greatest for a low-Earth satellite) and inclination (maximum for an equatorial satellite, zero for a polar orbit). This phenomenon is used to create a 'sun-synchronous' orbit for a photo-reconnaissance satellite. The object is to maintain identical lighting conditions from day to day as the satellite passes over points on the Earth. It turns out that a low orbit inclined at 98 degrees precesses just enough to compensate for the effect of the Earth's orbit around the Sun. On the other hand, there is no way to make a satellite's orbit precess quickly enough to follow the Earth as it turns; no satellite will follow a fixed line of longitude as it orbits.

The irregular shape of the Earth also causes the satellite's orbital shape to rotate along the orbit. If the orbit is elliptical the apogee and perigee will gradually move around the orbit. Rotation is zero at an inclination of 63.4 degrees – which is why that inclination was selected for 'Molniya' type orbits, the object of which is to keep the apogee fixed high above the northern hemisphere.

In general, to stay within a desired orbit, a satellite needs some form of intermittent propulsion. Similarly, to change orbit (for example, to get a better look at some target of interest on the ground), the satellite must use up fuel. In either case, fuel capacity may largely determine the overall lifetime of the satellite.[7] There is also air resistance. A few hundred miles above the Earth, the atmosphere is very thin, but at its very high speed the satellite still feels it. Eventually it loses enough speed to fall out of orbit altogether; it is said to have 'decayed'. Decay lifetime depends on the height of the satellite. At about 200km, it may be only days to months; at twice that altitude, it is several years (at 420km, Skylab lasted six years). A satellite down to a 110km circular orbit is likely to hit the Earth within the next half-revolution (and it typically breaks up at about 80km). At much greater altitudes, air resistance is no longer measurable.

It often matters just how precisely the satellite's orbit can be measured. Using a satellite to locate objects on the Earth entails an important complication. The satellite locates objects in relation to itself, but its data must be used by those on the ground. Its own location must, therefore, be known precisely. That amounts to knowing the time elapsed since the satellite passed over a reference point while following a precisely-known orbit. Thus, for any role requiring precision, the satellite must be placed in a very precisely-planned orbit, and maintained there by periodically firing correction thrusters. In addition the angle between the satellite's antennas and the Earth-its attitude – is also important. Land reconnaissance satellites do not suffer these problems, whether they use cameras or imaging radars. In either case, there are landmarks against which the satellite's data can be registered. There are even known features on the ground against which photographs can be scaled. The sea, however, is essentially trackless.

To do anything useful, the satellite generally needs some onboard source of electrical power (the only exception is a passive reflector, an idea tried early in the Space Age but long discarded). Most satellites use solar panels, which in theory last indefinitely. However, to get maximum performance they must maintain their orientation towards the sun; that generally costs fuel. For a low-flying satellite, big panels can cause appreciable drag due to atmospheric resistance; they cannot be used at altitudes much below 175 miles. Satellites flying lower can use solar panels shaped to their bodies, but these are less efficient. Furthermore, because a satellite spends much of its time in the shadow of the Earth, where solar panels are useless, it needs storage batteries to maintain power continuously. Nuclear batteries or reactors are an alternative. Nuclear batteries powered some versions of the U.S. Navy's Transit, and the Soviets used nuclear reactors to power their radar sea-surveillance satellites.

There are also other threats to satellites in orbit. They are subject to damage from radiation in space, for example during solar eruptions. However, much more damaging radiation may be man-made. Throughout the Cold War, it was feared that the Soviets would attack U.S. satellites with electromagnetic pulses (EMP) created by nuclear explosions in space. These bursts create what amounts to a violent thunderstorm when their radiation knocks electrons out of the atoms of the atmosphere. The resulting pulse can knock out electronics over a wide area on the surface.[8] The pulse also travels up from the atmosphere, so it can damage and perhaps disable satellites, even those at high altitude. In addition to the direct EMP attack, nuclear radiation hitting the skin of the satellite produces its own destructive effects, called TREE (Transient Radiation Electronic Effect) and SGEMP (System Generated EMP, EMP due to radiation hitting components), which can damage internal components. Because of TREE and SGEMP, simply covering the innards of the satellite in shielding material may not be enough protection. Something more subtle is needed. One important Cold War argument against using commercial communications satellites for vital military purposes was that they were not hardened against EMP and TREE, and therefore would be unlikely to survive the opening stages of a major war. Many U.S. military communications satellites (DSCS, Fleet Satellite, UFO) are hardened to JCS Level 1, 'one bomb per satellite', so one burst cannot destroy an entire constellation. This degree of protection adds about 7 per cent to the cost and 100lbs to the weight of a satellite.

The satellite may also be hit by space debris, such as meteorites, which can tear through its skin. Again, there is a man-made equivalent. During the Cold War both sides worked on anti-satellite weapons. These would generally have produced debris through which the satellite would fly at high speed, destroying itself. The alternative, a high-powered laser, seems never to have reached sufficient maturity to endanger satellites, at least not during the Cold War.

THREE

Getting into Space: Boosters

For a given launch rocket, the lower the orbit, the heavier the weight which can be placed there. The power of available boosters can determine whether or not a satellite project is viable. For example, early proposals for a U.S. military space communications system using geosynchronous satellites were dropped because the satellites in question were too heavy for available boosters, while the Soviets had to redesign their ocean reconnaissance satellites because their heavy lift rocket was cancelled – as it happened, for political rather than technical reasons.

There were two key elements to lifting heavier payloads into orbit. One was raw rocket power at lift-off, which in turn depended on the size of the engines which could be built. The other was staging. A rocket engine produces a set thrust, which it uses to lift its body and its fuel. As the fuel is burned, the weight of the body still has to be carried aloft. If, however, much of that weight can be jettisoned, a second stage with a smaller engine can accelerate a payload to much greater speed. Further stages can do better. As the example of the U.S. Atlas Centaur (see below) shows, given a very powerful second stage a rocket of limited performance could match one with a much more powerful first stage.

Most U.S., Soviet, and Chinese launch vehicles were adapted from liquid-fuelled ballistic missiles. The major U.S. exception, the liquid-fuelled Saturn, was developed specifically for the Moon programme, was never used as a satellite launcher, and is no longer in service. The United States also developed a few small solid-fuel boosters.[1] Because they have small throw weights, modern solid-fuel ballistic missiles generally have not been adapted as satellite launchers. That is why the French, who developed only solid-fuel strategic missiles, led a European effort to develop a separate series of space boosters (which currently compete commercially with U.S., Russian, and Chinese launchers).[2] On the other hand, some new U.S. commercial boosters use solid-fuel engines (often adapted from boosters added to earlier liquid-fuel rockets). The Japanese have developed some large solid-fuel space boosters, in addition to their liquid-fuelled ones.

The standard early liquid propellant combination was kerosene or alcohol fuel and liquid oxygen as the oxidiser. Neither was really storable; early liquid-fuelled ballistic missiles (and their space-booster equivalents) had to be fuelled just before take-off. Preparing for launch could be a lengthy procedure, yet it is often crucial to launch a satellite within a narrow time window, in order to place it in the intended orbit. Lengthy delays also made it difficult to maintain a reserve launch capacity in the event satellites were destroyed or boosters failed.

Second-generation ballistic missiles had storable liquid propellants, so that they could be maintained on alert in protected silos, and fired instantly. The Soviets used UDMH (Unsymmetrical Di-Methyl Hydride) fuel and nitric acid or nitrogen tetroxide (N_2O_4) as oxidiser.[3] These liquids react when they meet (and are therefore called hypergolic), and they can be extremely poisonous, but they are storable. The U.S. equivalent, in Titan II and its successors, and in Delta upper stages, was MMH (MonoMethyl Hydrazine, or Aerozine-50) fuel and nitrogen tetroxide oxidiser (the other two U.S. missiles which became space boosters, Atlas and Thor, never progressed to a second generation). Another direction of development was towards higher energy, using liquid hydrogen as the fuel. Because it and the oxidiser, liquid oxygen, must be stored at extremely low temperatures, this combination is called cryogenic. The first rocket engine to use cryogenic fuels was in the U.S. Centaur upper stage (1966), and this combination also fuels the main engines of the Space Shuttle.[4]

U.S. Space Boosters

When the U.S. programme began, both the Air Force and the Army were developing powerful liquid-fuelled engines (the Navy's Polaris lacked the power needed to orbit any substantial satellite). Thus the Navy's contribution to early satellite launching was a specially-designed research rocket, Vanguard, based on existing scientific rockets.[5] The Air Force developed two competing ICBMs, Atlas and Titan, and an IRBM, Thor. All three were developed into space boosters (Thor evolved into the Delta series). The Army group, under Wernher von Braun, was transferred in 1958 to the new National Aeronautics and Space Agency (NASA); its Juno series rockets (based on the Jupiter IRBM) evolved into NASA's Saturns. They received high national priorities at the outset because it seemed that they would be needed to compete with very powerful Soviet space boosters based on the R-7 ICBM (see below). It turned out, however, that much less powerful U.S. ICBMs, combined with very powerful upper stages (such as the NASA-developed Centaur), and often supplemented by solid-fuel boosters, were quite adequate. For each U.S. rocket, staging and strap-on solid-fuel boosters were used to increase payload. The two main upper stages are Agena and Centaur. Agena, which was developed for the Air Force, was based on a restartable rocket engine originally developed for the payload of the B-58 bomber. Centaur is a later (and much more massive) rocket developed for NASA, and first flown in 1966.

The most powerful current U.S. space launch vehicle is the Lockheed Martin (formerly Martin) Titan, adapted from the two-stage Titan II intercontinental ballistic missile. A reconditioned Titan II ballistic missile can place 4800lbs (2177kg) in a 100 mile orbit. Development of a Titan III launch vehicle based on Titan II was proposed in November 1961, the missile forming a core onto which building blocks could be added; the heaviest version would be able to lift 25,000lbs (11,340kg) into a low earth orbit. Titan IIIA (1964) used a simplified version of the missile. It could lift 5800lbs (2630kg) into a low (100 mile) circular orbit. Titan IIIB added a powered upper stage (Agena or Ascent Agena). Titan IIIC/D added two massive solid-fuel motors strapped to the missile's first stage. Unlike those used in Atlas and in Delta, they are about as large as the first liquid-fuel stage of the missile. The segmented motor technology used allows for motors of variable size (*ie* length, hence burn time). Titan IIID was used to launch heavy (Big Bird) reconnaissance satellites. In effect its lifting capability imposed a limit on the size and weight of such satellites, until the much larger cargo bay of the Space Shuttle became available.

Titan IIIE Centaur carried the massive Centaur upper stage, which has a larger diameter than the Titan main body. The Titan-Centaur combination was placed back in production in 1982 specifically because the Space Shuttle was unlikely to meet its schedule for launching heavy reconnaissance satellites; and without the Centaur upper stage Titan could not do so. At the time, the combination was the most expensive rocket in history, at about $100 million each. The stretched-core Titan 34D, first launched in 1982, was intended as a back-up to the Space Shuttle. It has larger solid-fuel boosters than Titan IIID, and several alternative powered upper stages. Titan 34D can place 32,000lbs (14,514kg) in a low earth orbit (115 miles) or 11,000lbs (4990kg) in a geosynchronous transfer orbit.

The follow-on is Titan IV, development of which was ordered in February 1985. It is designed specifically to match Shuttle capability, using a Centaur upper stage. The first and second stages are lengthened, for increased fuel capacity (longer burns). The solid-fuel boosters are enlarged (from 5½ to 7 segments). As in Titan III, the standard upper stages are the Boeing Inertial Upper Stage (IUS) and the Centaur-G. A Titan 402 (Titan IV/IUS) fired from Cape Canaveral can place 38,784lbs (17,592kg) in an 80 × 95nm low earth orbit. Titan 401 (Titan IV/Centaur) can place 10,000lbs (4536kg) in a geosynchronous orbit. The entire rocket weighs up to 1,900,000lbs at launch. Each of its two solid-fuel boosters provides 1.5 million pounds of thrust. The first-stage rocket motors (paired Aerojet LR-87s), which ignite after the missile takes off, provide another 548,000lbs of thrust, and the LR-91 second-stage motor provides 105,000lbs. The Centaur-G liquid-fuel motor provides 33,100lbs of thrust. This version can have a 66-, 76-, or 86ft payload fairing. Titan 402 has an Inertial Upper Stage; Titan 403, 404, and 405 all have no upper stage (NUS). Of these, 403 and 405 are comparable Vandenberg- and Canaveral-launched versions with 56- and 66ft payload fairings. Titan 404 is a Vandenberg configuration used to launch classified satellites, probably with a 50ft payload fairing. Versions with an 'A' suffix use the original solid fuel motors, which had been developed about 25 years earlier for the abortive Manned Orbiting Laboratory (Titan IIIM, to support MOL). Versions with a 'B' suffix use a newer three-segment Upgraded Solid Fuel Motor. This version seemed particularly important after the Shuttle *Challenger* was destroyed on 28 January 1986 due to a motor segment joint failure. This version first flew in 1997. Overall, compared to Titan III, Titan IV offers 25 per cent better performance; it can place a 22-tonne payload in low earth orbit.[6] The recovery plan adopted after the loss of the *Challenger* included purchase of thirteen more Titan IV (in addition to the ten planned in February 1985); all twenty-three were ordered in December 1987. Another eighteen were ordered later. The first Titan IV was launched on 14 June 1989.

By the early 1990s, over 300 Titan II, III, and IV had been fired, mainly for the Air Force, *ie* to launch military satellites. Recently Titan suffered three launch failures in a row (12 August 1998, carrying a Mercury ELINT satellite; on 9 April 1999, carrying a DSP satellite; and 30 April 1999, carrying Milstar satellites). The only previous failure had been on 2 August 1993; as of early 1998, Titan IV had been launched successfully twenty-three times. There was concern that key operating capability was being lost, perhaps due to rapid turnover of personnel. Titan is not part of the Evolved Expendable Launch Vehicle (EELV) programme (described below).

Besides Titan, the other U.S. liquid-fuelled ICBM was the more sophisticated General Dynamics (now Lockheed-Martin) Atlas. It was an elegant solution to the perennial problem of missile design, which was how to cut propulsion and fuel tankage weight. The missile's fuel tanks were, in effect, balloons supported by the pressure of the nitrogen gas used to force the oxidant and propellant into the combustion chamber. When the missile was designed, there was some question as to whether a rocket motor could be ignited in space. Atlas therefore fired its boosters (which were identical to its main motor: in the final Atlas F version, all were Rocketdyne MA-5s) on take-off. It was a 'one-and-a-half-stage' rocket. All three motors shared the same propellant tanks, but the two booster motors were dropped well before the main motor burnt out. All of this ingenuity limited the missile's growth potential, because there was no easy way of increasing its payload (Titan could be improved simply by using a more powerful second-stage motor). Titan was, therefore, selected for continued development and Atlas was withdrawn from military service in the early 1960s. Some had already been used as space launchers; for example, John Glenn rode an Atlas into orbit in 1962.

As a satellite launcher, Atlas can lift standard upper stages (such as Agena and Centaur [as Atlas-Centaur]) as well as an additional upper stage (Burner II), which carries its own guidance and control section. Centaur, development of which was authorised in 1958, proved particularly important. Its cryogenic fuels were so efficient that it transformed Atlas from a mediocre satellite launcher into a near-equivalent to the massive Soviet R-6: with only about 33 per cent of the Soviet rocket's first-stage thrust, Atlas-Centaur could lift 83 per cent of its payload (94 per cent in developed form).

Various versions of Atlas have tapered forward tanks (to accommodate small-diameter upper stages) or a cylindrical forward tank (to mate with the large-diameter Centaur). Atlas G/Centaur first flew in 1984. It can place about 13,300lbs (6033kg) in low earth orbit or about 5100lbs (2313kg) in geosynchronous orbit. An improved version (Atlas K), with 10 per cent more booster power and longer tanks (in both Atlas and Centaur), can lift 16,150lbs (7325kg) to low earth orbit or 6100lbs (2767kg) to geosynchronous transfer orbit. Atlas I is a commercial version announced in 1987. It was followed by Atlas II (a modified Atlas G), used mainly to support the DSCS III communications satellite programme, and by the commercial Atlas IIA. None of these boosters employs the strap-on solid-fuel boosters used in Titan or in Delta (see below). However, a commercial Atlas IIAS has four solid-fuel strap-on motors (Castor IVAs, which are also used in Delta II). It can lift 7000-8000lbs (c3630kg) into geosynchronous transfer orbit, or about 19,000lbs (8618kg) into a low orbit. Atlas III (see below) is part of the EELV programme.

Delta (a modified Thor IRBM) is the lighter-weight complement to the Titan and Atlas series. It began as a single-stage Douglas (now Boeing) intermediate-range ballistic missile. In this as in other U.S. launchers, the missile programme provided the first-stage motor and tankage; guidance was moved to the upper stage. Over time the thrust of the main motor was increased and tankage stretched (to form the Long Tank Thor and later the Straight Eight, with constant body diameter). Strap-on solid boosters (three, six, or nine of various sizes) were added. The core motor was an MB-3 (also designated LR 79-7).

The first upper stage was Able, descended from an early U.S. sounding (sub-orbital) rocket, Aerobee, and from the second stage (Delta A) of the Navy's Vanguard satellite launch rocket. It was first launched on 24 April 1958. Thor Able 1 added the Vanguard third stage. An improved version of Able, Able Star, was the first U.S. rocket which could be restarted in space, so it could correct its orbit as required. It was also more powerful; it could place 904lbs (410kg) in a low orbit, compared to 298lbs (135kg) for Thor Able. This combination launched Transit satellites as well as classified (*ie* intelligence-gathering) Air Force and Navy satellites. Thor Ablestar was first launched on 13 May 1960, with an Echo I satellite on board.

Meanwhile, in January 1959 NASA awarded Douglas (which made Thor) a contract for a further evolved version, an intermediate-weight satellite launcher. It was called Delta.[7] The second stage, Delta 104, had a multiple-restart version of the Aerojet AJ10-series engine used in the Able upper stage. The third stage was the solid-fuel Altair, the upper stage of the Scout solid-fuel rocket. Delta was first launched on 13 May 1960. Although the military called this rocket Thor-Delta, the new Delta name stuck. The Delta B upper stage (1963) had longer propellant tanks and a more powerful AJ10-series engine. Payload therefore increased to 825lbs (374kg), and this version placed the first U.S. Syncom satellite (86lbs) in orbit on 14 February 1963. Later Delta second stages incorporated digital flight controls and had more powerful versions of the same AJ10-series engine.

Agena was a more powerful upper stage. It was developed specifically to support a series of Discoverer reconnaissance satellites (and the MIDAS early warning satellite). It had to be able to orbit with the payload. Thor Agena carried the 1300lb (590kg) Discoverer I reconnaissance satellite, a payload twice as heavy as that of any earlier U.S. space booster, on 28 February 1959. This was also the first satellite to go into polar orbit. Agena B had enlarged tanks (compared to Agena) and, like Able Star, it could be restarted in space. It launched later Discoverer-series satellites, beginning with Discoverer 17 (on 12 November 1960). Agena D was a further improved version, often used as a spacecraft, with a motor capable of multiple restarts. Unlike Agena A and B, which had to be tailored to their payloads, it was modular, adaptable to any of several payloads. Its engine was derived from the AJ10 series used in Able and Delta. Thor Agena D was first launched 28 June 1962 to launch an Air Force reconnaissance satellite. In addition to Agena, Altair and a Boeing solid-fuel Burner II were used as Thor upper stages.

The next step, the Thrust-Augmented Thor (TAT), had three Castor I solid-fuel boosters strapped to the basic Thor body. They increased takeoff thrust from 78 to almost 152 tonnes. This version first flew on 28 February 1963. It carried Agena B and Agena D upper stages. TAT-Agena D could place 2500lbs (1134kg) in a low orbit (57 per cent more than a basic Thor-Agena D).

The solid fuel boosted version of Delta, Thrust Augmented Delta (TAD), using the Delta D upper stage, put the Syncom 3 satellite into geosynchronous orbit. Delta D could put 1280lbs (580kg) into low earth orbit, or 230lbs (104kg) into a geosynchronous transfer orbit. This and later versions also had uprated main engines (MB- 3 Block 3, 79.4 tonnes of thrust, rather than 68 tonnes). Delta E (Thrust Augmented Improved Delta) used a completely redesigned upper stage capable of multiple restarts in space, and it had a new upper stage; it could place 1620lbs (735kg) into low earth orbit or 1210lbs (549kg) into transfer orbit. There were also Delta G (two stages) and Delta J (new upper stage).

Then the core rocket was lengthened (Thorad, or Thor addition) to form the Long Tank Thor (with 43 per cent more propellant, first flown on 9 August 1966, to launch a classified satellite, using an Agena D second stage). The upper-body taper of standard Thors is eliminated. This rocket was used between 1966-72 mainly to launch classified satellites. It was sometimes fired without solid-fuel boosters (as Long Tank Thor) or with those boosters (as LTTAT – Long Tank Thor Augmented Thrust).

The equivalent development in the Delta series was Delta L, with long tanks and Castor II boosters. In this version not only the first stage but also the upper stages were enclosed in a straight-sided body (nicknamed 'Straight-8' for its 8ft diameter). Delta M used an alternative third stage, and Delta N was a two-stage version (which was also built under license in Japan). Later versions (M-6 and N-6) had six Castor IIs, only three of which ignited at lift-off (the other three fired later; total payload was increased to nearly 2850lbs [1293kg]).

In 1972 Douglas replaced the series of letters with four-digit numerical designators. The first number indicated the type of augmentation and the first stage; the second, the number of augmentation motors (3 or 9); the third, the type of second stage; and the fourth, the type of third stage (0 for none). There were already 900 and 1000 series. Delta 900 (1972-3) had nine Castor boosters (six of which would ignite on lift-off), improved electronics, and a better second-stage engine. Delta 1000 (1972) was a lengthened 'Straight-8' which could place 4050lbs (1837kg) in a low earth orbit. Delta 2000 (1974) introduced an entirely new second stage whose motor was based on that used in Moon landings as well as a new main engine (Rocketdyne RS-27, 93 tonnes thrust, derived from the Saturn IB engine). This version was first used on 9 November 1973, to carry a Canadian satellite into geosynchronous orbit. The Delta 3000 series used Castor IV solid-fuel boosters (longer than Castor II) and an extended long tank body. Major examples are Delta 2914 (1600lbs [726kg] in transfer orbit) and Delta 3914 (2100lbs [953kg] in transfer orbit), each with nine solid-fuel boosters. Delta 4920 (1989-90) has an extended long tank and Castor IVA boosters, placing 2650lbs (1202kg) in a transfer orbit; Delta 5920 (1989) has a standard-length first stage and Castor IVA boosters, to put 3095lbs (1404kg) into a transfer orbit.

Delta 6000 (Delta II) uses Castor IVA boosters, and has an extra-long extended tank. Delta 6925 was ordered by the Air Force in January 1987 specifically to launch Navstar (GPS) satellites. The first was launched on 14 February 1989, and last launched on 24 July 1992. The first-stage fuel tanks are extended by 12ft, and the payload fairing is extended. The second stage uses an improved version of the AJ10-series which powered earlier Delta second stages. There is a new solid-fuelled upper stage. Delta 6925 can place 8776lbs (3980kg) in a 115-mile orbit, or 3188lbs (1446kg) in a transfer orbit.

The Delta 7000 (Delta II) series has GEM (graphite-epoxy motors) lightweight boosters offering 40 per cent more thrust; six motors ignite on lift-off, and the other three later. It also has first-stage tanks 12ft longer than in earlier versions. Delta 7925 was first fired on 26 November 1990; through 1998 it had been launched fifty-six times (with two failures). This version can place 11,200lbs (5080kg) in a 115-mile orbit or 4000lbs (1814kg) in a transfer orbit. Like Delta 6925, Delta 7925 is used to launch Navstar (GPS) and communications satellites (such as NATO IVA, 8 January 1991). There are also related Delta 7325 and Delta 7425.

The Space Shuttle

In 1969 the U.S. government decided to build the reusable Space Shuttle (the core of a Space Transportation System) instead of a new-generation expendable booster. Its main element is a large manned spaceplane (a vehicle launched vertically into space but landing like an aircraft). A manned vehicle had to be recoverable, and there was an obvious weight limit to anything which, like the early manned capsules, descended by parachute. It would take a winged or at least an aerodynamic vehicle to land at a conventional airfield. Such spaceplanes were proposed for the initial U.S. manned spaceflight (Project Mercury), but the space capsule approach was chosen instead because it required so much less development. The tight timetable was set by the intense race being run with the Soviets. Space capsules were essentially manned versions of ballistic missile re-entry vehicles, which already existed. Similarly, capsule technology was used in the Moon programme because of the tight deadline.

In the United States the spaceplane seems to have originated with the Air Force, which wanted manned space reconnaissance and strike systems. The key perception was that a properly-shaped body lifted into space could achieve vast range by skipping off the upper atmosphere, like a stone across a pond. This kind of spaceplane had been conceived by a German, Eugen Saenger, as early as 1933, and it had been adopted by Bell Aircraft after the Second World War. Eventually it led to a project for a military lifting body, the X-20 Dynasoar, conceived in the late 1950s.[8]

By the late 1960s the Air Force's need to place very heavy weights (reconnaissance satellites) in orbit had led it to revive interest in spaceplanes. At the same time NASA was interested in a heavy lifter, to support its own ambitious plans for a large space station and for lunar and planetary exploration. In 1969 the new Nixon Administration stopped production of the big Saturn 5 booster and ordered NASA to develop a more cost-effective means of placing heavy payloads in orbit. To do that the system had to be almost completely reusable, which meant that whatever carried the satellite into low orbit had to be recoverable. It turned out, moreover, that a ballistic capsule would be a very poor cargo-carrier. What was needed was a 'space truck', which translated into a stretched-body spaceplane (as opposed to the relatively short X-20). It was also clear that a larger spaceplane could handle the heat stress of re-entry better than a small one, because the heating rate (due to friction) is a function of the skin area of the vehicle. As the vehicle is made larger, its weight increases faster than its skin area, but the weight determines how well it can absorb the heat of re-entry.

To hold down launch cost, the high cost of developing a new Space Transportation System had to be spread out over as many launches as possible, perhaps even one per week. By 1969 the United States had reached very nearly this rate (thirty-eight launches in that year, which would fall to twenty-nine in 1970). Conversely, to provide the Shuttle with enough business for frequent launches, NASA had to shut down the expendables. It therefore ended production not only of Saturns but also of the smaller Atlas, Delta, and Titan. NASA and its contractors were placed under heavy pressure to maintain a very high launch rate – a historian likens the change to turning NASA into an airline – which may have contributed to the mistakes which caused the loss of the *Challenger*.

Although the concept of a reusable Shuttle was chosen in 1969, it took three years to develop a design. Many of the required elements just did not yet exist: thermal insulation which did not burn off (ablate) as the vehicle re-entered the atmosphere, aerodynamic controls with which to land the spaceplane, even a reusable rocket (which could be throttled over a wide range of thrust) to boost it into orbit. At first it was hoped that the spaceplane would ride atop a large reusable booster which would fly back to base after firing. Then it turned out that the booster would have to continue firing until the Shuttle reached high altitude and very high speed. In effect it would become an immense hypersonic aircraft. Development costs would be comparable to that of the hypersonic airliner which had just been abandoned as too costly. Therefore, the design finally selected was a compromise, only partly reusable. The powered spaceplane (using cryogenic rocket motors) is fuelled by an unpowered external tank (ET) flanked by two solid rocket boosters (SRBs) much more powerful than the Titan strap-ons (the failure of one of these destroyed *Challenger*). The spaceplane returns to earth to be refurbished for reuse. The SRBs are jettisoned at an altitude of 28 miles and are recovered by parachute for refurbishment. The ET, however, cannot be released until the Shuttle engines shut down, once it is outside most of the atmosphere. It burns up on re-entry.

Although NASA was apparently the prime mover in developing the Shuttle, it needed support from the Air Force, the agency responsible for orbiting all U.S. military satellites. It had, in fact, to convince the Air Force to abandon expendable boosters in favour of the Shuttle. The Air Force demanded that the Shuttle be able to orbit a satellite twice as heavy and three times as large as anything the existing Titan could handle (30t, 18m long, as against NASA's projected 12t, 12 × 4.5m). The Air Force also wanted the capability to fly the quick single-orbit reconnaissance missions it had conceived for its own spaceplanes, and the ability to return the Shuttle to its launch site after such missions. That required that the Shuttle manoeuvre 1550 miles crossrange through the Earth's atmosphere, since the Earth would have turned while the mission was in progress. This requirement explains why the Shuttle has a large delta wing, and why it hits the atmosphere at a low angle of attack.[9] As it is, the Shuttle never executed the single-orbit mission, and it generally lands at Edwards Air Force Base in California, far from the launch site at Cape Canaveral, Florida.

The Air Force was also particularly interested in orbiting geosynchronous satellites. The Shuttle would ascend only to a low earth orbit, so a new two-stage unit Inertial Upper Stage (IUS) had to be developed to power satellites ejected from its cargo bay. As insurance against delays in Shuttle development, the IUS had to be compatible with the existing Titan space booster, and in fact it was first used on board a Titan in October 1982.

Finally, because the Shuttle was to replace all U.S. space boosters, the Air Force gained the right to requisition Shuttle flights whenever necessary. Military launches were far too important to have to wait for NASA's commercial business. Several reconnaissance satellites were built specifically to exploit the size of the Shuttle's cargo bay, and they had to be stored while the Shuttle moved towards service. After the first commercial flight (December 1982) the Air Force requisitioned a 1983 flight. It discovered that there were further delays, and early in 1983 it ordered twelve Titans as insurance; it also wanted expendables so that it could launch satellites on an emergency basis. That was fortunate, because the IUS encountered problems of its own, failing on its first (1983) Shuttle flight.

Shuttle development was ordered in March 1972, at which time development was expected to take five years, but many unexpected problems were encountered. The Shuttle first flew on 12 April 1981. The first operational (satellite-launching) flight was the fifth, November 1982. The last of four initial shuttles was delivered in 1985 (a fifth was ordered after the *Challenger* disaster). Plans called for a lifetime of 100 missions per Shuttle (twenty-six per year), but in fact turn-around time was always far more than the expected fourteen days. For example, in 1985 it averaged thirty-two days. Nor did the Shuttle manage to achieve the dramatic cost cuts envisaged. It is only roughly competitive with expendables such as Titan and the Russian Proton, at an annual flight cost of about $245 million.

A published list of three baseline military Shuttle missions shows delivery of a 65,000lb (29,843kg) satellite into a 150nm circular orbit (or placement plus retrieval of a 32,000lb (14,515kg) satellite) (Mission 1); insertion of a 32,000lb satellite into a near-polar orbit at an altitude of at least 100nm (Mission 3); and delivery and retrieval of a 32,000lb modular satellite in a sun-synchronous 150nm circular orbit (Mission 4: a 29,000lb satellite was to be deployed within two orbits of lift-off, and a 22,500lb satellite, presumably with fuel used up, retrieved). Mission 1 was presumably from Cape Canaveral; Missions 3 and 4 were from Vandenberg. Mission 4 envisaged a seven-day flight. Apparently the 32,000lb satellite was the heaviest existing U.S. reconnaissance type, KH-12, whose design was begun in 1977. As it and the Shuttle programme developed, both became heavier; the Shuttle was no longer able to lift the satellite as planned. By August 1978, moreover, the satellite's mission plan had changed to call for launching a satellite and retrieving another in a single Shuttle flight: two KH-12 flights would be made each year. Shuttle Mission 3 was largely superseded, therefore, by Mission 4. By August 1985 the Air Force hoped to build up to four flights per year, with an emergency capacity for five.

Vandenberg was a key facility because satellites could be launched directly from it into polar or near-polar orbits, which could not be done from Cape Canaveral.[10] Moreover, as a military base rather than a NASA facility, it was more secure and thus could better accommodate the very secret reconnaissance programme. However, while it was being developed the Shuttle gained weight, which meant that it could not perform as advertised, particularly from

Vandenberg, with its poor location. By 1979 it seemed that the Shuttle, as designed, could lift only 24,000lbs (10,886kg) into polar orbit from Vandenberg, even when operating at 109 per cent of rated thrust and using a special lightweight external tank. Some form of thrust augmentation was needed. In 1981, NASA reported that it hoped to achieve the desired 32,000lb capacity by December 1987; by August 1985 the first Vandenberg Shuttle launch was set for March 1986. At that time hopes were that using filament-wound casings for the boosters would cut take-off weight by 65,000lbs (29,483kg), to increase payload by about 6000lbs. Even then the Shuttle would only be able to place 28,000lbs (12,700kg) in the desired polar orbit. Late in 1985 structural tests showed that the new boosters probably could not withstand the required loads. The Air Force still seemed determined to use them, and they were assembled at Vandenberg. The Shuttle dwarfed the other U.S. boosters, and the largest U.S. military satellites were tailored specifically to its large cargo bay.[11]

Detractors pointed out that, once expendable launchers had been abandoned, the four Shuttles would become prime wartime targets. Once they had been destroyed, the United States would be entirely unable to launch satellites. A combination attack on satellites already in orbit and on Shuttles orbiting to replace them would be devastating. Then *Challenger* was lost, and Shuttle operations were suspended. It was soon evident that the planned Vandenberg Shuttle pad was unsatisfactory (as of June 1986 it seemed that no launch could be made until late July 1989). No Shuttle flights were ever made from Vandenberg. Instead, classified Shuttle payloads – including KH-12s – are launched from Cape Canaveral, the only Shuttle launch site.

The loss of the *Challenger* revived U.S. interest in expendable launch vehicles. With the Shuttle system as a whole out of action until the cause of the accident could be determined, American companies were allowed to use Chinese launchers – a practice which led, eventually, to the accusations against Loral and Hughes in the 1990s of assisting a potentially hostile power (see below). Meanwhile a new U.S. Expendable Launch Vehicle programme was begun, the aim being a capacity of at least 100,000lbs (45,000kg) at a cost of only 50 per cent of that of existing vehicles. In particular, expendables could provide more launch capacity at low cost, whereas to extend the badly-stretched Shuttle fleet entire new Shuttles would have to be built.

The very considerable development of the three main U.S. space boosters concealed the fact that their main liquid-fuelled motors had improved very little since about 1962, when attention shifted from liquid- to solid-fuel ballistic missiles. By the late 1980s it seemed that every U.S. rocket had been surrounded by solid-fuel boosters; there seemed to be little more room for growth. A new programme for an Evolved Expendable Launch Vehicle (EELV), announced in December 1994 (the formal Request for Proposals was issued in May 1995), called for the development of a new expendable system (both medium and heavy) based on existing technology (hence evolved). A major goal was to halve the cost of a Titan IV launch, on the basis of producing seventeen to twenty vehicles per year. Eventually the Air Force seems to have settled for an expected 25 per cent cost reduction.

Although the announced goal of the EELV programme was to buy a single type of core launcher with strap-ons for variable payloads, in fact in December 1998 the Air Force chose to spread the contract between two bidders, Boeing and Lockheed-Martin. The winning Boeing (ex McDonnell-Douglas) proposal was to use variants within a new Delta IV family built around a 650,000lb thrust cryogenic (liquid oxygen/liquid hydrogen) RS-68 common core engine powering a Common Booster Core (CBC). Delta IV Small uses the new CBC topped with Delta II second stage and an optional Delta II third stage. Delta IV Medium uses the cryogenic second stage of a Delta III, and can lift 9100lbs (4128kg) into a transfer orbit. Delta IV Heavy uses two more core engines as strap-ons as well as a stretched Delta III second stage with a Titan IV nose fairing; it can lift 28,700lbs (13,018kg) into a transfer orbit. The initial EELV contract to Boeing was for nineteen launches by the medium and heavy versions of Delta IV.

Lockheed-Martin uses Atlas III, a core powered by a U.S.-made version (RD-180) of the Russian RD-170 engine used in the Zenit space booster (see below). Compared to earlier versions, this one has lengthened tanks. It eliminates the earlier stage-and-a-half configuration. The medium version (Atlas IIIA) uses a stretched Centaur upper stage to place 11,595lbs (5260kg) in transfer orbit; a heavier version (IIIB) uses a two-engine Centaur to place 14,500lbs (6580kg) in transfer orbit. The EELV contract calls for nine launches using Atlas IIIs.

Alternative Launchers

Still, it is inherently expensive to throw away a massive booster each time it is used. One possibility is to use an aeroplane instead of the big first rocket stage. Quite aside from reusability, it uses air as its oxidiser, and thus needs far less total tonnage of propellant than a big rocket. The first known application of this idea was Notsnik, a satellite launcher developed by the U.S. Naval Ordnance Test Station (NOTS) at China Lake. When Sputnik was launched, NOTS naturally asked whether an aeroplane might not make an effective first stage for a satellite booster. Early in 1958 NOTS engineers devised Project Pilot, in which a five-stage solid-fuel rocket carried aloft by a fighter would loft a small (1.05kg) satellite.[12] The project was approved with a four-month deadline (and a budget of only $300,000), and it was soon nicknamed 'Notsnik'. The launch platform was a stripped-down Douglas Skyray (F4D-1) capable of reaching Mach 1.05, and thus offering a much greater boost than the Air Force's B-47 later did in 1959. On the other hand, the fighter had much less ground clearance; the booster could be no more than 14.4ft long, and maximum weight was only 2090lbs (950kg). It was to be launched at the peak of a 2G toss-bomb manoeuvre (at an up-angle of 50 degrees) at Mach 0.9 at 41,000ft. The first two stages were modified Asroc solid-fuel boosters; the third stage was a modified version of the Altair rocket used as a Vanguard third stage. NOTS provided the small fourth and fifth stages, which together were expected to place the satellite in an orbit with a 1400-mile perigee.

In this incarnation Notsnik was a miniature imaging satellite, conceived as the forerunner of full reconnaissance craft. It carried an IR detector, which was scanned by the combination of its own spin and its forward motion. In theory it would build up a television-type picture a line at a time. Data would be downlinked to the Minitrack stations already in place for the Vanguard scientific satellite programme. Because of the connection to imaging (and, presumably, because it had never been approved at a very high level), Notsnik was very secret; its existence was first officially revealed in 1994.

It appears that the rocket orbited a short-lived satellite on its first attempt, on 25 July 1958 (a Minitrack station at Christchurch, New Zealand, detected a signal; the rest of the network had shut down after the launch pilot thought – apparently incorrectly – that he saw the rocket explode after launch). After a failure, there was another apparent success on 22 August of the same year, the New Zealand station picking up signals on the expected first and third orbits. In both cases, probably the last stage failed to ignite, the fourth stage having given the satellite too low a perigee (60 miles) to last for very long. At present the only application of this idea is the U.S. Pegasus, a rocket which is carried aloft by a Lockheed L-1011 airliner. First launched on 5 April 1990, it can place 827lbs (375kg) in a 124-mile orbit. Pegasus has two solid stages and a liquid-fuel third stage. There is also a commercial conventionally-launched derivative, Taurus, which uses a Castor 120 solid-fuel booster as its first stage, surmounted by the three Pegasus stages. First launched on 13 March 1994, it can place 3004lbs (1363kg) in a low earth orbit or 950lbs (431kg) in a transfer orbit.

For some years there have been proposals for hypersonic aircraft capable of reaching very high altitudes. Any of them could, in theory, function as a reusable first stage. The Soviets tried a combination hypersonic first stage and spaceplane in their Spiral project (see below). Probably the most extreme version of this idea is a hybrid jet/rocket capable of taking off like an aeroplane, burning fuel in the atmosphere, and then switching to rocket operation to reach orbit. Because it can land like an aeroplane, such a craft is reusable. For example, NASA is currently sponsoring X-33, an unmanned prototype of such a craft, which is expected to reach Mach 15. The abortive British HOTOL fitted into much the same category. The Boeing X-37 is to demonstrate technology for low-cost access to space; it will be placed in orbit by a Shuttle, and it will then manoeuvre to a conventional landing (the current X-40 is a smaller-scale version to test autonomous landing techniques). X-37 is seen as the upper stage of a future two-stage orbiting system.

One other technique deserves mention here. A small satellite can, in theory, be placed in orbit by a gun. This idea was first raised in 1961 by Dr Gerald Bull, the Canadian ballistics expert. In experiments which ended in 1967 the U.S. Defense Department sponsored HARP (High Altitude Research Project), a programme in which a gun made out of two 16in barrels (joined end to end) fired small payloads into space, mainly to test missile re-entry shapes. Bull evacuated most of the air from the gun tube to reduce internal resistance, and he used a lightweight saboted finned projectile (which he called Martlet) 8in in diameter. He achieved an acceleration of about 25,000G. Bull's Mark II projectile weighed 475lbs (215lbs). On 19 November 1966 he fired a 185lb version to a record altitude of 111 miles, little short of orbit. He planned a powered Mk IV, which he estimated would have achieved orbit, but the HARP project was cancelled in 1967. That did not stop Bull from advocating gun launchers for lightweight satellites.

The great advantage of the gun is that it is reusable. In effect it replaces several rocket stages. Bull's gun imposed very high acceleration on the projectile (as in a conventional gun), and terminal velocity was far below orbital speed, about 3.6km/sec (7km/sec is needed to reach orbit). However, during the Second World War the Germans built a long-range gun, which they called V-3, with multiple combustion chambers which could gradually accelerate a projectile to very high speed. This device was, naturally, fixed, so that it could attack only a single target (most likely London), but a satellite-launcher could also be fixed in direction. Bull actually designed a German-type gun for the Iraqis (under Project Babylon), and was building a full-scale space launcher when he was killed in March 1990. It would have used a 500ft (150m) barrel 39.4in (1m) in calibre, and would have weighed 2100 tons; a model was displayed at the May 1989 Baghdad arms show. The device was intended either to bombard Israel or to launch small satellites (or both). Particularly for bombardment, the super-gun would have had the advantage, over rockets, of being able to maintain a very high rate of fire (on the other hand, it was a large fixed installation, and therefore vulnerable to attack).

Bull was not unique in his interest in such guns. In the early 1990s, the U.S. Strategic Defense Initiative Office considered using a gun to launch the lightweight Brilliant Pebbles satellites it planned to use as space sensors. It had to place about 4000 small (100kg) satellites in specified orbits at the rate of one every 30 minutes. Somewhat later a commercial version, the Jules Verne Launcher (after the French writer, in whose novel a large gun launches a spaceship to the Moon) was proposed, based on the launcher planned for the Brilliant Pebbles project. In 1998 the Applied Physics Laboratory of Johns Hopkins University, which is responsible for many U.S. Navy missile projects, estimated the characteristics of a gun capable of firing 100kg satellites into 700km orbit, something like the requirement levied to maintain a system like the Iridium series of communications satellites. It turned out that a 1.5km barrel could boost a projectile to 7km/sec muzzle velocity. Once in the atmosphere the shell would quickly lose speed, so that it would leave the atmosphere at about 6.6km/sec, and it would then continue to lose velocity (due to gravity), reaching a minimum of about 5.4km/sec. Then its rocket would boost it to orbital velocity, in this case 7.5km/sec.

Soviet Space Boosters

As in the West, Soviet interest in ballistic missiles (which were the basis for space boosters) can be traced back to captured German Second World War technology. Stalin himself showed considerable interest in the V-2 ballistic missile, and special research institutes (NII) were formed in 1946-7 to study and exploit it. NII-88 studied missiles; its Department 3 (from 1950 an independent design bureau, OKB-1), headed by Sergei P Korolyev, concentrated on ballistic missiles. A design bureau, OKB-456, was set up under Valentin P Glushko to develop liquid-fuel rocket motors. In 1946 the Soviets seized numerous German rocket specialists in their sector of Germany, moving them into a NII where they developed a series of longer-range derivatives of the wartime V-2. Meanwhile Korolyev developed the first Soviet ballistic missile, R-1.

The Soviet system was racked by politics. In 1946 Stalin purged the leadership of the Soviet aircraft industry. The ballistic missile project therefore went to the Ministry of Weapons Production, which was concerned with army weapons development. Its factories were unaccustomed to precision production. Once missiles were really important, in the late 1950s, it was difficult to accelerate production. It turned out to be possible to produce shorter-range missiles

(MRBMs and IRBMs) in reasonable numbers, but ICBM production lagged, to an extent not appreciated by Western intelligence. It now appears, for example, that Khrushchev was impelled to place shorter-range ballistic missiles in Cuba in 1962 mainly because that was the only way he could bring any substantial number of missiles within range of American targets. By that time the United States had a large ICBM force in being – the production of which, ironically, had been accelerated by Khrushchev's boasts about his own ICBM programme. To Americans, the latter's existence seemed proven by the string of Soviet satellite launchings using modified R-7 ICBMs.

Like Wernher von Braun, Korolyev was interested in space flight at least as much as in military missilery; but his patron, Stalin, wanted missiles. By 1953 Korolyev was working on an IRBM (intermediate-range missile, with a range of about 1500nm). The next step, the ICBM (range of about 5000nm) would require a much more powerful rocket engine. Korolyev had an inspiration: if IRBM engines were clustered together (in what he called a 'bouquet'), he could obtain the necessary thrust using the existing technology. Although a clustered engine would be much heavier than a more powerful single-chamber type, and might well be less reliable (all the engines, including the four booster clusters, would have to ignite simultaneously), these disadvantages were more than balanced by the prospect of achieving an ICBM within a few years. Korolyev sold his idea to the Soviet authorities, and in 1954 development of the R-7 ICBM was approved. By January 1956 the ICBM project explicitly included a satellite-launching version, to orbit both scientific and military spacecraft.

When Korolyev was assigned the ICBM project, the IRBM was still considered quite important. A second missile design bureau (OKB-586) was formed in 1954, headed by Mikhail K Yangel.[13] It developed the R-4 and R-5 MRBMs (and eventually the R-12 MRBM and R-14 IRBM), which had storable liquid fuels. By 1958 Yangel was developing storable-fuel ICBMs in competition with Korolyev, who preferred his semi-cryogenic (liquid oxygen plus kerosene) rockets.

R-7 (8K71; NATO SS-6) was massive not only because it used clustered rockets, but because it was designed to lift a very heavy warhead (U.S. H-bombs were substantially lighter).[14] As a consequence, it could orbit very considerable weights even at the outset, without a powered upper stage (development of which presented problems, because it had to be ignited after the rocket was airborne). Western designators for satellite-launching versions of R-7 were SL-1 (8K71PS Sputnik), SL-2 (8A91 Sputnik), SL-3 (8K72 Vostok), SL-4 (11A57 Voskhod; 11A511U Soyuz), SL-5 (11A510 Vostok), and SL-6 (8K78 Molniya and 8K78M Molniya-M), and SL-10 (11A59 Polyot). Of these, SL-1, a relatively unmodified test version of the R-7 missile without a powered upper stage, could place 50 kg in a 200km orbit. The modified SL-2 was used to place Sputnik 3 in orbit (1327kg at 217km).

The next step was to add a variety of powered upper stages, beginning with a lunar probe (8K72 Luna). SL-3 used an enlarged Luna stage to launch Vostok manned spacecraft. An upgraded second stage (8K72K) was used both for Vostok spacecraft and for the first attempt to launch a Zenit reconnaissance satellite. It could place 4550kg in a 200km orbit. A modernised version (8A92) launched Zenit-2 series reconnaissance satellites.

Voskhod (11A57) used a large second stage designed for a planetary probe (8K78) to increase payload into low earth orbit; without any third stage it was used to launch the Zenit-4 reconnaissance satellite. This version could place 5900kg in a 200km orbit. Soyuz (11A511) is a standardised two-stage version intended to replace the numerous earlier ones and is the most widely-used version of the launch vehicle. Compared to Voskhod, it has lighter-weight telemetry, and its engines are individually chosen for maximum performance. This rocket can place 6450kg in a 200km 51.6 degree orbit. An upgraded 11A511U, first launched in May 1973, uses a chilled higher-density fuel in its core; it can place 6855kg in a 200km orbit. An 11A511M (6600kg in a 200km 65 degree orbit) was designed to support the abortive Soyuz 7K-VI manned spacecraft.[15]

Molniya, first fired in 1960, added a third stage atop the two-stage lunar vehicle adapted from the standard R-7A. It could place 900kg in an interplanetary trajectory. Then the Molniya upper stage was mated with Soyuz. First launched in 1964 as Molniya-M (8K78M), this version can place 1800kg in an 820km orbit or 1600kg in a geosynchronous transfer orbit. There are also two minor versions, each used only twice in place of the UR-200 space booster: SL-5 and Polyot. SL-5 was used to launch prototype radar ocean surveillance satellites (1964-6), the satellites' planned launcher, UR-200, having been cancelled (see below). Polyot was a two-stage version of 11A57 Voshkhod. It launched prototypes of the Chelomey ASAT in 1963-4, as the planned UR-200 booster was not yet ready (it was later cancelled).[16] Derivatives of Vostok, Soyuz, and Molniya are still in use. Once development had been completed, the R-7 series proved remarkably reliable. Between 1970 (when the series reached maturity) and 1998, of 1296 launched, only 21 failed (as of 30 September 1998, at least 1571 R-7 series launch vehicles had been fired).

In the early 1960s Korolev's cryogenic-fuelled missiles were superseded by weapons with storable liquid fuels, developed by the Yangel (Yuzhnoye) design bureau. Its initial space boosters were the Kosmos series based on shorter-range ballistic missiles, hence equivalent to the U.S. Delta series: 11K72/73 (NATO SL-7, based on the R-12 [NATO SS-4] missile), and 11K65 (NATO SL-8, based on the R-14 [NATO SS-5] missile). SL-8 was credited with the ability to place 1700kg in a low earth orbit.[17] Yangel then developed a storable-propellant ICBM, R-16 (11K64; NATO SS-7), which was moderately successful as the successor to R-7. Its space booster version was called Tsyklon (NATO SL-11); it could probably place about 3000kg in a low earth orbit.[18]

From 1960 on a third major Soviet ballistic missile designer appeared: Vladimir N Chelomey. He had been responsible for long-range naval cruise missiles, and he apparently saw ballistic missiles and space systems as the wave of the future. He tried to take over these programmes, in what Russians have called the 'War of the Designers'. Chelomey had hired Khrushchev's only surviving son, Sergei, as an engineer in 1958, and he became Khrushchev's favourite. Unlike the two earlier rocket designers, he was connected to the aircraft production ministry. In April 1960 he proposed exactly the military-oriented missile and space programme Khrushchev wanted.[19] In April 1962 Khrushchev was told that the Soviet ICBM programme had stalled; for about two years Chelomey had been offering the resources of the aircraft industry as a solution. He had already conceived a variety of space systems, including a satellite to locate naval targets (see Chapter 8). He proposed a series of 'universal' rockets: UR-100 (NATO SS-11 'Sego'), UR-200, UR-500, UR-700. UR-100 was universal in the sense that a change in payload could make it an ICBM, an MRBM, or an ABM (in each case the guidance system was the same). Chelomey also proposed that UR-100 take over the anti-ship roles of his cruise missiles, using a homing warhead (this project was assigned to the Makeev group developing naval ballistic missiles).[20] UR-100 was broadly equivalent to the U.S. Minuteman. Because it used a lightweight warhead, it lacked the power to be an effective space booster.[21] Designed so that it could be fired within three minutes, UR-100 was part of the first generation of fully-satisfactory Soviet long-range missiles powered by storable liquid propellants. Because UR-100 was both an ICBM and a medium-range missile replacing earlier MRBMs and IRBMs, it was deployed in greater numbers than any other ICBM. For this project Chelomey beat out Yangel (R-26/8K66) and Korolyev (who offered a solid-fuel RT-2/8K98, NATO SS-13). As an index of producibility, the CIA later estimated that an SS-11 (UR-100) cost about a third as much as an SS-9 (Yangel's much heavier R-36).

UR-200 (8K81), using some UR-100 components, was conceived to meet the requirement for a 'global rocket' (GR-1). The same booster would function as either a heavy ICBM or as a space booster to launch the Soviet ASAT (IS), the ocean reconnaissance satellite (US), or a sub-orbital warhead (FOBS). The 'global' phrase probably referred mainly to FOBS, since a FOBS warhead would be launched into a partial orbit to hit anywhere in the world. The Soviets called FOBS a combat re-entry vehicle, the idea being that by using aerodynamic manoeuvring and by coming from an unexpected direction (over the South Pole) it could overcome expected U.S. ABM defences. The competitors were Chelomey's UR-200, Yangel's R-36, and Korolyev's 8K713. Korolyev's rocket was cancelled before being flown (a mockup was later paraded in Red Square, misidentified as an operational ICBM, and designated SS-10 'Scrag' by NATO). UR-200 development was officially approved on 16 March and 1 August 1961, and the missile was tested between 4 November 1963 and 20 October 1964. No NATO code name or number was ever assigned to UR-200, presumably because its tests were never clearly distinguished from those of other rockets.[22]

Chelomey's assault on the other missile designers generated enormous bitterness. After Khrushchev was ousted (13 October 1964) all of his projects were reviewed by a commission headed by a senior Soviet scientist, M V Keldysh. UR-200 was cancelled and Yangel's R-36 (8K68; NATO SS-9) was built instead as the heavy Soviet ICBM. Its FOBS version was R-36-O; its satellite launcher version was Tsyklon 3 (11K68; NATO SL-14). Provided with a specially-built third stage, Tsyklon 3 can place 3600kg in a low earth orbit (200km altitude). When R-36 was selected in place of UR-200, the ocean reconnaissance satellite had to be redesigned, since Tsyklon could not accommodate it.[23] Tsyklon 2 (R-36M/11K69; NATO SL-11) was an analogous ASAT (anti-satellite) launcher based on the 8K69 version of the R-36 missile. It could place 2800kg in a low earth orbit. Eventually it was used to launch surveillance spacecraft.[24]

UR-500 was intended as a larger 'global rocket' (GR-2). It would function as either a super-heavy ICBM (to carry the new 100-megaton warhead, which was tested in scaled-down form in 1961), to launch a military spaceplane (Raketoplan), or as a heavy-lift space booster for proposed military space stations and for planetary exploration. The piloted Raketoplan could collect intelligence, inspect satellites, and destroy them; it had an orbital manoeuvring engine, target detection systems, rendezvous systems, and space-to-space weapons.[25] Work on UR-500 began in the fall of 1961, and full development (to be completed within three years, *ie* by the end of the 1959-65 Seven Year Plan) was authorised on 24 April 1962 (the unsuccessful competitor was Yangel's R-46). UR-500 was conceived as a three-stage rocket, the second and third stages of which were modified versions of the first and second stages of UR-200. This combination was to have had five times the lifting capacity of UR-200 (about 12,500kg into low earth orbit). The first stage was new. UR-500 (Proton) was designed so that its tooling could be used to build a projected UR-700 powerful enough to launch manned flights to the Moon or to Mars. Thus the abortive UR-700 design greatly affected details of UR-500.

This Proton (8K82/NATO SL-9) rocket, far more massive than the R-7 series, survived Khrushchev's fall to become the heavy workhorse of the Soviet space programme (the Raketoplan, however, was cancelled, as was the ICBM version). On 16 July 1965 a two-stage version of UR-500 successfully placed a Proton scientific satellite in orbit. Payload capacity was 8400kg, about 24 per cent more than Soyuz (UR-500 was, however, 75 per cent larger; the published figure, 12,200kg, included the casing around the satellite). The rocket came to be known by the name of its first payload, although apparently it was originally to have been called Gerkules (Hercules).

The full four-stage version was UR-500K (8K82K/11S824 Proton-K; NATO SL-12). It had a longer second stage. The spacecraft payload provides guidance to the Block D (11S824) fourth stage. A later version uses the 11S86 (Block DM) fourth stage, which has a self-contained guidance system. It was developed specifically to launch geosynchronous military communications and early warning satellites (Raduga, Ekran, Gorizont, Potok, Luch). A later version (with 11S861 fourth stage) was introduced in 1982 and is the most successful current Russian commercial space booster. There is also a 17S40 fourth stage, used for heavier payloads launched into lower orbits; it has lofted the Arkon reconnaissance satellite as well as Iridium communications satellites. Equipped with the current DM4 (11S861-01) fourth stage, UR-500K can place 20,600kg in a 200km orbit, or can lift 2500kg directly into a geosynchronous orbit. Another version, with a D-1 (11S824M) upper stage, was used for planetary probes; it could place 4740kg in a transvenusian trajectory (11S824 had been designed to place a manned vehicle in orbit around the Moon). Proton-KM (8K82KM) is a further development with a Briz M fourth stage to replace the earlier Block D; it can place 22,000kg in a 185km 51.6 degree orbit.

Development of a three-stage version of UR-500K/8K82K (NATO SL-13) was authorised on 3 August 1964. It was intended to support the Almaz manned military spacecraft (approved 12 October 1964) and the LK-1 manned lunar orbiter (approved 11 November 1964). Compared to the original design, it had more efficient engines and considerably enlarged tankage.[26] This rocket was first launched successfully on 10 March 1967. However, there were so many failures that it was not considered acceptable until its 61st launch (on 29 September 1977). This rocket can place 20,600kg in a 200km orbit at 51.6 degrees inclination.

In the super-heavy category, comparable to the U.S. Saturn, UR-700 – intended for the Soviet flight to the Moon – was cancelled in favour of Korolyev's N-1 (NATO SL-15, which was itself cancelled after four launch failures). The other failed competitor was Yangel's R-56, which was abandoned about 1966 because it would take two R-56 launches to provide the lift of one N-1 or UR-700; thus to use it R-56 would require an additional complication, a rendezvous in orbit to assemble the desired Moon landing craft. However, R-56 was significant because it apparently led to the design of Zenit, which was built.[27]

While Chelomey was selling his Raketoplan to the Soviet military, the MiG design bureau was developing its own aerospace system, Spiral. Advanced development was officially approved in 1965, and the full project was formally approved on 26 June 1966. At that time the first unmanned orbital flight (of a spaceplane lifted by a Soyuz booster) was planned for 1970. Spiral consisted of a reusable hypersonic launch aircraft, GSR, which carried on its back a two-stage rocket (RB) carrying a spaceplane (OS).[28] An OS prototype was actually built, and is currently displayed at Monino as MiG-105 (actually MiG article 105). In theory OS would have been inserted into a 130km orbit. GSR construction was to have begun in 1970, and the hydrogen-fuelled version was to have been rolled out in 1972; the first complete flight was scheduled for 1977. With the beginning of the Buran space shuttle project (1975), Spiral was shelved. Only the prototype OS spaceplane survived, as a test vehicle.

By 1974 the United States was developing a Space Shuttle. It had important military applications and the Soviet military wanted an equivalent, which it designated OK. Buran (Snowstorm) seems to have been conceived to help build and service the Mir-2 and KM military space stations, on which would be based a variety of military combat spacecraft. It could deliver projected Almaz BKA military spacecraft armed with lasers or rockets, and in wingless form it could form a hangar in the space station for rocket interceptors (both anti-satellite and anti-missile). In this way it could fulfil its primary requirement, to help deny space to the enemy. Like the U.S. Shuttle, Buran would also orbit large reconnaissance satellites, which it could service periodically.

Korolyev's design bureau, merged in 1974 with Glushko's rocket engine bureau and renamed NPO Energiya, had just lost its major programme, the N1 lunar rocket (its chief, Korolyev's former deputy, V P Mishin, who had taken over in 1966, was fired over this failure). The other rocket designers were, by this time, deeply involved in ballistic missile development. Energiya was thus the natural developer for a booster for the projected Soviet shuttle equivalent. Unfortunately the Soviets lacked some key U.S. technologies, namely large segmented solid-fuel engines and big reusable cryogenic engines. The choice, then, was to use expendable engines which did not, therefore, have to be housed within the orbiter. Instead of riding a big fuel tank, as in the U.S. Shuttle, Buran would ride a big cryogenic rocket, Energiya. The Buran orbiter was conceived as a direct aerodynamic copy of the U.S. Shuttle (alternative configurations were considered and rejected) with a slightly greater payload and with the additional capability of operating unmanned. It was developed by the MiG design bureau, which had previously worked on the Spiral programme. Development of the Buran-Energiya reusable space transportation system (MKS) was authorised on 12 February 1976, the first flight being scheduled for 1983 (and the first unmanned Buran flight for 1984); manned flights were to be routine by 1987.[29] NATO designated the Energiya-Buran combination SL-18. In fact Energiya was not tested until 15 May 1987, and Buran first flew (unmanned) on 15 November 1988. Then the programme collapsed. Energiya never flew again, and the entire programme was formally cancelled on 30 June 1993. Buran had never carried pilots.

Energiya (11K25; NATO SL-17) thus became a very powerful booster in its own right. With two strap-on boosters (using Zenit first-stage RD-170 engines [see below]) Energiya can launch 65,000kg into a 51 degree orbit (200km altitude) from Tyuratam; with four strap-ons that increases to 105,000kg, and with eight strap-ons, to 200,000kg (170,000kg into a sun-synchronous 200km orbit). All of these figures are about 35,000 to 40,000kg less than Western estimates.

It appears that as the Buran-Energiya programme developed, the Soviet military also became interested in a means of shooting down the U.S. Space Shuttle, which it became convinced was a major military asset. To do that, it conceived a spaceplane, Uragan (Hurricane). There were apparently two parallel programmes, one by Chelomey (LKS spaceplane, and Proton booster) and one by Molniya (OK-M spaceplane, and a new Zenit booster developed by Yangel). Test craft designed for the MiG Spiral programme were used for the first few test launches. Full development was apparently authorised in September 1978. The one really speculative element of the MiG programme, the reusable first stage, was eliminated in favour of a conventional expendable booster.[30] On 1 and 28 August 1987 the Soviets flew what looked like boilerplate models of Uragan atop Zenit rockets. Uragan seems to have been cancelled in September 1987, perhaps because the U.S. government cancelled planned Shuttle military missions from Vandenberg Air Force Base after the *Challenger* disaster. After the project was cancelled, five cosmonauts who had been training for it were transferred to the Mir space station project.

Quite possibly as part of the Uragan programme, Yangel developed a new Zenit three-stage booster (11K77; NATO SL-16). It flew in

1985. Zenit can place 13,740kg into a 200km orbit from Tyuratam (51 degree inclination), 11,380kg into a 200km sun-synchronous orbit, or 5180kg into a geosynchronous transfer orbit. Zenit differed from earlier Yangel boosters in using the semi-cryogenic engines (liquid oxygen and kerosene) which Korolyev had favoured over toxic storable liquids.

With the end of the Cold War, Russian technology could be used in combination with its Western rivals. The most dramatic example was the use of the RD-180 motor in the U.S. EELV programme. Other examples were Boeing's project to sea-launch Zenit and a Franco-Russian consortium (Starsem) which adds a French upper stage to the existing Soyuz.[31] Several Russian strategic missiles are being offered as space launchers.[32] There is also a project for an 80-ton air-launched rocket capable of placing 2000kg in a low earth orbit.

Chinese Space Boosters

Chinese space launchers (the Long March, or CZ [Chang Zheng], series) are all derived from their own ballistic missiles. Sometimes designations are given with the first letters LM instead of CZ. Thus CZ-2C is also styled LM-2C. The series began with CZ-1, which launched the first Chinese satellite on 24 April 1970. Based on the DF-3 IRBM, it could place 300kg in a low earth orbit (440km altitude). A CZ-2 series was based on the two-stage DF-5 ICBM; the initial CZ-2A could place 3200kg in a low (200km) orbit or 1000kg in a geosynchronous transfer orbit. Like the early Soviet ICBMs, this employs clustered engines (in this case, four YF-20A; CZ-1 used four engines of an earlier type) to provide sufficient thrust. CZ-3 is a purpose-built space launcher whose first stage is derived from that of the CZ-2 series. All of these launchers were considerably improved. CZ-2C (1975) is very similar to the ICBM, but uses the more powerful YF-20B engine; it can place 2800kg in a low earth orbit. It has often been used to launch Western satellites under contract. CZ-2D (or LM-2D) of 1992 uses modified versions of the first and second stages of CZ-4 (see below); somewhat shorter than CZ-2C, it can place 3500kg in a low (200km) earth orbit. CZ-2E (1990) is a much larger rocket in the Proton or Titan category, capable of placing 9200kg in a low orbit or 3370kg in a geosynchronous transfer orbit. It employs four clustered YF-20B engines in its core, and four boosters (each with a single engine of the same type) clustered around the core. Compared to CZ-2, CZ-3 (1984) added new upper stages, and could place 1500kg in a geosynchronous transfer orbit. It had the new main engines (YF-20B). CZ-3A added a new cryogenic upper stage; it could place 2700kg in a geosynchronous transfer orbit. CZ-3B (1996), the most powerful rocket in the series, has enlarged tanks and four clustered boosters like those used on CZ-2E. It can place 5000kg in a geosynchronous transfer orbit. CZ-3C is a -3B core with two rather than four boosters; it can place 2800kg in a geosynchronous transfer orbit. CZ-4 (1988) employed longer first and second stages, and added a new third stage. It could place 1650kg in a sun-synchronous orbit, an orbit often used by photo reconnaissance satellites (and by weather satellites). Alternatively, it could place 4680kg in a low (200km) earth orbit. This launcher last flew in 1990, and it is no longer advertised.

The close connection between missiles and space boosters means that improvements in boosters may lead to improvements in the missiles on which they are based. In the late 1990s several Chinese space shots (carrying foreign satellites) failed. Two U.S. companies, Loral and Hughes, helped the Chinese solve the problems revealed by the failures – and were accused of having materially improved Chinese missile accuracy. Because some of the shots involved multiple satellites, there was also a suspicion that the assistance could have helped the Chinese develop a viable MIRV (multiple warhead) capability. The companies involved protested that they had merely been solving a commercial problem – which was also true.

Conclusion

Until the 1980s, for both the United States and the Soviet Union, space booster development was driven mainly by military requirements. It was vital to be able to launch heavier and heavier satellites, into a combination of low and more or less geosynchronous orbits. For a short time there was also a vital prestige mission culminating in the Moon landing (for the United States). Once Americans had landed on the Moon, the Soviets also considered missions to the outer planets, probably primarily as a way of regaining lost prestige. There was also intense Soviet interest in a military space station. For both countries, development of reusable spaceplanes (culminating in the Space Shuttle and Buran) was intended to meet military demands.

None of this is to deny the importance of civilian space activities. However, in the Cold War Soviet Union, the civilian sector came a very poor second to the military sector. Civilian research was almost always justified by its military applications, and the Soviet military was in the habit of co-opting civilian organisations such as Aeroflot. Thus in the account above it has been assumed that military considerations were always paramount in Soviet space booster and spaceplane conception.

By the 1980s, the commercial use of space was expanding rapidly enough to begin to drive booster development, at least in the United

States and in China; the European boosters (Ariane series) had always been commercial. After the fall of the Soviet Union, the Russians and Ukrainians concentrated largely on commercial work, to the extent of modifying existing ballistic missiles for this purpose, and also of collaborating with the Americans. Some time in the 1990s (just when depends on how costs are measured) commercial investment in space overtook military investment. It seems likely that commercial concerns, particularly the need to drive down the cost of placing payloads in orbit, will dominate future booster development.[33]

FOUR

Polaris and Precise Navigation

The scene is familiar: out of a calm sea a streamlined missile suddenly erupts. Its motor ignites above the water, after which it streaks off to fall onto a distant target. Yet this picture is very incomplete. If it is to hit the intended target, the missile, Polaris, has to be launched from a known position. But the sea is trackless. When Polaris was conceived, ships could find their positions only to within a few miles. That was hardly good enough. To be useful, Polaris had to hit within a mile or so of its target. If the navigational error of a few miles was added to the error inherent in its guidance system, it was unlikely even to hit a Soviet city a few miles across. Indeed, Polaris was developed almost as an act of faith, since no satisfactory submarine navigation technique yet existed when it was conceived (and approved) in 1956-7. Very fortunately, while the missile was being developed, a space technique, Transit, solved the problem.[1]

A ballistic missile like Polaris is aimed at a point on the Earth's surface. The missile cannot correct its course as it approaches its target, because it has no onboard sensor to distinguish the target from the surrounding territory. Rather, it is fired to a fixed range and bearing from its launcher. Guidance is exerted only so long as the missile's engine burns. To this extent accuracy depends heavily on how precisely the location of that launcher (and factors such as the speed and roll of the submarine) can be determined.[2] In the late 1950s an official U.S. writer likened an ICBM to a gun with a 200-mile barrel and a 5400-mile range – which should be much more precise than a naval gun, since these figures were comparable to a 16in naval gun (800in barrel) firing to a range of only 600yds. On this basis the desired accuracy of the missile, to within (say) half a mile, would be comparable to placing the 16in shell within about half a yard at 30,000yds. The great difference between the shell fired against a fixed shore target and the missile fired against a fixed target was that the battleship could see the target (or at least a reference point near it). Thus it could very easily translate the target position into co-ordinates centred on itself, and it could aim the gun on that basis. However, the missile had to be fired at a position found by a system entirely separate from the firing submarine. Moreover, this position was defined relative to the Earth, not relative to the firer.

Polaris was not the beginning of the U.S. Navy's efforts to achieve accurate submarine navigation. In the late 1940s it had begun work on a submarine-launched cruise missile, Regulus. The launch submarine tracked it and commanded it onto a course. Once it flew beyond that submarine's radar horizon it was taken over by a forward-based submarine which commanded it to dive onto its target. Both submarines had to know where they were in order to guide the missile. Conventional navigational methods, depending on the stars and chronometers, were unlikely to be sufficiently precise, particularly in very high latitudes near the Soviet Union. Since Regulus's range was relatively short, it was vital to be able to navigate precisely near the Soviet coast.

Electronic navigational alternatives had been developed during the Second World War. Navigation at sea (or, for that matter, above clouds) is difficult because both the sea and the upper air are essentially trackless. Electronic beacons, however, could lay down what amounted to paths and other landmarks. For example, in the 1930s electronically-indicated airways were set up in many countries. In 1940 the Germans used this Lorenz technology to set up paths along which bombers flew to attack British cities (two such paths crossed at the aim point).

In the case of the most important Second World War electronic navigational aid, Loran, master and slave stations were set up around the world. The master station sent out periodic pulses. The slave stations sent out their own pulses when they received master pulses. A ship could distinguish the two, and could measure the time difference between them. That equated to a difference in the distance from the ship to the two stations. A constant difference in distance defined a hyperbola. Maps were issued showing sets of hyperbolas associated with pairs of stations; a ship could fix her position at the intersection of two hyperbolas from two sets of stations. Unfortunately, due to the frequency used, Loran range was initially only 450-800nm by day, so successful operation demanded a world-wide network of radio stations, which would naturally make attractive wartime targets. Moreover, the Soviets were most unlikely to allow Western Loran stations near their territory. They had their own equivalent, but it was unlikely to be available in wartime. By the late 1950s there was a longer-range version of Loran, Loran C, operating at lower frequency (100 kHz, rather than the 1.85 MHz of Loran A), which offered somewhat greater range and precision, at the cost of greater complexity.[3] More importantly, because its frequency was low, Loran C could be received by a submarine at periscope depth; it therefore became important for Polaris submarines. Loran C coverage matched that of Loran A, except for additional coverage in the Mediterranean. In the 1950s a modified system, REDUX, was credited with a range of about 2500nm. Another possibility was to use signals the Soviets might inadvertently produce. For example, power lines inevitably radiate at low-frequency. In the early 1950s there was U.S. interest in using this radiation for navigation near the Soviet coast.

In the 1950s the U.S. submarine force considered RAFOS, or inverse SOFAR (sound fixing and ranging), a new technique used to locate downed airmen. It was known that sound could travel extremely long distances via the deep channel of the sea. Airmen could drop small explosive charges into the channel, and their sounds could be located by shore listening stations.[4] A submarine could suspend a hydrophone into the same channel to hear timed explosions set off at fixed places; in the Atlantic, it offered an accuracy of 5nm at a range of 950nm. That was hardly good enough for Polaris, but it gives some idea of just how difficult submarine navigation could be.

The alternative was to find landmarks in the sea. For example, the Earth's gravity varies from place to place. The U.S. Navy carefully measured that variation. An alternative was to use the carefully measured shape of the bottom (precise water depths: bathymetry). Neither was entirely satisfactory. For example, bathymetry is difficult in very deep water. Moreover, it was unlikely that the entire sea, out to (say) 2500 miles from the Soviet targets, could be mapped. Instead, specific areas would be very carefully surveyed. In fact, when the U.S. Navy began surveying for Polaris operations, the Soviets began to trail the survey ships. There was a very real possibility that in wartime Soviet submarines would lie in wait at the pre-surveyed points.[5]

It was, therefore, a considerable relief to find a solution in space. When the Soviets launched Sputnik in September 1957, it proclaimed its presence by the regular beeps of its radio. These beeps seemed to solve the submarine navigation problem, because the satellite followed a circular and entirely predictable orbit. Anyone with an accurate enough clock would know where it was at any given time. American scientists at the Applied Physics Laboratory of Johns Hopkins University, responsible for many U.S. naval air defence systems, decided to measure Sputnik's orbit by using the Doppler shift in the received signal. Then they realised that inverting this process would give the submarines the navigational fix they so badly needed.[6]

Although the satellite moved through its orbit at constant speed, the speed at which it seemed to move along the line connecting it to an earthbound receiver would vary considerably. The apparent speed depended on the angle between the line to the receiver and the line along the orbit, which changed as the satellite approached the receiver and then receded from it. That apparent speed determined the Doppler shift in the frequency of the signal from the satellite. The closer the receiver to the satellite's ground path, the larger the Doppler shifts between when the satellite first and last appeared (the shift passed through zero at the point of closest approach). A satellite thus mapped out a series of lines on what might otherwise seem to be a trackless sea. One set was satellite ground tracks; the clock and the zero Doppler shift would determine where the fix was along such a line. The size of the shift determined how far away the receiver was, along a parallel line. The satellite was a navigational beacon, as good as Loran – but it covered the whole Earth. It did have some limitations. The user had to track the varying satellite signal for 10-15 minutes, which meant the same period of mast exposure for a submarine.

Successful position-fixing requires that the user know the satellite's orbit precisely (it may change over time) as well as the exact timing of its passage. Thus the system incorporates ground stations which periodically measure satellite orbits and pass corrections to the satellite's onboard transmitter. Normally that information is provided in a message sent by the satellite as it passes overhead. The main source of error is refraction in the atmosphere, and to overcome it, satellites normally transmit at two frequencies; the receiver then compares Doppler shifts to cancel out most of this error. There is also an error associated with the movement of the receiver (course and speed), on board a submarine, during the fix.[7]

The concept was so successful that the United States launched its own navigational satellites, under the programme name Transit (as in a surveyor's instrument), mainly to support Polaris. The vital elements to the system were very accurate clocks and frequency standards, so that the satellite pass could be precisely timed and the small Doppler shift accurately measured. A single satellite pass provided a useable fix, allowing the submarine to update her inertial system every few days. Proposed in 1958, Transit entered service in January 1964, becoming the first operational U.S. satellite system.[8] The first experimental Transit 1A satellite was launched on 17 September 1959, but it failed to orbit; Transit 1B was orbited on 13 April 1960, and lasted 89 days. The first prototype operational satellite, Transit 5A-1, was launched on 18 December 1962.[9] The system was made commercially available in 1967, although low-cost receivers did not become available until the early 1970s. By 1968 three separate series, a total of twenty-three satellites, had been placed in circular orbits, mostly at about 805km altitude. In the late 1980s the Transit system consisted of six small (100lb) solar-powered satellites in 600nm circular orbits, each carrying a precise crystal clock and regularly emitting signals. Accuracy at a fixed site was about 100ft, but rated accuracy, 200m, was much worse.[10] By way of comparison,

Loran C was credited with an accuracy of 180m, but it was not available world-wide.[11]

Much depended on how stable the receiver was during the 10 to 15-minute fixing procedure. Eventually it was estimated that the system would give an error of 370m for every error of 0.5m/sec in ship speed. On the other hand, if the user was static, it could monitor pass after pass, averaging them out. In that case net accuracy (after twenty-five passes) might be as good as 5m. A receiver tracking a single satellite offered an accuracy of 80-100m; tracking two, it could eventually reach 15-25m.

Transit could not work effectively until the shape of the Earth (actually, the shape of the Earth's gravitational field) was understood in detail; conversely, Transit could be used to measure the Earth's shape – which would affect not only satellite orbits, but also the flight paths of ballistic missiles. Before the advent of Transit, it was known that the Earth was slightly flattened at the poles, but it was assumed that otherwise its mass was evenly distributed. Transit measurements showed otherwise. The Earth is somewhat pear-shaped, and the equator is slightly elliptical. There are 'lows' centred on the Carolina coast, about halfway between California and Hawaii, near the tip of India, and in Antarctica, and 'high' spots in Peru, near New Guinea, in the Iles Crozet, and in Ireland.

The system required considerable support from ground stations. Four stations in the United States (including Hawaii) collected Doppler data from the satellites as they passed overhead. These data went to a computing centre in California, which calculated each satellite's position and velocity, and estimated how these data would change over time. Each satellite received corrected orbital data from the computing centre, to rebroadcast it to ships using the Transit system. Given this data, navigators on board ships could calculate their positions.

Transit was not a complete solution for Polaris. To fire Regulus, a submarine had to surface and remain there for some time, so the 10-15 minute Transit fix time imposed little additional vulnerability. However, Polaris was fired by a submerged submarine. There was little point in putting up an antenna, and thus advertising that firing was imminent. Instead, Polaris submarines would periodically use Transit to fix their positions. Between such fixes, they would depend on a new internal navigational device, SINS (Ship Inertial Navigation System). It sensed the movement of the submarine and estimated its position accordingly. However, like any other inertial device, SINS drifted as time passed. It had to be reset periodically. Hence the key importance of Transit, or of some equivalent.

In fact the U.S. Navy combined all the available alternatives. Polaris submarines were equipped with three SINS, a Loran-C receiver (using a trailing wire, so that it could operate while the submarine was submerged), a Transit receiver, a terrain-matching sonar, and even a Type 11 periscope to take star sights for conventional navigation. Outputs from all the systems were integrated by a pair of navigational computers (NAVDACs) which fed the missile fire control system.[12]

In the 1960s a competing earthbound system, Omega, was added. It was a form of Loran using very low frequency (VLF, 10.2 kHz) radio, which penetrated water to periscope depth, so it could be used by a submerged submarine. Moreover, given its low frequency Omega could reach out to 6000nm, so that only eight stations were needed to cover the world (the last, in Australia, was completed in 1982). Due to its long wavelength, about 30km, Omega could not be as precise as Loran C; it was credited with an accuracy of 2200m (about 1nm). Omega was declared operational in 1968, when the first four stations were ready.[13]

Omega roughly coincided with a dramatic change in missile guidance. Polaris and its successor, Poseidon, relied entirely on the submarine's navigational system and any errors in that system would translate directly into errors in hitting a target. No matter how precise the missile's inertial guidance system, it could not get much closer than within a mile of a target. Polaris and its successor, Poseidon, were therefore always considered effective only against soft targets such as cities. Some strategists considered that a virtue. The missiles would be used only after the Soviets attacked the United States, and it seemed appropriate that retaliation would be exercised against the Soviet Union's cities and its population. In fact it could be argued that the only Soviet targets worth hitting were the hardened bunkers in which the Soviet leadership – which would have authorised the attack in the first place – was sitting. To get sufficient precision to attack such small, hard targets the U.S. Navy had to resort to real stars. The next-generation missile, Trident, used stellar-inertial navigation to make up for the errors inherent in Transit and SINS – and, for that matter, Omega. In this system the missile uses a star position, which it observes as it rises, to correct its estimated position.[14]

Accurate navigation has other profound consequences for naval warfare. Fleets at sea use a combination of on-board and off-board sensors; naval aircraft are often a sort of bridge between the two. For example, in 1941 the U.S. Navy operated a network of radio direction-finders around the Pacific. In theory, they would detect a Japanese fleet as it steamed towards action (in fact, the Japanese frustrated the U.S. HF/DF net by maintaining strict radio silence as they steamed towards Pearl Harbor). At the other end of the scale, U.S. warships were beginning to be fitted with radar. If, as the architects of the direction-finding net hoped, the Japanese did disclose their position and course by using their radios, then a U.S. force would be dispatched to deal with them. Interception would of course be tricky, not least because the fleet would not know exactly where it was relative to the detected Japanese ships. However, it could launch search aircraft. They could find the enemy force, overcoming navigational errors.

In modern terms, the HF/DF net (and, for that matter, code-breaking) produce time-late data. At best the Japanese would use their radios only intermittently, providing the U.S. net with a series of more or less precise fixes. Assuming that the Japanese were following a steady course, the fixes could be used to estimate their course and speed and thus to predict where they might be intercepted. The time-late data had inherent errors, and there was thus always some uncertainty as to just where the Japanese fleet was headed. Even the value of code-breaking could be limited. Much of the success of the U.S. submarine force against the Japanese was due to interception and interpretation of Japanese messages, many of them ordering important warship movements. In some cases, however, U.S. submarines could not intercept those ships, due to navigational errors – either theirs or the targets'. In their case, there was no way of overcoming inherent errors, since a submarine could not launch any search device, nor could it run fast enough to intercept a fast target. The lesson was that global sensors, which would generally be time-late, would be increasingly useful if navigation could improve, *ie* if a submarine at sea could know with certainty where it was when the global system decided where the enemy was.

Navigation also had important implications for the evolving style of naval air defence. In the aftermath of the Second World War, it became obvious that existing techniques for controlling air defences (particularly fighters) could handle only a limited number of targets. For example, in 1948 British trials in HMS *Illustrious* showed that her Action Information Centre could handle only about twelve raids (*ie*, separate air targets) per hour. Each stage in dealing with a raid entailed some delay. The radar operator could not instantly report the contact. He also generally inserted a small error in his report, not least because the blip on his scope was somewhat larger than the size of the radar beam demanded. The plotter entering the contact on a vertical plot would also add a small delay and some slight error. Deducing target course and speed added further delays and minor errors. Up to a point, all the errors were tolerable, because the plot was used to vector a fighter, and the pilot could make up for errors simply by using his eyes. But clearly there was a degree of error beyond which the pilot had little chance of intercepting a target, and this set the limit on how many raids the system could handle. Perhaps the worst lesson of the 1948 trials was that those inside the Action Information

Center (CIC, in U.S. parlance) were unaware of the extent to which they were falling behind. The solution, adopted initially by the Royal Navy and by the U.S. Navy, was automation. Radar operators could insert target data directly into a computer, which would do the rest. Its sheer speed drastically increased the number of targets (tracks) it could handle, typically to 128 or 256 in first-generation systems of the 1960s.

When CICs and Action Information Centers were first set up aboard ships, it was quickly appreciated that those within a fleet had to be linked together. The ideal was to offer each ship in the fleet much the same tactical picture (plot). For example, in the late 1940s and early 1950s the Royal Navy and the Royal Netherlands Navy experimented with television-transmitted plots, a camera recording the plot on board one ship so that it could be seen on board another. Only in this way could tasks such as fleet air defence be divided up efficiently. Computers made the task easier, since they could reduce a plot to a stream of computer commands. For example, in the standard Link 11 net used by many NATO navies, each ship transmits updates to the common plot. The computer on board each ship keeps track of the updates, to display the plot as needed.

None of this required space-based navigation. Yet the evolving system had a major navigational vulnerability. Each ship's sensors reported (to its own command) data in terms centred on that ship. For the common fleet plot to work, ships had to report in terms of a common grid, otherwise it would have been impossible for one ship to plot, say, an aeroplane detected by another. If ships were *not* properly oriented to the common grid, two of them might easily report the same object as two distinct targets. A fleet with inevitably limited overall resources could assign too many of them to the same target, leaving itself open to attack from some other quarter.

The Soviets began to deploy naval nuclear weapons just as the modern computer-based command systems first became operational. NATO naval forces had to spread out, so that one bomb could not destroy an entire fleet. Yet the fleet still had to be mutually supporting. That was possible, using long-range defensive missiles and naval fighters – but it greatly complicated the grid problem. Moreover, mutual support demanded quick and reliable communication among very widely separated units. The solution to the grid problem was not, in the event, space-based. Instead, each ship's local picture was matched to that of other ships, a process called self-consistent gridlock. It became possible as computer power increased, so that a ship's combat direction system could compare several slightly different pictures (in the U.S. Navy, as UYK-43 computers became standard in the 1980s).

FIVE

Passing the Word: Reliable Communications

Polaris was part of a nuclear revolution in warfare. Suddenly an initial blow could be decisive. It became essential that national commanders be able to maintain not merely reliable but also continuous contact with deployed and very distant forces. For example, U.S. national strategy in the 1950s was based on deterrence. For that to be credible, the Soviets had to know that as soon as their own attack was detected, U.S. forces – including deployed forces such as those aboard aircraft carriers – would strike back. The situation became acute as the Soviets began to deploy ballistic missiles, which could reach bases in the United States in only 30 minutes (warning time was likely to be 15 minutes or less). Missiles could reach bases in Europe and off the Asian coast even more quickly.

In each case, bombers (which carried the bulk of U.S. nuclear weapons even in the late 1960s) would have to be launched on warning of a Soviet attack, since otherwise they would be destroyed on the ground. Yet it was unacceptable to attack the Soviet Union simply on warning, which might be in error. The Air Force therefore developed a 'fail-safe' system, in which bombers took off on warning and flew to positions from which they could attack the Soviet Union. They would not carry out those attacks unless they received the critical 'go' message – which had to be sent over thousands of miles.[1] From a radio point of view, this was much the same as demanding that authorities in the United States always be able to send a 'go' message to, say, the Sixth Fleet strike carriers in the Mediterranean.

The Navy was in a somewhat similar position. Of all the forward-deployed nuclear-armed aircraft in the late 1950s, those aboard carriers had the best chance of taking off before their bases were destroyed (partly because the mobile carriers could not be hit by missiles, partly because some nuclear-armed aircraft were always ready to go). Yet the only existing means of long-haul communication to ships at sea, HF radio, could be unavailable for hours at a time due to atmospheric problems. The Air Force found itself relying on much the same technology.

During the Eisenhower Administration, then, the key issue in strategic communication was always to be able to order forces into action. That was not a trivial problem, because HF radio could not always be used. Something better was urgently needed. Space systems ultimately provided an answer, but not before considerable effort had gone into exotic alternatives.

In the 1960s reliable long-range communications became even more important to the U.S. military. The new Kennedy Administration adopted a doctrine of flexible response, in which control over the escalation of a nuclear war was intended to limit the damage the war would do. In this view war, even nuclear war, was imagined as a sort of bargaining process. As during the previous administration, it was assumed that defence against a Soviet nuclear attack was impossible. However, damage to the United States could be limited if the war were somehow controlled. To do that, the United States might offer to limit its initial strikes against the Soviets (eg to clearly military targets), in hopes that the Soviets would show similar restraint. Such attacks would certainly kill many civilians, but not as many as the sort of unrestrained attacks contemplated earlier. To convince the Soviets to limit their attacks, the United States had to retain the ability to make further attacks on the targets not hit initially. That demanded both enduring forces (such as Polaris) and the ability to command forces during a nuclear attack, implying both a capacity to plan further attacks and to communicate with the forces intended to execute those attacks.

Obviously naval forces were likely to have greater survivability than those based on land.[2] In particular, the flagships of the numbered fleets (eg Sixth in the Mediterranean, Seventh in the Far East) were likely to survive to plan theatre nuclear strikes. These ships needed particularly good long-haul communications, and thus became early users of satellite communication systems.

Kennedy Administration strategists went further. They gradually came to believe that virtually all military actions could be (and would be) read by the Soviets as messages. If they were misinterpreted, war might easily erupt. Thus the Administration became convinced that it needed extremely tight and continuous control over all U.S. forces.[3] That went even further in the direction of continuously reliable communications. To the Administration, the Cuban Missile Crisis dramatised this point. The Administration's thinking was encapsulated in one of its favourite books, Barbara Tuchman's *The Guns of August* (1962), which explained the outbreak of the First World War largely on the basis of a series of unthinking moves by the combatants. All such thinking was probably more fantasy than reality. As for nuclear bargaining, it seems unlikely that in the 1960s or even much later a President could actually tell whether a large nuclear attack was focused on military or civilian targets. Nor was there much evidence that the Soviets understood or cared for the restraint envisaged. For that matter, the difference in terms of people killed, at least in the United States and Western Europe, between military and all-out attacks was never very large; too many key military installations were always too close to too many people. President Eisenhower was probably right in thinking that deterrence boiled down to the threat of massive damage, and that it was unlikely to be tested in any more detailed way.

As for tight control of forces in all circumstances, the threat that a minor crisis would blow up into nuclear holocaust was surely grossly exaggerated. Again, as a former military man, Eisenhower had said that no one was likely to be so stupid as to unleash nuclear forces over some minor incident. Crises which looked as though they would escalate towards nuclear war would almost always be resolved at a lower level.[4] As for Cuba, in retrospect it is not nearly so clear that the world was at the brink of nuclear war. It seems that President Kennedy tended to inflate the crises in which he was involved, and it was certainly in his interest to make the rather poor bargain he extracted from Khrushchev seem more attractive, as a way out of a potential disaster.

In any case, the mix of theory and perceived reality made reliable communication extremely important. The need to control non-nuclear forces demanded not only reliability but also an unprecedented capacity, since such forces need far more than the order to execute some aspect of a preordained nuclear strike plan. Conversely, before Kennedy it had been assumed that naval commanders would have the good sense to do the right thing – and that nothing they could do would likely be disastrous (apart from launching nuclear strikes).

The effects of tight control were particularly evident in President Johnson's handling of the Vietnam War. Virtually all initiatives taken in the field had to be reported back up the chain of command, to C-in-C Pacific in Hawaii and then to the White House. Unfortunately, about 1968 the Soviets gained access to U.S. codes through a spy, John Walker.[5] Probably as a consequence, several attempts to spring surprises in Vietnam, such as at least one major air battle and the raid on the prison complex at Son Tay, failed. As experience on both sides during the Second World War showed, the victims of a code failure rarely realise what has happened, and the code-makers (in this case, the National Security Agency) are remarkably resistant to any suggestion that their products are at fault. U.S. coding techniques were not changed until Walker was caught.

Tight control meant that message traffic was so dense that information often moved extremely slowly. The loss of the intelligence ship USS *Pueblo* was a case in point. Her distress messages just did not reach the relevant authorities quickly enough, because they had to pass through too many systems before getting there. The degree of control exercised by the White House became notorious. For example, during the October 1973 Middle East War the White House reportedly sent the Sixth Fleet rudder orders, much to the fleet commander's disgust. In 1975, when the Cambodians seized the U.S. merchant ship *Mayaguez*, the White House demonstrated some of the problems which tight control could cause. When the carrier USS *Constellation* reported that she was approaching the area, a White House staffer ordered her to 'orbit at 35,000 feet' in the mistaken understanding that she was a Constellation radar aeroplane (an EC-121 – which could not have reached the ordered altitude in any case). To some extent, then, the availability of better communication supported a trend already underway. Only under President Reagan did the trend towards tight control seem to falter; he understood that there were few mistakes on the ground which could lead directly to a devastating thermonuclear war.

HF Radio

Before the advent of satellites, long-haul radio meant high-frequency (HF), which was developed in the 1920s.[6] Until the 1950s, HF was the only viable means of overseas voice communications, because undersea cables (which were used for telegraph/teletype) could not carry voice signals without unacceptable distortion. Thus it was HF radio which carried wartime conversations between President Franklin D Roosevelt and Prime Minister Winston S Churchill. Because HF signals were easily received away from their intended recipients, the Germans easily tapped into this circuit; hence the vital importance of wartime voice scramblers to protect it. Even when the first transatlantic telephone cable became operational in 1956, each circuit was extremely expensive. Thus the demand for high capacity point-to-point telephone service (both military and civilian) became an important driving force in the development of satellite communications. Only much later did the advent of fibre-optic cables revive undersea cables as a viable alternative.

Moving naval forces still needed some form of radio. HF achieves its very long range because its signals (sky waves) bounce off the ionosphere, each jump covering thousands of miles. HF signals also travel by surface wave (out to about 300nm). The earlier alternatives, low (LF) and medium (MF) frequency radio, generally did not travel nearly so far.[7] Compared to LF and MF, HF can also carry more information. As in all other forms of radio, the information is superimposed on the radio carrier signal. The higher the carrier signal frequency, the higher the frequency of the signal which can be imposed on it. A given amount of information could be transmitted in a shorter time via HF. The most extreme case was the German wartime use of burst transmitters (Kurier), which sent specially compressed signals to frustrate attempts at direction finding (HF/ DF).[8]

From a naval point of view, HF had some important failings. It was inherently somewhat unreliable. The ionosphere varies on a day-by-day (sometimes hour-by-hour) basis. The range of optimum frequencies, which depends on the vagaries of the ionosphere, is quite limited, and operators may find it difficult to find the appropriate frequency to reach a given station. Reception is virtually impossible in the shadow area between bounces to the surface, an area which varies as the ionosphere changes.[9]

Because the ionosphere consists of multiple layers, a signal can travel by several different paths ('multipath') and, at short ranges, also by surface wave. Copies of the signal arrive at the receiver at slightly different times, interfering with each other; the signal is said to 'fade'. The shorter the signals sent, the greater the chance that they will fade altogether. Typically individual signals (dots and dashes) are prolonged so that they do not fade out: reliable HF transmission has long been limited to 75 symbols per second (75 baud), often rendered 75 bits/sec, and often equated to 100 words/min.[10] This is teletype speed. Voice transmissions are often badly distorted (ideally they require more bandwidth than HF offers). Given the drastic limits on any one HF circuit, as demands increased, the number of separate HF circuits had to increase dramatically. Had capacity been greater, several messages could have been squeezed together (multiplexed) in a single channel.

HF required large antennas. Wavelengths were so long that sending and receiving antennas could easily interfere with each other.[11] For example, an HF antenna near another could act to reflect or block signals from some angles. Changing the frequency used by one antenna would often affect the operation of all others on board a ship with many antennas.[12] At the least, placing numerous antennas in close proximity drastically reduced their efficiency. The demand for greater numbers of independent HF circuits grew rapidly after the Second World War, but it was difficult to provide ships with enough separate antennas. The U.S. Navy developed special broadband antennas which could support multiple circuits simultaneously, but even they presented problems.[13] For example, the popular antenna position in the bows blocked an important gun arc. For a particular transmitter and receiver, frequency, and direction, different shipboard antennas will have different efficiencies. Establishing an HF link (or restoring it after loss of connection) means matching the receiver to the transmitter. It can be a laborious process. Moreover, HF wavelengths are so long that multiple transmitters on a ship can interfere with each other. Signals bounce off parts of the ship in ways so complex that they are very difficult to predict. For example, the U.S. Navy generally builds brass models of its ships in order to estimate the extent of interference between different emitters.

Antennas compact enough to fit on board ships were relatively inefficient, so under poor conditions they often failed to receive vital messages. Navies thus came to rely heavily on shore stations, where much larger (hence more efficient) antennas could receive signals more reliably, to retransmit them to nearby ships. Thus really good HF coverage demanded the use of secondary stations around the world: a navy's reach was determined to a considerable extent by its access to radio stations. This was not really a new problem; in fact HF required fewer radio stations than its predecessor, MF.[14]

For the U.S. Navy, the problem became more acute as it became obvious, in the mid-1950s, that the future focus of operations would be in what is now called the Third World. U.S. naval strategists concluded that, once the Soviets had enough nuclear weapons, the Cold War would largely stalemate in Europe, and move to the periphery of Eurasia and, perhaps, to other continents, such as Africa. The base structure built up in the aftermath of the Second World War had included vital HF relay sites. However, large ocean areas were described as 'radio deserts'. The U.S. Navy secured communications stations on British-owned islands such as Diego Garcia, Socotra and Ascension, located in some of these 'deserts'. However, 'deserts' persisted in areas off South Africa, around most of South America, in the mid-north Atlantic, and in parts of the Pacific. Moreover, even when stations existed, they could not easily be enlarged to provide for a higher tempo of operations or to cater for larger forces. Furthermore, some key stations were in politically unstable areas. If they were lost, so would crucial communications capability – at just the time it was most needed. Examples included Tripoli in Libya and Asmara in Ethiopia.[15]

The U.S. Navy wanted a radio station which could be as mobile, on a strategic level, as its fleet, to deal with the sort of unexpected crises which now seemed likely. The station would collect radio traffic from ships and force commanders (ship-to-shore), transmit messages back to the ships and the commanders (broadcast and individual shore-to-ship); and also relay traffic to and from the area of operations (medium- and long-haul point-to-point communication).[16]

Two solutions were advanced. One was a ship (or ships) with antenna efficiencies comparable to those on shore. The other, which some preferred to a ship, was an Air Transportable Communications Unit (ATCU) site housed in vans, with erectable masts. In both cases, the central issue was real estate: to be efficient, transmitting and receiving antennas had to be well separated. ACTUs could be coupled together to gain capability, but the result could be quite massive. For example, in 1962 a planned naval communications station at Souda Bay in Greece with approximately the planned capacity of the ship required forty-two vans, spread over about 800 acres.

The antenna interference problem was so severe that at first it seemed that the best shipborne solution would be to use one hull for transmission (AGMT) and another for reception (AGMR), the two being connected by microwave links.[17] A tentative long-range shipbuilding programme prepared in 1960 called for two pairs of ships, one for the Atlantic and one for the Pacific. The best hulls for the purpose would be laid-up Second World War escort carriers, because their decks were uncluttered and because they offered large open hangar spaces, which could be used without going to the expense of cutting open the ships. Alternatively, a single ship could be converted, offering about half the capacity and efficiency of the paired ships at a ship-to-ship range of 1500 to 2000nm. Later it was pointed out that the microwave connection between two separate ships would be both expensive (estimated at $1 million in 1962) and unreliable. By the spring of 1962 estimates showed that the escort carrier flight deck was long enough to separate transmitting and receiving antennas. Two separate ships would cost twice as much as a single AGMR capable of doing much the same job.[18] As a measure of just how much radio equipment had to be packed into a single hull, the U.S. Navy's 1970 communications manual warned that when all transmitters were radiating at full power, an AGMR could create a powerful enough electromagnetic field to set off exposed fuzes on board ships (*eg* missile ships and carriers) up to 4 miles away.

Quite separate from the long-haul issue is communication within a naval formation. Until the late 1950s, ships within a group operated operate more or less within sight of each other, using radios working at the very high frequencies (VHF, then UHF) at which propagation is quite reliable and signals generally do not travel much beyond the horizon.[19] That is why U.S. warships during the Second World War communicated relatively freely on their VHF Talk Between Ships (TBS) sets, whereas HF signals were fairly strictly controlled. But, by the late 1950s the Soviets were credited with naval nuclear weapons. The U.S. Navy therefore decided to disperse its fleets so as to limit the damage that any one such weapon could do, so it had to adopt HF radio for its vital tactical data link, Link 11, which fed the new computer-driven combat direction systems. The group's shared tactical picture was built up as a series of updates transmitted via Link 11, so the link had to be used more or less continuously.[20] The effect of the change to computer-driven combat direction systems, then, was to demand virtually continuous HF operation within a group of ships, despite the risk of interception. Ships in the group lacking Link 11 required constant formatted updates in the form of Link 14 teletype messages, so a group flagship had to broadcast nearly continuously on HF. Because ships would not be too widely dispersed, and because the vagaries of the ionosphere were all too well known, Link 11 relies on surface wave propagation. During the late 1970s, as it became more evident that the Soviets hoped to track U.S. formations from their Link 11 emissions, special measures were designed to limit propagation range.[21] Such limitation also helps protect one Link 11 net from interfering with another.[22] There is also an alternative UHF version of Link 11, for use when ships are within about 30nm of each other.

In this system, which is now standard in NATO, a master ship periodically calls the roll; the pickets (the others in the Link 11 net) send their updates. Link 11 data are used to control fleet aircraft and to cue missiles against enemy aircraft and missiles. Both the aircraft and the defensive missile systems can tolerate a certain amount of inaccuracy in their initial data, but only so much. That inaccuracy is due, in the end, to how stale the data are, which is largely a question of how long the Link 11 cycle is. Thus the need for accuracy limits the number of participants in a Link 11 net.

Link 11 posed special problems because it had to pass information more quickly than the usual 75 baud limit imposed by HF. The solution, which provides 2250 bits/sec, is to stack thirty different signals (each at its own frequency, carrying one bit of a 30-bit word). Thus, instead of transmitting 75 bits per second, the link transmits 75 *frames* (words) each second. Even so, six of the bits in each word are used merely to detect and correct transmission errors. In effect messages are sent in parallel.[23] The price is complexity and, often, unreliability, since it may be difficult to distinguish the full range of frequencies. Using a computer and periodic test messages, an HF channel can achieve this sort of performance while sending simple digital (on/off) signals. The computer regularly checks the test characters to adjust reception and thus to overcome multi-path. This technique is used in a current version of Link 11 and in the new Link 22, both of which operate at 2250 bits/sec. A similar technique has been used successfully for underwater communications, which also suffer from fading due to multipath.[24]

The major exception to naval reliance on HF was submarines. HF signals do not penetrate water, so to send or receive them a submarine must put a substantial antenna above water. Before the Second World War the U.S. Navy (and others) found that compact loop antennas could receive LF signals at periscope depth. The low data rate associated with LF was tolerable because the submarine did not expose herself while listening.[25] A submarine could not, however, send signals while submerged, so she generally used HF for that purpose. For Second World War submarines, which spent most of their submerged time at periscope depth, LF was perfectly adequate. However, for post-war submarines, which preferred to operate more deeply submerged, both LF and HF required either a periodic break in operations simply to receive signals, or some way of placing an antenna in position at or near the surface while the submarine stayed much deeper. The search began for some alternative. It turned out that VLF signals (at Omega-like frequencies) could propagate almost around the world in something like a surface mode; in 1959 signals from Annapolis were received by a submarine in the Eastern Mediterranean, 6000 miles away.[26] Later it turned out that ELF signals could penetrate much deeper into the sea.[27] In each case, reduced frequency meant reduced data rate, to the point that ELF is largely a bell-ringer to inform a submarine that she should rise to receive a message via a more effective medium. As in the case of LF, in neither case can a submarine be equipped to send messages to distant land stations.

The submarine communication problem was particularly critical for Polaris submarines and their successors. For them to present an effective deterrent, they always had to be on call to fire at the Soviets. All the water-penetrating forms of radio required very large, distinctive antennas, and clearly they could easily be destroyed by nuclear weapons. For many years the most survivable means of sending a 'go' code to the submarines has been a miles-long VLF antenna which could be extended beneath an orbiting TACAMO ('Take Charge and Move Out') aeroplane.[28] The aircraft in turn needs some link back to national command authorities – which is another radio problem, best solved, it seems, by satellite.

Another difficult situation arose with submarines assigned to work with a battle group. The U.S. *Los Angeles* class was justified specifically as a battle group escort. It therefore had to share the tactical picture used by the rest of the group, a picture normally transmitted by HF radio. The proposed solution was for the sub-marine to tow an HF receiving antenna on a kite which could project it out of the water. The kite was soon abandoned because it would have required too massive (and hence too draggy) a submarine sail to house it, but considerable work was done on towed antennas incorporating mini-hydrofoils.

HF Radio and the Transformation of Naval Warfare

When HF did work, it was far too easy for an enemy to intercept. From its beginning, naval radio had been very much a double-edged sword. It provided its users with timely communication, but when a fleet sent radio signals, those signals could generally be picked up by its enemies, and their sources could be located by a net of radio direction-finders (DF). Furthermore, in many cases their contents could be revealed by code-breakers. In either case the natural recipient of the mass of radio-derived intelligence data was the central directing brain of the navy, not the commander of a deployed fleet. Indeed, a headquarters ashore might well have better current information about an enemy fleet's movements than a commander afloat, at least until the fleets came into contact. Radio made it possible, in turn, for the directing brain ashore to provide that information to a distant fleet commander, so that he could bring his force to within horizon range (detection range) of the enemy. This was the beginning of a transition to the new kind of war which exploits satellites and over-the-horizon sensors.

What had in the past been considered intelligence data increasingly became operational data. For example, during the First World War the Admiralty used captured naval codes to read intercepted German naval radio messages. Given that data, it could plot the position of the German High Seas Fleet. Just such intelligence was used to send the British Grand Fleet into position to intercept the High Seas Fleet at Jutland on 31 May 1916. The Grand Fleet scouts could make up for the imprecision of the data (not least because neither fleet could navigate exactly). On the other hand, the British found decoded U-boat information far less useful. They had no way of searching the area around an expected or reported U-boat position, bridging the gap between time-late data and real-time location.

As is now known, the code-breaking success of the First World War was eventually exceeded in the next war. Code-breaking then provided the Admiralty with so precise and detailed a picture of enemy operations that it was no longer entirely clear how to divide responsibility between itself and the deployed fleet commander. For example, in June 1942 the British ran a convoy, PQ 17, to Russia past the Norwegian base of the German battleship *Tirpitz*. The convoy's escort was far too weak to stand off the battleship. If she appeared, the best defence was to scatter the merchant ships, so that only a few would be sunk by gunfire. On the other hand, any decision to scatter would make them vulnerable to U-boats and torpedo bombers, which the convoy escort was able to counter. As the convoy approached Norway, German radio signals suggested that the *Tirpitz* was about to sortie. At the Admiralty, the First Sea Lord, Admiral Sir Dudley Pound, decided that his information was significantly better than that possessed by the convoy commander. He ordered the convoy to scatter, against the advice of the local commander. The result was disastrous. *Tirpitz* never came out (her sailing orders, which had been given, were cancelled), and the U-boats and bombers destroyed most of the dispersed merchant ships. Sometimes what seems to be compelling information is misleading, intentionally or otherwise. The lesson may well be that the new style of warfare is quite vulnerable to error or deception; the central authority does not always know best.

Satellite Relays

The problems of HF unreliability and of limited capacity were so serious that some quite exotic techniques were seriously tested during the 1950s. All involved some alternative to the unreliable ionosphere, some alternative against which signals could be bounced. For example, radio signals can be reflected by meteor trails. It happens that meteor showers are so common that they form a reliable reflector, albeit not so high that they can carry signals around the world.

The U.S. Naval Research Laboratory (NRL) experimented with bouncing signals off the Moon itself. In 1946 the U.S. Army had successfully bounced a radar signal off the Moon, and the NRL had discovered that a really large radio telescope could detect Soviet radar signals inadvertently bounced off the Moon (see Chapter 7). The next step was to use Moon bounce for communications. The NRL bounced its first radio signal off the Moon, using the world's largest parabolic antenna, in October 1951, and made the first voice transmission over the Moon radio circuit in July 1954; the first two-way communication was conducted in 1956. For a time a Moon relay system was operational on a small scale. This technique was used to collect data from ELINT ships such as the USS *Liberty*; in fact, one of the first Israeli hits on the ship during the 1967 war knocked out her Moon-bounce antenna.[29] Remarkably, Moon relay is now used by radio amateurs.[30]

Another approach, tried in 1960, was to orbit a large entirely passive satellite, Echo-1 (a 100ft aluminium-coated balloon). Yet another was to seed space with metal needles which would reflect higher-frequency signals, forming a kind of artificial ionosphere.[31] Because the ionosphere reflects HF signals, reflector systems worked at higher line-of-sight frequencies (VHF, UHF, SHF, EHF) which would pass through it.

In each case, the idea was to provide a more reliable reflector than the ionosphere. As long as nothing but the chosen reflector affected the signals, there would be no multipath problem. Against that, signals would be travelling over much greater distances, so they would be far weaker at the receiver. A weaker signal would encounter more noise at the receiver, so it might be more difficult to distinguish signals; the data rate would suffer. If the signal-weakening problem could be solved, then the space system would offer a much higher data rate, since signals would be at much higher frequencies, and they would not have to be spread out in time to overcome multipath.

The successful solution was an active satellite relay, in which the satellite receives a up-lined signal, amplifies it, and sends it back down via a separate link. A combination of very high frequencies and amplification would reduce the required size of any shipboard antennas. As is now well known, a British radio engineer, Arthur C Clarke, later a very successful science fiction author, pointed out in 1945 that a single satellite in geosynchronous orbit is visible from 40 per cent of the Earth's surface; three such satellites, evenly spaced, cover virtually the entire planet. Radio relays on board such satellites would provide reliable high-capacity service over much of the world's surface. Clarke patented his idea, and later commented wryly that exactly such satellites entered service just after his patent ran out.[32]

Only near the poles is coverage poor, since a radio signal sent towards the satellite grazes the earth. That is one reason the Soviets used lower-altitude satellites in very eccentric (Molniya-type) orbits. During the latter stages of the Cold War, when the U.S. Navy expected to operate at very high latitudes, it planned a series of special Articsats to supplement its high-altitude satellites.

The satellite's great advantage is that it always operates in line of sight. There is no multipath problem to demand long pulses and thus to limit the rate at which data can be sent. Satellite operating frequency is set by the 'windows' in the ionosphere, through which signals can pass unhindered. They turn out to be at radar frequencies: VHF, UHF, SHF, and EHF, *ie* at 30 MHz to 300 GHz, all above HF.[33] Thus it takes only a relatively small dish, similar in size to that used by a fire-control radar, to form the beam for transmission or to gather it for reception. The dish sends a very narrow beam skywards. Although, inevitably, the dish has side lobes, they are relatively small, so hostile interception at the Earth's surface is difficult.

As in radar, the higher the frequency, the smaller the dish or other antenna needed to form a beam. Thus, although VHF signals can pass through the atmosphere, generally they are not used by naval satellites; the necessary antennas would be too massive. Similar considerations would apply to the earth stations. Also, the higher the frequency, the more signal it can support. On the other hand, the higher the frequency, the more subject a signal is to rain and other atmospheric interference. Atmospheric loss rises steadily at frequencies above 10 GHz, reaching a peak at about 25 GHz, with another peak at 50-60 GHz and a trough (still higher than the rate before the low peak) at about 35 GHz. These figures explain the choice of frequencies in EHF satellites such as Milstar. Loss due to rain rises steadily with frequency, exceeding atmospheric loss at about 3-4 GHz. In heavy rain (25mm/hr), attenuation at 10 GHz is about 500 times as great as at 1 GHz, and that at 20 GHz is about six times that at 10 (all expressed in dB/km). All of these losses could of course be made up for by increasing power and antenna gain (size), or simply by exploiting the greater bandwidth inherent in a higher-frequency signal. To further complicate matters, the ionosphere through which the signals must pass is quite variable. Geomagnetic storms affect the ionosphere and change its transparency at satellite frequencies. Because the changes are sudden and rapid, data can be lost or corrupted, as the ionosphere affects the signal passing through it.

The figures for absorption in the atmosphere mean that UHF (about 225-400 MHz) is essentially immune to weather. It is also immune to interference from foliage (UHF signals are, however, affected by the ionosphere). Moreover, UHF technology is simple and inexpensive, and UHF may also be peculiarly well suited to mobile users, because a satellite will generally produce a very broad UHF beam, covering a wide area (EHF and SHF beams are often much narrower, typically focused on a small area within the satellite's footprint). A satellite can produce only a limited number of separate beams, so this narrow focus may deprive many users within the footprint of simultaneous access to the system. That is a problem, for example, in the new U.S. Global Broadcast System (GBS). These considerations make UHF satellites attractive not only to naval users but also to an increasingly dispersed army.[34] On the other hand, the lower the frequency, the less the bandwidth available to carve into channels, and competition for channels becomes fiercer. With limited bandwidth it is difficult to provide a UHF system with very good anti-jam or stealth (LPI) performance. Systems may suffer from self-jamming. In the past, moreover, the cost of a terminal rose with frequency, which is why UHF is so widely used by the United States. Now that higher-frequency devices have come down in cost, there are so many UHF systems in service that it would be difficult to shift to some alternative.

Overall, communications satellite performance tradeoffs are defined by a link equation: so much signal strength goes into the link, and so much is available at the receiver. For a satellite there are actually two connected equations, one for the up-link and one for the down-link. In each case, a key point is that the amount of signal available depends on the product of three factors: the gain of the transmitting antenna, the gain of the receiving antenna, and the strength of the signal itself. As in radar, gain measures how much of the signal the antenna receives, or how sharply it defines the transmitting beam. In each case, it depends on the area of the antenna in square wavelengths: the larger the antenna, the higher the gain. Because what matters is a product, each factor can be traded off against the others. A bigger receiving dish can make up for a smaller transmitting dish or, for that matter, for a weaker transmitted signal. In either case, if the bigger receiving dish points its sharper (higher-gain) beam more directly at the transmitter, it will receive less of the noise which surrounds the real signal.

In a transponder satellite, the signal is received and amplified – and somewhat distorted. In effect the satellite adds both power and noise. The down-link adds more noise. Ultimately what matters is how well the receiver, which has its own antenna, can distinguish the signal from the surrounding noise. That is the content of the down-link equation. In some cases, such as Milstar, the satellite is designed to recognise the signal before transmitting it in very different form. In that case the satellite may be able to strip off some of the noise added during transmission by up-link.

All of this is true whatever the character of the up-linked signal may be. In fact the signal's originator may not even want it re-transmitted. As long as the satellite antenna has sufficient gain, it may pick up quite weak signals for amplification and re-transmission (or for recording). That is, there is no basic difference between a communications satellite and an electronic intelligence (ELINT) satellite. The only real difference is that the ELINT satellite picks up signals not intended for transmission into space. They may be line-of-sight radio continuing into space beyond the horizon, or the inevitable top lobes of line-of-sight transmissions. Satellite capability will ultimately depend on the required level of signal to noise for detection, and thus on the gain of the receiving antenna. For example, reportedly current U.S. ELINT satellites have receiving dishes about 100m across, while a typical communication satellite operating at the same altitude uses a dish a metre or two across. Thus the ELINT satellite would enjoy as much as 10,000 times the gain of the conventional satellite. That would make up for a signal 10,000 times fainter. Note that in radars the top lobe is often about 10,000 times weaker than the main lobe, and that this sort of ratio might also apply to line-of-sight communications systems on the ground.

As in any other radio system, the up-link transmitter applies a waveform, carrying information, to a radio signal. In typical active repeater (transponder) systems, the satellite strips away the up-link radio frequency, amplifies the signal, and applies it to the down-link radio signal. The down-link signal thus preserves the details of the signal sent up to the satellite. If (as is nearly inevitable) the satellite moves over the ground, it adds some Doppler to the signal. Operation is very simple, because the satellite need not apply any logic to the signal passing through it.

All signals contain some unintentional modulation in the detailed shape of their pulses. In theory, this modulation may be characteristic of the particular transmitter which creates the signal. This is the basis of radar fingerprinting. A satellite transmitting dish is not too different from a radar and, at least in theory, it too can be fingerprinted. Just how much of the satellite's apparent motion is reflected in Doppler depends on where the transmitter is, compared to the ground track of the satellite. Frequency (*ie* Doppler) can be measured very precisely. The down-link signal from a simple (analog) repeater satellite contains the detailed shape and Doppler of the up-link. In theory, both are measurable.

Even though the up-link may be extremely difficult to intercept, a satellite often down-links over a broad area. That is particularly the case if it has to serve mobile receivers, *eg* on board ships, since the satellite is unlikely to move its own beam to track them. Any receiver within its broad footprint can receive the down-link. In theory it is possible to build up a library of satellite transmitter fingerprints. Doppler measurements should indicate roughly where the transmitters are, though not with the precision achieved by HF/DF in the past (frequency-hopping may make such tracking difficult or impossible).[35] Given this combination, it should be possible to plot the positions of the transmitters. In 1991 the Soviets claimed that they had been able to track the build-up of U.S. forces for the Gulf War by 'traffic analysis'. That was unlikely in any literal sense, since for many years U.S. communicators have known how to counter such analysis.[36] Indirect tracking of satellite up-links would have been a very different proposition.

Then there is jamming. Most satellites use simple linear transponders: the stronger the uplink signal, the stronger the relayed downlink. There is only so much available power on any one transponder. A very strong uplink signal – a jammer – can sop up so much of the available transponder power that little or nothing is left for weaker signals at other frequencies. Furthermore, the transponder is not altogether linear. If enough power is fed in, it saturates, simply sending out a steady signal, rather than a varying one reflecting the details of the uplink. Thus a strong enough jamming signal, added to a weak uplink signal, may be enough to saturate the transponder and thus destroy the uplink signal altogether. A variety of methods have been devised to overcome these threats. The simplest, conceptually, is to avoid receiving the jamming signal in the first place. The narrow beam produced by a dish, which the satellite can manoeuvre, may be pointed away from a known jammer (at the least, this technique requires the jammer to lie close to the uplink signaller). Phased-array antennas, which replaced most dishes, can manoeuvre their narrow beams electronically. However, they cannot shape their beams to leave holes (nulls) where the jammers are. That requires an active array, in which each element of the array can be controlled separately. This is the satellite equivalent of an active-array radar such as the British Sampson or the German-Dutch APAR. Compared to a phased array, it offers an additional degree of control for each element (signal strength as well as phase). Other anti-jam techniques involve the structure of the signals and (in the case of Milstar) recoding within the satellite.

During the Cold War, satellites seemed to offer an alternative to surface communication systems in the event of nuclear warfare. A nuclear burst disrupts the ionosphere, providing signals with multiple paths rather than their usual single line-of-sight path. Signals following several paths interfere constructively and destructively at the surface; they seem to fade in and out. This flickering is called scintillation. Immediately after the burst the atmosphere is particularly turbulent, resulting in 'fast fading'; as it settles, there is 'slow fading'. Fading can be considered a burst error, like the effect of a pulse jammer; one countermeasure is interleaving, so that the burst is spread over several channels. Because a slow fade lasts longer, it is difficult to overcome. Another countermeasure is frequency-hopping, since scintillation effects vary with frequency. Both Milstar and the Universal Modem, which were designed to overcome nuclear effects, use frequency-hopping SSMA coding as an anti-jam and anti-scintillation measure. From a nuclear point of view, the higher the frequency the better. Because the absorption and scattering effects of high-altitude nuclear explosions are inversely proportional to the square of the frequency, nuclear interference would strongly effect a UHF system, moderately effect SHF (like DSCS), and only slightly effect SHF (like Milstar). Note that some natural effects, such as solar flares, cause nuclear-like problems on a lesser scale. VHF through SHF (30MHz to 30 GHz) signals are affected by solar radio noise. In some cases the satellite may mistake a burst of solar radio noise for a real signal, which is far weaker.

Information Capacity

Very crudely, the information content of a signal is proportional to its bandwidth. The information-carrying signal (baseband) modulates a carrier wave. The bandwidth of the imposed signal is a fraction of the frequency of the carrier; as a rule of thumb the information-carrying bandwidth can be no more than about 10 per cent of the carrier frequency.[37] Often it is much less, closer to 0.2 per cent. The amount of bandwidth available depends on how much noise the signal has to overcome. Noise can be thought of as an alternative signal spread out over a bandwidth. The narrower the bandwidth occupied by the desired signal, the smaller the fraction of noise mixed into it. That is why satellites generally use far less than the maximum 10 per cent of the carrier signal.

In any case, the higher the carrier frequency, the greater the bandwidth, so the more information the carrier signal can carry per unit time. Conversely, a higher-frequency carrier can transmit a given amount of information in a shorter time.[38] One way to see why is to imagine a signal running at the same frequency as the bandwidth, switching from positive to negative at the bandwidth frequency. The signal coded in this way cannot carry any more information than that supplied at each peak or trough of that signal. Actual information capacity is a fraction of bandwidth frequency, depending on the signal to noise ratio at the receiver. For example, for years the U.S. Navy's satellites used 25 kHz channels to send information at 2400 bits/sec.[39]

Better receivers and transmitters might considerably increase the amount of information sent via a channel. In the simplest system, a receiver distinguishes a 1 from a 0 by the strength (amplitude) energy associated with each (a 0 would probably be a negative square wave). If there were more energy available, it would be possible to send signals at two distinct strengths within the same duration, *ie* within the same bandwidth. That would double the amount of information sent. Given even more power, more distinct signal levels could be sent within a fixed bandwidth. A scheme using sixteen signal levels is twice as efficient, in its use of bandwidth, as one using only four – which is twice as efficient as one using only two (1 and 0).[40] To achieve such efficiency, the sending device must collapse a signal (consisting of a string of 0s and 1s) into a much shorter signal each character of which can have any value between 0 and 15. That requires a computer (in effect, translating a string of numbers from binary to hexadecimal – base 16 – form).

Performance depends partly on the technique used to represent the satellite's message. As in other forms of radio it can be by amplitude (AM), by frequency (FM), or by phase (PM: in effect, timing). Phase-shift keying (PSK) is now quite common, as it does not require more energy per character, hence does not overload an existing satellite transponder. For example, QPSK, Quad PSK, can give twice as many characters per bandwidth as conventional two-character PSK. An alternative is a form of FM, MSK, minimal (bandwidth) shift keying. In the actual design of a system much depends on the tolerable error rate (bit error rate, BER). For example, signals may be error-coded, allowing the receiver either simply to detect that there is an error (and thus to request a repeat) or even to correct the error. Such schemes reduce overall signal capacity because some of the bits flowing through the channel do not carry information directly.[41]

Overall, a satellite communication system can be power- or bandwidth-limited. The power-limited rate is given by the power which reaches the terminal, and which equates to a signal rate at a tolerated signal-to-noise ratio. However, the receiving modem has a finite efficiency in terms of bits/sec per Hz. Given a fixed bandwidth, that may actually be less than the level set by available power (transmitted power divided by transmission losses, multiplied by the gains of the transmitting and receiving antennas). Once a satellite is in orbit, the bandwidth it offers cannot be changed. Nor can the power of its transponders, which amplify the signals it receives from the ground for retransmission. If the system is power-limited, a bigger receiving dish (more gain) may improve system performance. Receiver noise may be reduced (to improve the signal-to-noise ratio). As for the transmitter on the ground, its noise level can be reduced, and its gain increased to reduce the overall level of noise mixed into the uplink signal. Alternatively, if the data rate is reduced, energy per bit will increase, which will also improve the signal-to-noise ratio. Ultimately what counts is the ratio of signal to noise, not the separate strengths of signal and noise. If the system is bandwidth-limited, the problem is in the coding system. Better coding (as described above) can drastically increase the data rate associated with a given bandwidth. The two approaches are complementary: a higher-gain receiver (with a bigger dish, or with better electronics) may be able to distinguish more subtle power differences within the signal it picks up. That better ability can be exploited by using a more complex signal. Typically a satellite has one or more broad channels which break down into sub-channels used for communication, separated sufficiently in frequency to make them distinguishable. For example, a 500 kHz channel (used in an early U.S. satellite) did not equate to twenty 25 kHz channels.

Satellites vary in design; some use a single transponder to cover several channels, whereas others use one transponder per channel. The single transponder solution is simpler, but in effect it must divide its power among the channels. A particularly strong signal in one channel (*eg* from a jammer) will swamp all the others. Smaller signals in other channels may not be transmitted at all. Thus, as launch capacity has grown and as electronics have become more compact, satellites have generally shifted to multi-transponder designs. For example, DSCS I satellites had a single 3-Watt 26 MHz transponder. DSCS II satellites had four channels and two 40-Watt transponders. DSCS III satellites have six transponders (originally two 40-Watt, two 16-Watt, and two 10-Watt), each with its own channel.

The standard HF data rate, the slow teletype rate, is 75 bits/sec. A typical fast teletype or computer rate is 2400 bits/sec, which in early

U.S. satellites required 25 kHz. Voice also requires the equivalent of about 2400 bits/sec for narrowband encrypted (secure) transmission, which may badly distort what the listener receives. The more faithful the reproduction, the more bits must be transmitted each second (up to 50,000) and hence the more bandwidth is needed. Facsimile circuits generally need voice quality. These considerations explain why the U.S. Navy's standard satellite channel is 25 kHz wide.

Once numerous pictures are involved, data rates rise very sharply. Television transmits one fairly crude picture every 2 seconds. The standard for a single colour television channel, the T-1 rate, is 1.544 Mbits/sec, more than 6000 voice channels (which may require as much as 2 GHz bandwidth in a satellite system). A black and white channel needs only about 4.5 MHz (128,000 bits/sec). On the other hand, a single high-quality image (*eg* from a reconnaissance satellite) may equate to as many as 100 million bits (100 Mbit). Some new communications systems are therefore designed for higher rates: T-3 (about 45 Mbits/sec) or OC-3 (155 Mbits/sec).

These information rates can be provided in many different ways. The simplest is a single channel represented by a single signal (single carrier per channel, SCPC). That is what a radio or television receives when the dial is turned to a particular station, which means a particular carrier wave frequency. However, it may be possible to send several different streams of information via the same physical channel, using multiplexing. There are three basic methods, frequency-domain multiple access (FDMA), time-domain multiple access (TDMA), and Code Division or Spread Spectrum Multiple Access (CDMA or SSSMA).

In FDMA, the simplest and most common technique, the channel is divided up into several sub-channels, each operating over a narrower part of its bandwidth, at a slightly different carrier frequency. If the differences are not too great, all of the sub-signals can be processed without serious distortion. The main problem is that the power used for each sub-channel must be balanced. The usual countermeasure to FDMA is a tone jammer sweeping across those frequencies, subjecting each channel in turn to maximum attack.

In TDMA, time is broken down into frames, one user per frame. Each signal is chopped up into short sequences, and they are interposed, so the channel carries first a bit of one signal, then a bit of a second, etc, until the cycle repeats. To make this technique work, the network demands very precise timing. To squeeze significant amounts of signal into each frame, the terminals feeding the channel (and fed by it) must run at high burst rates. On the other hand, there is no need to balance power among several sub-channels. The usual countermeasure to TDMA is a pulse jammer which imposes a burst of errors on one frame at a time.

SSMA or CDMA is the most sophisticated multiplexing technique of all. There are two techniques. As in TDMA, time is broken up into short intervals. As in FDMA, the basic channel is broken down into sub-channels at different frequencies. Each information channel has its own frequency-hopping scheme. During any given time interval, then, a receiver knows that (say) frequency X is carrying channel Y. Then all the channels hop, so that frequency X is now carrying channel Z. Channel Y is now on frequency W. A related technique is direct sequencing, in which a signal is 'multiplied' by a rapidly-changing key of different carriers. Hopping schemes can be uncoordinated, in which case signals may sometimes collide, or they can be maintained by some central controller so that they do not coincide (such schemes are called orthogonal, because in mathematics orthogonal functions are completely independent of one another). In current satellite practice, the simpler hopping schemes are exemplified by Milstar LDR and by the JSRC channel in the DSCS satellite. Examples of orthogonal coding are Milstar MDR and the Universal Modem.

SSMA is also an anti-jam technique. It counters the swept-tone jammer by evading its jamming signal: unless the jammer can match the frequency-hopping pattern, it will only rarely coincide with the target signal as it sweeps across frequencies. Unless the jammer has some way of anticipating the hopping scheme, it has to cover all possible hopping frequencies continuously – and the power it can apply to any one frequency will be drastically reduced. As for the pulse jammer, the TDMA solution is to chop up signals more finely and interleave them within each frame (*ie* within the pulse time of the jammer). That requires powerful computers at the transmitting and receiving ends. The burst of noise imposed by the jammer is spread among all the users, but none is very heavily affected, because the short-term burst is spread out over several de-interleaved frames. The main drawbacks to this technique are the computing power it demands and the significant delays in information transmission that it may entail. Overall, the greater the attention that must be paid to the threat of jamming, the greater the sacrifice in channel capacity. Thus an enemy need not use jammers in order to disrupt a satellite system; he only has to threaten their use.

From the point of view of a ship, a satellite dish imposes only a few demands. It has to be stabilised and it needs some means of tracking the satellite (*ie* the usual target-following electronics developed for fire-control radars). The dish must, moreover, have a wide sky arc, free of interference by superstructure or masts. Finally, the satellite signal is weak, so the dish needs powerful amplifiers. As electronics has developed, these demands have become easier to meet, and more and more ships have been fitted with satellite antennas.

The satellite in turn broadcasts a signal down over a wide area, often at a frequency different from that of the up-link: a single satellite can cover much of an entire ocean. It requires very few ground stations, so it can function whether or not there is some local agreement. Several equatorial countries have tried to tax or limit satellites overhead, but to no avail, since they lack any practical means of enforcement. No one has to guess in advance that a radio station should be put in place to handle some expected crisis; the satellite's radio footprint often covers so wide an area that it can deal with a great variety of possibilities. Factors such as the power output of the satellite determine just how elaborate the receiving antenna must be.

Clearly a single satellite cannot cover the entire world; really long-range communication requires something more. In most systems the satellite transmits down to a ground station, which retransmits to a second satellite for ultimate transmission as needed. In some later systems, satellites are provided with cross-links so they can dispense with intermediate ground terminals. However, they cannot be entirely independent of the Earth's surface, because generally they still require control, to maintain them in the appropriate orbits and attitudes.

Carrying the signal up and down imposes a definite delay between transmission and reception, because the signal follows a much longer path than point to point, as on the Earth. For example, a geosynchronous satellite 23,500 miles above the Earth imposes a path length of at least 47,000 miles, equivalent to a delay of something under a quarter-second. That can be significant, for example, if the satellite is used as part of a data-link net (as in the case of the standard NATO Link 11).

Much of course depends on how high the satellites are. If they are in anything like a geosynchronous orbit, then often a satellite will have both a sender and a receiver in sight at the same time. If not, then an intermediate ground station (repeater) is needed, as in the Vietnam case described below. The alternative, which the Soviets used, is store-and-dump via a lower-altitude satellite. The satellite receives a message for a terminal it cannot currently see, stores it, and then dumps it as it passes over the receiving terminal. This sort of arrangement becomes complicated if the position of the planned receiver (a moving ship) is not definitely known when the message is sent. In the geosynchronous or high-altitude case, that problem is taken care of by using a fairly broad beam for the down-link.

Early U.S. Programmes

By 1957 it was clear that, if an active relay satellite could be built, it would be far superior to the passive devices then under investigation. That year NRL proposed a relevant research programme. By the time its proposal had been adopted by the Navy (1959), the Defense Department's ARPA (Advanced Research Projects Agency) was already interested; in 1960 it set up a joint service satellite programme. The resulting ICDSP satellite was later abandoned when it proved too heavy for existing boosters. Indeed, in 1958-60 it was by no means clear that sufficient weight could be boosted into geosynchronous orbit to support the desired communications system.

A satellite in a much lower orbit would not long be in sight of sender and intended receiver. The interim solution, which the Soviets apparently later adopted for agent communications, was a store-and-dump delayed repeater. The first such satellite was the U.S. Army's SCORE (Signal Communication by Orbiting Relay Equipment), an entire Atlas missile which carried a tape-recorded Christmas greeting from President Eisenhower.[42]

In 1958 the Defense Department ordered development of a Clarke-type synchronous satellite system (Decree) backed by a medium-altitude repeater (Steer and Tackle). All three projects were merged into a single project, Advent, in February 1960.[43] Meanwhile all the strategic communications systems of the services were ordered merged into a single Defense Communications System run by a new Defense Communications Agency.[44] Advent was initially a synchronous satellite, but that proved too advanced, and instead a two-phase plan was proposed. Advent would employ groups of 1250lb stabilised satellites, the first in inclined 5600-mile orbits (launched by Atlas Agenas), others in synchronous orbits (launched by Atlas Centaurs). Technology advanced so quickly that by 1962 it seemed that 500lb (or even lighter) satellites could serve just as well. Advent was cancelled on 23 May 1962.

Now the Air Force proposed an alternative, a series of 100lb satellites placed randomly in polar orbits at about 5000nm altitude. The system was called the Initial Defense Communications Satellite Programme (IDCSP); a later geosynchronous system would be the Advanced Defense Satellite Communications Programme (ADSCP). Meanwhile Congress established a new civilian Communications Satellite (Comsat) Corporation, which would provide similar services. Before approving IDSCP, Secretary of Defense McNamara considered simply leasing Comsat service. That proved difficult, partly because Comsat envisaged international service. On 15 July 1964 McNamara decided to go ahead with a military system, and President Johnson approved it in August of that year. By that time the low-altitude concept had been abandoned in favour of a near-synchronous satellite system. Apparently the key development was the advent of a much more powerful launch rocket, Titan III, which was expected to carry several small satellites together into synchronous orbit.

Meanwhile the first civilian communications satellites entered service. AT&T launched its low-altitude (592 × 3500-mile) Telstar 1 on 10 July 1962, using ground stations at Andover, Maine, Goonhilly Down (UK), and Pleumeur-Boudou (France). The great impetus to this development was the desire to transmit television signals; existing underwater cables just did not have sufficient capacity (HF radio was altogether inadequate). It and subsequent experimental satellites could relay telephone and television signals, but only briefly, while the satellite was in sight of ground stations at both ends. However, on 26 July 1963 Comsat orbited its first effective Synchronous Communications satellite (Syncom 2, Syncom 1 having failed). In 1963 boosters could place just 35kg in a high (geosynchronous) orbit; that was just enough. Syncom 2 lay in a 28-degree orbit, and therefore was not really synchronous; but Syncom 3 (orbited 19 August 1964) was directly over the equator. It carried live pictures of the opening of the 1964 Olympics in Tokyo. In August 1964 an international consortium, Intelsat, was formed, launching its first satellite, Intelsat 1 (Early Bird), on 28 June 1965.

Because Comsat was a U.S. entity, its Syncom satellites were available for early Defense Department tests. For initial trials (1963) the U.S. Army provided two shore stations (Fort Dix, New Jersey, and Camp Roberts, California) with 60ft antennas (FSC-9 terminals). The ship terminal used a huge (30ft) steerable antenna on board the test ship USNS *Kingsport*. In several tests, *Kingsport* acted as a relay, for example supporting two-way voice communication between an aircraft in flight and a distant ship. The system used a 7360 MHz up-link and an 1815 MHz down-link.

At the start of the Vietnam War, the Syncom system was handed over to the Defense Department. Syncoms 2 and 3 were retired in April 1969, having exceeded their design lifetimes. For civilian purposes they were replaced by Early Bird (later Intelsat 1). It retired early in 1969, but was brought back temporarily to transmit pictures of the Apollo 11 Moon landing.

DSCS (Defense Satellite Communication System)

Meanwhile IDSCS was being built. At an international radio conference in Geneva, frequency ranges for communications satellites were allocated (to avoid interference); DSCS would use SHF frequencies, 8 GHz for its uplink and 7 GHz for its downlink.[45] These frequencies were attractive because they offered very wide bandwidths, hence large numbers of channels (each satellite operated several 50-, 60-, and 85-MHz repeaters). Given more bandwidth, a satellite can hop from frequency to frequency to evade a jammer. Moreover, SHF can support a higher data rate than lower-frequency alternatives (mainly UHF). SHF propagation is also more stable, so the system suffers less from fading and scintillation. At a higher frequency (shorter wavelength), an antenna of a given size offers higher gain and a narrower beam, to better resist both interception and jamming.

On the other hand, in the 1960s a satellite was unlikely to generate much power at SHF frequencies. The downlink would be relatively weak. Ideally, then, it would have to be concentrated in a relatively narrow beam pointed directly at the receiver. The receiver, moreover, would need a large antenna (with high gain) and a powerful amplifier. The high-gain receiving antenna would have a very narrow beam, so it would have to be pointed fairly precisely at the satellite. These considerations suggested that SHF communication would probably be limited to fixed ground stations or, at the least, to those with very limited mobility.

In all, twenty-six small (100lb) DSCS I satellites were launched between 16 June 1966 and 13 June 1968 (twenty-eight were made). They were placed in orbits (at 18,300-mile altitude) slightly below geosynchronous ones, drifting eastwards 30 degrees each day.[46] This system was used to prosecute the Vietnam War. For example, President Johnson and Secretary of Defense McNamara personally approved target lists. When DSCS I became available, it could transmit high-quality target photographs back to Washington. The Naval Research Laboratory set up the necessary satellite system in 1967. A DSCS I satellite over the Pacific transmitted pictures from Vietnam to a terminal on the U.S. West Coast (later, in Hawaii), whence it was re-transmitted, using a second satellite, to Washington (under Project Compass Link). This system had the highest encrypted data rate achieved up to that time.

DSCS I power was relatively limited, so there was some question as to whether a shipboard terminal would be practical.[47] Tests with Syncom 2 and 3 showed that a 6ft antenna could work. The first two SSC-2 terminals were installed on board the cruiser *Canberra* (1964) and the carrier *Midway* (1965). Early in 1965 the cruiser communicated with the test ship *Kingpost* via Syncom 3 satellite. Then the cruiser and carrier were able to communicate at a range of 6000nm. These experiments were so successful that seven improved SSC-3 terminals were bought in FY67, nine more following in FY68. They were intended for fleet flagships, carriers, and special command/control ships. The first was mounted above the bridge of the fleet flagship (cruiser) *Providence*. It was used in a major fleet exercise in November 1966.[48] As with SSC-2, this receiver used a 6ft parabolic dish. The system was fairly elaborate, requiring a shelter (7 × 7 × 12ft) containing the dish positioning and control elements, the transmitter power supply and control, and the operating console.

DSCS became the main high-volume Defense Department/State Department system, used by U.S. defence agencies for long-haul communication between fixed sites. As such, it was expected to provide a stable level of service. By way of contrast, the needs of a tactical user would vary dramatically over time. Thus at first this system was not to be used by tactical units in the field.[49]

The geosynchronous DSCS II, work on which was authorised in July 1968, followed DSCS I.[50] The first was launched in November 1971, and sixteen in all were launched through 1989. One was still operational late in 1998. The full system comprises four working satellites and two in-orbit spares; the first full constellation was completed in December 1978. Designed lifetime was 5 years (later 10). This system supported the Desert Storm campaign in 1991. DSCS II was provided with four channels so that it could handle multiple user communities: Channel 1 (125 MHz), Channel 2 (50 MHz), Channel 3 (185 MHz), and Channel 4 (50 MHz). It was much more sophisticated than DSCS I, with a multiple access capability, a command subsystem, attitude control, and station-keeping ability. However, it was not hardened against nuclear effects, nor did it have any anti-jam features. Each 1350lb satellite has one wide-beam (global, or earth coverage horn) antenna and two steerable antennas, to form three kinds of beam: global (18 degree wide), regional (6.5 degree wide, footprint 4100km diameter), and spot (2.5 degree wide, footprint 1600km wide). The narrow beam antenna is steerable (by ground command) within +/−10 degree. The Earth Coverage Horn was intended for embassies, distributed fixed users, and for ships and other mobile users. The spot beams were intended for fixed users; they saw some tactical use during Desert Storm. The satellite can handle 1300 duplex (2-way) voice channels or 100 Mbit/sec of data. It can be moved within two days to a new geosynchronous position to handle a contingency. Frequencies are: 7.9-8.4 GHz uplink, 7.25-7.75 GHz downlink. There are two transponders. The system was designed to work mainly with big fixed terminals such as the 60ft FSC-78 (in service 1975; nineteen built). The associated shipboard terminal was WSC-2, with a 4 or 8ft dish. While DSCS was being developed, work proceeded on a series of experimental satellites, Tacsat and LES-5/6, intended to support tactical users. They led directly to the Fleet Satellite described in Chapter 8.

DSCS III, the current system, began development in 1975 (the development contract was awarded in February 1977). The first was launched in October 1982. In all, fourteen were bought, the last of which was completed in 1991 and put into long-term storage. Plans called for a minimum of five operational satellites; originally DSCS IIIs were to have been launched in pairs by the Space Shuttle, but after the *Challenger* disaster, the Air Force decided to launch them singly on

Atlas IIs. Two were launched by Titan 34Ds in 1982 and 1989, two by Shuttle in 1985, and six by Atlas II in 1992-7. Two more are scheduled for launch by Atlas IIA, and the last two on Evolved Expendable Launch Vehicles (EELVs). The operational control sites are Camp Roberts, CA; Fort Detrick, MD; Fort Meade, MD; Landstuhl, Germany; and Fort Buckner (Okinawa); note the vulnerability of overseas cover to the loss of overseas sites.

Principal DSCS III users include the Defense Information Support Network (DISN), for which DSCS is the backbone, the Diplomatic Telecommunications System (DTS), the White House Communications Agency (WHCA), and Ground Mobile Forces (GMFs). Currently DSCS also services Special Operations Forces. These satellites also carry command links for the Air Force Satellite Control Network (AFSCN). Initially DSCS terminals were also installed on board Navy flagships. In the 1980s another role was added when the U.S. Navy began to deploy ocean surveillance (SURTASS) ships towing very long passive arrays. At the time, there was little expectation that data from the arrays could be processed on board. Instead, it was to be beamed back to the United States, where deductions from it would be fed into the ocean surveillance system. SURTASS data is transmitted to shore stations via a DSCS secure voice link (32,000 bits/sec). DSCS is unusual among satellite systems in that it serves so wide a variety of users, including those (*ie* White House and Special Operations) using very small low-gain antennas. In the case of the White House, capacity must be set aside to provide instant service, even though that service is not usually needed, while Special Operations requires special measures to avoid interception by an enemy.

Major drivers in the satellite design, which was considered advanced for its time, were the need to protect against jamming and nuclear and laser attack, and the desire for more (and more widely distributed) terminals of a wider variety of types. Because each of the six channels has its own limiter and transponder, it can have its gain set to accommodate a particular user: low gain for a large terminal, to reduce re-radiated noise; high gain for a small tactical terminal; and saturated for anti-jam terminals (USC-28 SSMA modems). However, there was no great interest in increasing capacity over DSCS II; DSCS III has the same bandwidth (405 MHz) and only 50 per cent more power. To counter jamming, the system has an encrypted uplink and is controlled via an X-band nulling antenna. Nominal satellite lifetime, set by fuel capacity at launch (600lbs of hydrazine), is 10 years. The satellite weighs 2143lbs.

In place of the dishes used on board earlier satellites, it uses three multibeam lens antennas (one 61-beam receiving antenna and two 19-beam transmitting antennas) controlled electronically from the ground, so that coverage can be tailored to a situation. The satellite can evade a ground-based jammer by ignoring (nulling out) a beam pointed at its uplink antenna. In addition to the lens antennas, the satellite has four wide-beam horns pointed at the earth (two transmitting, two receiving) and a high-gain steerable dish. There is also a UHF antenna (for the Single Channel Transponder, SCT), a one-way transmitter of the Emergency Action Message (the nuclear go-code), which uses an SHF or UHF uplink; it is part of the Air Force AFSATCOM system. The last four DSCS III have a second moveable antenna specifically to support increasing naval use of the SHF band.

Of the six channels, Channel 1 (50 MHz) was intended specifically for strategic control, providing Jam Resistant Secure Communications (JRSC) for low-data rate voice and data communications (32,000 bits/sec) for the Tactical Warning/Attack Assessment net, for the National Command Authority and for the C-in-C Network. It uses nulling to resist jamming. Associated equipment was the GSC-49 ground terminal (8 or 20ft in diameter; thirty-two bought). This mission is now shifting to Milstar (which was designed specifically to support it), and in any case it has become much less important with the end of the Cold War. Channel 1 is being 'destressed' and is to be used mainly by the Navy. This handover should be complete by 2003. Channel 2 (75 MHz) is intended mainly to work with small terminals which need more satellite power and high gain (using the steerable dish for its downlink). It is used by the Navy and by GMF (ground-mobile forces). Channels 3 and 4 (85 MHz) work with large fixed antennas, for users such as DISN and AFSCN. Channels 5 and 6 (60 MHz) work with the Earth coverage horns, providing full Earth coverage but low gain and low data rate; they also provide no jam resistance because they have no ability to null out a jammer. They are used for widely distributed users, including the Diplomatic Telecommunications System, the Navy, and Special Operations.

When the Gulf War broke out, only the four fleet command ships (USS *Blue Ridge*, *Mount Whitney*, *Coronado*, and *La Salle*, serving the Numbered Fleets or equivalent commanders) were fitted to receive DSCS messages, because the system was conceived as a channel between high commands. They relied on the 32 kbits/sec anti-jam Channel 1, which could not really cope with the demands of the war. It used a single or dual 4ft (1.2m) radome-enclosed dish. The SURTASS ships could send data via DSCS but could not receive DSCS messages. Plans called for DSCS terminals on board the carriers, but they were not yet in place. As an emergency measure, Air Force/Marine TSC-93 SHF vans carrying 4ft dishes were placed aboard large-deck ships (carriers and amphibious carriers) in a 'Quicksat' programme (adding 64 kbits/sec of data rate). Although

Quicksat was a war emergency programme, it proved impossible to get the terminals aboard the carriers during the Gulf War. The first ship to get a terminal (after the war was over) was the carrier *Abraham Lincoln*. During Desert Storm (1991), only the SHF satellites offered really high volume: a pair of DSCS II and III satellites transmitted more traffic than the US-to-Europe satellite communications used during the whole Cold War.

Post-war, the Navy decided to make SHF a primary channel for large-volume communication, *eg* of images for real-time retargeting. SHF terminals were ordered for every carrier and large-deck amphibious ship. As of 1994, forty-four systems had been bought, and options were open for fourteen more.

To accommodate the new tactical users, DSCS operation had to change: DSCS SHF shore stations were upgraded. The limited-capacity anti-jam channel, which operated on a spread-spectrum (spread-frequency) basis, was changed to a series of parallel channels with narrow separations in frequency (*ie* FDMA). Each ship is assigned its own channel. New large (7ft) WSC-6 dishes increase gain and thus quadruple signal strength without requiring more satellite power. Maximum capacity is now better than T1 (reportedly 2 Mbits/sec), although normally data is transmitted at 550-650 kbits/sec. However, typically only 64 kbits/sec to 1 Mbits/sec are assigned to a carrier or LHA/LHD. Full fleet SHF capability, as described in a 1992 Operational Requirement, extends to all cruisers, Aegis missile destroyers, and Tomahawk-capable *Spruance* class destroyers (cruiser installations began in 1999). The new large-dish version of WSC-6 and the new modem provide information rates, for surface combatants, of 256-640 kbits/sec, depending on antenna size (*ie* gain).[51] Current fleet SHF capacity (1999) is given as 128 -1024 kbits/sec.[52]

At the same time, the last four satellites in the series, not yet launched, were upgraded in a DSCS SLEP (Service Life Extension Programme). The first was scheduled for launch in July 1999 (actually launched 20 January 2000; the others are to be launched in 2000, 2002, and 2003; a fifth has not yet been funded). Major upgrades include increased bandwidth, less-noisy amplifiers, and higher power for all six transponders (50 W in each rather than two 40 W and four 10 W). This combination is expected to increase capacity by 700 per cent. Average lifetime should increase from 7.5 to 10 years. Upgrades to the amplifiers in the smaller terminals were expected to increase their data rates by about 30 per cent. All of these improvements make it possible to squeeze non-tactical users into two rather than three channels.[53] As of February 1998 estimated capacity of a single DSCS satellite was 109 Mbits/sec; DSCS SLEP increased that to 190. The last DSCS, to be launched in 2003, will probably go out of service about October 2012.

The stage beyond is the Wideband Gapfiller, the first of which should be launched in 2004, with two more to follow in 2005. To minimise cost, this is to be a commercial satellite, requiring no military research and development. The design is to emphasise capacity (estimated at 716 Mbits/sec) over protection, so nulling and hardening will be provided only if they do not drive up the cost. Gapfiller is conceived mainly as a tactical satellite; initially it was not to have provided any service between fixed points (later that was changed to 'no more than is on DSCS today'). The satellite will operate in both SHF (like DSCS) and EHF (Ka-band, like Milstar). Like DSCS SLEP (or DSCS IV), it will not have any anti-jam channel; instead it will have dual SHF (Ka-band) channels, for higher data rates. The Ka-band capability is valued because it provides the necessary wide bandwidth even with the sort of small dish a destroyer can carry. A baseline design, for estimating the cost of the satellite, had four Ka-band uplinks: one 1 degree spot, two 1.5 degree spots, one 3 degree spot. SHF uplinks: one 2.2 degree spot, two 6 degree spots, one 17.5 degree Earth Coverage antenna. Ka-band down links: 1 degree and 1.5 degree spots. SHF downlinks: 2.2, 6, 17.5 degree beams.

Skynet

The Royal Navy watched U.S. experiments with DSCS with great interest. In 1968 it tried its own satellite receiver, NEST, on board a frigate; like the U.S. Navy, it found that a small dish could be perfectly effective. A British Skynet system equivalent to DSCS had been begun the previous year, growing out of a 1966 U.S. offer to the British to participate in IDCSP. The British were interested, but they needed a small-ship capability in addition to the large-terminal capacity built into the U.S. system. Late in 1966 the U.S. and British governments signed an agreement under which the United States would develop a suitable satellite interoperable with IDCSP, initially called IDCSP/A (augmentation). At the same time NATO decided to participate directly in IDCSP, operating two IDCSP ground stations between 1967 and 1970, in preparation for its own NATO satellite programme (see below). The British Skynet design was derived from that for the U.S. satellite, but it was placed in geosynchronous orbit and was designed for station-keeping. It also had a despun antenna (the satellite itself spun for stability) for greater gain, a command system, and two channels (2 and 20 MHz).[54]

The first Skynet was launched in November 1969 to provide strategic communication between Britain and the Middle and Far East, the British having just lost their naval base (presumably with HF facilities) in Aden, at the mouth of the Persian Gulf. The satellites were built by Philco-Ford. Since DSCS I was not quite geostationary, and the first DSCS II was not to be launched for another two years, Skynet 1 was the first geostationary communications satellite system intended from the first for military use. It provided 23 voice circuits or 250 teletype channels.

A second-generation Skynet 2B (2A was lost on launch) replaced Skynet 1. It was launched on 23 November 1974. This system was built by Marconi Space and Defence Systems (with help from Philco-Ford), and thus became the first communication satellite built outside the U.S. and the Soviet Union. Like Skynet 1, these used SHF (X-band) up- and down-links, and they had paired transponders and receivers (one standby in each case). Channel bandwidth was the same as in Skynet 1 (2 and 20 MHz), but maximum power increased enormously, to 20 Watts (typically operating at 18). The system was set up so that the narrowband (2 MHz) channel, which was used by small (*ie* shipboard) terminals, received 80 per cent of the available power. The Skynet 2B command system failed early in 1977, so the satellite began to drift in longitude (following a two-year cycle, between 0 and 150 degrees East). It remained in use until 1987, and when tested in 1994 was still useable. Apparently this satellite did not provide adequate capacity during the Falklands War, the British having to rely on the U.S. DSCS system.

Skynet-3, which began project definition in 1972, was cancelled in 1974 as a Defence Review argued that, with the continued withdrawal from East of Suez, it was not needed. Presumably Royal Navy operations in the expected area of greatest interest, the Norwegian Sea, would not require anything more than HF (or perhaps even LF/MF). In fact, however, the Royal Navy continued to operate on a global scale, and increasingly it was using U.S. and NATO satellite communications.

The decision to build a Skynet 4 was apparently taken in 1978. This system began project definition in 1980, and production (by British Marconi and British Aerospace) began in 1982. The programme is sponsored by the RAF, but the Royal Navy is the chief user (and paid 80 per cent of first-phase cost). Like DSCS, Skynet uses SHF frequencies (7-8 GHz) to communicate with surface terminals. The satellite has three SHF transponders providing four SHF channels (60-135 MHz bandwidths), with beamwidths down to 3 degrees.[55] These channels are compatible with the U.S. DSCS channels. Two are jam-resistant. Skynet-4 also has two UHF transponders compatible with the U.S. Navy system; they are reportedly used mainly to communicate with British submarines. There is also a single experimental EHF up-link. There is apparently no direct equivalent to the U.S. Navy's Fleet Broadcast (see below). There are four beam patterns: earth coverage, wide area (North Atlantic), narrow-beam (Europe), and spot (Central Europe, for the army's small terminals). Like DSCS III satellites, Skynet 4 is nuclear-hardened.

Phase I provides global coverage in the form of three satellites; Phase II replaces the first two after five years. The initial trio of satellites, Skynet 4A, 4B, and 4C, were launched in December 1988, January 1990, and August 1990. These launches had been considerably delayed by the *Challenger* disaster.[56] Their replacements were 4D (January 1998), 4E (February 1999), and 4F (due for 2000). A proposed Anglo-French- German system was abandoned in favour of the future Skynet-5, the first of which is to be launched in 2003 (see below).

The associated ship's terminal is SCOT. Frigates and destroyers use a SCOT 1 (1.1m dish); larger ships use SCOT 2 (1.8m dish). An enhanced SCOT 1A provides a 1.2m dish for both classes of ships and the equivalent of four secure-voice communications channels. Bandwidth is 500 MHz, compared to 50 MHz for SCOT 1; the 1A terminal can be tuned to any 50-MHz channel. In the Falklands War, the use of radar-like frequencies for satellite communication had a very unfortunate consequences. On 4 May HMS *Sheffield* was escorting the carrier HMS *Hermes*. The British were well aware that satellite antennas had detectable side-lobes. To avoid giving away the carrier's position, they assigned *Sheffield* to communicate back to the United Kingdom by satellite. The antenna of her satellite dish unavoidably spilled some of its signal into her ESM receiver. To avoid false alarms, the latter was disabled for the duration of the communication. Naturally, that was when the Argentine Exocet-firing aircraft appeared. The ESM receiver would have picked up the attacking aircraft – but for the need to use the satellite dish.[57]

After the war, the Royal Navy found a solution. Communication was limited to bursts, during which the ESM set was blanked out. Unfortunately this practice drastically limits the capacity of the satellite channel. Worse, there is now much more interest in making communications stealthy. That is generally done by stretching out the signal, spreading it over a wide bandwidth, which makes bursting impossible. The problem remains.

NATO and Other Western Systems

While the British developed Skynet, a very similar NATO system was built. NATO I was the pair of IDCSP ground terminals. NATO IIA/B were similar to Skynet I but had a different antenna pattern, covering only the NATO countries (Skynet had a relatively uniform pattern pointed at the equator). The NATO satellite system was intended to provide reliable transatlantic and transeuropean service. Because of NATO's geographical limitations, only a single active satellite was needed (the second was a spare). NATO-2A and -2B were launched on 20 March 1970 and 2 February 1972.

Three third-generation satellites (NATO III) built by Ford Aerospace were launched between April 1976 and November 1978; each provided three SHF channels (several hundred voice and teletype channels), and could be served by shipboard SCOTs or by the U.S. WSC-6.[58] A fourth satellite, NATO-3D, with twice the RF power, was launched in November 1984.

The fourth generation NATO-4 is a modified British Skynet-4. It provides two 25 kHz UHF channels, presumably mainly for naval use, as well as four 60-135 MHz anti-jam SHF channels for high command links, which are compatible with DSCS. Each satellite has an area coverage transmitter for the Atlantic (to reach ships in indefinite positions) and a narrow-beam transmitter pointed at Europe.[59] As in the earliest series, one satellite provides coverage, the other being an in-orbit spare. NATO IVA was launched on 7 January 1991, NATO IVB on 7 December 1993; the system was fully operational in 1994. These satellites serve the NATO Integrated Communications System (NICS), serving national command authorities and designated NATO CINCs.

France placed her own military communications system, Syracuse, employing five SHF transponders (2400/75 bits/sec) on board the two French Telecom 1 satellites under a 1980 agreement; coverage extends from the West Indies to Reunion. Syracuse I became operational in 1987. The associated shipboard system used a pair of 1.5m stabilised antennas in radomes. Syracuse II, which entered service in December 1991, employs transponders on board three satellites: Telecom 2A (launched 1991), Telecom 2B (launched May 1992), and Telecom 2C (reserve). As in the Telecom 1 system, each carries five X-band (SHF) military transponders. Dishes are 0.90m diameter for surface ships and 0.40m diameter for Atlantique Mk 2 aircraft. Telecom 2B and 2C incorporate a laser cross-link to the Helios photo reconnaissance satellite (they down-link to surface units). A follow-on Syracuse III will be needed about 2004-06.

In December 1997 France, Britain, and Germany all agreed to build a joint next-generation military satellite system, Trimilsatcom. It would replace Skynet 4 and Syracuse 2 and it would provide the Germans with an initial capability (which they expect to need about 2007). However, in August 1998 the British abandoned the programme in favour of a national Skynet 5, on the ground that Trimilsatcom might well not be available in time.

Spain operates her own pair of Hispasat hybrid military-civilian geosynchronous satellites (1A and 1B), launched September 1992 and July 1993), which link her fleet flagship *Principe de Asturias* to her national command centre (Spain is not part of the integrated NATO military command).[60] The satellite terminal is a SCOT, so the ship can also receive NATO signals. The satellite's X-band payload is based on that of Skynet 4. The link to the command centre is the Sistema Espanol de Comunicaciones Militares por Satelite (SECOMSAT), which carries two voice, three teletype, and two data channels (1200 bits/sec Sacomar, 9600 bits/sec Simacar) plus a computer data channel. All channels are encrypted.

All Japanese destroyers are equipped to communicate using the Superbird B communications satellite, which was launched on 27 February 1992. The Superbird series is operated by the Space Communications Corp. of Japan, which is largely owned by Mitsubishi. It competes with the other Japanese space communications company, JSAT. The Superbird satellites were built by Space Systems/Loral (FS-1300 type). Superbird B1 was launched with a guaranteed 10-year lifetime. It employs Ku- and Ka-band links: twenty-three (plus eight backup) 50 Watt Ku-band transponders (36 MHz bandwidth) and two (plus a backup) 29 Watt Ka-band transponders (100 MHz bandwidth). There is also an identical Superbird A1 (launched 1 December 1993). The earlier Superbird A was launched on 5 June 1989, but most of its station-keeping oxidiser was lost in December 1990; commercial operations stopped. Superbird B was destroyed on launch, 22 February 1990. A third satellite, Superbird C, using the Hughes HS-601 bus (as in the U.S. Navy's Leasat), was launched on 28 July 1997.

In mid-1998 the Royal Australian Navy took over the U.S. Leasat-5 satellite when the U.S. Navy replaced it with a UFO satellite. The Australian Department of Defense also leased two transponders (L-band and Ku-band) on the Cable & Wireless Optus Ltd. Optus-B1 satellite launched in 1992. As of 1999, it plans to own the military section of a shared Optus C1/D satellite, to be launched early in 2002, under Joint Project 2008, to be in geostationary orbit at 156 degrees East. Links are to be UHF, X-, and Ku-band, with terminals on board frigates and submarines. The ground station is to be at the naval signal station near Canberra.

Two of the three Brazilian Brasilsat B commercial satellites carry military traffic. These Hughes-built satellites were launched (by Ariane) on 10 August 1994 and on 28 March 1995. Italy currently plans to launch a satellite in 2000, as the beginning of a Sicral (Sistema Italiana de Communicazione Rizervante Allarmi), which has been under study since the late 1980s. In addition to these countries, in 1999 Vietnam announced that it planned to obtain a military communications satellite.

SIX

Finding Targets: Reconnaissance

As the United States began to deploy ballistic missiles in the late 1950s, it faced an unrecognised problem: the U.S. targeters literally did not know where their targets were. Even after the advent of the U-2 spyplane in 1956, photo coverage of the Soviet Union was grossly incomplete. Intelligence might know that some key factory was near Omsk, but that was hardly enough. For example, about 1950 the Joint Chiefs of Staff, recalling Second World War bombing experience, ordered the Air Force's Strategic Air Command (SAC) to plan to attack the Soviet electric power plants if war came. Many were hydroelectric installations far from the cities. SAC demurred, not because it preferred to kill Soviet citizens in cities, but simply because it doubted its navigators could even find the Soviet installations. Moreover, this attack problem was substantially easier than finding factories and other targets whose location was even less fully known. To compound the problem, the Soviets carefully labelled their really secret installations, such as nuclear labs, with the names of cities up to a few hundred miles away. Arzamas-16, the main nuclear weapons laboratory, was a prime example; it was located in an old and very isolated monastery in Sarova, about 80km from Arzamas. The Soviets also deliberately distorted published maps of their country, to make air navigation far more difficult.

In the beginning, U.S. Air Force and Navy nuclear targeters had to rely largely on captured German aerial photographs dating from the Second World War. Naturally they did not show many of the important new targets, such as missile sites and nuclear and missile plants, which had been built up since the war. Nor did they extend throughout the Soviet Union. The Air Force and the CIA did overfly the Soviet Union beginning in 1952, but flights were relatively rare, and coverage was very fragmentary. Aircraft went only to places that were expected to be worth seeing, presumably based on signals and other intelligence. Resources could not possibly support any overall photographic survey of the Soviet Union, to find installations which might be of interest. The situation was so bad that in the late 1950s the CIA was reduced to releasing high-altitude balloons fitted with cameras, in the hope that prevailing winds would take them all the way across the Soviet Union.

As an illustration of just how serious the situation was, in 1960 U.S. targeters developed the first Single Integrated Operational Plan (SIOP), a combined Air Force-Navy strategic strike plan to be used in the event of general war. The Navy objected that it involved far too many weapons, and that the analysis supporting the SIOP had been rigged to demand the largest possible force for SAC. What no one seems to have realised for several years was that the plan could not have been executed, simply because many of the targets had not been accurately located. The necessary reconnaissance (for target location) simply had not yet been done.

The advent of missiles made the situation even worse. A bomber crew could be sent to the approximate location of a target. Provided with reasonably good photographs, the crew could make up for inaccuracies and it might well find a slightly mis-located target. A missile, however, had no such capacity, being aimed at a set of geographical co-ordinates on the far side of the earth. The miracle was that the missile arrived at all anywhere near its aim point. Something else had to ensure that the aim point was accurate.

Moreover, aiming the missile entailed something more than knowing the latitude and longitude of the target. The missile flew along an orbit set not only by its motor burn but also by the mass – and the shape – of the Earth. It hit at a point set by that orbit. Since the Earth is not quite spherical, connecting all these factors to ensure a hit in the desired place is not a trivial issue.[1]

Space satellites helped solve all these problems. The public still sees satellite reconnaissance systems as the primary military use of space. Quite as much as Transit, these systems put the teeth into Polaris and its successors – and now they are key to naval strikes against shore targets.

U.S. Reconnaissance System Origins

It was well understood by 1945 that big rockets, which could be developed from the wartime German V-2, could place satellites in orbit. By 1946 the then U.S. Army Air Force (which would become the U.S. Air Force a year later) was elbowing the Navy out of a putative satellite programme. At the time, it was interested mainly in attacking targets from space; reconnaissance was a secondary possibility. Studies continued through 1953.

By that time the focus of U.S. interest was fear that the Soviets might mount a surprise nuclear attack. President Eisenhower convened a blue-ribbon panel, the Killian Commission, to consider what could be done. The panel included a Technical Capabilities Panel (TCP), which suggested that reconnaissance coverage of the Soviet Union could detect any preparation for a surprise attack. High-altitude aerial reconnaissance could provide an interim solution; in due course the U-2 was designed and built. For the moment it could fly above any Soviet fighter or anti-aircraft missile. However, it was clear from the beginning that no aircraft could remain immune for very long. The TCP therefore recommended construction of a satellite monitoring system, as a longer-term solution to the surprise attack problem. Its proposal roughly coincided with an Air Force decision to press ahead with a satellite system, which it called WS-117L.

The system would have two complementary aspects. One was an ability to monitor Soviet airfields and, later, missile pads. Changes indicating preparations for attack could be detected well before the attack could be mounted. The other aspect was a complementary 'Ferret' system to monitor Soviet air defence radars which American bombers might face. If the satellite took photo after photo of key areas of the Soviet Union, successive photographs (say, on successive days) could be compared. Differences could be analysed at a ground station. A television camera on board the satellite could convert each photo (on film, in the satellite) into a radio signal, as it does on Earth. The satellite would radio this signal down to earth. The signal could be fed through a computer, and differences would pop out. But as it turned out, available technology was not up to this challenge.

The data link was the key problem. The low-flying satellite would not remain in range of its ground station for very long. Existing radio links could carry only so much information per unit time. Each picture contained an enormous amount of information, and there was no way for the satellite to edit it down. In modern digital terms, a picture might be equivalent to a megabit of data – a million bits. Yet early satellite links operated at teletype speeds, say 75 or 100 bits per second. At that rate, it would take hours to transmit a single picture. Even the 9600 bits/sec of the 1970s would not have solved the problem, since the satellite would be in range of the ground station for only a few minutes, long enough for perhaps a single frame. This is quite aside from the problem of storing multiple electronic frames in a pre-digital era.

Ideally, there would be some way to distil the data the satellite scooped up so that only the valuable information was sent down. Everything else is waste, clogging a limited communication channel. Unfortunately a satellite lacks the intelligence to decide what matters and what does not. Hence the intense interest, in both the United States and the Soviet Union, in manned military spacecraft during the 1960s and 1970s. Astronauts and cosmonauts found that they could easily identify many objects from space. They could boil down the complex images they saw into simple data: such-and-such a ship at such-and-such a position, for example. The bulk of what the astronaut saw was irrelevant, and he could filter it out in his reports to the ground.

In a few cases, filtering can be automatic. A raw radar display, for example, is a mass of different-size blobs, some of which represent real targets. The system in which the radar is embedded can be set to treat only echoes beyond a particular strength as real targets (the threshold is generally chosen to hold down the rate of false alarms). This is called automatic detection, detection being the act of deciding that a particular signal represents a real target, *ie* extracting it from the surrounding noise. It is quite common in modern radars, producing clean digital displays of air and surface targets. For the Soviets, one of the major advances required to develop a radar satellite for ship detection was automatic detection, not at that time part of typical Soviet earthbound radars.

Applying a similar simple technique to images would wipe out a great deal of important detail. In recent years, however, there has been intense interest in finding ways to compress images so that they can be sent via relatively narrow radio channels – often, channels provided by communications satellites. That is, in theory an image is described by breaking it down into its picture elements – pixels – which are small squares covering it. If the pixels are small enough, the eye cannot tell that the picture is grainy. It is enough to say how bright each is (or, for a colour image, colour and brightness). That is how a television works. A high-quality image will equate to millions of pixels. The simplest form of compression is to concentrate on the difference in brightness from pixel to pixel, so that no effort is wasted in transmitting the details of a large patch of uniform colour or brightness. To do this requires a computer to analyse the pixels, and another to reconstruct a picture from the series of differences which are actually transmitted.

Applying more computer power at each end can make up for a narrower channel. For example, a more sophisticated analysis of a picture might be to examine the variation not from pixel to pixel, but across its whole face. That variation can be expressed as a combination of some set of mathematical functions (wavelets are a current favourite). Then the channel need only transmit the number of wavelets of different kinds which add up to something close to the original picture. Pictures can also be described as combinations of mathematical objects called fractals. Again, very powerful computers for analysis and for reconstruction make it possible to send pictures via very narrow radio channels.

These techniques do not actually make sense of the information in the picture; they merely make it easier for a human user to get the picture in the first place. In that sense they are no closer to automatic (radar) detection. However, it is possible that reducing a picture to simpler mathematical form also makes it easier to recognise particular objects in that picture. In that case, picture compression really would be a close relative to automatic radar detection. For example, ideally a photograph of a field should be reducible to the positions of, say, tanks in that field.

But of course none of this technology was available in the 1950s. There was an alternative. Instead of radioing back its data, the satellite could expose film and periodically eject a capsule, which could be picked up in the atmosphere. As early as 1956, the Air Force's RAND research institute estimated that this approach would offer at least two orders of magnitude more data than the television link. However, it could not offer the kind of warning that was needed. The Air Force therefore awarded Lockheed a contract for the television system, WS-117L. However, it was soon clear that no such system could be ready before 1960. The alternative film system seemed attractive as an interim measure. Because the satellite would be much lighter than the television type, it could be placed in orbit by a Thor with an Aerobee 75 second stage (developed for Vanguard); soon the more powerful liquid-fuel Agena booster-satellite (developed for WS-117L) was substituted. Planned resolution was 60ft from an altitude of 135-140 miles. It was estimated that the television system would give a resolution of 144ft. The interim programme was designated Program IIA within WS-117L.

WS-117L was a normal classified programme, run by the Air Force. Although details were secret, overall the programme was not. In February 1958 Program IIA was officially cancelled, its rocket to be used to test 'biomedical capsules' in a new Discoverer scientific programme. In fact Program IIA was resurrected as a totally black CIA-run programme, Corona (named after the brand of the typewriter on its director's desk). Meanwhile the television-readout satellite programme was allowed to continue, although it never succeeded.

The lesson of that failure was that to radio vital tactical information down to Earth via a narrow link required some intelligence on the part of the satellite. It had to pick essential features out of the rich, complex picture visible from space. That real-time intelligence could be provided only by astronauts. The U.S. version of this concept, in which the U.S. Navy participated, was the proposed Manned Orbiting Laboratory (MOL), approved for development on 10 December 1963. Manned satellite reconnaissance, both visual and photographic, was tested during the June 1965 Gemini 4 flight, when MOL was on the point of receiving final approval for development.

Despite their limitations, satellites could cover far more of the Soviet Union than could aircraft like the U-2. Thus they could detect and locate potential targets, often for the first time. They could also monitor Soviet progress in weapons development and deployment. Their first great contribution to U.S. security was to demonstrate just how few of the massive R-7 (NATO SS-6) ICBMs were being deployed, c1961, *ie* that the 'missile gap' which had figured so heavily during the 1960 presidential election was no more than a fiction.[2] In the mid-1960s President Lyndon Johnson said that the United States had fired billions of dollars of hardware into space – and that it had all been worthwhile for the peace of mind space reconnaissance had provided him, the priceless ability to know what was actually happening in the closed Soviet Union.

Although it was officially approved in 1955, the U.S. reconnaissance satellite programme was kept quite secret. As a prerequisite, the U.S. government sought to establish a principle in international law of the 'Freedom of Space'. To this end in April 1955 President Eisenhower approved a plan to launch a U.S. scientific satellite as part of the International Geophysical Year (IGY), 1957-8.[3] Even its launch vehicle was deliberately made non-military (it was the unfortunate Vanguard). The Soviets made a similar announcement. There is some evidence that the Eisenhower Administration wanted the Soviets to launch first, in the expectation that their feat would establish the 'freedom of space' even more firmly than the U.S. scientific satellite could.[4]

When the Soviets launched their own satellite, Sputnik, in October 1957, President Eisenhower was undismayed because it was far too small to accomplish the one key military role, reconnaissance. He was surprised to discover that very few others, either in government or even in Allied countries, drew similar conclusions. Instead, they deduced that the United States, hitherto considered the leader in world military technology, was perhaps fatally behind. The Air Force responded by leaking an account of its reconnaissance system. A furious Eisenhower cancelled it, transferring the project to more secret CIA auspices. This became Project Corona. Because the project was so urgent, the booster was switched from Atlas, which was a few years from readiness, to the already-developed shorter-range Thor, whose Agena upper stage would become the satellite body.

By the end of 1959, three reconnaissance satellites were being developed: Discoverer (the black CIA film-return satellite), SAMOS (Satellite And Missile Observation System, the Air Force readout satellite), and MIDAS (MIssile Detection Alarm System). The first Discoverer launch was attempted on 21 January 1959; the first successful launch (albeit without a capsule) was on 28 February 1959. The first capsule (without film) was successfully retrieved on 12 August 1960, shortly after the U-2 incident in May of that year had ended the only other U.S. programme capable of photographing the

Soviet Union. Later that month Discoverer XIV launched a capsule with exposed film on the basis of which objects 35ft across could be identified. The programme continued through 1972, 145 satellites being launched.

SAMOS was the surviving version of the Air Force's original electronic-readout concept, which had been renamed Sentry in 1959. Under this name it included not only the read-out satellite but also Corona and MIDAS. Then the read-out satellite was renamed SAMOS. Samos 101A was high-magnification read-out; Samos 101B was high-magnification film recovery. Both were cancelled in 1961. Samos 201 was another high-magnification film recovery system, cancelled in 1962. Much heavier than Corona, Samos was launched by an Atlas booster. Of two known Samos-A launches, one failed in October 1960, while another succeeded in January 1961. Of eight Samos-B launches, only two (in 1962) succeeded. The programme was never declassified but, given the technology of the time, it is unlikely to have been very successful. Ironically, it was generally described in the press as the key U.S. reconnaissance satellite, and in this way its existence covered that of the real programme, Corona. Even after cancellation, the name Samos was widely associated with space reconnaissance, Discoverer being reported as a cover for it.

MIDAS, too, was a failure. It was intended to use IR sensors to detect missile launches, but again existing technology was insufficient. However, the idea was revived about a decade later in the current Defense Support Program (DSP) system, which had an important naval application, Slow Walker, in the 1980s (see Chapter 10).

In Paris in May 1960, in the aftermath of the U-2 shoot-down, General de Gaulle told Khrushchev that he might resent the fact that sputniks regularly flew over Paris. Khrushchev assured him that did not matter; he would not care what satellites flew high over the Soviet Union. Eisenhower took Khrushchev's statement, which the French reported, as the prized agreement on free use of space. In fact, the Soviets cared deeply. Despite Khrushchev's offhand comment, they protested each time the United States launched a reconnaissance satellite – which, by 1961, was no longer particularly secret. Certainly by 1962 the Discoverer cover story had been thoroughly penetrated. In response the United States adopted the policy, which is still in force, of refusing to acknowledge the function of reconnaissance satellites at launch time.

The U.S. photo satellites received a series of codenames; their camera payloads were Key Holes (KH). Corona itself was KH-1 through -4 (including -4A and -4B), carrying 40lbs of film. KH-4 had two panoramic cameras pointed fore and aft, their axes 30 degrees apart, to produce stereo images. Resolution was initially 25ft (later it was 9 to 25ft). Total mission life was six or seven days, rather than the three or four days of earlier satellites. KH-4A was the first in the series to carry two film capsules ('buckets'). That was possible because the launch rocket was much more powerful, using three strap-on boosters. Mission length more than doubled, to fifteen days. Typically the first capsule was recovered half way into a mission. Typical resolution was 10ft (sometimes 7ft). KH-4B had a better camera with 6ft (sometimes even 4.5ft) resolution; typically it photographed a 150 × 150nm area. Mission life extended to nineteen days (the camera cycle rate could be controlled more precisely than in the past). Argon (KH-5) was a single-camera geodetic (mapping) system, necessary to support ballistic missile targeting by providing an accurate map of the Soviet Union. Its camera had low resolution (460ft). Like the Coronas, it carried 40lbs of film. Lanyard (KH-6, July 1963) was an unsuccessful attempt to achieve higher resolution. All of these satellites were orbited by Thor-Agenas.

Eventually these satellites mapped the Soviet Union. On that basis U.S. missiles such as Polaris could finally be targeted. That was not quite enough, since it was necessary to register the targets relative to launch sites in the United States, or to submarines at sea. Transit fixes were apparently used to relate the satellite-generated map of the Soviet Union to the appropriate co-ordinates.[5]

Gambit (KH-7) was a high-resolution (reportedly 18in) satellite operating in parallel with the low-resolution KH-4 (1963-7). Like KH-4, it carried a single film capsule, and mission duration was about six days.[6] The new camera must have been far heavier than that on KH-4, since KH-7 needed an Atlas-Agena to place it in orbit. KH-8 (1966-84) was a larger (6615lbs as opposed to 4410lbs) version of Gambit, begun in parallel with the two-capsule KH-4B. It was launched by a more powerful Titan IIIB Agena booster. This satellite had an improved high-resolution camera, and it probably carried 160lbs of film in its two capsules. Mission lifetime was thirty days.

KH-8 was, in effect, the ultimate development of the original Corona. About 1965 the decision was taken to develop a wholly new system which would combine in a single satellite the wide-area (low resolution) coverage of Corona with the high-resolution coverage offered by Gambit. Its lifetime would be months rather than a few weeks. About 1966 this concept, which became Hexagon or 'Big Bird' (KH-9) was adopted in place of an Air Force proposal for incremental improvements to Corona (Valley). Much heavier than KH-8 (22,800lbs vs. 6615lbs), it carried four rather than two film-return capsules. Typical mission lifetime was 138 days (275 days by the end of the programme), and reported swath width was 80 × 360 miles. Part of the size growth may be ascribable to the use of a much larger main camera mirror (60in diameter rather than the 45in of KH-8). Most satellites carried two cameras, but on five missions a single lower-resolution mapping camera was carried. Reported resolution was 2ft. Because 'Big Bird' would fly at low altitude, it would be affected by air resistance. It needed a rocket engine (which added more weight) to boost it periodically, to keep it from inadvertently re-entering the atmosphere and burning up. Hexagon was considered so important by the U.S. intelligence community that a 1969 decision to cancel it (due to the cost of the Vietnam War) was reversed. Delays in camera development deferred the first launch from 1970 to 1971. Between then and 1986 nineteen satellites in this series were launched (there was one failure).[7]

Satellite data was eventually used to frame the clauses of the Strategic Arms Limitation Treaties; the U.S. government also relied on the satellites to verify Soviet compliance. In the treaties the imaging satellites were called National Technical Means (NTM), and the name has stuck.

KH-11 and Lacrosse

By the mid-1970s the United States had solved the television/data link problem. The great advantage of television is its timeliness, very nearly as timely as the satellite's pass over a target can be, and furthermore there is no longer a finite film supply to exhaust. Now the satellite's lifetime is the lifetime of its electronics and of the fuel supply needed to maintain it in orbit. However, there is still a rub. If the satellite reports only to a ground station, then it has only a limited time within sight of the station during which it can send images down. How much it can report depends on the available time and also on the bandwidth of the link.

The U.S. solution was to orbit additional communication satellites via which the imaging satellite could report: the Satellite Data System (SDS). At least two prototypes and eleven operational SDS satellites (some of which were apparently actually electronic reconnaissance satellites) were launched in 1976-87 into Molniya orbits, with extremely high Northern Hemisphere apogees.[8] In 1997 it was revealed that there are also SDS satellites in geosynchronous orbits. In addition to imagery, SDS almost certainly transmits electronic intelligence data. Some satellites, such as the Lacrosse radar satellite, apparently transmit instead through the geosynchronous NASA Tracking and Data Relay Satellite System (TDRSS).[9]

The television satellite, first launched on 19 December 1976, was Kennan (later renamed Crystal, when the name Kennan was compromised in 1986: KH-11).[10] According to a 1995 Russian summary, in all nine such satellites have been launched (a tenth was lost in a launch failure). Some writers class the last in the series (launched 28 February 1990) as an improved version (see below). It was described as incorporating both digital imaging and signals intelligence receivers for the CIA (*ie* NRO) and NSA.[11] All were launched by Titan boosters. The first, launched on 19 December 1976, was retired in 1979. Its first products, near real-time pictures, were shown to President Jimmy Carter on the afternoon of his inauguration in January 1977. Numbers 6 and 7 were retired in November 1994 and in June 1992. In 1995 two satellites remained in service. Design lifetime is five to eight years. The limiting factor in lifetime is the amount of fuel on board, since that is needed to manoeuvre the satellite (to some extent) to change coverage as required. Fuel is also needed to keep the satellite's cameras and solar cells pointed as necessary. Until the *Challenger* was lost, there were plans to make these satellites refuellable by Shuttle.[12] Reportedly KH-11 is similar to the Hubble space telescope; they use the same shipping container and the same specially-modified aircraft.

Because their lenses have limited resolution, these satellites must operate at relatively low altitude (a typical orbit is 1020 × 270km). The orbit is sun-synchronous: the satellite appears everywhere at the same local time. A pair of satellites can make morning and afternoon passes (reportedly at 1000-1100 and 1300-1400 hours). Differences in shadows on the two passes make it easier to interpret photographs. The satellite produces both large-scale (surveillance) and detailed images. According to a 1995 Russian article on the KH-11 programme, the surveillance image is a 90 × 120km strip (resolution 0.5-1.7m); the detailed image is a 2.8 × 2.8km strip (resolution 0.15-0.18m). Reportedly the cameras use a 2.3m (90.6in) mirror to form their images. The resolutions given here can be achieved only under ideal weather conditions. The satellite can also be used to photograph other objects in space (*eg* KH-11-4 was used to inspect the Space Shuttle *Columbia* in 1982 for damage to its heat shield).

At least initially the television system apparently did not offer the sort of image quality provided by film. Reportedly the satellite offers even better resolution than its predecessors, but probably the scanning process is not as consistent as capturing an image on a tightly-stretched film. Dimensions on the image, then, may not be as consistent as those on film. Through the early 1980s, therefore, the United States continued to launch film-capsule satellites of two types, KH-8 and KH-9. That both were used in parallel suggests that 'Big Bird' imagery did not duplicate that of KH-8. Typically two television satellites (KH-11) were maintained in orbit at any one time. A KH-8 would be launched early in the spring, to remain in orbit for a few months. A KH-9 would be launched late in the spring, to remain until about the end of the year. Given their short effective mission lifetimes, KH-8s in particular were presumably retained to photograph objects picked up by KH-11s.

KH-9 and KH-11 drastically changed American satellite policy. Instead of placing short-lived satellites in orbit on almost a weekly basis, the United States would follow its practice in other types of satellite design, of working for long operational lives. This practice has two drawbacks. First, even four decades into the space age, launches are not always successful. The loss of a billion-dollar military satellite is doubly embarrassing if no replacement can be ready for months or even years, because none seemed likely to be needed. Second, a long projected lifetime can be cut short if the satellite is attacked in orbit. Again, the lack of ready replacements can be devastating.

Soviet practice was quite different. They apparently tried to build long-life satellites, but they were not too successful, so therefore they accepted the need to launch replacements very frequently. Because their economic system favoured mass production of relatively short-lived equipment, this practice did not seem any more wasteful than most other Soviet operating practices. Certainly it would have been well adapted to wartime needs, when satellites might well have been destroyed or disabled in orbit. However, when the Soviet Union collapsed, mass production of boosters and satellites ended; the Russians had to make do with what was left over from the Soviet past. Since their satellites had only short design lifetimes, their space reconnaissance systems largely collapsed during the first post-Soviet decade. In February 1999 Yuri Koptev, director of the Russian Space Agency, said that 75 per cent of the 131 Russian spacecraft in orbit, sixty of them military, were effectively dead. Of those still operational, half were on the point of breaking down; three-quarters were past their planned lifetimes.[13]

An Advanced KH-11 (incorrectly called KH-12) reportedly carries considerably more manoeuvring fuel, which it can use either to reach a desired objective or to evade ASAT attack. Most of the growth from about 13,300kg to 19,600kg is apparently fuel (the KH-11 carries about 8 tonnes of it). This satellite reportedly produces better detailed (2.8 × 2.8km) views: a Hubble-type optical camera compensates for atmospheric distortion (resolution 0.8-0.1m in the detail view, 0.5-1.0m in the surveillance view). An IR camera (CCD array) has a resolution of 0.6-0.8m in the detail view and may operate over a broader spectrum than in KH-11. This satellite may have 'real-time' television capability. Advanced KH-11 also apparently carries ELINT antennas. There is a trade-off between resolution and the area the satellite covers. The satellite can store and transmit only so much data – so many pixels – per frame. In 1995 it was reported that the next satellite to be launched in this series would have its resolution reduced so as to increase its data transmission rate and also to provide the larger-area coverage that tactical commanders needed. It was no longer nearly so important that, for example, they provide crisp photographs of new ships on the slip in Russian shipyards. It was far more important, for example in Iraq, that they could provide information needed to plan Tomahawk missile flight paths so that they would avoid Iraqi radars and anti-aircraft batteries. Advanced KH-11s were reportedly launched on 28 November 1992, on 5 December 1995, and on 20 December 1996.

The same technology which made it profitable to build a really massive photo satellite with a long lifetime limited the number of such satellites the United States built, so that very few were ever held in reserve. That in turn increased the impact of the Soviet ASAT system. Presumably the lull in ASAT tests after 1972 convinced those responsible for the big U.S. satellites that the Soviets had given up on this sort of weapon, an unfortunate error which might be attributed to the excessive belief in detente then current.

The optical satellites cannot penetrate clouds or, for that matter, foliage or camouflage nets. The latter would hide Warsaw Pact armoured forces. The solution was a high-resolution side-looking (imaging) radar, carried by a Lacrosse satellite in a 704 × 676km orbit. The programme (Indigo) began in 1977, and Lacrosse-1 was launched in 1988 (Lacrosse-2 followed in 1991). This system was so important that the first was launched by the first military Shuttle flight after the *Challenger* disaster (had it not been for that, the satellite would probably have been launched in 1987). Using a centimetric radar, Lacrosse may achieve resolution comparable to that of the big photo satellites, reportedly about 1ft. This is *not* the sort of radar satellite which might be used for real-time aircraft or ship or cruise missile detection; that would require many more satellites in low orbits, for continuous cover. However, it should be effective against large armoured forces, which move slowly and which spend much of their time waiting under cover. Moreover, the area which the satellite would have to search for such vehicles would be relatively small, so the satellite could be cued. Allied operations during the Gulf War were reportedly supported by three imaging satellites (KH-11) and by the first Lacrosse. They also exploited civilian imaging satellites with lesser resolution: U.S. LANDSAT and the French SPOT-1 and -2.

The Impact of Imaging Satellites

The advent of the photo satellites drastically changed the U.S. intelligence community. Suddenly there was a flood of images, which had to be interpreted. Photo interpreters came to dominate intelligence simply by their sheer numbers. Moreover, imagery seemed to offer objective information. Traditional kinds of intelligence, even signals intelligence, were far more ambiguous and required far more interpretation. It had always been preferable to have hard information about an enemy's equipment and strength, but now that sort of information in effect seemed to drive out more traditional efforts to discover enemy intentions and the quality of enemy forces.

The legacy of this development can be seen in the approach to war in the Gulf in 1990. The Allies had excellent information on the Iraqi order of battle, most of it undoubtedly provided by imagery and SIGINT satellites. The Iraqis clearly had very powerful forces. That these forces were quite hollow, from a human point of view, emerged only after the war had begun. Traditional kinds of intelligence would probably have concentrated more on the latter aspect, perhaps mainly because it would have been so difficult to obtain hard information about enemy strength.

In a way this was the beginning of the new kind of warfare. Satellite imaging systems were much more like time-late sensors than like traditional intelligence sources. To the extent that they were bought for strategic targeting, they were the sensors directing the U.S. strategic attack force. That this force was targeted in very slow motion, and that it was never actually used, did not change the relationship between the sensors, the attack planners, and the weapons themselves. That the satellites were often operational sensors was obscured by their bureaucratic status as part of the U.S. intelligence community, and by the frequent use of their imagery for traditional intelligence functions.

All of these imaging satellites concentrated on land targets. The reason is simple. It would take too long to photograph the world ocean (or even a substantial part of it), and while the satellites were scanning they would probably miss moving objects such as ships. Moreover, it would be quite laborious to review the mass of satellite photographs to find small ships on a vast sea. By way of contrast, a satellite scanning a land area for large fixed objects can often find them as it makes multiple passes. Even so, imaging satellites had considerable impact on naval operations, as distinct from strategic operations involving long-range missiles or bombers. In 1960, when the satellite programme began, the West had very little knowledge of what the Soviets were building. Attaches passing the yards at Leningrad often could see ships under construction (the big nuclear cruiser *Kirov* in the late 1970s was a spectacular case in point), but they could hardly visit the main submarine yards at Severodvinsk in the north or in the closed city of Gorkiy. Nor could they visit the big-ship yard at Nikolaev on the Black Sea, where Soviet carriers were built, or the small-combatant yard at Zelenodolsk on the Black Sea. Imaging satellites could visit and revisit all of these sites. Even in the case of Leningrad, where something could be seen at street level, an overhead satellite offered a much better view.

This capability was undoubtedly far more important to the West than to the Soviets, since the Soviets never publicised their ongoing programmes, nor did they ever display sketches of planned or authorised ships. To a considerable extent, then, U.S. imaging satellites helped protect the West from unpleasant surprises in the form of new submarines and surface ships. There were, of course, limits on this capability. At Severodvinsk, submarines were built under cover (partly to protect builders from the cold). Furthermore, the titanium-hulled submarines had to be built in enclosed areas because titanium welding demanded a special atmosphere. In each case, however, much could be learned from parts and materials visible before delivery or use. In that case much depended on interpretation. For example, in the late 1970s some massive reduction gears were delivered to the building at Severodvinsk; they were sometimes described as 'carrier-sized'. The building itself was large enough to accommodate a nuclear carrier. As it turned out, the project was the 'Typhoon' class missile submarine rather than an aircraft carrier (which was being built at Nikolaev).

There was also an operational aspect. Satellites might be unable to detect and track Soviet warships at sea, but they certainly could periodically monitor Soviet naval bases. At the least, a satellite pass could reveal that Soviet ships had or had not sortied. This is not a new idea; in 1941 the fact that the *Bismarck* was at sea was revealed by photo coverage of a German-controlled anchorage. The difference is that in the 1970s and 1980s a satellite could easily be assigned to check a given port every so often, despite whatever air defences the Soviets might have assigned there.

Electronic Reconnaissance

The programme Eisenhower approved in 1955 included not only imaging but also signals intelligence, the 'Ferret' mission. This was largely a matter of detecting and localising radars, so that bombers could attack effectively. It was split between Air Force and Navy, since the two services would likely attack different areas.[14]

Probably the first U.S. space ELINT device was a receiver/recorder aboard the Discoverer 13 photo satellite (launched August 1960), called Scotop; it recorded Soviet radars tracking the satellite. However, satellites could also detect radio and radar signals not directed into space. The beam from an antenna on the surface, looking horizontally, extends past the horizon into space. The antenna also leaks radiation to the sides and straight up (side- and top-lobes). A satellite which detects this radiation can also locate the radar. On the

Earth, one directional cut is not enough to locate anything, but the satellite has an important advantage. The source of the radiation it picks up is on the surface of the Earth, so the line down to that source cuts the surface in a very definite place.

U.S. naval interest in space interception of radar signals actually predated the Space Age. In 1946, Donald Menzel of Harvard, a former naval reservist, suggested using the moon as a radio reflector; he was undoubtedly well aware of the Navy's need for better communication. In 1948 a paper in the *Proceedings of the Institute of Radio Engineers* suggested that the Moon might have an ionosphere of its own, in which case reflection might be quite efficient. That led James Trexler of the U.S. Naval Research Laboratory to calculate the characteristics of the radio system involved. To his surprise, it turned out to be much like current long-range radars. Their signals were probably already bouncing off the Moon. A big enough Earth antenna might collect them, and thus might measure the characteristics of Soviet air search radars, some of which (at 500 MHz) had already been detected near Alaska. In June 1950 the U.S. Navy made its first investment in Project PAMOR (Passive Moon Relay), also known as 'Joe'. Initial tests of a big radio dish sunk into the ground at Stump Neck, Maryland, showed that the Moon was an excellent reflector, much better than expected.[15] This experience ultimately led to the Moon Relay described in Chapter 5 above. The Stump Neck bowl (220 × 263ft) turned out to be too small (with insufficient gain), so early in 1955 NRL proposed building a 600ft steerable (so as to follow the Moon) radio telescope specifically to intercept Soviet radar signals. Eventually the project died of cost overruns.[16]

The Navy 'Ferret' programme, supported by the Naval Research Laboratory, was code-named Project Tattletale, then GRAB – to grab what it could, although the name was turned into an acronym with a scientific-sounding cover meaning, Galactic Radiation And Background.[17] Grab was the first U.S. electronic intelligence satellite, conceived by Reid D Mayo of the NRL Countermeasures Branch. In 1958, when Vanguard was first successfully launched, the branch had just developed a new submarine ECM (actually, ESM or ELINT, for S- and X-band) antenna, a wideband spiral inside the glass at the top of the periscope. Late in March 1958 Mayo proposed placing one on a satellite (in fact Grabs were launched piggyback with other satellites). A rough calculation suggested that useful signals could be picked up at a range of over 600 miles.

There were two key factors to this kind of performance. One was that a radar had to pick up signals which had made the round-trip from antenna to target and back again, whereas the signals picked up by the satellite receiver would make only a one-way trip. For example, a big air-search radar might be designed to detect targets 200nm away.

At a distance of 2000nm its out-bound signal would be 100 times weaker than at 200nm. However, the reflected signal would be more than 40,000 times weaker than that hitting the target (how much weaker would depend on the echoing area of the target). The radar would have to pour enough energy into space to get back enough of a signal to detect; and that energy would be much more detectable well out into space. Thus a relatively unsophisticated receiver at 2000nm distance could probably detect the radar signal. The other was that Grab would detect the powerful main lobe of the radar, not only the weak top lobe, because the main lobe would continue beyond the horizon, out into space. This also made it likely that a very small (low-gain) antenna in space could effectively detect powerful surface radars. Eventually it became obvious that the omni-directional antennas on the satellite would pick up all S-band radars except those directly under it. Using this blind area, the 'hole in the doughnut', Grab could locate radars, by correlating their times of disappearance and reappearance with the satellite's orbit. Because the antenna was so compact, Grab could be quite small, 0.51m in diameter (weight 18kg).

NRL formally proposed Project Tattletale, to detect S-band signals, in July 1958; the satellite would be based on Vanguard. President Eisenhower approved full development on 24 August 1959.[18] The objective was to define the characteristics and location of Soviet air defence radars which U.S. strategic bombers would face in wartime; in 1959 the first U.S. integrated strategic war plan (SIOP) was being written. At an altitude of 500nm and an inclination of 70 degrees, Grab swept out a 3500nm wide intercept swath, covering virtually the whole of the Soviet Union in a single pass.

The first dummy subsatellite was launched on 13 April 1960, and the first operational one followed on 22 June 1960 (Solrad 1, in a 596 × 935km orbit inclined at 66.7 degrees, launched by a Thor Able Star). In all, eight were launched, of which at least two became operational. These satellites were launched piggyback on Transit navigational satellites. Grab needed a good cover story, not least because it would broadcast continuously from orbit. The ground stations also had to be explained. NRL chose a solar radiation study, SOLRAD; the laboratory had already used detectors in V-2s to measure solar X-rays. There was also GREB, Galactic Radiation Energy Background. Unlike other intelligence systems, which also had scientific cover stories, this one had substance; Grab actually carried X-ray detectors, and it actually down-linked scientific data. The cover was apparently successful. The Soviets shut down radars when other satellites were overhead, but never for Grab.

Grab was a transponder, in effect a lower-altitude equivalent to the Moon intelligence-gathering system NRL had developed some years earlier. It might be thought of as a communication satellite – perhaps the first successful one – but for radar rather than radio signals. Grab received main-lobe pulses from target radars and repeated them beyond the horizon to ground stations, where they were recorded. The satellite downlink provided radar frequency, pulse repetition rate, and scan rate as well as location (from Doppler data). Typical Soviet radars intercepted were Gage and Token, main-lobe signals from both of which could be picked up at altitudes up to about 1000nm. At an altitude of 500 miles, Grab would detect a radar about 1750 miles away. Its achievements included the first detection of a Soviet anti-ballistic missile radar.[19]

Because of its low altitude, Grab could relay the radar signals only a limited distance, about 1750 miles (*ie* to its horizon), so the system's ground terminals had to be abroad, as were its ground control stations. The terminals (mainly the naval intelligence station at Bremerhaven, Germany) taped Grab signals, and the tapes were then couriered to the United States for analysis. The relay or transponder concept made for the simplest possible satellite; all the complexity of recording and analysing data was kept on the ground. The alternative, for a low-flying satellite, would have been to record data and then dump it to a selected ground station, a technique the U.S. Army tried in its first communications satellite, Courier, at about the same time.

Most current electronic intelligence satellites probably also use the transponder technique. They are, then, closely related to civilian communications satellites, albeit probably with much higher-gain antennas. As in the case of Grab, the transponder technique minimises signal processing and storage on board a satellite which is, after all, of limited size. Any store-and-dump ELINT satellite can deal with only a limited range of signals. In the Soviet case, Lourdes, Cuba, was probably the ground station for satellites picking up signals over the United States (it is also apparently within the footprint of the downlinks of some commercial communication satellites). On the other hand, Soviet passive naval reconnaissance satellites operated in store-and-dump mode – which was practicable because they picked up only a narrow range of radar signals.

The Navy operated the Grab system until August 1962. On 14 June 1962 Secretary of Defense Robert S McNamara established the National Reconnaissance Office (NRO) by a top-secret directive. NRO then took over all satellite reconnaissance, including Grab. Probably later Solrad satellites were also used for ELINT. The last two in the series, Solrad 7A and 7B, were launched on 11 January 1964 and on 9 March 1965, into near-circular orbits at roughly 900km (inclination 70 degrees). The programme continued through 1967. In its later stages it was apparently code-named Poppy and then Siss Zulu.

Grab provided data to both the Air Force (for SAC) and NSA, the codebreaking National Security Agency. Since it was optimised for radar frequencies, the latter use of data suggests that it was picking up communications at radar frequencies. Indeed, it may have discovered that the Soviets were shifting to such links from earlier HF radio systems.

The corresponding Air Force system was Ferret, Program 102 (also Programs 698, 706, and 770), a series of 900 to 1500kg satellites ('heavy ELINT') launched by Thor Agena-B and -D rockets.[20] The Air Force SAINT (satellite inspection and neutralisation) ASAT programme may have been a cover for this ELINT programme. The first in the series, which apparently failed to enter the desired orbit, was Discoverer 37 (21 February 1962). Ferret 2 (18 June 1962) was launched successfully into a 347 × 377km (82.1 degree) orbit; later vehicles went into higher orbits (*eg* OPS 1439, 9 February 1966, 507 × 511km, 82 degrees). This programme apparently ended in 1971 (OPS 8373, launched 16 July 1971), fifteen craft having been launched successfully. Satellites were reportedly 5.0ft in diameter and about 38.5ft long. Reportedly they fell into three generations: three in the first, nine in the second (beginning January 1963, ending January 1968), and four in the third. When these satellites were designed, there were no automatically deployed space antennas (reflectors), so probably they relied on strip antennas spread along the length of the vehicle. Experience with these satellites apparently showed that it would be worth while to switch to geosynchronous and Molniya-type (12-hour) orbits in order to obtain continuous target coverage.

Apparently also as part of Program 102, beginning in August 1963 small ELINT satellites were carried piggyback with many of the KH-4 and -7 Corona satellites, then boosted into higher working orbits at altitudes of 300 to 800km (KH-8s, however, carried no sub-satellites). These satellites were usually placed in constellations of four for maximum direction-finding effectiveness. One Corona-type satellite (launched 21 December 1964) had no camera at all on board, just a receiver and a tape recorder; its capsule carried the tapes back to earth. The mission lasted four days. Further Ferrets were launched with the big KH-9s and also with Jumpseat SIGINT satellites (see below).[21]

On 5 September 1988 the Air Force launched the first of a next-generation series of 1700kg Ferrets (possibly code-named Singleton) into a 500-mile 85-degree orbit, using a Titan II. Others followed on 5 September 1989 and on 25 April 1992. Apparently three more planned launches were cancelled, the Titan IIs reserved for this 'classified use' being reassigned to the Strategic Defense Initiative Office (now BMDO) in 1993. It now appears that this shift reflected a decision to use White Cloud naval reconnaissance satellites (see below) for the Air Force Ferret mission of locating foreign radars. That interpretation would explain the assignment of an Air Force intelligence detachment to the White Cloud ground facility in Guam in 1995.

Signals Intelligence

The radar-locating satellites described above are to be distinguished from SIGINT satellites, operated by the NSA to detect communications traffic. Satellites offered NSA a valuable capability. Until the mid or late 1960s it had probably exploited the HF radio the Soviets used for internal communications. As in many other areas, the Soviet Union had not invested in infrastructure projects such as long-haul telephone wiring. Soviet memoirs of the Second World War often refer to the special high-frequency (HF) radio telephones which senior commanders used to communicate with Stalin. Surely similar circuits were used throughout the Soviet Union, to cover vast largely uninhabited – and undeveloped – areas. They represented a major opportunity for eavesdropping from beyond Soviet borders, since HF signals propagate over long distances, well beyond their intended recipients. Obviously NSA never said whether it exploited this opportunity, but it would seem extraordinary if it did not. Any success it achieved was probably reported by a series of NSA traitors who defected to the Soviets in the late 1950s and early 1960s.

Certainly the Soviets seem to have been aware of the problem. At least one retired Strategic Rocket Forces commander has acknowledged that, until about 1970, it might take hours or days to contact remote missile sites. That implies that HF radio was unacceptable, so in some cases couriers had to be used (the U.S. government was aware of this, and up to the late 1960s this and a generally sluggish Soviet responsiveness meant that for a few years the United States probably could have executed a successful first strike). Similarly, the Soviets replaced HF links with expensive submerged telephone cables in areas such as the Pacific. We know as much because an NSA traitor, Ronald Pelton, eventually revealed that the U.S. Navy's submarines had tapped the cable between Kamchatka and Vladivostok, which carried much of the Soviet Pacific Fleet's most sensitive traffic.

Cables are very expensive, and they have limited capacity. Throughout the world, from the 1950s on, the favoured alternative was microwave relays. Repeater towers began to appear, receiving a stream of signals and then repeating it for transmission to the next tower. Such towers were far easier to build than massive buried cable nets, and they had considerable signal capacity. Unlike HF, they sent signals only – it seemed – along their lines of sight. They certainly did not inadvertently send signals thousands of miles by bouncing them off the ionosphere. However, like radar signals, theirs inevitably continued into space; their antennas also inevitably had top lobes. Their signals, then, were subject to interception by satellites.

Given the Soviet obsession with standardisation, it is likely that their microwave links used radar components. Thus at some point in the 1950s or 1960s Ferret radar intelligence programmes inevitably began to merge with radio intercept programmes. There were still important differences: a radar produces a series of sharply-defined short pulses, whereas a communication channel produces a more or less continuous signal, its average power far less than the peak power of the radar (whose average power may approximate that of the communication channel). Even so, whatever efforts NSA was making to detect microwaved messages would also have picked up the top lobes of microwave radars, including shipboard ones.[22] To some considerable extent the Russian statement, noted below, that U.S. ELINT satellites were intercepting terrestrial UHF communications confirms this speculation.

Once it was clear that the Soviets were using radar frequencies for communications, NSA began to sponsor ELINT (SIGINT) satellites. By this time the U.S. Navy no longer planned to fly its own bombers deep into Soviet territory (it would use Polaris instead), but the Air Force still badly needed data on Soviet air defences. Since the Air Force was the agency responsible for launching NSA satellites (details of which are secret), it is not certain which satellites, if any, were devoted to the Air Force mission. It is also possible that some of the satellites typically listed under the Ferret programme were actually intended for NSA use.

Reportedly the first NSA satellite was the 700kg Canyon (TRW 'Spook Bird'), first launched on 6 August 1968, with a second on 13 April 1969. It was intended primarily to monitor the radio communications of command posts and the staffs of higher-echelon commands, most importantly the Soviet strategic rocket forces. Signals are acquired using a 10ft diameter umbrella-type reflector with multiple feed horns at its focus. Approximate dimensions are length 28ft and diameter (extended) about 30ft. These were the first U.S. intercept satellites in so-called quasi-stationary orbits, not quite geostationary (which would be 36,000km altitude and 0 degrees inclination). The quasi-stationary orbits are 30-33,000km × 39-42,000km at an inclination of 3-10 degrees. A ground observer sees such an orbit as a closed intersecting loop, elongated along the horizon, about 30 degrees long, covering about 5-6 degrees in elevation. The satellite views a broad area, and can see the same emitter almost throughout its orbit. Because coverage is nearly continuous, a target cannot avoid interception simply by turning off an emitter during a satellite's transient passage. Satellite lifetime is about 5-7 years; data is down-linked by radio (reportedly using a pair of dishes). The ground station is at Bad Aibling, Germany.

The first two Canyons placed in orbit were, according to a Russian article, used to follow the build-up of Soviet forces in the Far East during the growing Sino-Soviet crisis. Among other things, they intercepted communications nets controlling Soviet bombers in the Far East, which were then undergoing intensive training.[23] The success of these satellites led to series production, the last of seven such satellites (one failure) being launched on 23 May 1977.

Chalet (also called Program 366 and Vortex), was the successor series, using 820kg (also reported as 1200kg) satellites. It began with a 10 June 1978 launch into a quasi-stationary orbit (29,929 × 42,039km at 12 degrees).[24] Probably the main change, compared with Canyon, was a much larger deployable antenna, which could be built of composite materials. Like Canyon, this satellite was intended primarily to intercept UHF communication systems (the 1993 Russian article describes its main antennas as being oriented in the direction of the stationary orbit, to pick up satellite links). The main (foldable) antenna was 30-45m in diameter, and the satellite carried even larger whip and Yagi antennas. These satellites probably have onboard digital signal processing and beamforming. Reported lifetime is 5-7 years. After Rhyolite was compromised, this system was reportedly modified to intercept Soviet missile telemetry. If the satellite is essentially a transponder, this sort of modification occurs at the ground station. Thus it is possible to reorient a satellite already in space, much too high to be reached for modification. Reportedly the frequency coverage of the series grew over time, requiring heavier satellites (1400 to 1600kg).

After the name Chalet was compromised in 1979, the programme was renamed Vortex; the latter may also have indicated a modification to the satellite. The first modified Chalet was launched on 1 October 1979 (as Vortex). At the height of the programme three satellites were reportedly maintained at all times, one to cover Eastern Europe and the western Soviet Union, one to cover the central Soviet Union, and one to cover the eastern Soviet Union. The same satellites would presumably cover the Middle East. The primary ground station is at Menwith Hill in the United Kingdom. It received a unit commendation in 1989, probably for its assistance in supporting the U.S. naval effort in the Persian Gulf in 1987-8. The associated ground processing system is code-named Runway.

After the name Vortex was compromised in 1987, the programme was renamed Mercury. The last Mercury was destroyed in a 1998 Titan launch failure. Mercury is also described as Advanced Vortex, presumably with an even larger, higher-gain antenna. The first satellite in this series was launched on 27 August 1994. Reportedly British Aerospace was contractor for a British equivalent to Vortex, Zircon, launched 4 May 1989 (however, there were also many reports that Zircon had been cancelled).[25]

There are also 700kg Jumpseats in Molniya orbits, to cover the northern Soviet Union (21 March 1971 – 31 July 1983, with one failure; the last one, typically, was 1028 × 39,321km at 63.4 degrees inclination). Two more such satellites were reportedly orbited in 1985 and in 1987. They are apparently similar in size to the SDS satellite-data relays. A 1993 Russian account of the U.S. SIGINT programme suggests that Jumpseats were intended mainly to intercept communications via Soviet Molniyas, but that seems unlikely, given the geometry of the situation. When paired with Ferret satellites (989) launched by KH-9s, the system is called Yield (renamed Willow in 1982). The Jumpseat ground processing system is code-named Ruffer (it is also used for the follow-on Trumpet). Presumably the ground station is Buckley Air National Guard Base, Colorado (which serves Trumpet).

Parallel to Canyon was a series of quasi-stationary Rhyolite (Project 720) satellites intended primarily to intercept Soviet missile telemetry. That probably meant operation at higher frequencies than Canyon and its successors. Thus Rhyolite could acquire centimetric radar and even radio-telephone signals.[26] The dish was reportedly 15-20m in diameter (presumably it had a higher gain that that of Canyon because it operated at shorter wavelengths). The contractors were TRW and E-Systems. The first 700kg Rhyolite was launched on 19 June 1970, and the last on 6 March 1973. A second series of four improved Rhyolites (Argus) was launched beginning on 18 June 1975; the last was launched on 7 April 1978. At least the first was code-named Argus. Compared to Rhyolite, Argus had a higher-gain antenna, perhaps as much as twice the diameter of the Rhyolite antenna. Of the four, one may have failed to get into its high transfer orbit. One Rhyolite ground station (servicing a satellite located above Borneo) is Pine Gap, Australia. The processing equipment is code-named Rainfall.

Rhyolite was compromised in 1976 by spies at TRW; reportedly the Soviets were then led to encode their missile telemetry, beginning about 1978 (the U.S. delegation to the SALT II missile arms control talks then wrote an anti-encryption clause into that abortive treaty). Plans for more Argus satellites were reportedly abandoned in 1975 to release money for more imaging satellites. However, this programme was revived in 1979 when the United States lost access to missile tracking facilities in Iran.

Argus was updated to become Aquacade (Program 472: this name may also have been applied to Rhyolite after the 1976 compromise).

Later the programme was renamed Magnum; by 1985 it had been renamed again as Orion. This new 4500kg (10,000lb) satellite exploited the extra lift capability of the Space Shuttle (it later had to be redesigned so that it could be launched by a Titan booster, since the Shuttle failed to meet the required schedule). Reportedly Magnum/Orion uses a 425-426ft (129.5-130m) diameter deployable dish antenna. This large dish focuses multiple beams on several sets of horn antennas, so the satellite can simultaneously focus on several quite distinct areas. As an example, a 1993 Russian article suggests that one array might point at Kuwait, another north of Kuwait, a third south of Baghdad, a fourth at Baghdad, others to Moscow, Poland, etc. According to a 1993 Russian article on the U.S. SIGINT programme, these satellites intercepted transmissions via Soviet communications satellites (Raduga and the later Gorizont) in adjoining areas of the synchronous orbits they occupied, presumably by exploiting side lobes.

Reportedly the first Magnum/Orion was launched by Shuttle on 24 January 1985. The last of three was launched on 15 November 1990. All three were reportedly operational early in January 1991, at the outset of the Gulf War. An additional satellite was held in reserve at Vandenberg Air Force Base at least through 1993. At that time the Titan booster was suffering failures, and there was some fear that this extremely valuable satellite might be lost in a launch accident (as had just happened to a Navy reconnaissance satellite). The ground stations for this system are at the Buckley Air National Guard base, Colorado, at Bad Aibling, Germany, and at Pine Gap, Australia. The processing equipment is codenamed Roster.

Mentor is an advanced (next-generation) Magnum/Orion, also for quasi-geosynchronous-orbit SIGINT. The first Advanced Orion was launched 14 May 1995. Another class of new-generation SIGINT satellites reports to a ground processing system code-named Ramrod, one of which is located at Buckley, Colorado (responsible for 75 per cent of the information reported). The other is apparently at Pine Gap, Australia. The first two satellites were launched on 25 August 1994 and on 8 May 1998. These satellites are described as concentrating on specific geographic reporting areas, which suggests that they are geosynchronous. All of these new satellites used Titan IV boosters.

Trumpet (Advanced Jumpseat) is apparently a corresponding satellite using a high Earth orbit or perhaps a Molniya orbit (like Jumpseat). Reportedly it uses a 490-492ft (149.4-150m) diameter deployable dish, and weighs 11,500lbs. The programme began in 1984, and the first satellite was completed in 1991, but it was not flown until 3 May 1994. Two more were launched (in 1995 and 1997) before the programme was cancelled. The sole ground station is at

Buckley Air National Guard base, Colorado. This system uses the same ground processing system, Ruffer, as its predecessor, Jumpseat.

A planned fifth generation system, for deployment after the year 2000, is Intruder. It apparently is to replace both the Chalet series and the Rhyolite series. An earlier plan for a replacement for Chalet/Vortex, Vega, was cancelled in the 1990s after it was criticised for its narrow concentration on Russian communication systems, now no longer the only important U.S. SIGINT target. As of 1993, it was estimated that the United States had three to five Rhyolite and successor satellites in service, two over the Indian Ocean and two or three over Africa and the Atlantic Ocean.

Reportedly these satellites were first used for combat purposes during the 1973 Middle East War, when they (and Chalets) monitored Arab air defence radars as well as command links, including the voice radio chatter of Arab pilots. Other reports of U.S. exploitation of SIGINT satellites include intercepts of strategic missile telemetry during the 1970s and of SA-5 tests against ballistic missiles at Sary Shagan (1973-4). Reportedly satellites were used to track SS-24 mobile ICBMs by intercepting the coded communications between the launch vehicles and their control centres. U.S. satellites also reportedly provided the first news of the Chernobyl disaster in 1986, intercepting radio (presumably microwave-borne telephone) communication between Kiev and Moscow. The news was confirmed by the IR satellites (DSP) usually used to detect missile firings.

In addition to all these specialised satellites, there are persistent reports of ELINT packages aboard other satellites, such as the big imagery type. As in other ELINT operations, the point of placing such packages aboard satellites otherwise not considered ELINT types would be that electronics are often turned off when an ELINT satellite or aircraft is known to be in the area.

National vs. Tactical Systems

The U.S. government held both satellite imagery and ELINT data very closely. For example, it appears that none of this data was released to tactical forces during the Vietnam War. Clearly the fear was that wide dissemination would compromise valuable sources of information.[27] If the Soviets knew just how good (or bad) U.S. photo satellites were, they could camouflage installations accordingly. If they knew exactly what NSA was intercepting, they could shut down key circuits, or encrypt ones they imagined were secure. For that matter, if they knew exactly which low-altitude satellites carried NSA's receivers, they could shut down key emitters when the satellites passed overhead.

Alternatively, one might say that in the 1960s and 1970s the satellites were reserved largely to support the central strategic war – Washington's war – that might have been. The government saw tactical operations, even a big war in Vietnam, as a secondary proposition, on which invaluable national assets ought not to be wasted. The great movement since the late 1970s has been to release space assets to tactical users, under a series of TENCAP (Tactical Exploitation of National Capabilities) pro-grammes. Perhaps they became possible because many in Washington came to doubt whether the big strategic war, the war for which the satellites had been conceived, would ever be fought. If non-nuclear, often local, war was all there was, then it deserved all the support which could be mustered. This change is an important theme in later chapters.

The Soviet case was different. There was no real distinction between national and tactical systems. On the other hand, there were two semi-competitive intelligence organisations, both of which may have controlled space assets, the KGB and the GRU (reportedly the GRU was the primary authority for ELINT satellites). In theory the KGB, the Party's intelligence organ, roughly paralleled the U.S. CIA and NSA, feeding strategic data to the Soviet leadership. The GRU was the intelligence arm of the Soviet General Staff. In Russian, intelligence and reconnaissance go together, so the GRU was also responsible for targeting. For example, the Soviet ocean reconnaissance system was a Navy, hence GRU, asset. It is less obvious who controlled other Soviet ELINT satellites, and thus whether their product was generally available to the Soviet Navy. Similar questions apply to the imagery satellites.

Weather Observation

Although not directly tied to targeting, satellite weather observation was an essential space imaging service supporting carrier strike operations. Satellites offered something quite unique: weather observation over an enemy's territory. Those familiar with the story of the Normandy invasion will recall how important the crucial forecast was, and to what extent it succeeded because the weather over France came from the west, from areas the Allies could monitor. Carrier air operations, indeed the manoeuvring of fleets, were critically affected by weather. The classic examples are the two disastrous Pacific typhoons of 1945, into which the U.S. fleet blundered, due in part to poor weather predictions. Quite aside from making air operations more or less practicable, weather observation affected the imaging satellites, since there was little point in wasting valuable film over an overcast area.

Weather also had more subtle effects on naval operations. For example, in much of the world radar 'ducts' form over the sea, acting as waveguides. A radar in a duct can see far beyond its rated range; conversely, signals from a radar outside a duct generally bounce off the top or bottom of the duct. The extent of ducting, and the heights of the ducts, can be calculated from the profile of temperature and humidity – which can be measured from space. Observation on the spot would be more precise and, thus, preferable. However, it would also be far more expensive, since each surface or subsurface measurement covers only a very limited area. In some places it would be altogether impossible: an enemy or potential enemy controlled the area of interest.

The importance of satellite weather observation is indicated by the prominence of antennas aboard U.S. aircraft carriers dedicated to receiving satellite weather data. Data are processed directly on board the carrier, but they are also transmitted back for central processing in the United States, by a system originally called NAVWEPS (the Naval Weather and Environmental Prediction System).

The initial U.S. space weather system was TIROS (Television and Infra-Red Observing Satellite), developed by NASA for the U.S. Weather Bureau. TIROS I was launched on 1 April 1960 into a 450-mile orbit. The Defense Department then proposed a single system which would meet both its needs and those of the Weather Bureau. When that proved impossible, work on a Defense Satellite Applications Program, specifically to support strategic bombers and satellite reconnaissance, began in 1963 (Air Force Program 417). A decade later the programme was declassified and renamed the Defense Meteorological Satellite Program (DMSP). DSMP became a tri-service programme in 1977.[28]

Soviet ASATs

Given the existence of the U.S. satellite systems, by 1960-1 Soviet anti-satellite (ASAT) systems enjoyed a very high priority; a special anti-satellite command (PKO) was established in 1963-4 (at about the same time a special anti-missile command, PRO, was formed). There were two alternative ASAT trajectories. The simplest was direct ascent, the ASAT being fired from the surface directly at the passing satellite, but this requires considerable energy and very precise homing, since it is easy for the ASAT to miss. The alternative was to launch the ASAT into the same orbital plane as the target. In such a co-planar interception the ASAT manoeuvres in only two dimensions. The ASAT is launched when its site is directly beneath the orbital plane of the target satellite. The ideal launch site is at the equator, since in some cases a northern site may never be under a satellite's orbital plane. The Soviets generally favoured co-orbital interception. In 1961-3 they considered and rejected (after flight trials) an air-launched direct ascent missile with a 50kg warhead. Note that neither technique was really suited to destroying a satellite in a very high (*eg* geosynchronous) orbit. Ascent would be lengthy, and it would be difficult to command the interceptor into position to destroy so distant an object.

Several alternative techniques were considered. One was to use a large ballistic missile to place a large nuclear warhead near the target satellite. Radiation effects could disable a satellite as much as 1000km away. It was discovered in 1962 that a large bomb exploded in space would create a massive electrical storm, the electromagnetic pulse (EMP), when the particles and radiation it created struck the upper atmosphere. After that much attention was devoted to protecting military electronics, particularly micro-electronics, from high-altitude blasts. What was less well appreciated was that a mirror pulse would travel up to devastate satellites in orbit. This kind of EMP may have been very nearly the only ASAT measure which could disable a high-orbit satellite. One reason for using military rather than civilian satellites during the Cold War was fear of a Soviet ASAT explosion at the outbreak of war. Although the Soviets considered specific missiles as nuclear ASAT platforms of this type, in fact any large ballistic missile would do. In their 1961 test series the Soviets exploded several nuclear weapons in space in tests U.S. intelligence interpreted as anti-missile experiments; they also presumably served as ASAT tests.

The favoured alternative was to place an explosive-laden satellite close enough to the target for the fragments from a blast to destroy or disable it. The necessary electronics were developed by KB-1, the avionics organisation founded in 1947, and responsible for most Soviet missile guidance systems. Plans initially called for the booster to be Korolev's big R-7 (NATO SS-6); the satellite was to have been developed by the MiG design bureau. However, in the early 1960s V N Chelomey took over the project, offering one of his 'universal rockets' (UR-200) as the booster. Chelomey's design bureau, OKB-52, took over the design of the ASAT vehicle itself; it became the 'Istrebitel Sputnikov' (IS), the satellite interceptor. The IS used a radar seeker and a 300kg warhead, which would break up into twelve groups of shrapnel. Its effectiveness depended on the direction to the target: lethal radius was 400m head-on, 2km tail-on, and probably 1km for a beam aspect. IS had a multi-start rocket motor to manoeuvre it into position. Apparently the satellite was ultimately to have used an optical (probably IR) seeker. This system was publicly named Polyot. It is probably most important to this naval account for Chelomey's attempt to use the same satellite 'bus' for an ocean reconnaissance satellite (see the next chapter).

Tests of IS began with a 1 November 1963 propulsion/control test. After his patron was ousted about a year later, Chelomey's UR-200 rocket was cancelled in favour of the alternative Tsiklon, based on the R-36 (NATO SS-9) ballistic missile. The IS had to be redesigned (similar considerations affected the ongoing naval space reconnaissance programme, as recounted in the next chapter). It was first tested (as Cosmos-185) on 27 October 1967. The gap between the two Chelomey tests and the first Tsiklon test may have made it appear to Westerners that the Soviets were hesitant in pressing this programme ahead, for fear of encouraging U.S. ASAT efforts. In fact the programme had apparently been transferred to another design bureau. After tests in 1968-72, the IS system was accepted for service in 1972. The trials had shown that it was effective from 230km altitude up to 1000km. Tests then stopped. Since 1972 was also when the SALT I agreement was signed, Westerners tended to associate the apparent abandonment of the programme with the new treaty. SALT made repeated references to 'national technical means', *ie* satellites, as the keys to verifying compliance with the treaty. However, IS was clearly a wartime or near-war measure, hence had nothing to do with a peacetime treaty. Moreover, nothing in the treaty banned the development or deployment of an ASAT system. Interceptors were apparently based at Baikonur, where each of two launch pads could support several missions per day.

A modified IS-A was tested in 1976-8, at which time it was probably placed in service (in 1981-2 missiles in storage were tested). Tests were interpreted as attempts to demonstrate quicker interception and to evaluate new attack sensors (optical or IR, which apparently failed). The initial tests required that the satellite make two orbits before attacking, but in four tests (two successes) the satellite destroyed its target after making less than a single orbit. In this case there was an excellent chance that U.S. space surveillance systems would not even detect the interceptor; the target would simply disappear. At the least, there would be no opportunity to order the target to manoeuvre away from the interceptor. The tested altitude limits of the system were expanded to a minimum of 159km and a maximum of 1600km. During the Cold War the U.S. Defense Department estimated that by 1972 the Soviets had an ASAT system capable of destroying a satellite at an altitude of 5000km. The launch location probably limited the systems to satellites in inclinations of 45 degrees or more. In addition, an air-launched direct-ascent ASAT missile was developed during the 1980s, to be launched from a MiG-31. It was analogous to the U.S. missile launched from an F-15.

The Soviets also experimented with ground-based attack techniques, both the use of high-powered lasers and what they characterised as radio-electronic combat (REC). Work on lasers and particle beams began in the 1970s; ultimately estimated range was about 1000km, although IR satellites in high orbits might have suffered damage.[29] It is not clear to what extent laser systems were ever successful to the point that they could operate effectively in wartime. REC may mean little more than jamming or even sending the satellite false control commands.

The U.S. Space Shuttle presented both a great challenge (it could, in theory, manoeuvre in space) and a great reward (attacks would have devastating consequences on overall U.S. space capability). To deal with it, the Soviets began work on a 'space interceptor', Uragan, specifically to deal with Shuttles in space. This project was cancelled after the *Challenger* crash which re-oriented the U.S. military satellite programme back to expendable rockets.[30]

Soviet Imaging Satellites

The Soviets appreciated the value of photo satellites because they had their own programme, proposed in 1956 by Korolyev's OKB-1 (which was then developing the necessary booster, R-7 [SS- 6]).[31] The bureau proposed a reconnaissance satellite with a film-return capsule. After Korolyev's design bureau began a manned space project in 1958, clearly it lacked the resources for two entirely separate series, so they were combined as Vostok (East).[32] Of four types, Vostok-1K (11F61) was the basic manned (one week duration) vehicle, Vostok-2 (11F62) a reconnaissance satellite (one month endurance), Vostok-3KA (11F63) a one-week manned vehicle, and Vostok-4 (11F64) another one-month reconnaissance vehicle. Yuri Gagarin rode Vostok-3KA to the first manned space flight in 1961, his craft being renamed simply Vostok.

The reconnaissance craft became the parallel Zenit-2 (11F62) and -4 (11F64, later 11F69). These heavy (about 4600kg) satellites carried only a single massive (2400kg) pressurised return capsule, containing both cameras and film. Zenit-2 was a wide-area satellite, Zenit-4 its narrow-area companion. Mission endurance was initially set by the capacity of the satellite's storage battery (eight days), and later by film capacity; in the mid-1970s it was typically no more than two weeks. Because the retro-rocket (which brought the capsule out of orbit) could not be restarted, these satellites had no ability to manoeuvre in space.[33]

The decision to merge manned and unmanned systems slowed the Soviet programme, so that Zenit was first (and unsuccessfully) launched only on 11 December 1961. The first success was Cosmos-4, 1962. The system entered service in October 1963. Developed versions included Zenit-2M (11F690, 1968), Zenit-4M (11F691), Zenit-4MK (11F692M), Zenit-4MT (11F629, for stereo images for mapping), and Zenit-6 (11F645, for surveillance, *ie* low-resolution images).

Yantar (Amber) was a third generation (second, to Russians) system, intended to exploit the new Tsiklon boosters in order to extend orbital lifetime. Progress was initially slow because the design bureau involved was overloaded with ballistic missile work. Work on the new satellites began in 1970. The new satellite had orbital manoeuvre engines in an additional module, so that it could adjust its altitude for either wide-area (survey) or detailed imaging. Its return module could carry two to four cameras, and in some cases they operated in the infra-red. There was apparently also a separate ejectable film capsule, which could be jettisoned part way through a mission (as in the U.S. KH-4). Mission endurance of early versions was typically fourteen to seventeen days. A recent U.S. account describes the third-generation spacecraft as about 2.4 × 6.5m (6300kg), carrying film in a spherical 2.3m (2400kg) capsule, which also contained an explosive charge, to be used if the capsule could not be recovered on Soviet territory. Satellite lifetime was given as two to three weeks. Yantar-2K (11F624, 1974) had a design lifetime of thirty days. The first attempted launch (23 May 1974) failed, but a second satellite (launched 12 December 1974) remained in orbit for twelve days, and Cosmos-905 (launched May 1977) finally achieved the designed thirty-day lifetime. The following year the Yantar-2K system was accepted into service as Feniks. The typical orbit was 170-180 × 350-360km, the satellite being oriented not along the direction of flight but (like the U.S. KH-9) towards the nadir. Unlike the U.S. satellites, at the end of a mission a Yantar was not allowed to fall into the ocean, but instead it was landed on Soviet soil, so that its expensive reconnaissance equipment and computer could be recovered for re-use.

The next stage, Yantar-2K1 (11F693, first flown 1979), had a forty-five day lifetime. It became operational (as Oktan) in 1981. Compared to the earlier Yantar-2, this satellite flew an elliptical orbit with a low perigee (typically 170km) for better resolution near an important target. The satellite carries multiple film capsules, so it can exploit its longer lifetime. It was therefore possible to maintain one satellite always in place by launching eight or nine each year. Approximate dimensions are 2.4 × 7.0m (6.7 metric tons). Both Oktan and Feniks were withdrawn in 1984 in favour of a third generation of longer-endurance satellites (initially sixty, then ninety days).

In the mid-1970s typically between thirty and thirty-five such satellites were launched each year, but only for very short missions: total in-orbit time rarely reached 400 hours in one year (a year has 8760 hours). Launches of these satellites continued into the 1990s, becoming much less frequent as longer-lived craft came into service.

Some of the later launches were probably civilian Earth resources versions called Resurs-F1, used to map the Soviet Union itself.

As in the U.S. programme, the early Soviet programme included a manned reconnaissance satellite, Soyuz-R. It was part of a Vostok successor programme proposed by Korolyev in 1962, another version being a manned ASAT interceptor (another version was designed to orbit the Moon). Like the ocean reconnaissance system described in the next chapter, Soyuz-R was included in the space reconnaissance part of the 1965-70 five year plan (the plan was signed out by Minister of Defense Marshal Malinovskiy on 18 June 1964). As proposed, Soyuz consisted of two spacecraft which would have docked together to form a 13-ton space station. One carried imaging and ELINT equipment, while the other was a transport section to bring cosmonauts up to the station, and to return them to earth. As it happened, Soyuz-R fell victim to Soviet politics. Just before his patron, Khrushchev, fell in October 1964, Chelomey managed to get permission to develop a 20-ton three-man space station, Almaz, which would be launched by his UR-500K Proton booster. It would operate for two to three years, taking photographs throughout its lifetime. The crews would be rotated every ninety days. When Korolyev died in 1966, Chelomey managed to have Soyuz-R cancelled in favour of his Almaz (part of the Soyuz project survived for a time as a ferry to service Almaz). At that time it was expected that Almaz would be tested in 1968, to enter service in 1969.

While these projects were in the early design phase, the Americans tested manned reconnaissance in the June 1965 Gemini 4 mission. The Soviet government ordered a parallel experiment on an urgent basis, the existing Voskhod and Soyuz 7K-LOK spacecraft being modified. In their cases the missions envisaged were reconnaissance, both optical and electronic (ELINT); inspection of enemy satellites from orbit (with the potential to attack them); and early warning of strategic attack. Under a 24 August 1965 order a military research version of Soyuz was to fly by 1967. The code name for this 7K-VI was Zvezda. It carried a specially-designed recoilless gun for self-defence against enemy interceptor satellites, aimed by turning the entire satellite. The main reconnaissance instrument was an OSK-4 optical sight connected to both a camera and a radar illuminator, aimed using a special saddle. There was also a special (presumably IR) missile launch detector (Svinets) and the satellite carried directional antennas (for DF and ELINT) on long booms (for maximum base line). The satellite was powered by nuclear batteries, solar batteries having been rejected because of weight (due to the storage batteries associated with them), the limitation associated with keeping the satellite oriented towards the sun in daylight, and fear of the consequences if the solar panels did not deploy properly. Moreover, the military equipment required substantial power.

By 1967, the Soviet military planned to fly fifty Soyuz VI missions in 1968-75. Cosmonauts were in training. Then the programme fell victim to internal politics. Soyuz VI had been developed by the OKB-1 out-station at Samara, under Chief Designer Dmitri Ilyich Kozlov; the main OKB-1 office in Kaliningrad concentrated on other versions of Soyuz. In October 1967, however, the main office (Chief Designer Vasiliy Pavlovich Mishin) began attempts to take over the program. To do so Mishin began to attack the Soyuz VI design, hoping to substitute his own Soyuz 7K-S. He succeeded (his version had solar panels), but the whole programme was cancelled in 1969, no satellite having flown. By that time Kozlov was working on unmanned reconnaissance satellites; his Yantar-2K benefited from design work done on Soyuz VI.

Almaz barely survived. After the Americans got to the Moon, Brezhnev ordered Mishin to produce a space station in the shortest possible time, as a way of retrieving something from the propaganda disaster. Eight complete Almaz spaceframes existed, and one was orbited (as Salyut 2) on 3 April 1973, to beat out the U.S. Skylab. It depressurized before its crew could be launched to occupy it. Salyut 3 was successfully launched on 24 June 1974. It carried a telescope a meter in diameter, which the cosmonauts used to photograph airfields and other potential targets; it was aimed using an optical sight. Film could be developed and examined on board, the cosmonauts reporting important features by radio. There was also an IR camera, and the spacecraft had a Nudelmann cannon (originally developed for Soyuz VI) in the nose for self-defence against a feared attack by a U.S. Apollo spacecraft. There were also space-to-space missiles. The first film capsule was successfully returned to Earth on 22 September 1974. A second space station, Salyut 5, was launched on 22 June 1976 (there was no Salyut 4); its first film capsule was returned to Earth on 22 February 1977. The station itself was de-orbited (and thus destroyed) on 8 August 1977, its fuel having been used up.

Thirty military personnel were trained for the programme. However, the results of the two flights were apparently disappointing, and the programme was cancelled on 1 January 1978. It is possible that cancellation was due mainly to Chelomey's personal unpopularity within the Soviet defence establishment.[34] The military experiments survived on board other space craft (Salyut and Mir). Officially, the conclusion was that the higher cost of a manned station outweighed its advantages. Among other things, the Almaz spacecraft carried a side-looking radar, which was associated with naval experiments (see Chapter 8).

Like their U.S. counterparts, the Soviets also saw television and a data link as a desirable alternative to film capsules. Chelomey proposed such a television surveillance system (TGR) as early as 1963. As in the United States, it was not yet practical. Eventually they modified Yantar (Yantar-4KS1/Neman) as a television satellite. As in the U.S. system, the imagery can be transmitted either directly down to the ground station, or via relay satellites (two Geizer satellites, at 80 degrees E and 346.5 degrees E; Potok network). The first such television satellite, called the 'fifth generation' in the West, was Cosmos-1426 (December 1982); note that the first Geizer had been orbited in May 1982.[35] These satellites had a new propulsion unit for both re-entry and for manoeuvring in orbit, and they had solar panels to provide power for their long missions. Satellite lifetime was 6-8 months; afterwards the satellite was braked so that it would dive into the denser part of the atmosphere (and torn up by onboard explosives) to disintegrate, the pieces falling into the ocean. The longest lifetime of a satellite of this type was 419 days (Cosmos 2267). Given a long lifetime and a low launch rate, presumably there was no longer great interest in retrieving satellite parts for re-use. Three satellites of this type were launched in 1995-8, of which Cosmos 2359 (June 1998) was still in orbit in the spring of 1999.

This television type was apparently not considered effective for high resolution viewing, since development of film capsule satellites continued. The first of what the Soviets called their sixth generation (Yantar-4KS2/Kobalt) was Cosmos-2031 (18 July 1989), launched into a Yantar-type orbit. Three launched in 1990-2 lasted fifty-eight days each; the most recent, Cosmos-2242, lasted 124 days. A follow-on satellite (seventh generation, to the West: Orlets-1/Don) could carry ten film capsules in a ring around its 'equator'. When the satellite completes its mission, it is blown up in orbit. Six such satellites have flown in 1989-97. One lasted 124 days in orbit. Another, heavier, Orlets satellite (Cosmos 2290) was launched by a more powerful booster (11K77 Zenit-2; the others all used Soyuz U/U2). It lasted 224 days in 1994-5. This satellite may have combined film return and television techniques, and may have been part of a system called Kuban.

Work on a next-generation television system, Arkon (at NPO Lavochkin), began in 1984, the first launch being planned for 1988. By that time this programme, like many others, was in deep financial trouble, and work stopped in 1991. However, some satellites had been made, and by 1996 some of their components were running out of guaranteed lifetime. Moreover, by that time Russia had very few imaging satellites in orbit. The available satellite, Cosmos-2344, was therefore launched on 6 June 1997, by a four-stage Proton-K. It had been conceived as one of at least twenty spacecraft flying at different altitudes to provide 24-hour global surveillance. Each satellite would provide both area and detailed views. This satellite went into an unusually high orbit, 2749 × 1516km (inclination 63.3 degrees). Previous Soviet reconnaissance satellites had all flown in much lower orbits, typically 220-240 × 260-280km.

Soviet ELINT Satellites

The early Zenit imagery satellites carried a radio intercept receiver, Kust. Its success led to the creation of two series of ELINT satellites, Tselina (Virgin Soil) and US (see Chapter 7), the latter for naval purposes, beginning with experiments in 1965-7.[36] Given limited resources, there was some question as to whether two distinct series of ELINT satellites were needed. It was suggested that a single type of wideband satellite could supply all Soviet military customers. However, US survived because it was part of a larger system of naval target designation, linked to a centre which also controlled the associated radar satellite.

There were two Tselina series: Tselina-O (11F616, about 900kg) for wideband search, using ten modular receivers, and Tselina-D for detailed search. The first Tselina-O was launched on 30 October 1967 (a 26 June 1967 attempt failed). Tselina-O was accepted into service in 1971. There was a follow-on Tselina-OM (11F617). Typical Tselina-O orbits were 500 × 550km (74.1 degrees). These satellites were launched by SL-8 (Cosmos 11K65M boosters: modified R-14/SS-5 IRBMs with added upper stages). In all, thirty-nine Tselina-O were launched between 30 October 1967 and 31 March 1982 (there was one failure, 22 July 1971).

Tselina-D (Ikar, 11F619, 1640kg) was generally launched into an orbit similar to that of Tselina-O (sometimes somewhat higher), almost always at about 81.2 degrees inclination. Like Tselina-O, this satellite used a Tsiklon-3 booster. The first was launched on 18 December 1970 and the last of seventy-nine (a failure) on 25 May 1994. Tselina-D was accepted into service in 1976. After 1984 the wideband Tselina-O was abandoned in favour of full reliance on Tselina-D.

Tselina-D is a store/dump system, which collects its ELINT data and then dumps it to the appropriate ground station. Reported position-finding accuracy (as given in 1985) is 10km. Since several vehicles were often launched in close succession, presumably different versions were used together. The Soviets tried to maintain six such satellites in six orbital planes spaced 60 degrees apart (mean altitude 700km). In the mid-1990s this system was nearing the end of its operating lifetime, with three or fewer spacecraft in orbital planes 30 degrees apart at altitudes of 635-665km.[37]

A third-generation Tselina-2 (11F644) system (3500kg) began with a 28 September 1984 launch into an 828 × 865km orbit (inclination

71 degrees), using a Proton booster. Beginning with the third satellite (22 October 1985) this system was placed in orbit by the new Zenit-2 booster. Compared to Tselina-O/Tselina-D, it could handle a wider frequency range, it conducted more detailed analysis, and it was more reliable. The Soviets tried to maintain four such satellites in orbit at any one time, their orbital planes 40 degrees apart (a 1992 three-satellite constellation occupied planes 45 degrees apart). Besides downlinks, these satellites can report data via geosynchronous Geyser satellites. In all, seventeen of these Tselina-2 satellites have been launched, the latest on 28 July 1998. There were also five failures, on 28 December 1985, 4 October 1990, 30 August 1991, 5 February 1992 (three failed launches in succession), and 20 May 1997. The October 1990 failure wrecked the Zenit launch pad, and thus delayed any replacement launch.

These spacecraft were all described as interceptors of radar signals.[38] The Soviets presumably also had a SIGINT capability, perhaps in the form of their Strela-2 low-altitude (785-810km, inclination 74 degrees) military store/dump communications satellites. These craft are sometimes said to be intended to pick up low-power clandestine signals, store them, and then dump them while passing over Russia. They would seem well-adapted to detecting and retransmitting radio signals such as those from surface microwave transmission systems, a role which would probably be more important than clandestine communication.[39]

U.S. intelligence apparently first perceived a Soviet capability to tap into the U.S. domestic microwave telephone system in the late 1970s. At that time special efforts were made to convince Americans in sensitive positions that their telephones were anything but secure; the Soviets were said to have used eavesdropping to hold down the prices of American grain they were buying. The usual explanation was that the Carter Administration had disastrously allowed the Soviets to build their new Washington Embassy on a rise which offered a particularly good opportunity to intercept the main microwave beam into the city. It seems more likely that this was an attempt to avoid drawing attention to U.S. capabilities to intercept Soviet domestic microwave relays. Presumably the big ELINT station at Lourdes, Cuba was important not because it could pick up radio traffic from Florida (which was well over its horizon) but as a down-link from ELINT satellites. Lourdes in turn would have been crucial because in the 1970s Soviet ELINT satellites could not pass their product up to communications craft for direct transmission to the Soviet Union (the first Geizer data relay satellite was placed in orbit in May 1982).

There was also a geosynchronous ELINT satellite (two-year lifetime). Reportedly one was stationed over the Atlantic to monitor

Trident telemetry, down-linking to Lourdes, Cuba. Most likely this satellite was a geostationary Raduga (Rainbow: 11F643) communications satellite, working at SHF frequency.[40]

Again, ELINT satellites were not intended for naval applications (the Soviet EORSAT and the U.S. White Cloud/Parcae, described in later chapters, are a different proposition). However, often they could (and did) pick up naval communications traffic, and at least in the U.S. case (see Chapter 10) they could apparently add a valuable contribution to more purely naval systems for passively detecting enemy ships. Probably based on U.S. experience, the U.S. Navy assumed that Soviet ELINT satellites would be used to help locate U.S. warships. It is not clear in retrospect to what extent that was done, since ELINT satellites were probably subordinated to the KGB rather than to Soviet military intelligence (GRU).

Like the Americans, the Soviets had their own weather satellite system. First launched in 1969, it was Meteor. This was the system credited with finding large openings in the Arctic sea ice (polynyas) from which submarines could launch their missiles.[41]

U.S. ASAT Programmes

Just like the Soviets, Americans realised that satellites offered critical wartime advantages. Beginning in 1959, all three U.S. services proposed ASAT systems. The U.S. Air Force became the first (in the world) to test an anti-satellite weapon, using a Bold Orion missile fired from a B-47 bomber.[42]

In March 1961 the U.S. Navy proposed a homing weapon, Early Spring (initially a modified Sparrow), atop a submarine-launched Polaris booster.[43] The system was rejected, largely because Polaris strategic missiles were considered far more useful than ASATs. Its main advantage over a land-based system was that the launch point could be varied as desired. Another abortive proposal was Skipper, a modified Scout rocket to be launched from surface ships or submarines.

The ASAT issue became more critical when, in August 1961, Khrushchev said that he had to capacity to place H-bombs in orbit, so that he could hit any spot on the planet. In theory, an orbiting bomb was a worse threat than an ICBM, because it could attack with far less warning. There would be little or no indication that a particular satellite carried a bomb, and the bomb would hit its target a few minutes after having been released. The heavy U.S. investment in missile-detection radars would have been circumvented; it would no longer be possible to launch U.S. forces before they were struck. In practice orbital missiles were not likely to be very accurate. It would be extremely difficult to time attacks from orbit so that bombs hit more or less simultaneously. Failing that, the first explosion – even the first detection of something leaving orbit – would be sufficient warning on which to launch U.S. forces. Even so, the threat was underlined in September when the Soviets tested a 58-megaton bomb, the largest ever built (and in fact a derated version of their 100-megaton weapon). Khrushchev continued to threaten the use of orbital bombs through the autumn of 1961. That Christmas Secretary of Defense McNamara asked the Aerospace Corporation, which developed satellites and other hardware for the Air Force, for an urgent proposal for an anti-satellite weapon. He withdrew within a month, possibly because it seemed that the Army's developmental Nike Zeus anti-missile weapon could easily be adapted to the new requirement. In May 1962 McNamara secretly approved development of an ASAT version. It was first tested in May 1963, and became operational at Kwajalein on 1 August 1963. The programme was called Project Mudflap (the Bell Labs designation) or Program 505.

The U.S. Air Force was already developing SAINT, the Satellite Inspector. It was first proposed when the Air Force began developing its own intelligence-gathering satellites, in 1956. SAINT covered a variety of techniques for both inspection and interception, although inspection was emphasised. The programme was approved after a satellite detected in December 1959 could not be identified (it turned out to be part of the second stage of the U.S. Discoverer V, launched the previous August). An Atlas Agena B would have placed the SAINT vehicle in orbit slightly ahead of the target satellite; SAINT would then have manoeuvred to within 50ft for inspection by television and radar (and, in later versions, by IR and even X-ray). The results would have been downlinked. The Air Force expected that SAINT would also have the ability to destroy a satellite if the target turned out to be hostile. It was also interested in a manned version, the theory being that the pilot was needed to decide how to use the sensors and to interpret their readings. As of 1960, it was hoped to make SAINT operational by 1967, with the first development flights scheduled for March 1963. However, the Air Force cancelled the programme on 3 December 1962. It was too ambitious and too expensive.

Instead, the service modified its Thor intermediate-range ballistic missile as Program 437, to carry a nuclear warhead up to satellites at altitudes of up to 700 miles. By 1962 Thor was being withdrawn as an operational IRBM, and many surplus missiles were available. On the other hand, the Thor airframe and engine were still in production as a space booster, to place satellites in orbit at just about the same altitude the Thor ASAT would have to reach to destroy them. The Air Force had issued an Advanced Development Objective (ADO) for an ASAT on 9 February 1962; it was to use an existing booster. Probably the

ADO was issued because of the satellite scare of the previous Christmas. Studies were soon completed, and by October 1962 Air Force Systems Command could submit a plan to the Defense Department. Tentative approval was granted on 13 December 1962; Thor promised greater range and altitude than the Army's Zeus. Thus the new system was being approved almost exactly as SAINT was being abandoned.

The Air Force was told on 15 February 1963 to prepare for a Thor ASAT standby operational capability once tests had been completed. The Soviets were still claiming an ability to orbit bombs or missiles carrying large thermonuclear warheads. Deployment was approved (on Johnston Island in the Pacific) on 8 May 1963. It was justified publicly as a counter to a projected Soviet orbital bomb system. The problem was considered so serious that on 6 July 1963 this programme received the 'highest national priority'; it was tested in February 1964.

Compared to Nike Zeus, the modified Thor could reach higher altitudes, but could not react as quickly because it used non-storable liquid fuel. Given delays inherent in satellite tracking and the fact that intercept calculations had to be made at NORAD (which received satellite data) and then passed to Kwajalein, Thor could be expected to intercept satellites twice a day. Thor was credited with the ability to intercept targets at altitudes up to 200nm (maximum demonstrated Zeus altitude was 150nm) and at slant ranges up to 1500nm (it could attack higher targets at shorter ranges). It carried a 1.5 MT warhead. There was also considerable interest in an alternative satellite-inspection device, which was tested in 1965-6. Both Thor and Zeus were direct-ascent weapons, which had to wait until their targets almost passed overhead. Neither missile was likely to intercept its satellite target very precisely. Instead, each would carry a nuclear warhead. The Thor and Nike Zeus projects together explain U.S. interest in high-altitude nuclear explosions, which first demonstrated EMP effects off Hawaii.

Both systems were disclosed by Secretary of Defense McNamara in September 1964 (Congress had been secretly informed in 1963) after the U.S. government was criticised for doing nothing to deal with a reported threat of Soviet nuclear weapons in orbit. Zeus was more energetic, and thus could reach satellite orbit much more quickly; Thor could attack satellites in higher orbits. On 23 May 1966 Secretary McNamara ordered the Nike Zeus ASAT phased out, and that was completed the following year. Thor already provided the necessary capability. Furthermore, by this time McNamara was fighting hard to keep from deploying a related Army ABM system. He may well have feared the political effects of keeping a related ASAT system in service. He actually wrote that the proposed Nike-X (Zeus follow-on) would duplicate the Thor capability; there should be only one system (and not Nike-X, which he abhorred). When the decision was taken to deploy Nike-X (as Safeguard), it was credited with ASAT capability.

Finally, there was Hi-Ho (also rendered Hi-Hoe), developed by the Naval Ordnance Test Station (NOTS) at China Lake, which is best known for Sidewinder and the 'eye' series of air-to-ground munitions. After having built Notsnik (see Chapter 3), China Lake received permission to build an improved air-launched booster, Caleb. It was available for a Navy attempt to develop an air-launched ASAT, Hi-Ho (Notsnik II), in 1961-2. In the last of three flights (26 July 1962), fired from an F4D, Hi-Ho apparently reached an altitude of 1167.3km (725.5 miles).

By 1963, then, the United States had a limited ASAT capability which required the use of nuclear warheads in space. That was probably acceptable because the main threat against which ASAT had been bought was orbital nuclear weapons; it would be used only in a global war. It was by no means clear that an enemy would leave such weapons in orbit long enough for them to come within range of either of the Pacific Ocean ASAT systems. The weapon the Soviets actually fielded, FOBS, launched a satellite but de-orbited it before it made a complete orbit; orbital motion was used mainly so that attacks could be made over the undefended southern reaches of the United States (the big missile detection radars pointed north, as was logical). Moreover, the United States soon came to depend quite heavily on space assets. As the single 1962 EMP test showed, a nuclear burst in space would likely damage several satellites in its vicinity; it would not distinguish friend from foe. This consideration helps explain why efforts were made to harden U.S. military satellites against nuclear effects (similarly, U.S. warships were shock-hardened to protect them from the effects of nuclear ASW weapons they employed). In 1967 the Soviets signed the Outer Space Treaty, which outlawed keeping nuclear weapons in space. The U.S. government seems to have seen this agreement as proof that the Soviets had little interest in orbital bombs. In Congressional testimony members of the Joint Chiefs of Staff agreed that, even if the Soviets violated the treaty and deployed orbital bombs, they would be of little consequence; the Soviets already had enough quite conventional missiles to devastate the United States, if they could fire first.

Given its nuclear warhead, it seemed less and less likely, then, that the Thor ASAT could or would ever be used.[44] On this basis, on 4 May 1970 the Air Force was ordered to phase down the system as soon as possible; on 1 October its operational reaction time was lengthened to thirty days, which meant that it was no longer really operational. However, it was not finished. For example, damage done by

Hurricane Celeste in August 1972 was repaired at considerable cost (the system was not again operational until March 1973). What finally killed it was a need to use Johnston Island to support other research, particularly anti-missile tests. The ASAT system was terminated effective 1 April 1975.

In 1970 there was, moreover, little real interest in ASAT operation. Before a war, there would be little point in attacking an enemy's low-altitude satellites – *ie* reconnaissance satellites. That would merely be to start the war prematurely. Once war began, slow-motion reconnaissance using film capsules (the only kind yet available) would have little tactical impact. This situation seemed to reverse only when the Soviets deployed real-time ocean reconnaissance satellites which would be used, in wartime, to support attacks on U.S. naval forces. Suddenly ASAT was far more important. This revival is left to Chapter 8.

Space Surveillance

In all the ASAT systems, the missile was only one element of a larger system. Some means had to be found to detect satellites even if they did not radiate; to decide what they were; to determine their tracks; and only then (if need be) to attack them. The key element, the surveillance sensor, was already under development by the Naval Research Laboratory (NRL). The laboratory had been responsible for the Minitrack system used to track the satellites launched by the Vanguard rocket, but that had depended on the satellites' signals. Once Sputnik had been launched, ARPA, the Defense Department Advanced Research Projects Agency, began Project Shepherd, to detect 'dark' (non-radiating) satellites passing over the United States.

There were three alternatives. The Navy proposed the bistatic SPASUR radar 'fence', which was eventually adopted. The Air Force proposed netting together a number of existing radars, many of them conceived to detect ballistic missiles. Their data would be fed into a control centre (SPACE TRACK). The Army's proposed DOPLOC radar system was rejected altogether. Ultimately the choice was to make SPASUR a primary surveillance sensor, feeding its data into the space surveillance system. The existing radars would also feed data, but generally they would be cued by the SPASUR 'fence', because its chance of detecting a passing satellite was far greater. Once a satellite was detected and its track established, its path could be passed to one of a net of very long range telescope-cameras (Baker-Nunn optical cameras). Elements of the surveillance system had already been established to support the experimental Vanguard programme: a Minitrack system to track the satellite's radio signals and the Baker-Nunn optical net.

SPASUR was NRL's way to convert the Vanguard system into what ARPA now wanted. If the satellite did not choose to emit, then some signal would have to be imposed on it. A conventional scanning radar might easily miss a passing satellite, so NRL proposed a massive fixed bistatic system, amounting to a radar fence across the United States. In the first tests, the Minitrack receivers already in place at Blossom Point, Maryland were teamed with a big parabolic radar emitter at Fort Monmouth, New Jersey; the signals were bounced off the first Russian Sputnik. This experiment having succeeded, NRL proposed a national radar 'fence'. ARPA approved the project on 20 June 1958. The first parts of the system were placed in service on 29 July 1958, and the system was effectively complete by June 1961. By that time administration had been turned over by NRL to a new U.S. Naval Space Surveillance Facility at Dahlgren, Virginia. As completed this Space Surveillance System (SPASUR, WS-434) lies across the United States at 33.5 degrees North, between San Diego and Fort Stewart, Georgia.[45] Any satellite with an orbital inclination greater than 33.5 degrees must pass through the fence twice a day.

Almost forty years later, SPASUR is still the only uncued U.S. space surveillance sensor, the basis of all other kinds of surveillance (the other main sensors are narrow-beam devices such as tracking radars and long-range cameras, which must be cued to be effective). Its radar fence extends, in effect, 15,000 miles into space (maximum range is sometimes given as 22,000 miles). Because beams are narrow in a north-south direction but very broad in the east-west direction, the fence effectively extends well beyond the continental United States, to Africa in the east and to beyond Hawaii in the west. It makes about 160,000 observations each day, and is considered critical for sensing the breakup of spacecraft and the tracking of their debris.

The SPASUR fence does have some inherent limitations. It cannot see objects orbiting with inclinations of less than 33.5 degrees, which mainly means geosynchronous satellites. Also it cannot routinely detect very small objects. In the past, the U.S. Air Force has operated several radars to fill in some of these gaps.[46]

To some extent SPASUR and other U.S. surveillance systems can be cued by intercepted telemetry associated with a satellite launching itself. This process cannot be completely silent, since the launch control centre has to be able to monitor the rising booster to know whether or not to abort the mission. Once the United States could reliably intercept telemetry, particularly using satellites, then, it could at least know that a launch was proceeding. However, telemetry would not necessarily indicate the orbit into which the satellite was going, particularly if it went first into a low orbit and then manoeuvred away. It took something like SPASUR to solve that problem.

SPASUR is part of a larger Space Surveillance Network (SSN) directed by the U.S. Air Force Space Control (formerly Surveillance) Centre at Cheyenne Mountain. Once a target has been detected and tracked, a telescope camera can be cued to produce an image or a useable signature as well as a more precise track.[47]

Quite aside from cueing ASATs, this sort of surveillance can be a vital way of warning against an enemy's reconnaissance satellites. On land, that meant warning laboratories, for example, that an imaging satellite was coming within range. Warning against imagery satellites came in the form of a SATRAN (Satellite Reconnaissance Advanced Notice, also called Stray Cat) programme begun in 1966. The U.S. Navy formed its own countermeasures programme, for Fleet Support/Satellite Vulnerability, *ie* to warn the fleet against particular vulnerabilities to satellite detection.[48]

SEVEN

A New Kind of Naval Warfare

The Soviets pioneered the new information-centric style of naval warfare which is now identified with space systems. It began with the coast defence system erected by the new Soviet Union in the 1920s, a system similar to that the Germans had built during the First World War to defend the conquered Flanders coast.[1] In each case, radio added depth to conventional forms of coast defence. Offshore scouts detecting an approaching enemy fleet could instantly relay that information to a central commander ashore. He in turn could move defending craft into position to attack the approaching enemy, even before it came within sight. Given radio to provide information, and a radio net to command dispersed forces, a central commander ashore could fight a battle well beyond his horizon. This was far more than a traditional coast defence based on shore guns and on torpedo craft which could sortie once an enemy fleet came into view. To make it work, the central commander ashore had to be given full control over all of the craft assigned to him. Traditional naval practice, which emphasised the initiative of each ship's commander, was abandoned. The Soviet central control concept mirrored army tactical concepts; the army dominated the Soviet military.

The command net enormously increased the effectiveness of small attack units, such as torpedo boats, coastal submarines, and naval aircraft, because the shore commander could co-ordinate their attacks. For example, the radio net could co-ordinate the movements of boats or aircraft which were not in visual touch with each other, and thus could arrange attacks from many different directions, to saturate a target's defences. Combined-arms attacks (*eg* air and surface torpedo strikes) could be mounted. In theory, the command net could also preclude the usual bane of co-ordinated attacks, attacks against friendly craft. If the central commander knew exactly where the targets were, and exactly where all his attacking assets were, he could order firing only when that was appropriate. Co-ordinated or combined-arms attacks were particularly important to an impoverished Soviet state which could not afford massive quantities of any particular weapon. Central control was also attractive for political reasons. The new Soviet state had to depend heavily on ex-Tsarist officers whose loyalty was always open to question. Control was exerted through loyal Commissars, who had to countersign commanders' orders (but who lacked professional military competence, so they could not themselves command forces). The more centralised the control of naval forces, the less the risk of independent action.

From a tactical point of view, central control made it easier for the small coast defence craft to fire at maximum range, since the central controller could cue them, based on the information he was collating. That was attractive, because it reduced their vulnerability. Normally, the greater the range, the lower the hitting rate. However, in a co-ordinated attack, the net hitting rate might be quite acceptable. The Soviets badly wanted to fire from maximum range, because they could not easily replace even the smallest 'mosquito craft' – which, in the West, were generally considered entirely expendable. One way of limiting the vulnerability of the coastal craft was to limit their own radio transmissions back to the central commander – in effect, to limit any sort of feedback.

Torpedo bombers were a case in point. In the Soviet Navy's shore-based naval air arm, medium bombers formed Mine-Torpedo Regiments.[2] To attack at maximum stand-off range, the Soviets developed a technique of dropping torpedoes from high altitude. Clearly single torpedoes would be ineffective, since (like bombs) they would be dropped relatively inaccurately. Moreover, ships would have considerable time in which to evade. However, if formations of bombers dropped their weapons together, then in theory they could create patterns of torpedo tracks which would be almost impossible to evade. This concept combined two key elements in Soviet thinking, the attack from maximum range and collective (or co-ordinated) tactics. It also reflected a favourite Soviet habit, careful mathematical analysis of combat situations to discover what might be called 'scientific' solutions.

As it happened, such solutions often failed in practice, presumably because the analysts omitted many factors in real situations. When the Soviets disclosed their revolutionary wartime torpedo innovation in the 1970s, they omitted the usual box score, showing how many ships had been hit versus how many torpedoes had been dropped. Later it became clear that the tactic had been altogether unsuccessful. Yet so strong were the imperatives behind it (such as avoiding counterattack against the bombers) that it was retained and refined post-war. Americans found that the Soviets practised at a given altitude until they reached a required standard of accuracy, then tried to increase altitude and therefore stand-off.[3] They also developed a series of very fast torpedoes specially adapted to high-altitude attack, most prominently the rocket-powered RAT-52 (the number indicated the year of approval for service introduction).[4] Quite large aircraft, such as the

Tu-16 ('Badger') jet medium bomber, were adapted for long-range torpedo attack.[5]

Information was the system's vulnerability. If the central commander's information was corrupted, he might well order some of his platforms to attack others, rather than the enemy's. Moreover, much depended on the timeliness of information (not to mention of orders based on that information). In a rapidly evolving situation, mistakes might easily be made. The Soviet concept of operations was very obviously subject to the modern insight that much of warfare depends on cycles of information handling: Observation (of some enemy move), followed by Orientation (digestion of that information), then Decision, then Action – to which the enemy has to react. This concept, of what is now called (at least in the United States) the OODA cycle, was first publicised in the 1970s by U.S. Air Force Colonel John Boyd. Boyd argued that the Air Force's F-86 Sabres had succeeded so well in Korea because, due to their design, their pilots could react more quickly than could their enemies; they could get inside their enemies' OODA cycles. Boyd applied the same idea to the fall of France in 1940. He argued that the French General Staff had been unable to react quickly enough to German attacks. Because its OODA cycle was so slow, it found itself reacting to earlier German moves rather than to current ones. Ultimately its reactions were wholly irrelevant, and it collapsed in what amounted to a nervous breakdown.

Whether or not Boyd was right about the fall of France, the Soviet system of operations certainly implied an OODA cycle. Slowness might well cause a commander to order his units to attack the wrong targets, or even no targets at all. The Soviet system could also be contrasted with wartime British arrangements to control coastal forces in the English Channel. The British established what amounted to a Combat Information Centre (CIC) ashore, to collate all the available information on German movements in the Channel. In this they seemed to be following the Soviet pattern. However, the output of the CIC was used, not so much to order coastal craft to attack as to apprise their commanders of the situation around them, so that they could make intelligent decisions. That typified the difference between Soviet and Western naval practice.

The information-oriented tactics the Soviet navy developed before the Second World War meshed so well with the Soviet political system that one might imagine that they were inevitable. However, it was really best adapted to coast defence, because its elements all reported to a fixed commander. The further the action from the coastal headquarters, the more complex the command problem, and the greater the chance of confusion. Operating far from land, a blue-water fleet demands decentralised control. As in the West, a fleet commander could certainly benefit from externally-obtained information, such as that from HF/DF or from code-breakers, but command decisions would have been made on the scene, by the embarked fleet commander. No distant commander could possibly enjoy good enough information or complete enough communication with the many fleet elements.

Thus any Soviet attempt to build a blue-water fleet carried a real possibility of transforming command style, whatever the Soviet leadership may have imagined. As it happened, the Soviet navy very nearly changed its complexion both before and after the war, because Stalin decided in the mid-1930s that as a major power the Soviet Union needed a conventional blue-water fleet built around expensive capital ships. As it happened, Stalin's heavy-ship programme was stopped by the 1941 German invasion. It was, however, revived after the war and by the mid-1950s there was a real possibility that the Soviet Union would build up the sort of deployable concentrated fleet employed in the West. But Khrushchev quashed that plan, largely because he needed the resources involved to pay for the new nuclear-armed strategic forces which he believed would dominate future warfare.

The style of warfare the Soviets had been developing for several decades blossomed into net-centric warfare when they deployed ship-launched and coast-defence missiles beginning in the 1950s. They could be fired well beyond their horizons, and which aeroplanes could fire well beyond their own visual range. To use these weapons effectively, the Soviet navy needed some means of finding its targets, a means not based on the ships or aeroplanes or coastal batteries which actually fired the weapons. Ships armed with the new over-the-horizon missile weapons had to depend completely on external sensors. In the past, long-range sensors (such as HF/DF) had coached naval forces into positions from which they could find and then attack an enemy. In the new type of warfare, the long-range sensors went a step further and provided sufficient information for actually firing. The Soviets took this step first because their evolving style of naval warfare fitted so naturally into it. We can now recognise this new style of war as the precursor of the current U.S. 'net-centric' warfare.

By this time the concept of coast defence had changed dramatically. Before the Second World War, it was generally assumed that an enemy trying to land an army would have to seize a major port at the outset. Soviet coast defence could, therefore, be focused on a few key areas, such as Leningrad in the Baltic. However, wartime operations showed that forces could land on open beaches far from ports; in effect they could make their own ports. Moreover, carriers could attack targets ashore from hundreds of miles away. Within a few years they would be able to deliver nuclear weapons, so it would become vital to stop them before they came within air attack range of cities deep inland. Thus potential enemies had to be dealt with far offshore. Hence intense Soviet post-war interest in long-range coast defence missiles.

The Beginning of Missile Warfare in the Soviet Fleet

The Soviet missile development programme began in 1944, when the British provided the Soviets with the remains of a German V-1 cruise missile. After a senior Soviet aircraft designer, Polikarpov, died, his design bureau (OKB-51) was given to V N Chelomey, who was ordered (on 13 June 1944) to develop an equivalent. Chelomey would later be extremely important in both ship-launched missile development and in the Soviet space and ballistic missile programmes.

Like the V-1, Chelomey's 10-Kh (Lastochka), was intended to attack fixed targets. Essentially a copy of the V-1, it could be launched by an aeroplane (both Tu-2 light bombers and Pe-8 heavy bombers were used). It was first tested in 1945. In May 1947 this missile was ordered developed to arm Tu-2 and Il-4 bombers. A coast defence version, 10-KhN (Volna [Storm]), with an underslung booster, was tested in 1947. Chelomey went on to build two improved versions, both with twin pulse-jets on either side of their bodies, 14-Kh and 16-Kh (Priboye [Surf]).[6] Both were tested in 1947-53. In each case, guidance was by autopilot, as in the original V-1. Performance was unimpressive. Nominal range was better than 100km (*ie* comparable to the later KS-1), but the speed of 10-Kh was only Mach 0.3, about 200kts.

Like the V-1 itself, 10-Kh was extremely inaccurate. Even so, the Soviet navy showed interest in arming both a surface ship and a submarine with surface-launched versions of both 10-Kh and 16-Kh. At this time a large battlecruiser (Project 82) was in the design stage, and plans were drawn up to arm it with the new weapons. Similar studies were made of the new *Sverdlov* class cruiser and of the ex-German cruiser *Lützow* (Project 83). Sketches were made of alternative single, twin, and turreted launchers with armoured ports and projecting launch rails. One such launcher could replace a main battery turret. It is not clear how serious any of these plans were. Prospects for a submarine conversion were apparently more realistic. In 1952-3 plans were drawn up to modify a K-class (Series XIV) submarine, *K-56*, to carry 10-KhN, under a project code-named Volna. This name was later applied to the larger programme to take land-attack cruise missiles to sea on board submarines. A total of fifteen of these 10-KhN missiles were tested between 17 December 1952 and March 1953 at a range of about 240km. They flew at altitudes of 200 to 1000m.[7]

On 13 May 1946, Stalin ordered the creation of a missile-development programme.[8] Two special research institutes were created, NII-1 within the Ministry of Aircraft Production (for cruise missiles), and NII-88 within the Ministry of Weapons Production (responsible for army weapons) for both surface-to-surface and surface-to-air rockets. Associated with NII-1 was a new experimental design bureau, OKB-293 (created June 1946), under M R Bisnovat. At NII-1 itself were the developers of the wartime Soviet BI rocket fighter, Aleksandr Ya Berezhniak and Aleksei M Isaev.

The Soviets had captured German-developed air-launched anti-ship guided missiles in 1945: the Hs 293 glide missile and the FX-1400 radio-guided bomb. Both had been effective wartime anti-ship weapons.[9] Such missiles fitted very naturally into Soviet coastal defences, both those ashore and those centred on aircraft. A glide missile was a very natural successor to a long-range high-altitude torpedo, for example. An equivalent to the FX-1400 bomb was apparently actually placed in Soviet production. The Soviet Ministry of Weapons Production briefly tested Hs 293 in 1948 (from a Tu-2D), and may have sponsored Czech attempts to produce it (or at least to assemble it from surviving parts).

The German missiles were command-guided. The operator had to be able to see the target throughout the missile's flight towards it. Typically he used a joystick to guide the missile, using a flare in the missile's tail to indicate where it was relative to the target. Moving the joystick caused a radio to send commands to the missile flying or falling away from the launch aircraft. This type of operation limited missile range and thus exposed the launch aircraft to possible counterattack – particularly once Western anti-aircraft missiles entered service. The Soviets wanted a guidance system with which an aeroplane could attack from beyond defensive range – at which its pilot could not possibly see the target well enough to identify it. The pilot would have to rely on electronics not merely to guide the weapon but even to detect and identify its target. A few wartime anti-aircraft systems had been able to fire in the dark ('blind-fire'), but in their case the identity of the target was more or less obvious, since it was heading for the ship on which they were mounted. A blind-fire *offensive* weapon was a very different proposition. It had to be linked to some sort of system which would differentiate targets from non-targets. In the new kind of warfare the mechanism deciding which were the appropriate targets became central. Much the same considerations applied to missiles which might replace existing guns and torpedoes, either coastal or shipboard. Again, the existing weapons were directed at a visible target, and visibility implied that the target could be identified more or less accurately. The new missiles, however, could fly far beyond the visual horizon.

In 1949-53 OKB-293 developed a coast-defence missile, Shtorm (Storm) to a specification set by a naval institute (NII-4). Like Chelomey's 10-KhN, Shtorm was also proposed for shipboard installation (in this case, on board *Skoriy* and 'Kotlin' class destroyers and *Sverdlov* class cruisers).[10] Shtorm competed with Beria's KS (see below); presumably to protect that programme, OKB-293 (and the Shtorm programme) was shut down in 1953.[11]

Meanwhile GSNII-642/KB-2 (of the Ministry of Agricultural Machine Building), which had tested the German Hs 293, developed a series of Shchuka (Pike) air-launched missiles to a 1948 naval specification, as superior equivalents to the wartime German Hs 293.[12] Shchuka-A (the '1948 torpedo') was command-guided; Shchuka-B (the '1949 torpedo') was radar-guided (with the first such seeker in the Soviet Union). Both were intended for air launch by light bombers (ultimately Il-28s) and so had much shorter ranges than the big KS described below. Shchuka was to achieve a range of 15-20km (up to about 11nm) at an altitude of 2km (about 6500ft). Rated speed was 320m/sec (1150km/hr) and range was 60km. These missiles were tested in 1951-5. They were unusual in having detachable warheads, which were intended to dive and explode under a target's keel. Shchuka was reoriented towards a ship-launched version in 1955. Eventually KB-2's experience in developing an active seeker for Shchuka bore further fruit: it developed the seeker for the very important P-6 and P-35 anti-ship missiles ('Shaddock').[13] Beside the air-launched variant, there was a coast-defence version. Having completed its work on Hs 293, KB-2 also developed Soviet equivalents of the wartime German FX-1400 anti-ship glide bomb. Eventually it was placed in limited production as Chaika (Seagull), serving alongside the RAT-52 stand-off torpedo.[14]

KS achieved early political success because its chief developer was Sergei L Beria, the son of Lavrenti Beria, the feared chief of the secret police and the head of the Soviet nuclear weapons programme. Beria's university (Leningrad Military Communications Academy) diploma project was Komet, a control system for an air-launched cruise missile. The specification for an air-launched anti-ship missile was issued on 8 September 1947. The carrier aeroplane, a Tu-4 (a Soviet-built version of the U.S. B-29) flying at 1500 to 400m would detect a target at a range of 100km, and the missile would be launched at 60km. It would reach a speed of 950km/hr (about 600mph).

A new Special Bureau (SB)-1 (later KB-1) was created by the Ministry of Weapons Production specifically to develop such missiles and their guidance systems. As its head, Beria was able to obtain both Soviet engineers (including senior designers) who had been consigned to prison camps and experts taken from East Germany. KB-1 came to concentrate on missile electronics, which was fortunate for the Soviets. During the Second World War the Germans had concentrated instead on missile airframes; in later parlance, they had neglected system engineering – they spent very little time on the guidance mechanisms essential to successful missile systems. In exploiting German developments, the Soviets formed separate establishments devoted to missile airframes and to the new electronics (which included missile guidance technology of various sorts). By 1950 it was clear that control systems, in which SB-1 specialised, were the key to the success of all such weapon systems. In August 1950 SB-1 was re-organised as KB-1, a fully-fledged design bureau. In addition to the guidance system of the KS anti-ship missile, it was responsible for the control system of the first Soviet surface-to-air missile, Berkut (S-25, NATO SA-1). Significantly for later events, KB-1 expanded into other electronic areas. In February-March 1955 it split into divisions, one of which (SKB-41, later the independent OKB-41) was concerned with airborne intelligence collection. Others dealt with ABMs and with surface-to-air missiles.

The air-to-surface system was designated Kometa (Comet). Kometa-I (production version K-IM) was the electronics on board the missile; the guidance system on board the launching aeroplane was Kometa-II (the production version was K-IIM). The missile airframe was Kometa-III. Chelomey's KB-51 seemed to have the necessary experience and on 8 September 1948 he was ordered to develop an appropriate airframe for Kometa III based on his evolved V-1. The desired combination of speed (950km/hr) and range (60km) demanded a new engine. Chelomey flew a version of his 14Kh missile, 14Kh K-1, with increased wing area, and with a new engine. A Pe-8 launched an autopilot-guided version during the first half of 1948. However, KB-1 disliked pulse jets, and it rejected Chelomey's design. It turned instead to the MiG design bureau (OKB-155), which offered a scaled-down version of its new MiG-15 fighter, powered by a time-expired RD-500K (copied Derwent V) engine. The missile was designated KS (Kometa Snaryad, or Comet Projectile). Its designer, Berezhniak, later left MiG to form his own design bureau, which is now named Raduga (Rainbow). The Kometa guidance system was reportedly developed largely by German engineers the Soviets had imported after the war (it became known to the West when the Germans were repatriated in the 1950s).[15] The system was first tested on board a modified MiG-9 fighter.

A bomber carrying KS missiles picked up the target using its own radar. It would lock its beam onto the target, and the missile would fly down the beam until it was 10-20km from the target. Then the missile would switch to semi-active homing, using the reflection of the guidance beam from the target. In at least some versions, at still shorter ranges the missile would turn on its own radar illuminator, using antennas in the wings to guide it into the target. Apparently plans initially called for the missile to cruise on autopilot until its onboard active radar detected the target. The Kometa system was apparently adopted because the autopilot allowed for too much drift. Overall, Kometa was far superior to the command system the Germans had used during the war, although it still exposed the bomber to counter-attack.

KS was first successfully fired at a ship target on 21 November 1952. It was accepted for service in 1953, on board Tu-4KS propeller-driven heavy bombers. They could carry either one missile under the port wing or one missile under each wing. System integration in the Black Sea Fleet began in June 1953, with system tests beginning in April 1954. In 1954-5 tests, of eighteen missiles fired, fourteen hit their targets. In one test a single missile hit sank the old cruiser *Krasnyi Kavkaz*, which had been fitted with huge radar reflectors to simulate a carrier (there was later speculation that charges on board the ship had helped). The first air regiment (124th Long Range Heavy Bombardment, later the 124th Long Range Mine-Torpedo Regiment) equipped with Komet was formed on 30 August 1955, with eight Tu-4 and twelve Tu-4KS. Units in the other fleets soon followed.[16]

The Soviets already knew that the Tu-4 was inadequate; the equivalent U.S. Superfortress had suffered badly in Korea. Given its short range, the missile would probably have to be launched from inside the CAP screen around a U.S. carrier. At least, something with much better performance than a Tu-4 was needed. The new Tu-16 (NATO 'Badger') medium jet bomber, already selected as a long-range torpedo bomber (Tu-16T), was chosen. A pair of missiles would be carried underwing, as on a Tu-4KS. Tests of a Tu-16KS version began in 1954, and in June 1957 the first operational aircraft were delivered to the 124th Regiment. The first operational missile was launched there in December 1957. Similar aircraft reached the Northern and Pacific Fleets the following year. In 1958 for the first time both missiles were launched simultaneously from a bomber. By that time about 100 Tu-16KS had been built. NATO designated KS, the first operational Soviet air-to-surface missile, AS-1 ('Kennel').

KS imposed considerable limitations on the bomber carrying it. The missile was quite bulky, so a Tu-16 carrying two of them was limited in speed (to 840km/hr against 1050km/hr unencumbered) and in range (to 3150km vs. 5700km). The missile had been designed for use by the much slower Tu-4, so to fire it the bomber had to slow down to 420km/hr (originally to 360km/hr), about 250kts, partly because it had to extend the radome of its Kobalt-N radar below its belly, to guide the missile, and descend to 3000 to 4000m (about 13,000ft). A bomber launched its missile at a range of 35-45nm (rated range was about 55nm, based on the need to capture the missile in the guidance beam). After firing, the bomber had to slow further (to 320km/hr) and had to remain on course for 3-4 minutes (13-17nm) to keep the guidance beam on target and then to provide the illumination the missile needed to home from 15nm range. From its launch, the missile approached its target in a shallow (3 degrees) dive. Within a group of attacking aircraft, at first only one could fire at a time, because the guidance radars lacked frequency stability: missiles would be unable to distinguish between them (simultaneous attacks were possible only if they were at least 90 degrees apart).[17] These limitations became evident in early tests, and in 1956-7 the system was considerably improved. An improved version (G) had extending wings (to reduce air resistance before launch) and more fuel (range increased to 140-160km) and could be launched at greater altitude (1km).

For its part, the U.S. Navy found even KS quite impressive. By about 1960 the U.S. Navy considered missile-armed 'Badgers' the most serious threat to carrier battle groups, whose main wartime task was to hit Soviet land targets.[18] The carriers could outrun diesel-powered submarines, but not bombers. U.S. naval intelligence did not yet consider the Soviet nuclear submarines, which were fast enough to intercept carriers, operational. The Soviets did have long-range missiles for their submarines (see below), but as yet they lacked the key element, the ability to find a carrier precisely enough for the missile to destroy her. The U.S. Navy therefore considered submarine-launched missiles mainly as threats to fixed targets ashore, such as naval bases.

Once KS had been proven, the MiG bureau proposed a coast defence version. Competing programmes were eliminated. On 19 February 1953 both Chelomey's 10-KhN and Bisnovat's Shtorm were cancelled. A projected coast defence version of Shchuka was cancelled at about the same time. Chelomey's OKB-51 and Bisnovat's OKB-293 were dissolved, and their personnel transferred to the MiG bureau. According to later Russian accounts, the choice of the MiG KS over the alternatives (and the dissolution of the other bureaus) was determined entirely by political factors. Not only was the elder Beria extremely powerful, but Artyom Mikoyan, the head of the MiG bureau, was the brother of Anastas Mikoyan of the Politburo. By the time Beria fell in July 1953, the other bureaus no longer existed, and KS was not particularly associated with Beria.[19]

On 21 April 1954 the Council of Ministers ordered the development of the coast defence system, Strela (Arrow), using the KS missile. Static Strela systems were commissioned in 1957 on the Black and Barents Sea coasts. The guidance system was designated S-1; the missile was S-2. There was also an army version, FKR-1.[20]

Strela was significant mainly for where it led. Nikita Khrushchev's son Sergei convinced him that conventional naval forces were unlikely to be effective against such missile weapons. On 30 December 1954 development of a shipboard version of Strela, to equip rebuilt *Sverdlov* class cruisers, was ordered. The shipboard version was designated KSS (Korabelnye Snaryad Strela [Shipboard Missile Strela]); it was fired from a newly-designed SM-58 ramp launcher, like that later installed to fire KSShch/SS-N-1 missiles on Project 57bis (NATO 'Krupny') class destroyers. This version had folding wings; before firing it was housed in a small hangar at one end of the ramp. It was fired at an elevation of 10 degrees. Tests began in the Black Sea in September 1955. The sketch design of the missile-armed version of the cruiser, Project 67, was ready in May 1956. These ships would have carried nineteen missiles (including four ready-use rounds in the hangars). To accommodate them all four 6in turrets had to be eliminated, and plans to provide a helicopter hangar (for targeting) had to be abandoned in favour of providing only a helicopter pad. Project 67 would also have had all secondary and anti-aircraft weapons replaced by six twin 57mm mounts (ZIF-75 type).[21]

The cruiser *Admiral Nakhimov* was converted as trials ship (Project 67EP), her two forward turrets being replaced by a missile launcher. She was also fitted with a stabilised antenna for the S-1 control system. Post-conversion trials began in October 1955 (two months later than planned), and missile trials began only on 26 April 1956. Missiles were fired at a landing barge and at a radio-controlled torpedo boat making up to 40kts; both were fitted with radar reflectors. At minimum range (15km) two out of three shots failed, because the missile did not have enough time to come into the guidance beam. However, seven of ten shots at maximum range (43km) were successful. On the basis of the tests, estimated hit probability against a destroyer was 0.8. There were also guidance tests using a converted MiG-17 fighter. Given these successes, in 1956 it was planned to begin Project 67 conversions. Three ships would be completed in 1959. Two more ships, to be completed in 1960, would have a more advanced system (Vector) in which the missiles would use active seekers, and thus would be effective out to 100km (54nm) range.

Second-Generation Missiles

By this time another long-range surface-launched missile was on the way. When Chelomey's OKB had been dissolved, he took refuge in the Soviet Navy's weapons development arm, NII-4. There he promoted ship-launched land-attack weapons, of which 10-KhN had been the first. They offered the Soviet Navy, which lacked aircraft carriers, a tremendous opportunity: surface ships or submarines, which it did have, could now strike targets deep inland. About 1954 Chelomey began work on a new turbojet-powered transonic missile with swept wings, P-5, suitable for submarine launch. About the same time the Beriev design bureau, usually associated with flying boats, began work on an equivalent, P-10. It may have been inspired by the U.S. Regulus I subsonic cruise missile.[22] Both projects paralleled the somewhat earlier effort to place a ballistic missile on board a submarine.[23] Ilyushin designed a much longer range (3500km) strategic cruise missile, P-20, also for submarine launch.[24] Development of all three submarine-launched cruise missiles was formally authorised by a Council of Ministers decree issued on 25 August 1955. The immense 21-ton P-20 was cancelled early in 1960.

P-5 won out over Beriev's P-10 because it was far more compact. It had folding wings and, unlike P-10, did not require a launch rail. Because they were powered by turbojets, both missiles had to be started on the surface. The success of P-5 launched Chelomey on a career as the main supplier of long-range cruise missiles for the Soviet Navy. First flown in 1957, P-5 became known to Westerners under the NATO code-name 'Shaddock' (SS-N-3). It became operational in 1959.[25] P-5D was an improved version with a Doppler airspeed indicator, for better autopilot accuracy; at its full range of 500km it had a dispersion of 4km. There was also an army version, S-5 (1960). All of these weapons were intended to hit fixed targets. To use them, the shooter needed nothing more than precise navigation.

These missiles were part of a post-Stalin revolution in the Soviet navy. By 1953, when Stalin died, plans were proceeding for the next Ten Year Plan (1956-65). Khrushchev himself apparently ordered a high-level review, beginning with a special commission of the Party Central Committee (March and April 1954). The existing plan was presented (and rejected), and a new programme was presented in November 1954. It was rejected again. The issue was raised again the following spring and summer. In October 1955 Khrushchev called a meeting at Sevastopol, a main fleet base, to discuss the future of the fleet. Those called to testify ranged from a destroyer commander up to Minister of Defence Zhukov. The majority of participants agreed that for the future the main forces of the navy should be submarines and naval aircraft armed with missiles. They disagreed on the role of large surface combatants, however armed, and the ten-year plan under discussion was rejected. Khrushchev, for example, specifically rejected a new light carrier (Project 85) then under consideration. Following the conference there were special meetings on the long-range shipbuilding plan (soon superseded by a 1959-65 Seven Year Plan) in January 1956 (by the Defence Council) and on 9-10 May 1958. The programme actually adopted came out of the May 1958 conference: missile-armed nuclear submarines and large missile-carrying aircraft would be the main weapons against aircraft carriers, large surface warships, and their escorts. A projected cruiser programme was cancelled; however, the Soviet Navy managed to extract permission to build four small missile cruisers (the 'Kynda' [Project 58] class), and the compact Chelomey missile fitted aboard them.

There was already a programme to adapt the air-launched Shchuka to shipboard use, ordered on 20 September 1955 (the programme was assigned to a team led by chief constructor M V Orlov of TsKB-34). The shipboard version (the only one to see service) was KSShch (Korabelnye Snaryad Shchuka, *ie* ship missile Shchuka). Just as Shchuka had been conceived as a lightweight air-launched missile, KSShch was seen as a lightweight shipboard missile, suitable for a destroyer rather than a cruiser. It was tested in September-October 1958. The first destroyer armed with the missile, *Bedoviy*, was accepted for service in October 1958.[26] The initial version was command-guided. A modified RG-Shchuka had the first Soviet active radar seeker, with an acquisition range of about 20km. The missile was commanded by the firing ship's gun fire control system, suitably modified. That limited it to horizon range, about 30-40km, although it could fly about 100km at 60m altitude. In common with many later Soviet shipboard missiles, KSShch could attack both land and sea targets. For the latter it had the detachable warhead of its earlier air-launched version. Tests were less than successful. For attacks against land targets the missile was commanded to climb to 1000m and then dive.

KSShch was a new kind of weapon. For the first time, a ship could fire at a target well beyond her horizon. The solution, incorporated in plans for the first Soviet missile-armed surface warship, the Project 57bis destroyer (NATO 'Krupny' class) armed with the KSSch missile (NATO SS-N-1), was a radar-equipped helicopter – which, in the event, was not ready in time. The radar would have reported down to the destroyer (the shooter), which would control the missile. The missile would fly most of its path at high enough altitude to remain in touch with the launching ship. This concept was simple enough in the case of a surface ship, which could accommodate its own helicopter; but a submarine was a different proposition. The airborne radar would have to come from shore, and there would have to be some way of co-ordinating its appearance with the submarine's.

The new missiles, then, demanded a new kind of command and control, which soon evolved into information-oriented warfare. For example, without the helicopter the missile destroyer was limited to horizon range. Thus to a considerable extent killing her helicopter would neutralise the destroyer. For that matter, the destroyer could not fly her helicopter continuously; it lacked endurance. She therefore had to be cued to attack, just as a bomber had to be cued to fly out to a likely target area. Again, without the cueing the ship was relatively ineffective.

On the other hand, whether or not the destroyer had some means of extending her horizon, the new missiles did offer enormously increased firepower. That applied to quite small combatants. During 1955 Berezhniak, who had designed KS, began work on a small shipboard missile, P-15, the famous 'Styx' (SS-N-2).[27] Unlike KSShch, it was fired only to horizon range, hence could be guided entirely by the firing boat. In this it was a direct successor to the torpedo, albeit with much more hitting power. Indeed, one type of 'Styx' boat, the 'Komar' (Project 183R) was effectively a converted P-6 (Project 183) torpedo boat. Like the torpedo, 'Styx' (and Western equivalents like Exocet) gave a small inexpensive boat the ability to destroy a large expensive warship, at least in coastal waters. To some this seemed enough to proclaim a revolution in naval warfare. When Egyptian-launched 'Styx' missiles sank the Israeli destroyer *Eilat* on 21 October 1967, they helped convince the U.S. Navy that its own destroyers needed an equaliser. The Harpoon anti-ship missile, already under development for air launch by P-3s, was ordered adapted for ship use. It could outrange 'Styx' – but only if the firing ship could obtain over-the-horizon targeting information. Work to solve this problem led into the over-the-horizon targeting system, which was the first important U.S. Navy tactical use of space assets (see Chapters 8 and 9).[28]

There were soon some longer-range anti-ship weapons. Even before P-5 had flown, under a 12 August 1956 Council of Ministers decree it was ordered developed into an anti-ship weapon for both submarine (P-6) and surface ship (P-35) use. Decrees were generally issued as the outcome of a developmental process, so the idea for dedicated anti-ship versions of long-range missiles probably goes back to 1955 or earlier. These missiles would fly well beyond the horizon of the ship or submarine firing them, and would probably be used to attack carriers or other high-value targets within formations. The Soviets accepted that it would take several hits to destroy such a target, and that, left to their own devices, missiles might well lock onto the other ships in a formation, and thus might be wasted. Chelomey's solution was to incorporate a radar video data link from the missile back to the firing ship. The shipboard operator could see the missile's radar picture, and could choose the ship onto which the missile seeker would lock. Using the link, the missile remained under the ship's control until it disappeared beyond the ship's radar horizon. Thus a submarine firing P-6 had to remain surfaced (in order to control the salvo of missiles) for some considerable time after the missiles had been fired. P-35 became operational in 1962, and P-6 in 1964.[29] Later there was a coast defence version, S-35.

P-6 and P-35 required more. The firing ship needed some way of detecting the target in the first place. For the Soviets this was the beginning of the new style of naval warfare. In this case the primary sensor was a big radar on board a converted 'Bear' bomber (see below), whose picture could be linked down to a ship or a submarine. The overall system, the combination of airborne radar, data link, fire control, and missiles, was called 'Success'.[30]

Meanwhile the MiG bureau developed a successor to KS, which it called Kometa-10 or K-10 (a contemporary strategic missile, for launch by Tu-95s, was K-20). The missile element of the system was the supersonic K-10S (NATO AS-2 'Kipper'). Considering the known limitations of the existing KS, the Soviet military wanted to double or triple the launch range, and to increase launch altitude closer to that of the ceiling of the missile carrier; to double missile speed, while allowing it to fly a complex path towards the target, including a final leg at low altitude; and to permit salvo launch. To meet the requirement for a complex path, beam-riding had to be replaced by radio control, the bomber ordering each change of missile course. Work on a KS follow-on was authorised on 3 February 1955, and the K-10 project was approved on 16 November. The missile was to be suitable for both the existing Tu-16 bomber and for its projected successor, then known as Tu-105 (it later became the Tu-22 'Backfire').

The overall system, including the bomber, would have an effective range of 1600 to 2000km. The bomber would detect the target at a range of 180 to 250km, and it would launch the missile 160 to 200km from the target, at an altitude of up to 11,000m (about 33,000ft) and at a speed of 700 to 800km/hr. The missile itself would fly at 1700 to 2000km/hr (up to about 1250mph) at 11,000m, carrying a 1000kg warhead. The system as a whole would allow the bomber to remain at least 100km from the target, beyond what was then standard CAP radius, which was about 30 to 40nm (the threat of such missiles in turn caused the U.S. Navy to push its CAP stations further out, ultimately to about 150nm from the carrier).

The new K-10 guidance system was built around a new airborne radar (YeN: NATO 'Puff Ball') and an active radar missile seeker (YeS). The new missile could be launched at high altitude, because it did not dive directly at the target. Instead, it initially dove about 1000 to 1500m before climbing back to the bomber's altitude, the bomber controlling its course. When it was 105km from the target, the bomber ordered it into a second dive (at an angle of 13 to 18 degrees), then levelled off at about 1000m altitude (3280ft). For the final 40km to the target it used its own active radar seeker, locking on at a range of 15 to 20km. Because the missile's autopilot maintained it at a constant altitude, the guidance beam did not have to be steady, as in the

Kometa system; after firing, the aeroplane could roll (within the stabilisation limits of its YeN radar) and could turn up to 80 degrees away from the target. By abandoning terminal semi-active homing, the missile could attack targets at greater ranges, 180 to 200km, too distant to reflect enough illumination energy back to the missile. The bomber could escape as soon as the missile ceased to require command guidance. Up to eighteen missiles could be launched simultaneously from different directions.

The K-10S warhead was twice as heavy as that of the earlier KS.[31] The combination of a new guidance system and much higher performance (which was useable because of the new system) made K-10S much more effective. Moreover, it exposed its launching aircraft to much less danger.[32] The missile first flew (in unguided form) on 28 May 1958. It and the associated Tu-16K-10 missile carrier were accepted into service on 12 August 1961.[33] In all, 220 Tu-16K-10 ('Badger C') were built.

While K-10S was being developed, the MiG bureau's earlier KS missile was being adapted for shipboard use, as already described. Council of Ministers resolutions dated 17 and 25 August 1956 ordered development of a ship-launched version of the missile, designated P-40. It would form the main battery of a projected nuclear-powered cruiser, Project 63. Missile range would have been 200 to 350km (one source puts it at 400km). There was also some interest in a shore-defence version of P-40. Both Chelomey's P-35 and MiG's P-40 far outperformed KS. The Council of Ministers therefore ordered (4 July 1957) that the new missile cruisers then in the planning stage (Projects 67bis and 64) should be armed with these weapons. However, there was some question as to how effective any of these ships could be against enemy aircraft carriers capable of striking from more than 400km (over 200nm) offshore. P-40 died in December 1958 when Khrushchev approved the 1959-65 shipbuilding plan, killing the big Project 63 cruiser. P-35 survived, possibly because it could fit on board a smaller ship, and because it was a submarine weapon.[34]

Khrushchev's decision to kill the big missile cruisers did not affect his more general preference for missiles over conventional naval weapons; quite the opposite. Initially plans called for each type of missile to be tested aboard a modified version of an existing ship, while a specialised missile ship was designed and built. By about 1957 Khrushchev was demanding that the programme be accelerated. The weapon test platforms were all weaponised. For example, P-5 was tested aboard a modified 'Whiskey' (Project 613) class submarine, 'Whiskey Single Cylinder', while a much more heavily modified version, 'Whiskey Long Bin' (Project 665) was developed. To get more missiles into service quickly enough, the test bed was slightly modified as 'Whiskey Twin Cylinder'. Hence the completion of apparently more austere conversions after the more sophisticated version was in production.[35] Later other over-the-horizon missiles were developed for smaller combatants: Termit-M (the second-generation version of 'Styx': P-15M/NATO SS-N-2C), Malakhit (P-120/SS-N-9), and Moskit (P-270/SS-N-22).[36] They, too, needed some special means of detecting targets before they could be fired – and some means of designating the appropriate target within a formation.

K-10S did not quite solve the problem of a KS-1 replacement. Because it was 50 per cent heavier than KS-1, a 'Badger' could carry only one, under its belly, rather than one under each wing. Moreover, it required a massive radar system in a specially-built aeroplane. Unless they could be armed with a simpler stand-off missile, the thousand or so Tu-16s in service were obsolete. The Kometa system was in fact simpler and lighter, but it too was obsolete. In 1957 an official report by the Soviet air force high command charged that it imposed too low a limit on the bomber's launch speed and altitude, its guidance system was unreliable, and it could not be salvoed. The main problem was the missile's turbojet engine. Despite considerable efforts, it could not be started at an altitude greater than 7000m (about 22,500ft).

The MiG missile designer, Berezhniak, was already working on a compact rocket-powered small boat missile, P-15. He chose to install the P-15 rocket engine in a modified KS, thus overcoming the launch height problem. In fact Berezhniak considered simply using a bomber launched version of the lightweight P-15. The shipboard missile made minimal demands on the launching ship, so an airborne version would make few demands on the bomber. In particular, it would not need the sort of massive fire control system employed with K-10S. The Chinese actually adopted this solution many years later, using C-601, their version of 'Styx'.

The Soviet air force wanted something more. Initially Berezhniak was authorised (2 April 1956) simply to modify KS with a rocket motor based on that of P-15. By eliminating the bulky turbojet, Berezhniak managed to reduce fuselage diameter. Flight speed increased to 1200 to 1250km/hr (about the speed of sound). It proved possible to provide the new version with a K-10S warhead (940kg). The missile was provided with a new autopilot, based on that developed for the contemporary Kh-20 (a strategic missile carried by Tu-95 'Bears'). Ultimately KSR weighed little more than KS; bombers typically carried two missiles. It had about twice the range of KS, 160 km.[37] Initial tests at Feodosiya (June-October 1958) showed that it had difficulties homing on its target, so it was modified (as KSR-2) with a larger-diameter radar dish and its tailplane was moved.

Maximum speed was Mach 1.18; range at high altitude was about twice that of KS-1, at 180-230km.

While KSR was being developed, it was decided to modify K-10S so that it could be delivered by 3MD ('Bison') strategic bombers. The control system would be tested on board modified Tu-16 bombers, carrying KSR missiles modified with the K-10S guidance system, including its active terminal seeker. The new associated aircraft radar was Rubikon, which had the same functions as the YeN on board a Tu-16K-10. This K-14 system was cancelled in 1960, but the KSR test version was placed in service as K-16, using the Ruby-1K (K-PM, NATO 'Short Horn') radar, which replaced the Rubidiy of the earlier Kometa system. The entire system was called Rubikon-1K. Typical detection range for a large warship was 200km. Missiles were launched at a speed of 700 to 800km/hr and at an altitude of 4000 to 10,000m, which was still slightly short of that for K-10S. Bombers armed with KSR-2 were designated Tu-16K-16.[38] K-16 was accepted into Soviet service on 30 December 1961, and the following February it was formally decided to re-equip Tu-16KS missile bombers with the associated guidance radar. Earlier, on 20 July 1957, development of an anti-radar version of KSR (K-11 system) had been approved. The associated version of the missile was KSR-11 (KSR-2P), accepted for service on 13 April 1962.[39] There was also a target version, KRM. The KSR-2/KSR-11 system became operational in 1962 as K-11-16, and 'Badgers' designed to deliver these missiles were designated Tu-16K-11-16 ('Badger G'). NATO designated KSR-2 as AS-5 'Kelt'.

Khrushchev favoured air-launched missiles much as he favoured ship-launched ones. In the late 1950s he cancelled all bombers (*ie* aircraft dropping gravity bombs) in favour of 'rocket carriers'.[40] Unlike ship-launched weapons, at least at first air-launched anti-ship missiles could be targeted by the carrier's own sensor, a big airborne radar. If it was powerful enough, it could reach out well beyond the CAP stations around a carrier. For example, at 30,000ft the radar horizon is about 270nm away. By the late 1950s missiles had entirely superseded anti-ship torpedoes as the weapon of choice of the Soviet naval air arm.[41] The mine-torpedo regiments were all eliminated when Khrushchev reorganised the Soviet military in 1960; on 20 March 1961 those equipped with missiles were renamed naval rocket units. All air force (Long Range Air Army) 'Badgers' carrying anti-ship missiles were transferred to the Soviet Navy. At the same time all naval fighters were all transferred to the Soviet air defence force (PVO). At this time, also, the Soviet coast artillery was renamed coastal rocket-artillery.

Bombers and sea-based missile platforms were complementary. Bombers could move from one base to another to cover the long

Soviet coastline. On the other hand, fixed air bases clearly invite attack, and bad weather over the base may make operations impossible.[42] Moreover, ships and submarines could keep the seas for sustained periods, always ready to attack.

Cueing

The submarines, ships, and bombers armed with the new long-range weapons had to be cued to fire them. For example, if bombers had to spend much time searching for their targets, they would alert defenders; their big radars might well act as beacons for the defending fighters. The missiles had to be aimed fairly precisely, because any protracted search would leave them open to detection and counter-attack. Cueing was very much in the spirit of existing Soviet naval tactics. The key sensors reported to a fleet command centre, which could command the shooters into position.

The first major cueing sensors were ground-based radio direction-finders. When they overran Germany in 1945, the Soviets captured not only missile technology but also a new means of HF radio direction-finding, a fixed-array system the Germans called a Wullenweber. The Soviets built their own version, which they called Krug.[43] They considered it safe to assume that any Allied naval force approaching their shores would have to communicate with its home base, and in the 1940s and 1950s the only means of long-haul naval communication was HF radio. The Krug net could pick up even fairly weak signals – and, just as importantly, short (burst) signals designed to defeat interception.

Apparently HF/DF was considered good enough to cue first-generation missile carriers. Given HF/DF data, they would use their own radars to find their targets, but K-10 was a different proposition. Regiments carrying this missile included specialised pathfinders, converted 'Badgers' equipped not only with long-range radars but also with ELINT equipment capable of homing on the enemy carrier group, exploiting its characteristic emissions.[44] Passive (ELINT) search was relatively safe, since the pathfinders could probably operate outside the detection range of the battle group's own radars. At least, there would be no emissions to alert the battle group. Once close enough, the pathfinders would switch on their own radars to locate their targets precisely enough to cue the bombers.

For example, as of about 1962, two pathfinders would have preceded a regimental attack by 'Badgers' equipped with K-10S missiles. Once cued, the bombers would attack in two waves, each of six to eight bombers (several regiments might attack together). By using the pathfinders, the bombers could minimise the time between turning on their radars and launching their missiles. However, because the missile could not be locked on at launch time, they still had to loiter near the target while they commanded their weapons.

The passive pathfinder role required that the Soviets collect electronic signature information in peacetime; otherwise, the pathfinders would not have been able to distinguish carrier formations in war (an HF/DF net collected its own signature data). It was not enough simply to be able to intercept electronic emissions. Attack resources were inevitably finite, so there had to be some way of distinguishing one potential target from another. The Soviets had to build a global system simply to collect vital identifying information. It included intelligence-gathering auxiliaries (designated AGIs by NATO), which first appeared at just about the same time as the K-10S carriers. Later there were specialised ELINT aircraft, such as converted An-12RR transports. Both were in addition to frequent sorties by naval and air force reconnaissance aircraft over Western naval formations.

Clearly the AGIs and the ELINT aircraft did more than collect electronic signatures. They observed peacetime movements of Western warships, and their data fed into the same ship-plotting system which used HF/DF to cue bombers and pathfinders. However, their signature-gathering role was extremely important. It illustrated an important characteristic of any passive ship detection/tracking system, the need to gather enough information to identify the waveforms collected by the system. Moreover, this need is permanent, not least because each ship's emitters change over time, as radars are modified or replaced. Almost certainly the Soviets limited themselves to identifying radars by type, *eg* distinguishing SPS-12, a common air search radar on board some carriers in the late 1950s. The stage beyond, fingerprinting, distinguishes one SPS-12 from another, using subtle unintentional differences in waveform, to identify, say, one particular aircraft carrier.

The pathfinder role became more difficult as the U.S. Navy learned to confuse Soviet radars. For example, screening destroyers were fitted with corner reflectors (later, with blip enhancers) so that the pathfinders' radars could no longer distinguish the carrier among them. At the least, they or the bombers would have to come closer and closer in order to distinguish the carrier, and thus would expose themselves to destruction by the CAP.[45] As carrier group defences improved, the Soviets developed escort jammers. They became so important that, at times, one of the three squadrons in a Soviet naval air regiment consisted entirely of jammers.[46]

In the 1950s, the main U.S. strike aircraft was the North American Savage, with a combat radius of about 400 miles (about 700km). That figure in turn defined the necessary range of the Krug/bomber system. However, by 1956 the U.S. Navy was flying the A3D Skywarrior, with about triple the range of a Savage, and was beginning to develop the A-5 Vigilante, which combined Skywarrior range with supersonic dash speed. The Soviets would have to move their defences further out to sea. They credited the K-10S (Badger/AS-2) system with a range of about 2000km (1250 miles), but that figure was apparently based entirely on the flight characteristics of the Tu-16 missile carrier.

The real limitation on the range of the Soviet anti-ship system was its ability to collect precise information. As with any other land-based HF/DF system, Krug lost accuracy as its targets moved further from its antenna arrays. The less precise the initial DF 'fix', the greater the area the pathfinders would have to search. Ultimately they might not have sufficient endurance to find the targets at all. At least, the longer the search, the better the chance that the search itself would alert the target – or that the pathfinder would be shot down by enemy naval fighters. Thus the key limitation on K-10 was not the speed or the flight profile of the missile, so much as the range limit inherent in the information-gathering and processing system on which it relied. Moreover, the Western navies were aware of Krug and thus of the key role of HF radio. In the 1960s the U.S. Navy began to abandon HF radio for long-haul communications, substituting satellites. Ultimately Krug and its successors would become virtually useless. As recounted earlier, the U.S. Navy did retain HF for its tactical link (Link 11), but its antennas limited the amount of energy it put into the sky waves which Krug picked up.[47]

Once within range, the shooters had to be able to detect and track the targets. The target would move an appreciable distance while the missile was in flight. The missile could search only a limited area; it had to be directed into a relatively small area near the target's expected position. For bombers, that was not too serious a problem. Flying high enough, they could detect and track their targets well before they had to fire their missiles. To this end the Soviets developed long-range sea-search radars: Kobalt and then Rubin ('Ruby') or Kobalt-N (NATO 'Crown Drum') for KS and KS-1, respectively, then YeN (NATO 'Puff Ball') for K-10S, and R-1 Rubin-1A for KSR-2 (AS-5). A larger version of Kobalt-N (YeN-D, also NATO-designated 'Puff Ball') equipped Tu-95-K-20s armed with the big Kh-20 missile. By the mid-1960s the Soviets expected Rubin and YeN-D (at an altitude of 10,000m [about 33,000ft]) to detect large surface ships at ranges of 300 to 400km (up to about 220nm). Western estimates were somewhat lower, about 175nm, probably based on measured pulse rates. Reportedly the big Soviet sea search radars were based on the U.S. APS-20 early warning/sea search set, widely used in the 1950s both in the United States and throughout NATO (in many cases on board Neptunes used for ASW).

'Blinder' and the Kh-22 Missile

The story of the Tu-22 ('Blinder') gives some idea of the interaction of aircraft, missile, and airborne sensor performance in the face of U.S. battle group defences. This supersonic 'Badger' successor was in the later stages of development by 1956, as the first jet missile carriers were entering service.[48] Like the earlier bombers, it was affected by Khrushchev's decree cancelling all pure bombers in favour of missile carriers. Production of the bomber version was stopped, although a reconnaissance version, Tu-22R ('Blinder C'), was produced both for the air force and for the Soviet Navy.[49] A missile version, Tu-22K ('Blinder B'), was designed, with an associated new supersonic rocket-powered missile, Kh-22 (NATO AS-4 'Kitchen'); the system was K-22. Development of the missile and its fire control system was authorised by a 17 April 1958 Council of Ministers decree. The joint proposal for the K-22 system and the Tu-22K missile carrier was presented early in 1960 and approved that March. The first Tu-22K was completed early in 1961; it carried a dummy Kh-22 missile in that year's Tushino air show. The prototype Kh-22 was completed in 1962. Unfortunately the Tu-22 had not been designed to carry a massive missile; even carried semi-submerged, a Kh-22 seriously affected its flight characteristics. So did the massive bulged radome needed to accommodate its guidance radar.

Unlike its predecessors, Kh-22 had an active seeker powerful enough to lock onto a target (at greater range than the launch range of a K-10S or KSR-2) before the missile was launched. The aeroplane could therefore turn away soon after launching it, instead of continuing on to could keep tracking the target. That made the Tu-22K/Kh-22 combination considerably more threatening than earlier systems.[50] Kh-22 was accepted for service use in 1964, before the Tu-22K tests had been completed, because it was also intended to arm the Tu-95 strategic bomber.[51] In fact the Soviet state commission declared the Tu-22K trials a failure, but Tupolev had sufficient political influence to save the programme. As for K-22, it was not fully operational until 1970.

The Tu-22K was unpopular within the Soviet air arms, to the extent that it was sometimes considered unflyable. To reduce cross-section (*ie* drag), it had only a single pilot, who had to struggle with unusually heavy controls (a defect eventually cured). Some crews even refused to fly it. Among other problems, it landed at much higher speed than its predecessor, the Tu-16. It was difficult to maintain, and its PN radar was unreliable. Because its ejector seats fired downwards, it could not safely be flown at low altitude; its only real protection against interception was its high speed. Undercarriage collapses (due to inadequate shock absorbers) were fatal, because a fully-fuelled Kh-22 crushed under the aeroplane's belly would explode.[52] Only seventy-six Tu-22Ks were built, not enough to equip both the air force and the navy. They were issued only to three air force regiments, which had a secondary anti-ship role for the North Sea and the Mediterranean.[53] An additional Tu-22K regiment may have operated in the Pacific. In addition to these aircraft, an escort jamming version, Tu-22P ('Blinder E'), was developed. Each Tu-22 regiment (three squadrons, ten or twelve aircraft per squadron) included a Tu-22P squadron.[54]

From a defender's point of view, Tu-22K was more threatening than its predecessors because it could attack from beyond the range of an early-1970s CAP orbiting at the usual 150nm and armed with short-range missiles (Sparrows). Even if it had to come within fighter range to launch, the supersonic bomber would spend very little time there, before turning away. Like earlier strike forces, a regiment of Tu-22K missile carriers (twenty-four to thirty aircraft, including four to eight Tu-22P escort jammers) would have been cued into position by four Tu-22R reconnaissance aircraft, which would find the carrier formation using ELINT and perhaps their radars. The Tu-22Rs were intended to locate the carrier(s) precisely enough that the Tu-22Ks would not have to search for them, and therefore not give themselves away by premature radar emissions. They could also perform stand-off jamming. Just before the attack, two of the four would climb to at high altitude, to jam enemy defensive radars. There was still the problem of deception, *eg* by blip repeaters. To defeat it, the other two Tu-22Rs would dive to 100m altitude and head for the heart of the battle group, to determine where within the group the high value target(s) were, by visual reconnaissance from a range of 10 to 15km. Their information would be passed back to the strike group via the high-flying Tu-22Rs. Because the Soviet Navy did not operate any Tu-22Ks, the U.S. Navy of the late 1960s and early 1970s was largely unaware of the threat they presented; Kh-22 seemed to be a strategic missile. The Navy was, however, quite aware of the threat presented by Tu-22R reconnaissance aircraft, and the Soviets considered their peacetime mission so dangerous that a crew could be decorated after only twenty reconnaissance flights over inter-national waters.

From the Soviet naval point of view, Tu-22K became obsolescent soon after it appeared in quantity. Once U.S. carriers were equipped with F-14 fighters, they would be able to destroy incoming Tu-22Ks beyond their attack range. The problem was the attacking force's need for information. The Tu-22Rs would, in effect, announce that an attack was imminent. If they could be destroyed in time, the entire attack would have to be aborted. Even if the bombers were coached into position, they still had to pop up into radar view (of the fighters and their E-2 radar control aircraft) in order to lock their missiles onto the targets. An F-14 armed with Phoenix missiles, orbiting 150nm from a carrier, could destroy targets 250nm from the carrier – beyond Kh-22 launch range (which was set by the acquisition range of the PN radar). Because the Tu-22 had very limited supersonic range (about 1000km), the reconnaissance aircraft generally could not approach their target indirectly. Once they were spotted, then, they would define the attack axis, and F-14s could be vectored out along it, to destroy the approaching bombers well before they could launch.

Kh-22 was too heavy to be carried by a Tu-16. Instead, a smaller equivalent, KSR-5 (NATO AS-6 'Kingfish'), was developed as a KSR-2 successor. Development was ordered in August 1962; a corresponding anti-radar missile, KSR-5P, analogous to KSR-2P/K-11, was ordered developed in February 1964. These missiles were analogous to the larger Kh-22. The associated fire control system was K-26, and the carriers were converted from Tu-16K-10s; they were designated Tu-16K-10-26 or Tu-16K-26 ('Badger G'). The associated radar was YeN-D. The new system could control four different missiles: K-10, KSR-2, KSR-5, and KSR-11. Bombers equipped with it sometimes carried a K-10S on the centreline and two KSR-5s underwing, but then only KSR-5s once K-10S had been retired. K-26 was accepted for service on 4 December 1969.[55]

Sensors for Surface Shooters

The problem of the surface shooters was more complex, because from the beginning they had to deal with targets beyond their horizons. In 1959-61 the Soviets investigated radar ducting and tropo-scatter in both the Atlantic and in the Baltic. Under some circumstances, radar signals are trapped in a layer above the sea, which forms a kind of natural waveguide. Signals in the layer can travel well beyond the usual radar horizon. Under some circumstances, this effect prevents surface ships from detecting aircraft, since signals do not penetrate above the duct. At longer wavelengths (such as L-band and UHF) the troposphere reflects radar signals, so that they travel well beyond the horizon. In the Soviet case, the key perception was that most radar signals (with wavelengths ranging from 3 to 200cm) would travel beyond the usual radar horizon, at least intermittently.[56] They could be picked up by a shipboard antenna, and used to determine the direction to the target. Ranges of about 400km (about 220nm) were achieved. The systematic use of such long-range propagation by the Soviets was apparently unsuspected in the West. The Soviets called the phenomenon DTR, meaning long-range tropospheric radar propagation. Western navies were quite aware of ducting (which was called anomalous propagation), but they tended not to use it very systematically, because it was often a seasonal phenomenon. There were, however, areas of the world in which it occurred consistently, such as the Baltic, the eastern Mediterranean, the Persian Gulf and the Arabian or South China Sea.[57] In those cases a radar operating at the right frequency (generally X-band, about 3cm wavelength) could consistently see well beyond the horizon. About 1970, for example, the U.S. Navy considered (but rejected) duct radar as an alternative to airborne early warning for its abortive Sea Control Ship. However, none of the Western navies seems to have realised that intermittent ducting was so consistent a phenomenon on a global scale that it could reliably be used to target long-range missiles.

Work on the first system of this type, Molniya (Lightning), began in 1962, and it was accepted into service in 1967. Depending on conditions, range was 150 to 350km. The passive reception antenna was combined with the antenna used to track and control the outgoing missile. This technique was applied only to surface ships. A submarine could certainly pick up radiation from a target, but just how precisely target direction could be found would depend on how big the submarine's antenna was. The Soviets considered that no submarine mast could accommodate a large enough antenna. The big missile-tracking antennas used by surfaced submarines certainly could have been used, but the submarine would have had to operate continuously on the surface to detect the appearance of the target in the first place.[58]

Ducting was incorporated in the Dubrava, Titanit (Dubrava successor), Monolit, and Mineral fire control systems (whose main antennas were housed in 'Band Stand' radomes) of Soviet (but not export) 'Nanuchka' and 'Tarantul' class missile boats and of *Sovremennyy* class destroyers, alongside more conventional techniques. These ships apparently used a variation on the P-6/P-35 technique, in which the ship could select a target on the basis of video supplied by the missile.[59] Some ships armed with the P-15M version of 'Styx', which could also go beyond the horizon, were equipped with another ducting radar, Garpun (NATO 'Plank Shave').

The Soviet Navy was, however, fundamentally a submarine fleet, so the submarines' need for long-range target information took priority. The work on radars for missile-carrying bombers had shown what great ranges they could achieve. A search radar (Uspekh) on board a long-range bomber could link its picture down to a missile shooter. The U.S. Navy already used exactly such a video data link between its APS-20 early warning radar and carriers and radar picket ships. If, as some have suggested, the big Soviet radars were derived from APS-20, the Soviets also had the AEW link they needed. Given such a link, a submarine or surface ship could see targets well beyond its horizon. The new radars were placed on board Tu-16RTs 'Badgers' and Tu-95RTs 'Bear-Ds' (of which only the 'Bears' became operational). The combination of aircraft, radar, and submarine-fired missile was called Success.[60] It was the first of what the Soviets called 'reconnaissance-strike complexes,' systems which would both seek out and attack targets. Its airborne targeting system was developed by KB-1. The Uspekh radar link was developed by NII-132 in Kiev, under Chief Designer I V Kudriavtsev. The bomber transmitted its radar picture down to a submarine or surface ship, which used the data to control its missiles. The missile transmitted back its own radar picture of the target area, and the launch ship locked it onto the chosen target. There was speculation that the missile downlink was inspired by one which had been proposed for an abortive U.S. hypersonic missile, which was to have been carried by the cancelled P6M Seamaster seaplane bomber. At 10,000ft the bomber's radar horizon was about 150nm away. The bomber itself could be about 100nm from the submarine, so the system could support an attack on a carrier as much as 250nm from the submarine – the sort of range needed to ensure safety against pre-emption by the carrier. Even so, the Soviets considered the radar aircraft quite vulnerable. Moreover, the presence of a specialised maritime reconnaissance aircraft revealed that a strike was imminent. NATO sailors, for example, learned that 'Bears in the morning mean missiles in the afternoon.' Destroying the bomber early enough would abort the attack. Furthermore, the submarine inevitably had to transmit to the bomber to set up the crucial targeting link. That in turn would give away its presence and, probably, its position.

The submarine commander would hardly want to fire when the bomber was in place. The bomber would indicate target position, course, and speed. Using such data, the submarine fire control system could project ahead where the target would be when a missile arrived, perhaps half an hour after it was fired. Even if the missile followed the path assigned to it, there was an excellent chance that it would miss the target, even the target group of ships. Even a large nuclear warhead (500 kT was typical of early anti-ship missiles) might well fail to destroy a carrier, since its lethal radius would be no more than a few miles.

It was soon obvious that the situation was even more complicated. The submarine had to know not merely where the target was, but also where it was within a formation of ships. Otherwise the missiles might attack the wrong targets. It was not that the Soviets feared hitting neutral ships, but rather that the submarine had only a limited number of missiles. Once they arrived at the carrier task force, they would be subject to numerous countermeasures, including both defensive missiles and jammers. Thus it was essential that all missiles be targeted on the appropriate ship in the group.

The Soviets were learning just how complicated the new style of naval warfare could be. Almost everything now depended on the shore command which would set up the anti-carrier strike. Using its own long-range sensors, that command had to detect and, at least crudely, track a target up to a thousand miles away. It had to coach a submarine into position *and* order a radar bomber into position to co-ordinate with the submarine. Only when both submarine and bomber came together could the attack be mounted. Nothing quite like this had ever been done before. For example, during the Battle of the Atlantic the Germans did coach their U-boats into position to attack – but then the U-boats relied on their own sensors. They did not have to be co-ordinated with aircraft, and they were firing at targets they could detect directly. Much the same applied to U.S. submarines coached into position (mainly by code-breaking) to attack Japanese targets.

Space-Based Sea Reconnaissance

Meanwhile Chelomey had a more radical idea. In 1959-60 he proposed a space-based reconnaissance and detection system (MKRTs) as an alternative to Uspekh. It would be developed by his design bureau (OKB-52) using the main electronics bureau (KB-1, led by General Constructor A A Raspletin) as the key subcontractor. At this time KB-1 was deeply involved in a programme to develop a Soviet anti-satellite system (ASAT). Recent Russian accounts suggest that the idea may have come partly from the Soviet Navy's Armaments Institute (IV-VMF) and from its Directorate of Missile-Artillery Armaments (URAV). A satellite designed to detect enemy ships so that they could be attacked by long-range cruise missiles figured in a navy display arranged for Khrushchev in 1961, and Captain 1st Rank K K Frants (of IV-VMF) briefed Khrushchev personally on the early progress of the programme. Reportedly Admiral Sergei Gorshkov, commander of the Soviet fleet, showed considerable personal interest in the programme as it developed.

The two vital considerations were immunity to adverse weather conditions and quick data transmission (a satellite would not be in sight of a ship for long). For the former, the satellite would use radar and its passive complement, ESM. For the latter, it had to limit the amount of information it transmitted to any ground terminal. To do that, it would store and send processed (automatically detected) data rather than raw radar data. It would be controlled by a single station in the Soviet Union, which would tell the satellite when to dump its data to a waiting receiver, which might be on board a ship or a submarine or on Soviet territory. That is, as in the case of Uspekh, a central command ashore would have to cue the shooter or shooters into position not only to attack but also to receive the satellite downlink. There was to be no provision for the shooter to interrogate the satellite as it passed overhead.

Much clearly depended on just how accurately the satellite could be placed in a known orbit, because without that precision the data it collected (relative to its own position) could not be translated into positions on the surface. The satellite had first to be launched into a precise orbit, which meant starting the launch sequence automatically on receipt of a time signal. A uniform time standard had to be developed, so that the single ground station could measure precisely when the satellite passed over it. Two daily passes over this single station sufficed for a precise measurement of the satellite's orbit, and the station could order the satellite to manoeuvre as required, as it passed overhead. The satellite itself would automatically control its attitude (its angle to the Earth's surface). The system included the shipboard station, which would receive and process information from the satellite, using it to select the main target and entering that target's position into missiles.

Chelomey's proposed system was formally approved on 16 March 1961 (a detailed decree of 3 June 1962 specified the roles of the prime contractor and subcontractors); Uspekh was still in the development stage. The system was designated US, a Controlled Satellite. A special Soviet Defence Ministry order of 18 June 1964 included it as part of a new five-year plan (1966-70) for space-based reconnaissance. It was given the codename Morya-1 (Seas-1). By the time it was operational, the codename was Legenda (Legend). Other programme names associated with the space system were Korvet (1975) and Korall-B and Kasatka (1981-2). Quite naturally Chelomey wanted the naval satellite launched by his UR-200 booster; UR-200K could place a 4-ton satellite in a low earth orbit. Raspletin proposed fitting the naval satellite with both an active radar and an ESM set to identify any target detected by the radar.

As Khrushchev's star set in 1963-4, so did Chelomey's. After nine shots his big booster was cancelled. KB-1 was made prime contractor for the space reconnaissance system, Chelomey's OKB-52 becoming simply the subcontractor responsible for the satellite bus. Under a 24 August 1965 decree, UR-200 was replaced by an alternative, Tsiklon (Cyclone), a modified version of the R-36 ICBM (NATO SS-9).[61] Because the operational version, Tsiklon-2, had limited capability (2.8 tonnes at 200km altitude, 65 degrees inclination), the system had to be divided in two, with separate active-radar and ESM satellites. To save development effort, both used a common bus similar to that of the contemporary IS. Furthermore, these satellites had to have additional rocket engines to insert them into orbit. They also carried medium-thrust rockets for orbital corrections and low-thrust motors for stabilisation.

The Soviets designated the two ocean reconnaissance satellites US-A (active radar) and US-P (passive radar). In the West they were called, respectively, RORSAT (Radar Ocean Reconnaissance Satellite) and EORSAT (ELINT Ocean Reconnaissance Satellite). For the best possible resolution, US-A had to fly relatively low (265 × 250km). At such altitudes conventional solar panels would encounter far too much air resistance. Chelomey's solution was a small nuclear reactor. He may well have been pushed in this direction by his organisation's somewhat earlier work on an electric (presumably ion-drive) rocket for interplanetary exploration.[62] The use of nuclear power raised an additional question: once the satellite decayed out of orbit, its reactor would fall somewhere on Earth, creating a possible international incident. To avoid such a problem, RORSAT was designed to boost its reactor section into a 900-1000km orbit after its 45-day operational life was over; it would not decay out of this orbit for about 400 years. By that time the fuel would have so decayed as not to present any serious hazard. The spacecraft would also boost itself into a high orbit in the event of some catastrophic onboard failure rendering it useless.

The US-A satellite bus was 10m long and 1.3m in diameter, with a weight of 3800kg. The reactor plus the boost system weighed 1250kg and accounted for 5.8m of the total vehicle length. It used an X-band side-looking radar developed by NII-17, with a semi-circular antenna about 4m long, extended from one end of the cylindrical satellite bus. At a slant range of 400km, the radar would see a spot about 4km long on the earth. That was far from satisfactory; the system therefore achieved the resolution it needed by using a synthetic aperture technique. It added up the returns it received as it moved along its orbit, so that in effect its antenna was greatly lengthened. To do that, the satellite needed an onboard computer. The radar, probably one of the first in Soviet service to provide processed rather than raw video, was developed by a Moscow scientific/technical institute (NII-17).[63]

US-A was designed to operate in pairs. Missile targeting demanded not merely a current target position, but target course and speed, so that target position could be projected ahead. Two positions, from satellites spaced 20-30 minutes apart, were therefore needed, and standard practice was to take another pair of positions on the next orbit, 90 minutes later, for confirmation. To get the required pair of plots, it was vital that satellite positions in orbit be carefully controlled; hence the need for manoeuvring rockets, to correct orbital spacing. It was also essential that the satellite orbit be precise in the first place. That required an automated launch system. The pair of satellites would not quickly revisit any given patch of ocean, yet data more than two hours late was considered inadequate for targeting. To get fuller coverage, several pairs would have to be placed in position at any one time. The CIA assumed that in wartime the Soviets would try to put up at least a second pair of RORSATs.

When RORSAT first entered service, in the early 1970s, the CIA estimated that under favourable conditions it could detect cruisers and destroyers, and that it could probably detect large ships (such as carriers) under most conditions.[64] That translates to an effective spot size, at the surface of the sea, about 100 to 150m long. Under good conditions, most of the signal sent back up to the satellite would come from a ship in that footprint. However, in rough sea conditions, much more of the return would come from the surrounding sea. Only a carrier (about 300m long) would be so large, compared to the footprint, that its echo would almost always be much larger than that of the sea. The CIA analysis hinted at a major weakness. Satellites follow predictable orbits and could be detected by SPASUR (see Chapter 6) and other elements of the U.S. space surveillance system. Special efforts were made to identify particular satellites. SPASUR and other U.S. operations in turn generated a space almanac, which provided deployed naval formations with notice that particular Soviet ocean reconnaissance satellites were due overhead. The best countermeasure would have been some means of inserting false targets into the satellite's side-looking radar. That was apparently quite difficult, because the satellite built up its picture from a large number of individual echoes. Thus, well into the 1990s, the development of a countermeasure against a space-based synthetic-aperture radar was a major goal of U.S. Navy electronic warfare research.[65] Even so, a great deal could be done. X-band radar is badly degraded by rain (which is why many naval radars operate at the longer S-band, despite their need for much larger antennas). Thus a ship could generally hide from US-A by turning into a rain squall, if one was present. As for the carrier, which the satellite surely could always detect, remember that the satellite was looking for large objects oriented parallel to its orbit. From its point of view, a carrier oriented towards the orbit might well appear to be a small ship (since the carrier's beam would be the dimension seen by the satellite). That would be even truer of destroyers and cruisers.

The US-A space programme began with seven test vehicles (launched 28 December 1965 through 25 December 1971). The first two (Phase 1) tested propulsion systems (for orientation and stabilisation). Since the Tsiklon-2 rocket was not yet ready, these satellites were launched by modified R-7s (NATO SS-6, the missile which had launched Sputnik and other early Soviet satellites). The next two (Phase 2: December 1967 to January 1969) tested the command system, and carried full control systems and simulated nuclear reactors. Of three satellites launched, one failed to reach orbit. The other two were initially placed in a low operational orbit (280 × 260km at 65.1 degrees inclination), and then boosted into a high orbit (950 × 900km). The CIA saw these experiments as part of a programme to develop a manoeuvrable satellite. Not until 1971 did it become aware that the Soviets were working on a space-based reactor. Phase 3, beginning in October 1970, was the full system test, comprising six launches. The State Commission evaluating the system was headed by Admiral N N Amelko, Deputy C-in-C of the Soviet navy. The first satellite, Cosmos-367, was boosted into a high storage orbit almost as soon as it reached operational orbit; it apparently failed catastrophically. However, the next two satellites, launched in April and December 1971, succeeded, remaining in low orbits for eighteen and ten days, respectively. Both were dummies, without radars. Cosmos-516 (August 1972) was the first to carry a radar; it remained in low (operational) orbit for thirty-one days. On the other hand, a sixth satellite, launched on 25 April 1973, failed to orbit and fell into the Pacific. Tests were considered complete in August 1973, and the next satellite, Cosmos-626 (December 1973), was operational. It lasted forty-four days. At about this time the CIA finally decided that the system was for naval radar reconnaissance, using a side-looking radar.

Meanwhile, in 1972 series production of US-A satellites had been ordered at KB Arsenal (Leningrad Factory No. 7). Its first satellites flew in 1978. The first pair of satellites (Cosmos-651 and -654) were launched in May 1974; they were boosted into high orbits after seventy-one and seventy-four days, far beyond the designed satellite lifetime. In April 1975 a pair of satellites was launched to support the Okean 75 naval exercise then proceeding in the Atlantic, Mediterranean, Pacific, and Indian Oceans, in effect as a system test. They lasted, respectively, forty-three and sixty-four days.

Presumably on this basis the system as a whole, including the shipboard terminal, was declared operational in October 1975. However, there were still problems, and tests continued as late as 1978. For example, in December 1975 the reactor on board Cosmos-785 failed immediately after the spacecraft entered orbit. The most embarrassing failure was Cosmos-954. In September 1977 its high-orbit boost system failed. It fell to earth in Canada on 24 January 1978. The Soviet leadership was appalled; the programme stopped while the spacecraft was redesigned. The reactor section was redesigned to burn up completely ~if the spacecraft re-entered the atmosphere, ejecting fuel elements at an altitude of 114-120km. This new design was proven when Cosmos-1402 experienced an accident in January 1983 (and fell to earth on 7 February 1983).

US-A satellites were placed in 250 × 265km orbit with 65 degrees inclination, the satellite repeating its ground track every seven days (110 satellite revolutions). However, the swath its radar swept out was wide enough so that the satellite would see a given target area much more frequently. Even so, US-A operations were suspended until Cosmos-1579 was launched in June 1984. It operated with a US-P passive satellite (Cosmos-1567), their orbital planes (150 degrees apart) intersecting over the North Atlantic, the satellites passing about 45 minutes apart. This kind of pairing was repeated in later launches, the active and passive satellites providing the necessary pair of target locations. Apparently the Soviets lacked the resources to launch each type of satellite in the desired pairs.

The last of twenty-four operational US-A, Cosmos-1932, was launched on 14 March 1988, apparently timed to support a naval exercise. Control was lost in mid-April 1988, and the reactor was boosted into a safe orbit on 30 September 1988. Apparently the Soviets hoped to resume launchings, since through 1988-90 they continued to maintain that nuclear reactors in space were essential.

Had the programme continued, it would have been based on a pair of longer-lived prototypes launched in 1987, equipped with a new TEU-5 Topol reactor, Cosmos-1818 (February 1987) and Cosmos-1867 (July 1987). Both were placed in 800km orbits, and they remained operational for, respectively, six months and a year. Designed reactor power output was no greater than that of the earlier Buk (and in fact output was only about half of what was planned), so these satellites presumably achieved longer radar range by using much better signal processing. By this time the United States was actively developing anti-satellite weapons; higher orbits would have made the Soviet satellites safer. The satellites would have been able to see further into the Arctic, an increasingly important area. Finally, the safety-orbit problem was finally solved, since a satellite in an 800km orbit would not fall to earth for 350 years. As it was, the programme died, most likely because by the late 1980s the Soviet economy was collapsing.

After the collapse of the Soviet Union, Chelomey's old design bureau, renamed Mashinostroyeniye, offered a lower-powered radar satellite, Almaz, for export (1991). Although it could not match US-A performance, Almaz characteristics give some idea of what the earlier radar satellite was like. Almaz was first tested as the Cosmos-1870 satellite. Its 1.3 × 15m slot antenna produced a beam 4 degrees wide in the vertical plane, but only 25 minutes of arc in the horizontal (S-band, 250 kW, 0.1 microsecond pulses, 3000 pps). This synthetic-aperture radar (SAR) produces its images either optically (Almaz) or digitally (Almaz-1). Data are stored on a magnetic recorder for playback on command. Orbiting at 250 to 280km (inclined at 73 degrees), the satellite radar sweeps out a swath 30 to 40km wide on the surface; each frame is 20 to 240km long along the satellite track. Range resolution is 15 to 30m. Presumably the US-As (RORSATs) swept out the much wider swath required to have a good probability of intercepting moving ships.

The passive ESM satellite, US-P, could afford to fly higher than US-A, where air resistance would be negligible. Moreover, its passive sensor needed far less power than the active radar used in US-A. Hence it could be powered by solar cells. Its X-shaped antenna, similar to that on the civilian Okean spacecraft, was pointed down towards the Earth. The radio direction-finding system was developed by the Kaluga Institute of Radio Technology branch of TsNII-108. The shape of the antenna suggests that direction to the target was found by interferometry, rather than (as in shipboard direction-finders) by amplitude comparison between beams. Compared to US-A, US-P used an elongated version of the same bus (up to 17m long, total weight 3000kg, framed by two large solar panels, about 50 per cent longer than the main body). Western analysts estimated that US-P could locate its targets within about 2km. Typically the Tsiklon-2 booster placed it in a 120 × 415km transfer orbit, from which its main engine (300-600kg thrust) boosted it up into a circular working orbit at about 435km altitude. Like that of US-A, it typically flew at an inclination of 65 degrees. The chief system limitation was the long gap between revisits to the same spot, six to fourteen hours in northern latitudes and twenty-eight hours at the equator. The CIA estimated in 1983 that continuous coverage, given the width of the satellite's footprint, would require at least four satellites.

From the beginning, US-P had been considered a complement to US-A; it was therefore tested by the same State Commission, headed by Admiral Amelko. A dummy test satellite, Cosmos-699, was launched in December 1974. The first full test satellite, Cosmos-777, was launched in October 1975. Typically one US-P flew between pairs of US-A, presumably to identify the target they detected and tracked. The usual orbit was 430 × 445km at 65 degrees inclination (like US-A), the satellite repeating its ground track every four days (sixty-one satellite revolutions). However, after the Cosmos-954 disaster, it was no longer clear whether US-A flights would continue. US-P might have to provide primary detection and tracking, which would require operation in pairs. The sixth and seventh satellites were launched 60 degrees apart in April 1979, flying one behind the other, to support a major Soviet naval exercise.

Cosmos-1735 (February 1986) introduced a new lower-altitude orbit (405 × 420km); the satellite returned to the same ground track every three days (forty-six orbits). Now three rather than four satellites could completely cover the Earth. Cosmos-1737 (March 1986) introduced a new 73.4 degrees inclination (with the same three-day period). The higher inclination gave better polar coverage, presumably associated with the new U.S. Maritime Strategy, which emphasised operations in the Norwegian Sea. In the past, satellites had generally been launched from Tyuratam. The higher inclination was better adapted to launches from the other Soviet space centre at Plesetsk, making it easier for the Soviets to launch on short notice.

From the 1970s on, KB Arsenal and TsNII Kometa (responsible for systems engineering) developed a modernised pair of satellites, US-M (US-AM [17F16] and US-PM [17F17]), with improved performance and longer lifetimes (increased by a factor of four to six). These versions incorporated micro-electronics. The US-M system was declared fully operational by an October 1979 Decree of the Council of Ministers; the first US-AM was Cosmos-1176 of April 1980. The lifetime of such satellites had increased to 120-130 days by the mid-1980s. Beyond US-PM, a further improved version was US-PU(17F120). It was probably associated with the new flight profile described above for Cosmos-1735 (1986). As an example of longer lifetimes, Cosmos-2122 (US-PU) lasted 775 days, a record for the system, before it died in March 1993.

By that time the US-A programme was dead. US-P survived; in 1989, for the first time, the Soviets had four US-P in two orbital planes separated by 172 degrees at the equator. In 1990 six US-P were briefly in service at the same time. One US-P was launched in December 1997, the system declining to only two operating satellites. By that time finances were so badly strained that it seemed unlikely that further ocean reconnaissance satellites would be launched. However, another US-P, Cosmos 2367, was launched on 26 December 1999, so as of 2000 the Russian sea surveillance system is still (barely) alive.

The combination of passive and active satellites, control stations, downlinks, and surface stations comprised the Legenda (Legend) system. Both satellites shared a common data link, the surface antenna for which received the NATO nick-name of 'Punch Bowl'. As a whole, the system was designed to minimise any need for ground stations. It used a single automated station (at Noginsk) to measure the satellite's orbit and thus predict its ground path, while issuing the necessary commands. These included the command to dump data to a specified location at a specified time. Alternatively, data could be stored for later transmission to Moscow, where it was incorporated in a central plot kept by the Soviet navy. To this end, both types of reconnaissance satellites had special antennas atop them for real-time or near-real time data dump via Molniya-2 and Tsiklon communications satellites. The ship and submarine systems were developed by NII-132 (Kiev).

Operation was not particularly flexible. Because the satellite had to be commanded to dump its data at a preplanned time and place, the intended recipient (submarine or surface ship) had to be ordered into position to receive the data. Thus a satellite-mediated attack would necessarily be a relatively protracted operation, the satellites first locating a prospective target for the planners. Then the planners would have to coach the attackers into data-reception positions. Once there, the submarine or surface ship would await the satellite, then take several minutes to receive its radar data. Only then would an attack be practical. This seems clumsy; on the other hand, for a submarine to contact the satellite to trigger a data dump might reveal her position to a distant enemy receiver. The Soviets considered the entire Legenda system vindicated by its performance during the Falklands War (1982). For example, satellite information allowed the Soviet General Staff to determine the precise moment when the British began to land troops in Falkland Sound. Some Soviet writers considered American interest in anti-satellite weapons (ASATs) indirect tribute to the success of their satellite-based ocean reconnaissance system.

Satellite ocean reconnaissance promised a revolution, in which Soviet ships and submarines could engage Western targets almost anywhere on what they called the world ocean. However, to make use of satellite data those units had to know exactly where they were: they had to be cued into position to attack. Moreover, the centre in Moscow, which would still run the operation – just as in the old days of coast defence – had to be able to contact the deployed units. Thus, once the idea of ocean satellite reconnaissance had been accepted, the Soviets also had to develop complementary means of navigation and long-haul communication, both of which were best accomplished via space. Thus in 1962 the Soviet Navy issued a specification (TTZ) for a communications-navigational satellite. A NII-695 proposal, Tsiklon, was adopted in favour of KB-1's Molniya-2, which had been proposed in parallel with the Molniya-1 communications satellite (see below).[66] Tsiklon operated in either data transfer or in direct transmission (navigational) mode. The navigational system was similar in concept to the U.S. Transit described above, using much the same frequencies (150 and 400 MHz). The corresponding shipboard system was Tsunami. The first satellite in the series was Cosmos-158 (May 1967); the system was tested on board a Project 641 ('Foxtrot' class) submarine. The system was declared experimentally operational in 1969; early installations were the cruiser flagship *Admiral Senyavin* and the submarine tender/flagship *Tobol*, as well as 'Whiskey' (Project 613) and 'Zulu' (Project 611) class submarines. Tsiklon was formally accepted into service in 1971. Begun with Cosmos-514, it used three satellites at about

750km altitude at 74 degrees inclination (beginning in 1973 with Cosmos-574, satellites were at about 1000km at inclinations of 83 degrees, so they could operate further north). Satellites did not last for very long; the last of twenty-eight launches was Cosmos-1027 (July 1978).

The first of an improved Tsiklon-M series, Cosmos-700, was launched on 26 December 1974. Elements of this Parus (Sail) system, which was declared operational in 1977, were Tsunami-AM on the spacecraft, Tsunami-BM (R-790) on board surface ships and submarines, and the Tsunami-VM ground station. Tsunami-BM in turn was part of the Molniya (Lightning) automated communications suite introduced in 1970 (distinct from the Molniya communications satellites). Tsiklon-M used six satellites at altitudes of about 950km at 83 degrees inclination, their orbital planes 30 degrees apart. The last satellite in the system, Cosmos-2266, was launched in April 1994. In the late 1970s, U.S. intelligence credited Parus with an accuracy of about 180m.[67] Civilian derivatives of this system (for navigation only) were Tsikada (Cicada, for its clicking emissions) and Nadezhda (Hope), the latter a package aboard the Cospas communication satellite.

The Soviets also fielded a unified (all-service) system, Kristal. The shipboard Kristal-K appeared in 1971, followed by Kristal-BK in 1984. This system may be the military version of the civilian GEO-IK geodetic satellite system, which operates at higher altitude (about 1500 km) than Tsiklon, and at an inclination of 73.6 or 83 degrees, carrying a Tsiklon-type 150/400 MHz Doppler navigation system operating up to twelve hours a day. The satellite bus is similar to that of Tsiklon. However, it has some added features, such as an onboard radar (9.4 GHz) to determine its height above the sea, a light signalling system (to indicate the satellite's position against fixed stars), a radar transponder (for precise ranging), and a laser reflector (also for precise ranging). Satellite lifetime is typically one to two years.

The Soviets also experimented with a system of small satellites, Strela, in 1964-5, and placed it in service in 1972-4. These satellites have no navigational element. They orbit at 1500km/74 degrees. Octets were launched by a single booster from Plesetsk. The series began with Cosmos 336-343, on 25 April 1970, and each satellite had a two-year operational life. About twenty-four were required for global coverage, using VHF and UHF links. The last octet was launched on 3 June 1992. The system became operational in 1973-4 as Strela-1M. The follow-on is Strela 3, six of which are launched together into a 1400km orbit inclined at 82.6 degrees. The first two were test-launched in 1985, and the first operational satellites followed in 1986. The timing of these satellites roughly corresponds to the reported timing of the Kristal system. Strela-2 is a much larger satellite, probably with a SIGINT role.

Molniyas are communications satellites in highly elliptical orbits (to which they have given their name: typically 450-600 × nearly 40,000km, inclined at 63 degrees, period 718 minutes), which, unlike Tsiklon, spent much of their time (up to eight hours at a time) in northern latitudes – where the Soviet fleet spent much of its time. Molniya-1 was part of the Korund strategic command and control system, which included the Rucheiy communications system. Korund connected deployed units, including aircraft, with the national command centre. It was a more automated development of the earlier Saturn command/control system. To increase communications reliability, Molniya-1 itself was partly modernised, tests beginning in 1970. The system as a whole required four space vehicles. Korund was accepted into service for the fleets, the strategic rocket forces, and the air forces in 1975. It included the Ministry of Defence communications centre (Rucheiy) and an automated control system (TsUKOS). It insured rapid and secure communication with deployed strategic forces and it also provided telephone communications with deployed forces in distant areas such as the Soviet Far East, which in the early 1970s was increasingly important. Eventually the system was expanded to eight satellites to provide high-level non-military government communications (the 'Surgut' system). This system was in service for about twenty years. The Molniya-1 satellite carried sixty duplex UHF channels or a single television channel, using a single transponder (1 GHz uplink, 0.8 GHz downlink). The Molniya-1 prototype was launched in 1964, and thirty-three production satellites were launched beginning in 1965.[68]

A Molniya-2 system (6 GHz uplink, 4 GHz downlink), *not* that initially rejected in favour of Tsiklon/Parus, was placed in service in 1971-7. Compared to Molniya 1, it had better solar panels for higher power (1 kW rather than 40 W) and thus could employ lower-gain (wider-beam) antennas (horns rather than dishes). It employed a total of fifteen satellites. Although no navigation element is generally associated with Molniya, its origin in the 1962 specification suggests that one was present, probably much like the Doppler system used by Tsiklon. Each satellite offered 2000 duplex SHF channels or three television channels.

Molniya-3 is a civil series for television transmission, begun in 1974. Molniya-1 satellites were initially placed 120 degrees apart (so that three satellites could cover the Soviet Union for nine hours each day); but Molniya-2 and -3 were 90 degrees apart, and Molniya-1 satellites were moved to positions between the Molniya-2 and -3 satellites in 1976.

The Soviets also fielded a military geosynchronous satellite system, Raduga/Potok, about ten of which were in service in 1992. It apparently provided the sort of reliable communication capability needed to replace long-haul HF, since many Soviet flagships displayed the necessary radome-enclosed dish (NATO-codenamed 'Big Ball') in place of the earlier pair of high-gain HF horns. Raduga first appeared in 1975; it uses nine positions to insure global coverage. Gorizont is a television relay system. Since the Russian fleet operates at very high latitudes, the Russians cannot abandon Molniyas in favour of Radugas: the latter cannot cover the polar region. Note that apparently none of these systems carried the down-link from the ocean reconnaissance satellites; for that a special antenna (NATO codenamed 'Punch Bowl') was needed.

Weapons Directed from Space

The new satellite systems provided targeting for two new generations of Chelomey long-range anti-ship missiles, Bazalt (NATO SS-N-12) and then Granit (NATO SS-N-19). The systems seem to have been conceived with Bazalt-armed submarines in mind. Thus the first 'Punch Bowl' radome was installed on board a Project 651 ('Juliett') class submarine between late 1966 and early 1969. The receiving system probably entered production in 1970. The first conversions of Project 675 ('Echo II') class nuclear submarines (which fired the Bazalt missile) were completed in 1972-3. Unlike the radar video link from a 'Bear D', 'Punch Bowl' could be used by a submarine at periscope depth. Space-based targeting was clearly far more important to submarines than to surface ships. Whereas a surface combatant might find it difficult to operate far from home waters (and friendly fighters), a submarine could go anywhere on the world ocean. Thus it is striking that surface warships armed with Bazalt (the *Kiev* and *Slava* classes) were not equipped with 'Punch Bowl'.

Bazalt was in effect the final version of the P-6 concept. Like P-6, it requires a tracker on board the launching ship. The great problem of such missiles had always been that the target might move out of seeker range while the missile flew out towards it. In Bazalt, Chelomey tried to minimise the problem by slashing flight time, giving it a new engine, probably a hybrid turbojet-ramjet, for very high speed. The flight profile was high-low, the missile detecting its target while at high altitude, then switching off its radar (to prevent counter-detection) and diving under the target's radar horizon. It then flies at low altitude, switching its radar back on only when it approaches the predicted target position. Using a new guidance system, the missile would choose one target (hopefully the designated one) among the array of ships it would approach. It also had a larger warhead. Work was authorised by a 28 February 1963 Council of Ministers decree. This date implies that some preparatory work began soon after the Chelomey design bureau began work on the satellite reconnaissance system.

The sketch design was completed in December 1963, and flight trials began in October 1969. Bazalt was accepted into service in 1975 as a P-6 replacement for 'Echo II' (Project 675) class submarines; there was also an abortive Project 688 to carry twelve to sixteen missiles of this type. In 1977 this missile was selected for the new aircraft carrier *Kiev*. It survives on board *Slava* class cruisers. Note, however, that some of these ships have been shown firing P-35 missiles, or at least P-35 airframes (perhaps adapted to Bazalt guidance systems). The missile is associated with an Argon shipboard command system. Range is 550km at a speed of 3000km/hr and an altitude of 50-50,000m. In at least some cases Bazalt was replaced in the late 1980s by another missile, Vulkan (P-1000), details of which are not yet available. This development suggests that Bazalt was less than successful.

The satellite system was even more closely associated with the next missile system, P-500 Granit (NATO SS-N-19), carried by 'Oscar' (Project 941) class submarines, *Kirov* class nuclear cruisers, and the carrier *Kuznetzov*. All have the 'Punch Bowl' satellite reception antenna. Work on Granit began in 1969. It was the first Soviet turbojet missile suited to underwater launch, and thus realised the goal of building a 'universal' missile for both surface ships and submarines. It is also a true fire-and-forget missile with a much more sophisticated guidance system than the P-6/P-35 family. A salvo of Granit missiles approaching a target communicate between themselves to choose the optimum target. As in Bazalt and earlier missiles, Granit follows a high-low flight path, but in its case there are two different active radars as well as a passive radar seeker. The powerplant is a high-speed turbojet (KR-93). Flight trials began in November 1975, and the missile was accepted into service on 12 March 1983. For Western navies, 'Oscar' was particularly threatening because, apart from a Punch Bowl antenna thrust briefly above the surface, it never had to show itself. Nor did it have to remain in place once it had given away its approximate location by firing from underwater. Moreover, given satellite data, in theory 'Oscar' could attack targets as much as 500km (about 270nm) away.

Once the satellite system was in service, the Soviets began to retire their 'Bear D' (Tu-95Ts) radar reconnaissance aircraft. Now all are gone, and a projected successor, Tu-142MRTs, was built only as a prototype. But now virtually the whole of the ocean reconnaissance satellite system is also gone. This decline drastically reduces the capabilities of surviving long-range missile shooters. To some extent the surface ships, many of which never had satellite receivers in the first place, can use onboard helicopters (Ka-25Ts and now Ka-31Ts) to extend their horizons. They still need cueing, since helicopter operation is likely to be very intermittent. They may rely on ducting to alert them to the presence of potential targets. Perhaps, too, the existing maritime patrol version of the 'Bear' (Tu-142MR, 'Bear F') can down-link the video from its own big radar (which it can use to target onboard Kh-35 [AS-20 'Harpoonski'] anti-ship missiles). The 'Oscar' cannot of course rely on a helicopter. It does have a towed acoustic array, which may detect surface ships out to several convergence zones, say out to 140 or even 185nm. That is still not nearly as good as what Legenda offered, and under some water conditions ranges it will be dramatically less. Moreover, the submarine still needs some form of cueing. HF/DF is unlikely to be effective against a navy, like that of the United States, which has largely switched over to satellite communications. The Russians may have considered the loss of the satellites acceptable on the theory that they could continue to exploit U.S. satellite transmissions (see Chapter 8), but presumably that gap in security has now been plugged.

The bombers also benefited from the new satellite technology. In 1959, just as Chelomey was beginning to think through the satellite concept, the Soviet air force issued an initial specification for a bomber broadly equivalent to the U.S. B-70 then in development. As in the past, it would be used against both land targets and aircraft carriers. The new bomber was to cruise at Mach 2.8 (maximum speed Mach 3), with a tactical radius of 2000km. Sukhoi won the ensuing competition with a bomber designated T-4. The new bomber would be armed with a new kind of missile, Raduga's 1500km-range solid-fuelled Kh-45 Molniya. Like the contemporary U.S. Skybolt, it would have followed a ballistic path up out of the atmosphere. Guidance would have been by command, rather than inertial, as in the U.S. missile. For the antiship mission, an active radar seeker would have provided terminal guidance. Given the missile's range, it would have been fired from well beyond the bomber's radar horizon. For example, even at 90,000ft the radar horizon is only 460nm (about 850km). Thus the new missile was viable in the anti-ship role only if the bomber were fed with targeting data from an external source, the satellite-based sea surveillance system, work on which began in 1961 (about when the specification for the bomber was finalised). By this time, too, there was less interest in a theatre bomber (a Tu-16 or -22 follow-on), a role largely taken over by ballistic missiles. The naval anti-carrier mission was more important (the balance tipped back towards theatre bombers after Khrushchev was ousted).

The competition for the new bomber was rigged to squeeze Tupolev out, for political reasons. Sukhoi won in 1962 with his T-4 design.

However, Tupolev had important friends in the Soviet defence establishment. While work on the winning T-4 proceeded, Khrushchev fell and the new Brezhnev regime favoured Tupolev. The Soviet air force supported his proposal (Samolet 145) for a Tu-22 follow-on, which he had conceived in 1962 as a much less expensive alternative to the costly all-titanium T-4. Where T-4 would have survived enemy defences by flying high at high speed, Tupolev proposed a low-altitude penetrator (Mach 1 at low altitude) capable of cruising at Mach 2 at high altitude, with a range of 5000km. That is, the bomber would follow a hi-lo-hi flight profile, using its high altitude supersonic speed to reach a target area (before targeting data became stale), then going to low altitude to avoid defending fighters, and then rising again to high altitude to escape at high speed. Tupolev's main selling point was that his new aeroplane would be an inexpensive upgrade of the existing Tu-22, a technologically safe step. That implied that it would use an upgraded version of the Tu-22 weapon system: K-22M, with a Kh-22M missile.[69] The new Molniya missile would be abandoned. To emphasise its relationship with the existing Tu-22, the new bomber was designated Tu-22M (NATO called it 'Backfire'). In fact Tupolev had overstated his case; Tu-22M was far from a simple derivative of the earlier, unsuccessful, Tu-22. Russian writers have dwelled on Tupolev's politics, but it seems fair to point out that the U.S. government, far richer than its Soviet counterpart, found it impossible to finance either the B-70 or Skybolt; Tu-22M may actually have been all that the system could have carried.

Because 'Backfire' could fly much further than a Tu-22 at supersonic speed, it could follow an indirect path towards a carrier battle group. The battle group defenders would no longer be able to estimate the axis along which a threat was likely to develop. The bomber's key advantage, however, was its access to satellite data, which enabled it to fire from well below the radar horizon of defending fighters. A 'Backfire' could not receive satellite data directly, but the centre plotting carrier movements on the basis of that data could communicate with the bomber by radio or by communications satellite. Presumably 'Backfire' attacks were to have been co-ordinated with recent satellite passes over the target area. A below-the-horizon attack protected the bomber from carrier-based fighters, but it demanded a new kind of missile guidance system. The missile could no longer be locked on before launch, as had been the case with Kh-22. Instead, a Kh-22M missile fired by a low-flying Tu-22M popped up into its high-altitude cruise trajectory. The missile had a new much more precise autopilot. It could be directed towards a distant target without turning on its seeker, which was not activated until the missile guidance system estimated that the weapon was about 80nm from the target. Moreover, there would no longer be penetrating pathfinders, to discover which ship within a formation was the target. As in the earlier 'Shaddock', missile video was downlinked to the launch platform, in this case the bomber. Using this data, the bomber could select the target within a formation, using an uplink to command the missile. This technique minimised the target's chance of detecting the missile seeker or jamming it. On the other hand, the missile still had to detect its target in order to lock on at a range of at least 70nm, to begin its dive. This late lock-on was similar to that used by the ship-launched Bazalt (SSN-12). During the 1980s, the U.S. Navy experimented with operating carriers in Norwegian fjords, on the theory that radar clutter would prevent the attacking aircraft (or missiles, as it turned out) from distinguishing the targets outside minimum range. At least, they would have to abandon their favoured attack technique, coming much closer to attack, and laying themselves open to counter-attack by land-based fighters.

The combination of very high-altitude flight and a steep terminal dive made AS-4 a particularly difficult target. The 'Backfire'-Kh-22M threat brought forth the space-related U.S. Navy Outer Air Battle programme in the 1980s (see Chapter 10).[70] Conversely, once the Russian (ex-Soviet) ocean surveillance system collapsed in the late 1990s, 'Backfires' were denied the ability to attack at low altitude; now they had to acquire targets themselves, at high (vulnerable) altitudes. Presumably they would pop up periodically (to search) while following a line of bearing towards the expected target position.

Bomber development proved protracted. The prototype Tu-22M0 flew on 30 August 1969, followed by the initial production version (Tu-22M1) in July 1971. It was unsatisfactory; the first version to enter service in large numbers, Tu-22M2 ('Backfire B'), appeared in 1972. Typically it carried a single Kh-22 under its belly (alternatively, at a cost in range, it could carry two Kh-22N underwing). This version was not entirely satisfactory, particularly when it flew at low altitudes (although that was the designed flight profile). Range was 3000 to 3500km, well below that Tupolev had promised.[71] Furthermore, the ECM system interfered with the aeroplane's automatic flight control system. It proved impossible to develop an escort jammer version, so that regiments of supersonic 'Backfires' had to operate with subsonic Tu-16P jammers. Eventually, in 1986, an escort jammer (Tu-22MP, using a Tu-22M3 airframe) was built, but it never progressed beyond the prototype stage.

Finally a new NK-25 engine was adopted; it was more powerful yet more economical than the earlier NK-144. It powered the final version, Tu-22M3, which flew on 20 June 1977 and which entered naval (and air force) service in 1981.[72] It could carry three Kh-22M anti-ship cruise missiles, and had a much longer range than Tu-22.

Unlike the earlier aeroplane, it could fly safely at low altitude (its seats ejected upwards), and it carried a crew of four rather than two (two pilots, two navigators). The radar was a modified version of that aboard the Kh-22K (PN-A), and the aeroplane had an inertial navigation system. It also had an electro-optical bombsight.

As a measure of Soviet reliance on indirect target detection, initially no reconnaissance 'Backfires' were built. At first 'Backfire' units depended on surviving Tu-22Rs, although these aircraft lacked the range of the new bombers. In 1984 a Tu-22M3 was modified for reconnaissance as Tu-22M3(R) or Tu-22MR, and twelve were built. They may have carried side-looking radars. Presumably the existence of these aircraft reflected dissatisfaction with the time-late data offered by the satellite system; there was apparently no possibility of providing the bombers with receivers for the reconnaissance satellite down-link. As the satellite system crumbled, reconnaissance aircraft became more important, and in 1994 conversions of Tu-22M2s began (as Tu-22MRs).

As for Raduga, the missile developers were still interested in aeroballistic weapons. In 1967 they began work on what they called Kh-2000 (later redesignated Kh-15).[73] It was likened to the U.S. SRAM (AGM-69), a relatively short-range weapon which a bomber could loft to hit stand-off targets. When the Soviet Union collapsed, only the inertially-guided version existed. However, it appears that plans had always called for a companion anti-ship version (Kh-15S), which would use a millimetre-wave radar for terminal guidance. Unlike earlier anti-ship missiles, this one would have been carried internally, on a rotary launcher. Kh-15S was first displayed, in brochure form, at the 1993 Moscow Air Show. In the brochure, the bomber launched it at high altitude. Perhaps that indicated that the satellite system had not quite sufficed to cue bombers.

In U.S. parlance, the Soviet ocean reconnaissance system was the SOSS, the Soviet Ocean Surveillance System. The parallel global naval communications system seemed quite formidable: in the 1970 Okean exercise, it was said, ships had manoeuvred in co-ordination even in different oceans. That made sense. If indeed the Soviets planned to win the 'battle for the first salvo', then they could not afford to attack in any one ocean area much before attacking in another, for fear that the first attacks would alert the U.S. targets. Co-ordination was, moreover, a natural implication of the sort of central control the Soviets prized.

Co-ordination would complicate the defence in another way. Usually a commander, for example of a carrier task force, would estimate the likely direction from which a threat might arise. In the eastern Mediterranean, for example, the most likely threat axis pointed up towards Soviet territory. Given this sort of perception, the commander did not have to surround his force with protection. For example, he could limit his CAP to stations placed in the direction of the threat axis. His early warning aircraft could loiter, not over the centre of the force, but perhaps 250 miles towards the assumed threat. That would increase warning time, so that more fighters could be launched in time to deal with the incoming missile bombers. Co-ordination could change that situation. The Soviets might time air operations so that large forces would arrive simultaneously from two quite different directions. For example, a carrier force in the Norwegian Sea might face attacks from both the Northern and Baltic Fleet areas, some bombers coming down the Norwegian Sea and others crossing Scandinavia from the Baltic. Now the early warning aircraft would have to be held back towards the centre of the battle group, for fear that displacing it towards one threat would leave it to miss part of the attack altogether. The force would lose much of its reaction time.

The Soviets were also known to be interested in co-ordinating attacks by different arms. As early as 1961 they had written about the need for bombers (or missile carriers) to use air-burst nuclear weapons so as to avoid damaging submarines striking simultaneously; the latter would use underwater-burst weapons to avoid damaging the bombers. Such ideas implied a very high degree of co-ordinated timing, in hopes of swamping defences.

It now appears that the 1970 exercise was not quite as successful as it seemed at the time. Apparently one lesson was that fully centralised control was unlikely to be effective. The decision loops from deployed ships back to Moscow may have been too long and too slow in operation. In 1970-2 the Soviets converted two *Sverdlov* class cruisers to flagships, equipping them with high-gain directional HF radio antennas (NATO nickname 'Vee Cones'), which became the mark of a flagship. The flag facilities on board may have been prototypes for those installed on board the next-generation flagships, the *Kirov* class cruisers and the *Kiev* class aircraft-carrying cruisers, both of which were fitted with similar long-haul HF radio antennas. Some submarine tenders also served as forward flagships – also marked by the characteristic HF antennas. All of these ships were later fitted with standard satellite antennas in radomes (NATO nickname 'Big Ball'). The use of flagships violated a central tenet of the new style of warfare, trying to avoid offering the enemy any specially valuable central target. Reportedly this new vulnerability was raised during the Okean 1975 exercise, in which the two converted cruisers took part, and the close-in armament of the new-generation flagships was accordingly strengthened.

EIGHT

Dealing with an Emerging Soviet Threat

Until 1967, the U.S. Navy had only a secondary interest in the movements of the Soviet surface fleet. The U.S. Navy was intended to attack the Soviet Union and to protect sea communications to NATO and to the Far East. The main threat to its carrier-based bombers was Soviet air defences, and the main threats to the carriers (and to sea communications) were Soviet submarines and land-based bombers carrying long-range missiles. The Soviet surface fleet might be fairly large, but it usually stayed home. The U.S. perception was that, without carriers, Soviet surface ships could not go very far beyond the protection of land-based aircraft, so that even though the ships might be considered ocean-going, in practice they were part of a coast defence force.

However, after the Six-Day War (1967) in the Middle East the Soviet surface fleet began to operate in numbers in the Mediter-ranean. The new 'Shaddock'-firing ships and submarines suddenly seemed quite formidable. Soviet destroyers ('tattletales') began shadowing U.S. carriers. Clearly they would help direct missile attacks once war began. For the first time in many years, the U.S. Navy faced a serious surface threat which could, moreover, attack from a considerable distance. The Soviets began to write about the virtues of a first missile strike, the 'battle for the first salvo'. Given the realities of international relations, it seemed certain that, were war to come, it would be begun by the Soviets, who would try to ambush the U.S. fleet, particularly the strike carriers in the Mediterranean.

The U.S. Navy faced a terrible dilemma. Those strike carriers provided a large part of the tactical airpower supporting the Southern Flank of NATO, and would be badly needed at the outbreak of war, when (it was assumed) Soviet and allied forces would head for the Dardanelles. Yet if the carriers were always on station, they would always be within missile range of the Soviet surface and submarine fleet in the Mediterranean. Without making any overtly hostile move, these ships could get into attack position and await the word to strike. This was a new problem. When the main threat to the carriers had been naval bombers based in the Ukraine, the bombers' runs towards the carriers would, at least in theory, provide some warning. In the case of the surface ships and submarines, the only warning (if any) would be the appearance of a 'Bear D' to set up a radar picture. Similarly, in the northern Pacific carriers were also the main U.S. tactical assets. The Soviet Pacific Fleet also began to operate regularly away from home waters. Again, there was a real fear that surface missile shooters might prove effective at the outbreak of war.

At this time the U.S. Navy planned to move carriers into the Norwegian Sea soon after the outbreak of war, to form the core of a NATO Strike Fleet Atlantic. Because they would not be within range of Soviet warships or bombers at the outbreak of war, the problem of their initial defence was somewhat simplified. Moreover, it appears that the Royal Navy was responsible for neutralising Soviet surface warships in the Norwegian Sea (in effect, to pave the way for the Strike Fleet).

Late in 1967 the Soviet threat was dramatised when Soviet-supplied Egyptian missile boats sank the Israeli destroyer *Eilat*. The U.S. Navy was already well aware of the existence of a variety of Soviet anti-ship missiles, and indeed it was developing counter-measures to those it expected (as it turned out, largely incorrectly) would turn up in Vietnam. Its efforts, however, were considerably accelerated after the *Eilat* incident. Within a few years they provided all the carriers with short-range defensive missiles (Sea Sparrows). They would form the inner layer of a complex missile defence. Yet there was widespread scepticism within the fleet that the missiles could react quickly enough to a surprise war-opening attack.[1]

The problem was so serious that some rather radical solutions were proposed. One possibility, mooted around 1974, was to substitute re-commissioned battleships for forward-deployed carriers. Presumably the Soviets would have to attack them as part of their initial strike. The battleships' advantage was that they would be relatively difficult to destroy, yet their loss would not enormously reduce NATO striking power in the eastern Mediterranean. In effect they would function as lightning rods.

A war game conducted at the Naval War College in the early 1970s demonstrated just how far some tacticians were willing to go. The U.S. side was well aware that whichever fleet struck first would probably win. It had the U.S. President send the Soviets a message via the 'hot line': the approach of any Soviet aircraft to within 50 miles of battle group centre would be considered an act of war. Since just such an approach (by a 'Bear D') was a prerequisite for a co-ordinated missile strike (the satellite system was not yet operational), the Soviets had to trigger exactly the U.S. attack which would destroy them. This was a very neat, if utterly unrealistic, solution. That anyone took it even slightly seriously illustrates just how difficult the situation was. As long as the war began with U.S. forces within range of the Soviet missiles, the Soviets were likely to attack quite effectively.

At the very least, the Navy had to keep track of Soviet surface warships in the Mediterranean. In 1968 the U.S. Navy convinced the NATO countries involved to create a joint surveillance operation. At about the same time work began on studies to determine just what information a task force commander needed in order to counter those surface forces.[2]

The U.S. Ocean Surveillance System

In 1969 the development was ordered of OSIS (Ocean Surveillance Information System), to collate the data the U.S. Navy collected into a coherent picture of world-wide Soviet naval operations. A Specific Operational Requirement was issued in September 1970, and the system became operational in 1972. OSIS was a larger-scale equivalent to the existing computerised combat direction system, NTDS. In NTDS, operators at consoles entered radar plots into the central computer. Given a sequence of three plots, the computer could derive target course and speed, and then project ahead target position. An operator would check periodically to see whether the target had veered off course. In OSIS, the operator at a console worked from a wide variety of sensor inputs, many of them passive, and some of them delayed in transmission. The use of passive sensors made it easier to identify the ships, and thus to apply known data such as typical cruising speeds. Given position, course, and speed, the computer typically projected ahead ship location. The OSIS centre at Suitland, Maryland, and the subsidiary centres (FOSICs) at Norfolk, London, and Pearl Harbor (for the Pacific) worked with the two Fleet Ocean Surveillance Information Facilities (FOSIFs), at Rota, Spain, and at Kamiseye, Japan, to support the two forward-deployed fleets, the Sixth (Mediterranean) and Seventh (Far East) Fleets.[3]

In 1968 the U.S. Navy had only one global sensor capable of tracking Soviet warships, Bullseye, a net of HF direction-finders similar to the Soviet Krug. Initially an extension of the Second World War HF/DF net used against U-boats, it was intended mainly to locate submarines transmitting back to base. Without satellite communication systems, Soviet submarines had to rely on HF radio for their communications, and those signals could be intercepted by the shore net and DF'd. The Soviets were undoubtedly aware of the vulnerability associated with HF radio, and they followed the wartime German example of compressing their signals into the shortest possible intervals.[4] For its part the U.S. Navy tried to reduce the minimum signal duration its HF net could intercept and DF. Bullseye was complemented by a net of long-range underwater detectors, SOSUS (Sound Surveillance System).

During the early 1950s the existing war-built net was modified so that it could intercept time-compressed transmissions as little as half a second in length (based on the wartime German Kurier). By 1960 the Navy sought the ability to detect and DF signals as short as a millisecond. The minimum time from intercept at the control station to measurement of bearing at the outstation (for triangulation) was twenty seconds; without prior knowledge of the time and frequency of a signal, it would be impossible to deal with a time-compressed signal. However, if stations kept records of their intercepts, they could be compared regularly and submarine positions determined soon after the fact. Netting was therefore useful even if stations could not exchange data during an intercept. The first new-generation HF/DF station was placed in service in 1960 on Okinawa; fourteen more followed in FY61-64. They in turn were replaced by the FRD-10 stations in the early 1970s. Although the Soviets had begun to use satellites, HF was still a primary means of communication, so HF/DF was still well worthwhile. At least some Bullseye installations were modernised in the early 1990s. Just as the U.S. Navy suffered from HF 'deserts,' it lacked Bullseye facilities in many important places. When the British withdrew from 'East of Suez', coverage of the Indian Ocean, which was becoming increasingly important, was lost. The Navy therefore opened HF/DF stations at Udorn, Thailand (December 1972), at Masirah in Oman (October 1974), and at Diego Garcia (August 1974).[5] They worked with an existing Royal Australian Air Force station at Pearce, Australia.

As the Soviet fleet went out onto the open oceans, Bullseye could pick up their long-haul HF traffic – which was particularly important because Soviet ships so often operated in solitary fashion. However, there was no guarantee that any particular Soviet ship would use her HF radio on even a daily basis.[6]

In the mid-1970s the U.S. Navy introduced a shipboard HF/DF system, Outboard, in effect a sea-based Bullseye. Typically each battle group was provided with two Outboard-equipped ships, to provide it with a local triangulation capability (of signals received mainly via surface wave). However, there was a much more interesting possibility. If Outboards could be linked with land-based Bullseyes in real time, then very distant (sky wave) signals could be triangulated. The new communications satellites provided exactly this possibility: a new Tacintel (Tactical Intelligence) broadcast conceived about 1974 soon linked the two kinds of HF/DF sites.

White Cloud

Something more was needed. Any surface ship had to operate certain radars more or less continuously, and if these could be intercepted, then the ship could be tracked. This was much the same logic the

Soviets had followed in their own passive satellite programme. To function effectively, the OSIS sensors had to be able to identify not merely the class of ship seen at a particular time and place, but also which particular ship had been detected. Otherwise it would be impossible to use pairs or triplets of sightings to establish target course and speed. The Soviets had solved the problem by using pairs of satellites (and hoping that both in each pair saw the same target, which was not always necessarily the case).

The U.S. developers had a more elegant solution, radar 'fingerprinting,' in mind. Major naval radars are often very nearly hand-made. Each radar in a series is slightly different and so are its pulses. There may be minute differences in the timing between pulses or in the length of a pulse, or they may be variations in the way the pulse rises and falls (unintentional modulation on pulse, or UMOP). The differences may arise from the radar signal generator or the antenna itself. The idea is much the same as using the lands on a gun barrel to identify the bullet fired from that particular gun. A satellite carrying a really precise receiver could, in theory, pick up pulse details subtle enough to enable analysts on the surface to identify a particular radar on a particular ship. To make theory into reality, very detailed radar signatures had to be collected. In 1971 the U.S. Navy placed a new automated ELINT aeroplane, the EP-3E, into service.[7] It carried a precision pulse analyser (ULQ-16), which was probably used for fingerprinting. The same analyser went on board a number of surface ships. In each case, the platform picking up the radar pulses could visually identify their source, either by overflying it or by coming within visual range. That data could go into the library OSIS needed. Radar fingerprinting was called HULTEC, hull-to-emitter correlation.[8]

The few EP-3Es and the specially-equipped surface ships could not be everywhere at once: they could not function as the search sensors of the system, at least not most of the time. For surveillance, the system needed a different class of sensors. Feasibility studies of a dedicated ocean surveillance satellite began in 1968. The Naval Research Laboratory and DARPA developed a passive satellite, White Cloud.[9] Like the Soviets' passive system, it detected specific naval radars. The overall system is called Classic Wizard, and White Cloud is reportedly code-named Parcae.[10] The first experimental satellite was launched on board a Thor-Agena on 14 December 1971. It used three sub-satellites. The difference in the time a signal arrived at the three indicated the direction of the emitter. The satellites were made by the Naval Research Laboratory, the lead U.S. Navy organisation for electronic countermeasures (including surveillance), as in the development of Grab (see Chapter 6).

By the mid-1970s the U.S. Navy looked ahead to a next-generation active radar satellite, Clipper Bow.[11] It would have used SAR techniques to produce images on the basis of which ships could be recognised. The radar satellites would detect not only ships but also sea-launched missiles, and thus would offer the fleet early warning. The satellites would combine radar with electronic surveillance to identify their targets. Tests showed that a SAR could detect a ship and her wake, and that analysis of the SAR data could produce an image by means of which a ship could actually be identified. Two satellites were planned for 1982-3. Congress killed the project in 1979 (FY80 budget), ordering the Navy to participate instead in similar Air Force/CIA projects (probably Lacrosse). By that time the Navy was less than enthusiastic, because Clipper Bow was clearly going to be far too expensive. Numerous satellites would have been needed for solid sea coverage.

White Cloud alone was pursued instead. To make use of the satellite information, the Navy mounted a radar fingerprinting project, HULTEC, while White Cloud was being developed. A track vector analysis capability (*ie* the ability to deduce the direction of a ship's course from a series of satellite sightings) was developed for OSIS specifically to exploit the resulting data. White Cloud had what turned out to be an important limitation. It could see only those ships using the sensors to which it was tuned. Where Clipper Bow would automatically have detected all the ships of a group, White Cloud might well miss ships using radars in which it was not interested. That did not seem terribly important in the 1970s, when attacks would be mounted by pilots entirely capable of distinguishing important Soviet warships from the unimportant ones around them.

Three clusters of operational satellites were orbited in 1976-80 (beginning 30 April 1976), followed by five clusters of improved White Cloud satellites between 1983 and 1987. Each cluster of three satellites (plus a mother) orbits at an altitude of about 700nm (inclination 63.5 degrees) and reportedly can detect emissions from ships 2000nm away. Sub-satellites use low-thrust rocket motors to maintain position at distances 30 to 240km from each other. They are most effective when the lines between the satellites are at right-angles, the satellites forming very nearly a right-angled triangle at the equator. As the satellites fly towards the poles, however, the sub-satellite orbits converge. To maintain DF capability, one is moved into an orbit 50 to 100km below the others. Since the distance between satellites changes as they move through their orbits, that distance must be continuously monitored. Also, to measure differences in time of arrival the receivers aboard the sub-satellites must be synchronised. Both tasks are accomplished by millimetre-wave radios communicating between the sub-satellites. A complete White Cloud system comprises four sets of satellites with orbital planes 60 to 120 degrees apart (at the equator) and a series of ground stations.[12] Satellites lasted seven to eight years in orbit. A 1993 Russian article, presumably based on official analysis, claimed that the sub-satellites are 0.3 × 0.9 × 0.4m, and that they intercept signals at 0.5-4 GHz, *ie* up through S-band. Second-generation (redesigned) satellites were launched beginning on 8 June 1990, further groups following in 1991, 1993, and 1996. The sub-satellite spacing is halved (to 30-110km), and the 1.4 GHz transmitters of the earlier system are eliminated (they interfered with radio-astronomy). According to the 1993 Russian article, the upper limit of intercepted frequencies is increased to 10 GHz (X-band).

Because the swath the satellites searches at the surface is so wide, coverage from one pass can overlap with the next, 108 minutes later. Four clusters can monitor any area of the sea between 40 and 60 degrees of latitude thirty times each day. Two clusters (the fifth and eighth) failed after only two or three years, probably because an entire cluster could be disabled if any one of its sub-satellites failed.

Classic Wizard (White Cloud) satellites reported to a Regional Reporting Centre (RCC), which in turn sent out their raw (uncorrelated) product in near real-time via the TRAP (Tactical Related Applications) broadcast, a one-way encrypted UHF satellite broadcast (at 4.8 kbits/sec). White Cloud reports were also carried on the separate Tacintel broadcast. OSIS obtained Classic Wizard reports via the TRAP broadcast and via separate record traffic. Because surface emitters other than ships are of great interest to the other services, White Cloud data are now distributed beyond the Navy.[13]

OSIS was subject to a major upgrade, OBU (OSIS Baseline Upgrade), in the 1980s and 1990s.[14] The object was to get within 95 per cent of real-time precision without buying new sensors. That would be achieved by increasing computer capacity, updating the OSIS data base with each satellite pass and each new sensor reading.

The style of ASW the U.S. Navy developed in the 1960s, using SOSUS, was not too different from the style of anti-ship warfare the Soviets were developing, using Krug to cue shore-based 'Badgers'. It provided a model in which a shore-based analysis centre, collecting information from a variety of sources, sent out long-range weapons to attack (actually, in this case, to re-detect) the targets it found. Probably this model helped inspire the creation of OSIS. Note, too, that both Bullseye and SOSUS could help track surface ships. The Soviet Union depended heavily on HF radio not only for submarine but also for surface ship communication. For example, its few flagships showed prominent long-range directional HF antennas. As for SOSUS, most of the sounds it picked up were generated by surface ships rather than by submarines. In the Atlantic, for example, surface surveillance was necessary in order to sift out the vast number of non-targets. Both Bullseye and a form of SOSUS were permanently deployed in foreign areas.

When the satellites were conceived, only the aircraft carriers and shore-based patrol aircraft offered a serious long-range capability against Soviet surface ships armed with 'Shaddocks'. The U.S. Navy had had little interest in open-ocean surface warfare, because it had faced little surface threat. However, aircraft, armed with weapons intended to destroy targets ashore, were likely to be able to hit warships in the open ocean. It was, moreover, known that unmodified air-to-air missiles could hit surface targets, so the fleet's aircraft already had some stand-off anti-ship capability, albeit a limited one. U.S. surface warships could use their own anti-aircraft missiles against surface targets, but only out to horizon range (because the missiles were semi-actively guided). They certainly could not outrange 'Shaddocks'. Thus, at least for the time being, the problem of pre-empting Soviet anti-ship attacks was mainly the problem of directing carrier aircraft. Effort immediately went into improving their ability to carry out this mission. In 1968 the U.S. Navy already had an anti-ship missile, Harpoon, under development. It was conceived mainly as a patrol aircraft weapon, to attack surfaced 'Shaddock'-firing submarines. The aeroplane had to fire as soon as the submarine was detected, in hopes of destroying it before the missiles had been launched (when the guided version of 'Shaddock' entered service, the submarine had to remain on the surface long enough to guide the missiles, providing the attacker with more time in which to act). Missile range, then, was set by the need to get to the submarine in time, before the submarine became aware of the attacking aeroplane (using its radar search receiver).

With the rise of the surface threat in the Mediterranean, Harpoon suddenly became the weapon of choice. Now it had to be fired from outside the envelope of whatever anti-aircraft missiles the Soviets might be developing. The rocket engine originally envisaged was replaced by a turbojet. Fortunately Harpoon was small enough that it could be carried on board existing U.S. attack aircraft. A carrier-based bomber carrying Harpoon could outrange 'Shaddock' itself. Until it was ready in the mid-1970s, the carrier bombers would have to make do with such existing stand-off weapons as Bullpup.

The OTH Picture

From the carrier's point of view, what mattered was the picture of Soviet naval activity beyond the ship's horizon, what became known as the OTH (Over-The-Horizon) picture. The carrier's horizon was defined by the battle group's (then known as the task force's) longest-range sensor, its airborne early warning aircraft. Stationed close to the carrier, an E-2 could see about 250nm. The carrier commander needed the OSIS picture to see ships any further out, beyond the range at which they could fire 'Shaddocks'.

This was not quite the same as the Soviet ocean surveillance system, because the U.S. Navy thought very differently. The essence of U.S. (and other Western) thinking was that virtually all decisions would be made by the local commander. The purpose of the ocean information system was to provide that commander with complete enough information for decision-making; the link with something beyond the carrier battle group was there mainly to provide an OTH picture. That is not to say that the new ideas of tight control, described earlier, were not having some effect.

Moreover, the U.S. system deliberately denied its potential victims any sort of warning. Unlike the Soviet system, it was not built around the requirements of an attack against a specified target. Even when the Soviets switched from using radar aircraft to satellites, they still had to wait for satellite data to be dumped to the shooters, who would be converging for an attack. The shooters would not normally be privy to the plot maintained at the main naval command (or commands) ashore, only to what they needed for the strike at hand. By way of contrast, the evolving U.S. system constantly tracked the Soviet fleet, and also constantly provided that data to deployed users. One reason why was that warning of a possible Soviet attack was at least as important as the ability to destroy particular Soviet surface ships – which, apart from their offensive capacity, were of little intrinsic importance. The carrier commander's main wartime role was to destroy targets ashore. He had to know where the Soviet fleet was in order to brush it off. Destroying Soviet surface ships might be a good thing, but it would not win the larger war, not even the larger sea war, where the main Soviet players were submarines and land-based bombers.

Warning would depend largely on signals intelligence. Presumably the Soviets had to send their fleet a warning signal, to prepare for their strike, some hours or even days before it was due. If their fleet did not attack in a co-ordinated way, then the initial attack would in alert the Americans, probably without damaging them badly enough. The attack alert would have to come early enough to allow the Soviet ships to disperse before striking (so that no single counterattack would destroy the fleet). Then there would presumably be an execute order. The hope was that enough study of Soviet naval exercises, and particularly of the sequence of signalling within them, would provide a pattern against which to compare events leading up to a real surprise attack. All of this made the first global Soviet naval exercise, Okean 70, extremely important to the U.S. Navy.

By 1972, with OSIS entering service, the U.S. Navy planned an additional step, the formation of a Fleet Command Centre (FCC) to fuse intelligence with other operational data. OSIS managed a picture of Red (enemy) tracks and White (neutral) tracks. The FCC was the shore data manager for Blue (friendly) tracks. It built them up from, among other things, its own orders to the fleet, and from its own message traffic. It formulated the combined Red/White/Blue picture for the fleet.

The Fleet Satellite

It was clearly essential that all of the tactical data now being assembled be instantly available to the fleet. The obvious solution was a Fleet Satellite. It would pass the picture compiled by OSIS and the FCC to carriers by a secure satellite link, eventually called TADIXS, using a dedicated channel. Eventually the computers on board ships were powerful enough to make direct use of intelligence data. The TADIXS link was split in two. TADIXS A carries the processed intelligence picture, in the form of OTH-T and, now, Tomahawk Mission Data Updates. TADIXS B carries real-time intelligence data, and requires a TRE (Tactical Receive Equipment, USQ-101). TRE also receives the TRAP broadcast. It was operationally evaluated in 1988, and it entered service in 1991.

TRAP and its ilk raised a problem. There was a great difference between sending the fleet a tactical picture based on intelligence data of varying levels of sensitivity, and sending out that data itself. Moreover, much of the transmission was being automated, as intelligence sensors were treated (under TENCAP) more like ordinary sensors. There had to be some way of screening intelligence data before it was sent. For a time that was done manually (the system was called ESPRIT). Ultimately an automated multi-level screener, Radiant Mercury, was placed in service. This type of automatic screening is also essential when some intelligence data are automatically released to allies as part of their ocean surveillance pictures.

In the late 1960s there was already considerable pressure to develop a tactical satellite communication system to replace HF. Beginning in 1965, the U.S. Navy and the Air Force prosecuted an air war against North Vietnam, under the nearly direct control of the White House. DSCS satellites provided a point-to-point high-capacity circuit between Saigon and Washington. Within Southeast Asia, communication between Saigon and fixed air bases was relatively simple. The carriers moving offshore were a different proposition. The Navy stationed a communications relay ship (*Annapolis* or *Arlington*, in rotation) offshore to receive messages from both Washington and Saigon, and then to send them out to the fleet.

Early internal U.S. Navy descriptions of the relay ships commented that they would have been extremely useful during the Cuban Missile Crisis, in which the naval force had been subject to extremely tight control. That had been true even for a force quite close to land-based communication centres; it would be much more the case for a crisis in a radio 'desert'. One clear implication was that, regardless of its investment in combatants, the U.S. Navy would find it difficult to negotiate two simultaneous crises unless it had more relay ships. Yet it experienced a two-crisis situation when the Six-Day War erupted in the Middle East in June 1967 while the Vietnam War was going on. Moreover, the special relay ships were expensive to man and to operate. The DSCS link to Saigon clearly showed that something better could be done.

The necessary technology was beginning to fall into place. In the mid-1960s two experimental tactical satellite programmes, LES-6 and Tacsat, were proceeding. Unlike DSCS, these satellites operated at UHF, for simplicity and also for relative immunity to weather and penetration of foliage. At UHF satellites could produce a powerful enough broad-beam down-link to work with simple broad-beam antennas, which did not need anything like the precision SHF demanded.[15]

The LES satellites were developed by MIT's Lincoln Laboratory, which supported Air Force space technology. Late in 1965 the Defense Department established Program 591, under which the three services were to evaluate the potential value of UHF satellite communication. Lincoln Laboratory was chosen to build the test satellites, which became LES-5 and -6. LES-5 would be built and launched as soon as possible, and LES-6 would incorporate the lessons learned. The three services would build their own terminals for test purposes. LES-5 was placed in a near-synchronous orbit with three IDCSP satellites (1 July 1967); it proved quite effective. Mobile users could receive its signals via simple low-gain helix antennas. Two days after being placed in orbit, it successfully connected U.S. aircraft, a submarine, surface ship, and army units.[16] LES-6 went into a synchronous orbit on 26 September 1968. It offered a single 500 kHz-wide UHF channel. It remained active through mid-1976.

The next step was a big geosynchronous UHF/SHF satellite with cross-links to other satellites and the ability to switch bandwidths as needed. The contract for this Tactical Satellite, the largest communications satellite yet (1690lbs), was awarded to Hughes in December 1966. It was the first U.S. geosynchronous satellite built from the first for military purposes (the first foreign geosynchronous satellite was the British Skynet I, launched 22 November 1969). The single Tacsat satellite was launched on 9 February 1969 into a near-synchronous orbit; it failed in December 1972, after 34 months in orbit.[17]

From the Navy's perspective, UHF operation was attractive because the fleet was already equipped with UHF radio for line-of-sight communication.[18] Task forces were already dispersing quite widely, mainly to limit the effects of single nuclear bursts on a group of ships (hence, for example, the use of HF for the vital tactical digital channel, Link 11). Thus an early objective in the Tacsat programme was to assess the capability of the mass of existing UHF equipment (in the other services as well as the Navy) to gain long-range coverage through satellite operation.[19] It seemed likely that antenna problems would be less severe at UHF than at SHF; simple UHF antennas could capture radio energy more effectively. For example, aircraft 'blade' antennas might be effective. Trials with LES-5 and -6 were encouraging, but existing UHF transmitters and receivers were inadequate. They were not stable enough, and the receivers were not sensitive enough. Ten WSC-1 ship terminals were bought, each supporting up to five voice or teleprinter channels (there were also single channel airborne terminals, ARC-146). A lower-cost ship terminal for smaller ships was also developed. Tests with Tacsat in 1969-70 showed that voice, teletype, and data transmissions were all feasible; aircraft could use the system for limited communication.

Tacsat was conceived as the first of a tri-service constellation of satellites. Thus, as of May 1970 the U.S. Navy expected to rely on this system for its future tactical communications. Plans called for equipping the fleet down to the level of minecraft with full-duplex terminals to handle all over-the-horizon traffic. In addition, every ship, submarine, and craft would receive the fleet broadcast via a simple UHF satellite terminal. Aircraft requiring data communications (presumably mainly P-3s and E-2s) would also be fitted with full-duplex satellite terminals.[20]

Perhaps the most important service the satellite offered was a reliable Fleet Broadcast, to replace existing HF service. Assembled at a NAVCAMS (Naval Communications Area Master Station) ashore by a Fleet CINC staff, the broadcast carried intelligence, logistics, and operational weather messages. It also carried ship-to-ship message traffic, minimising ship-to-ship HF communication subject to intercept and DF. Instead of sending messages directly to other distant ships, a ship sent her messages to the NAVCAMS for inclusion in the Fleet Broadcast.[21] All ships listened to ('guarded') the Fleet Broadcast, stripping off messages addressed to them. This procedure entailed delays in two-way traffic, which was why there were separate channels (using DSCS) for more urgent two-way high command communications. Parallel to the main Fleet Broadcast was a submarine broadcast. In addition to HF, which could not penetrate the water, it was sent by VLF transmitters, which a submarine could hear at periscope depth. In fact the interservice plan soon broke down, presumably because the services found their budgets stretched too tight by the war in Vietnam. The Navy soon had to propose a single-service solution, the Fleet Satellite. To provide interim capacity, in June 1970 the Joint Chiefs of Staff transferred available DSCS I systems and the Tacsat to the Navy. The Navy was also allowed to use LES-6. The Fleet Satellite would operate at UHF, for simplicity and to avoid weather problems. The key channel would run at 2400 bits/sec, about Link 11 speed, which required a bandwidth of 25 kHz.

To minimise vulnerability to jamming, the Navy decided to use a large SHF shore terminal transmitting a spread-spectrum signal. To minimise cost, it decided to use UHF (244-400 MHz) receivers on board ships. The satellite processed the received spread-spectrum signal to produce the necessary Fleet Broadcast down-link. Like its equivalent in HF, the receive-only satellite-transmitted Fleet Broadcast has sixteen channels, each 15-75 bits/sec to feed teletypes (total 1200 bits/sec).[22] It is computer-processed (originally by a system called NAVMACS, the Naval Modular Automated Communications System) on board the receiving ship.[23] The non-directional receiver is SSR-1, which uses an inconspicuous crossed vertical wire loop. The backup is a UHF (up- and down-link) channel normally used for secure voice communication.

A Fleet Satellite programme was approved in September 1971. In 1973, with Tacsat dead (as of 1972), and the Fleet Satellite unlikely to fly before 1977, the Navy leased a trio of Gapfiller transponders aboard each of three commercial Marisats (maritime communication satellites) scheduled for launch in 1976 (February 1976 for the Atlantic, June 1976 for the Pacific, October 1976 for the Indian Ocean and as a spare for the other two). Each had a wideband channel equivalent to that of LES-6 and two narrowband channels like the ones planned for the Fleet Satellite.[24] Gapfiller provided the Navy with the experience to operate its own Fleet Satellite when the full system became available. LES-6 was turned off when the first Gapfiller entered service (March 1976). Once the three satellites were in service, the leases were extended to 1979. With other U.S. capacity available, the British leased some Atlantic capacity in 1981. Gapfiller was phased out in 1990, but two more satellites were in use early in 1995. In addition to the fleet broadcast, both Gapfiller and the Fleet Satellites provided two-way communication for most ships, using the 25 kHz UHF up- and down-links originally conceived to service OSIS. Both also provided a 500 kHz strategic link for the JCS/CinC nets.

The Fleet Satellite provided a total of ten 25 kHz channels (two high- and eight low-power) in addition to the Fleet Broadcast channel and the wideband strategic channel. It also carried eleven low-power narrowband (5 kHz) channels for Air Force communication (AFSATCOM). Note that some of the Air Force channels could store and rebroadcast their Emergency Action Message (the nuclear go-code). The combination of Fleet Satellite channels can be turned into more than thirty voice and twelve teletype channels or twenty-two high-capacity data channels. UHF earth coverage is a 19-degree cone. Each satellite weighed 2250lbs and was nuclear-hardened (to JCS-1 standard). Of eight satellites launched between February 1978 and September 1989, six were successfully orbited. The system became operational in 1981. Design lifetime was seven years, but some satellites had to last twice as long; they began to retire from 1994 on due to lack of fuel for station-keeping. To conserve fuel, controllers allowed them to drift up to 10 degrees off their equatorial positions. In 1999, four are still available, mainly as in-orbit back-ups. Two are still partly operational. FLTSATCOM 7 (1986) and the last in the series carried an additional EHF package, the FEP (FLTSATCOM EHF Package), to provide an initial experimental EHF capability while Milstar (see below) was built.

WSC-3 is the UHF transmit/receive terminal. Its antenna resembles a shallow washbasin angled atop a pedestal. Some versions of WSC-3 were adopted as standard U.S. Navy line-of-sight radio transmitter-receivers, partly because the device's modular design made for simple modification for that purpose. WSC-3 is widely used by countries friendly to the United States, which presumably also use the Fleet Satellite and its relatives.[25]

When the satellites were first orbited, their 25 kHz wide channels were considered good for 2400 bits/sec (later the same bandwidth supported rates as high as 19,200 bits/sec, and some of the 5 kHz channels were considered good for 2400 bits/sec). This was used, for example, for CUDIXS text messages (see below) and for a Secure Voice Subsystem (SECVOX) which uses a voice digitiser.[26] CUDIXS is the Common User Digital Information Exchange System, processed on ship by the same NAVMACS used to handle the Fleet Broadcast. It is used to exchange classified data (but not signals intelligence) with shore stations, and it supports two networks each of fifty primary (one-way) and ten special (two-way) users. Formats match those of the Fleet Broadcast.[27] CUDIXS and a tactical intelligence broadcast (TACINTEL: see Chapter 9) were functional replacements for a coded 75 bits/sec teletype (ORESTES), which survived as a backup. These satellites supported the Maritime Strategy the Navy adopted for the 1980s. They carried the important tactical information exchange links, such as TADIXS, established at that time, described in Chapter 9.

Submarines posed special problems, as UHF radiation does not penetrate sea water. The frequencies which are effective for a submerged submarine require transmitting antennas far too massive for satellite use. Three approaches have been tried. One is to keep repeating a short message. When she can, the submarine puts up an antenna and picks up the message. This technique was used for the

SSIXS message sent by UHF (see below). The high satellite data rate limits antenna exposure, and makes repetition possible. The higher the frequency, the higher the data rate; hence the current interest in submarine SHF and EHF masts. Recent research has followed two alternative paths. One is to devise a larger submarine antenna, perhaps a conformal one which can be integrated into the sail. With much higher gain, it can support a higher data rate. A second is to devise a UHF-capable floating wire antenna, which the submarine can deploy while submerged, perhaps for two-way communication. The array would compromise the submarine to some extent, but far less than a mast-mounted dish. Typically it would be trailed on a 100ft cable. As of 1999, there is a DARPA-Navy programme to develop such an antenna. The array itself would float on the surface; the satellite beam would be formed electronically (the beam must be maintained as parts of the antenna are blanked out by sea water washing over them, or by waves shadowing them, and as the antenna itself snakes in the water). New types of antennas promise useful UHF performance in diameters far smaller than UHF wavelength (which is about a meter), perhaps 1 to 3in. Prototype antennas were first tested at sea (from a surface platform) in May-June 1999. A second approach is to use a low data rate system, such as the U.S. ELF transmitter in Wisconsin, to send a 'bell-ringing' message which causes the submarine to come to periscope depth and put up an antenna. Similarly, surface ship sonars can be modified to send special 'bell-ringing' signals (the U.S. version is Probe Alert). A third approach, which has been less successful, is to use a blue-green laser. It can penetrate the ocean to a substantial depths (it has been proposed as a sensor), and can carry considerable amounts of information. Effective information transmission is limited, however, because the laser must scan its relatively narrow beam over the wide area in which a submarine may be located, repeating the message frequently enough for any submarine touched by the beam to receive all of the information. The U.S. Navy and DARPA tested an airborne blue-green laser ('Project Y') in 1987; they proved that clouds did not prevent transmission to a submarine at operational depth. The Navy also tried 'Project Lambda' (*ie* inverted Y), a laser uplink from a submarine; but it would seem that any such link would be too easy to intercept, giving away the submarine's position. Reportedly plans for a laser-communications satellite (SLCSAT) died in FY89 budget cuts, but interest in such systems survived, *eg* in the form of a proposed two-way Tactical Airborne Laser Communication (TALC) system. A Tentative Operational Requirement (TOR) was issued, and NOSC San Diego ran tests with the experimental submarine *Dolphin*, a blue-green laser, and a small lightweight satellite. The House Armed Services Committee added $12 million in FY92 and $15 million in FY93 for this programme. Eventually it died, but the idea still holds interest. It may be particularly important if, as seems to be the case, submarines in a post-Cold War world must co-operate much more closely with surface forces.

These needs were not, of course, only limited to the U.S. Navy. A recent Russian account describes parallel Soviet-era programmes, including the development of an ELF system. Note that competing ELF systems might well jam each other, because of their very long range and their very limited channel capacity. At the very least, once the Soviets clearly became interested in ELF, the U.S. Navy had to take jamming into account. The Soviets were also interested in blue-green lasers, and experiments in this direction may have been interpreted as experiments in laser radar for submarine detection.[28]

There are also times when a submarine has to send a message without coming up. Submarines carry one-way UHF buoys (BRT-6) which carry a tape-recorded message. It is automatically dumped into the submarine information exchange system (SSIXS, see Chapter 9). To protect the submarine, the message is sent after a preset delay; and it can be repeated up to fifteen times, to ensure that it is received. This buoy was approved for full production in 1983. Work on a two-way tethered UHF satellite communications buoy began under the FY89 programme.[29]

The Fleet Command Centre and the Task Force Command Centre

Initially the FCC was intended to send out only the locations of likely targets or other high-interest units, a technique called a HITS (High-Interest Targets) broadcast. That limited the load on communications media and also on the receiver. However, it was soon clear that everyone needed a much fuller picture. In analogy with the NTDS/Link 11 combination, the picture would be passed as a series of updates. Lockheed received a contract for a computer to compile and display this picture, USQ-81. It differed from Link 11 computers in that the picture, of the sea area up to 1500 miles from the carrier, was likely to be far more complex – albeit changing more slowly – than that presented by Link 11. The carrier's OTH picture would be displayed in a new Task Force Command Centre (TFCC), later also called a Tactical Flag Command Centre. One of its tasks would be to correlate the time-late OTH picture assembled at OSIS and the FCC with the near-real-time picture provided, largely via Link 11, by battle group (or task force) assets such as early warning aircraft.

The combination of FCC and TFCC, with the dedicated satellite link, would form the basis for an automated Navy Command and Control System (NCCS). The FCCs were defined as the shore stations (and wartime command centres) of the unified Atlantic and

Pacific commanders, the commanders of the Atlantic Fleet (Norfolk) and Pacific Fleet (Pearl Harbor), and the deployed numbered fleet commanders (London, Naples, and Yokosuka), a total of seven sites. The combination of FCC and links was attractive to the U.S. political authorities because through them deployed (particularly nuclear-armed) naval forces could be controlled. Thus the FCC was also the naval node of the national-level World Wide Military Command and Control System (WWMCCS).

The FCC-TFCC idea found a resonance in the Royal Navy, which in the early 1970s was about to decommission its last aircraft carrier, HMS *Ark Royal*. For almost a decade, since the decision to abandon the carrier force had been made, the Royal Navy had warned that with the carrier it would lose its long-range anti-ship capability. The problem was not the ability to shoot at long range; the Royal Navy expected that it could develop a suitable missile at a reasonable cost. The problem was detecting targets, the same OTH problem the Americans were facing. The Gannet airborne early warning (AEW) aircraft on board British carriers were valued at least as much for their ability to detect and track surface contacts as for their value in air defence. With the demise of the *Ark Royal*, they were gone. While the Americans in the Mediterranean worried about what they might be missing 250 miles from the carrier, the British had lost even that much detection range.

The British built their own FCC, the naval headquarters at Northwood. Its computer information system was OPCON. Conceived in 1974, it became fully operational in 1985. Northwood maintained a shipping picture, based partly on British resources and partly on a version of the U.S. TADIXS broadcast from Norfolk, Virginia. The British satellite link to deployed flagships, which have display terminals, was designated Link R. As in the U.S. Navy of the early 1970s, the primary OTH attack weapon was still an aircraft, in this case a Sea Harrier carrying a stand-off missile, Sea Eagle. The OPCON system supports 150 local display terminals, with remote workstations aboard deployed flagships (mainly the British light carriers) and one at SACLANT headquarters in Norfolk, Virginia. As of early 1989 OPCON workstations or terminals were to have been installed in the German naval headquarters at Glucksburg, in the Allied Forces Northern Europe headquarters at Kolsas near Oslo, and in SACEUR headquarters at Brussels. An OPCON terminal was installed on board the U.S. flagship *Mount Whitney* for North Atlantic exercises. The Royal Netherlands Navy has its own OPCON computers at Den Helder, built under a September 1988 contract. They are linked to those at Northwood: a Dutch ship can talk to a British ship via radio to den Helder, the link to Northwood, and Link R to the British ship.

Later the French Navy adopted a similar system, SYCOM; the shipboard counterpart is AIDCOMER. The two were linked via the French military satellite system, Syracuse (ships can also link with SYCOM via the commercial INMARSAT). Initially SYCOM was apparently more focused on planning than on maintaining a shipping or other intelligence picture. A follow-on SYCOM NG is more like the U.S. and British systems. The staff requirement for its development was written in September 1989, and the system became operational at the end of 1992. Like the U.S. TFCC, AIDCOMER was apparently initially conceived only as a large-ship system, for the carriers in which French naval power is concentrated. Later it was extended to smaller ships as Mini-Aidcomer. Moreover, AIDCOMER is primarily a tactical advice system rather than an OTH picture-keeper. Unlike the U.S. systems, from the first AIDCOMER was integrated with the combat direction system of the new carrier *Charles de Gaulle* (one combat direction system computer is reserved to exchange data between SYSCOM and AIDCOMER). The naval staff objective for this system was issued in December 1987, and a prototype was tested on board the carrier *Foch* off Lebanon in September 1989, and the ship received the first production version in 1992. By 1996 AIDCOMER equipped the two French carriers and thirty-two more combatants, down to frigates and amphibious ships, as well as five naval headquarters ashore; one joint command centre; two maritime air operations centres; and three training and technical centres.

Fleet Satellite Successors

For a time FLTSATCOM handled about 90 per cent of Navy long-haul traffic. However, it could not suffice. A new programme began in 1976-7, as Congress directed the Defense Department to use more commercial systems. As Gapfillers were phased out the Navy leased four HS-381 commercial satellites (Leasats), orbited 1984-5, from Hughes. A fifth, built as a ground spare, was launched in January 1990. These satellites were effectively co-located with the Fleet Satellites, adding more channels. The 2915lb Leasat carried the Fleet Broadcast plus six wideband (25 kHz), five narrowband (5 kHz) and the single broadband (500 kHz) channels. The contract allowed for purchase by the Navy after seven years (the nominal lifetime), and the first three were bought in 1991-2. Lifetime set by fuel capacity was ten years, so the first two were retired beginning in 1994. Unlike military satellites, this commercial type was not hardened against attack (*eg* against EMP). However, its use indicates the number of U.S. communications satellites that can be turned to military use in an emergency.

Leasat and Fleet Satellite in turn were replaced by Hughes's UHF Follow-On Satellite (UFO), modified versions of its commercial thirty-nine channel HS-601 design, selected at the end of July 1988. With over 1500 UHF terminals in service, the Navy had little option but to continue using this frequency range. Unlike all other U.S. government satellites, UFOs were turned over to the government only when already in orbit and checked out; as in the commercial world, the manufacturer was responsible for launch services.

As in the earlier types, this is a UHF satellite with a supplementary SHF up-link for the Fleet Broadcast. As designed, it also had a 500 kHz channel, nine 25 kHz channels, and eight 5 kHz channels. Unlike the earlier Fleet Satellites, it is now dedicated entirely to tactical users: by 1988 strategic functions were moving to Milstar. The 500 kHz strategic channel has now been carved up into eight 25 kHz and thirteen 5 kHz tactical channels. The special regenerative Emergency Action Message channels were eliminated. The Army and Special Operations Forces are taking over the Air Force's 5 kHz channels for voice and data transmission. Under the new joint communications concept, all the services will eventually be able to use this (and all the other channels of the UHF system).[30] Beginning with satellite 4, an EHF capability (eleven transponders and a 5-degree steerable spot beam) is added; the first EHF satellite was launched 28 January 1995. The last three UFOs (EHF-8 through -10, 1998-9) have Ka-band GBS transponders (see Chapter 12). The UFO system consists of nine satellites (two each over the continental United States, the Atlantic, Indian, and Pacific Oceans plus one spare, so that two cover each footprint, dividing up the work). The system covers the earth between latitudes 71 degrees North and 71 degrees South. The first entered orbit in September 1993.

All of these systems suffer from the one great limitation of geosynchronous satellites: they cannot reach users in the Arctic. This problem first affected the Air Force, which wanted to use satellites to communicate with its strategic bombers.[31] The problem worsened in the 1980s, when the U.S. Navy's Maritime Strategy required the fleet to advance up the Norwegian Sea towards the main Soviet naval bases on the Kola Peninsula. The Air Force placed its narrow-band UHF transponders aboard two SDS satellites (described above), which were intended primarily to relay data from reconnaissance satellites, but whose Molniya-type orbits kept them above the Arctic for much of the time. The Navy planned a separate set of Arcticsats, which were never built, due to the end of the Cold War. Given the importance of the UHF satellite links, the Navy probably also placed transponders aboard SDS satellites. In addition to these links, there are apparently also DSCS-type links on board at least two SDS satellites, under a programme called Mission 22 (M-22).[32]

Countering Soviet Tactics

As the Fleet Satellite was being built, the U.S. Navy gradually became aware of Soviet attack tactics, as the bombers and missile-firing ships and submarines began to exercise. The sense of co-ordination over vast areas was nurtured by a global Soviet naval exercise, Okean 70 (1970). The new Soviet ocean reconnaissance satellites seemed to be a very natural extension of earlier practices, and in fact U.S. intelligence credited them with operational status (US-A in 1971, US-P in 1975) some years before they had been completely cleared for service.

What could the U.S. Navy do to counter the Soviet tactics? There were two major insights. First, the directing brain, using a plot somewhere in the Soviet Union to direct attacks, probably was not computerised. That almost certainly limited its capacity to deal simultaneously with multiple targets. Second, the directing brain depended on long-range sensors, such as Krug and the satellites. With limited attack resources, the Soviet commander had to be very sure that his attacks were being made appropriately. For example, there had to be a real fear of ambush, and even of corrupted data causing some missile shooters to attack Soviet warships. In the mid-1970s the Soviets would have had to track only a very limited number of targets in wartime: up to twelve carrier battle groups and, perhaps, a few major amphibious groups. In peacetime numbers would be even smaller. Yet no other U.S. surface warships were really worth tracking, because no others could, by themselves, directly harm the Soviet Union. Meanwhile the Nixon Administration signed the SALT I treaty in 1972. Both sides agreed to cap the number of ballistic missiles they had, leaving the Soviets with considerable numerical superiority. For Nixon, this result was about the best he could hope for. Previous administrations had largely abandoned work on new U.S. strategic systems.

Both sides had hoped for more. The Soviets wanted to limit forward-based U.S. systems, most notably Europe-based tactical aircraft and carrier bombers. The Americans wanted to limit Soviet sea-based cruise missiles, 'Shaddocks', which carried nuclear warheads and thus could attack U.S. and European cities near the sea. Neither class of system was limited, probably because neither side had any equivalent against which to demand cuts. In fact the two classes of weapons were linked, since the main impact of 'Shaddock' was to threaten the U.S. carriers. It is not clear to what extent that was appreciated by Nixon or by his chief foreign-policy advisor, Dr Henry Kissinger; more likely they thought only in terms of strategic weapons (by this time the carriers were no longer part of U.S. strategic strike planning).

Above: The new kind of space-based naval warfare is symbolised by the union of a long-range missile, Tomahawk, and surface combatants fed by space-based sensors. Here USS *Shiloh* fires a missile at Iraq, 3 September 1996. Note that *Shiloh*, an Aegis cruiser, had been designed as a fleet air-defence ship; her ability to hit distant land targets is almost entirely a function of her access to space systems. (U.S. Department of Defense)

Left: HMS *Marlborough*, a Type 23 frigate, illustrates a problem inherent in space communication. The big radome on her bridge roof, abeam her mast, encloses a SCOT satellite communication dish, operating in the same frequency range as the Type 911 Seawolf director radar located forward on the same platform. That will also be the frequency range of many hostile radars, which the ship's ESM equipment must be able to detect. Almost inevitably, some of the satellite up-link will spill over into the ship's masthead ESM antenna (for the UAF system), under her Type 996 radar. The current solution is to use burst satellite transmission, turning off the ESM set during brief bursts. But given rising demand for transmission time, the only solution may be a multi-purpose (shared aperture) antenna system which can replace not only the satellite dishes but also radars operating at similar frequencies – and the ship's ESM antennas. (Crown Copyright)

Above: HMS *Ark Royal* illustrates another issue in shipboard satellite systems. She shows a big radome for her SCOT antenna (large-ship version) abeam her mainmast. It connects her to the British Skynet system, to the NATO system, and probably also to the U.S. DSCS, which operates at much the same SHF frequencies. However, many important U.S. messages still go via UHF, and for that she needs a very different antenna, the 'washbasin' of WSC-3, visible atop her bridge, pointed to the right (it is also used for ship-to-ship communication, and its low angle of elevation suggests that it is set for that purpose). Many ships have other satellite antennas in addition to these two standard naval types. Given the very limited topspace available, and the demand for clear sky arcs for satellite dishes, placement becomes a real problem. Note, too, the vast difference in size. SCOT operates at a much higher frequency than WSC-3, yet needs a bigger dish to form a narrower beam. That so small a dish suffices for the UHF system helps explain why UHF is still the preferred choice for many applications, despite its low data rate. (MoD(UK))

Left: Ultimately space systems are viable to the extent that satellites can be reliably and efficiently orbited. Many existing boosters are based on the liquid-fuelled strategic missiles developed early in the space age. Here the first Delta II launches the first operational Global Positioning Satellite, 14 February 1989. Delta was based on the U.S. Thor IRBM. In the course of development, the booster added a ring of solid-fuel boosters (visible at the base of the rocket) and a new larger-diameter upper stage. Staging transformed Delta from a mediocre short-range missile into an effective orbiter. (U.S. Air Force)

Two of the earliest naval space roles are illustrated by these two satellites, arranged for a piggyback launch: reconnaissance and navigation. The smaller satellite, at the top, is the Naval Research Laboratory's GRAB, the first U.S. signals intelligence satellite. It rides atop the Transit 2A navigational satellite, also developed by the Naval Research Laboratory. This photograph was released in June 1960, at the dawn of the space age. (U.S. Navy)

DSCS II (Defense Satellite Communications System Phase II) was the first U.S. geosynchronous communications satellite. It was spin-stabilised, its solar cells placed around its cylindrical body so that they could catch the sun as it spun. An onboard motor despun the internal section carrying the antennas. There were two sets of antennas, low-gain (wide beam) horns for reception and transmission, and steerable 44in parabolic dishes (2.5-degree beams) for narrow-beam down-linking (on satellites 7 to 16, one of the dishes was defocused to 6-degree beam for wider-area coverage). (TRW Systems Group)

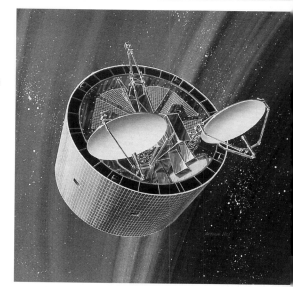

The U.S. Fleet Satellite (FLTSAT) was specially designed for the U.S. Navy after an attempt to develop a tri-service system failed. This 1974 sketch clearly shows the satellite's main antenna, its 16ft UHF parabola (which it deployed after launch) but not the 12ft UHF helix which operational satellites deployed to one side of the main dish, or the X-band horn used to pick up the Fleet Broadcast up-link. Momentum wheels inside the satellite stabilised it, and the two flat solar panels tracked the sun. (U.S. Department of Defense)

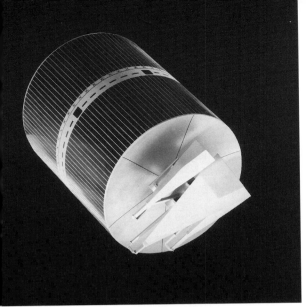

NATO developed satellites mainly for point-to-point communication. This NATO III satellite shows its three horn antennas: a big narrow-beam European coverage transmitting horn (5.4 x 7.7 degree beam) and two wide-beam antennas, one for transmission and one for reception (9.6 x 15 degree beam). The widebeam receiver was for Atlantic coverage. As in the case of DSCS II, this satellite was spun for stabilisation (hence its solar cells are arranged over its casing), the inner part carrying the antennas being despun. (Aeronutronic Ford Corp.)

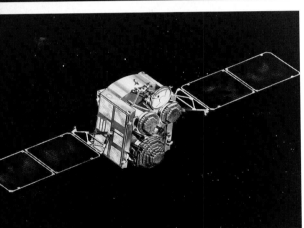

DSCS III illustrates some current satellite features. Instead of dishes, it uses multi-beam arrays: two transmitters, each 28in across (19 beams each) and one receiver (45in across, 61 beams). There is also a steerable 33in dish (3-degree beam) for transmission, visible above the two fixed transmitting arrays. Two pairs of horns (transmission and reception) are visible to the left of the fixed multi-beam arrays. The object resembling a small propeller is a crossed-dipole UHF antenna (there are two, one to receive and one to transmit). (U.S. Air Force)

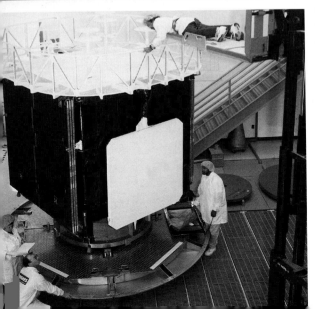

To be launched, a satellite must fit within a small space at the nose of a rocket, yet performance depends on just how large antennas and solar panels can be. This Hughes UHF Follow-On Satellite (UFO-2), folded for launch, shows how such problems are often solved. Its solar panels are folded accordion-style against two of its sides, and its white receiving antenna is folded against a third side. This satellite was launched on 3 September 1993. (Hughes Space and Communications Co.)

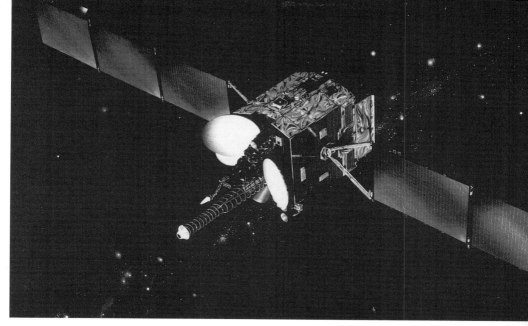

Skynet 4 is the current British military satellite system. This artist's view clearly shows its 8ft UHF helix antenna. The two dishes operate at SHF, using multi-horn feeds to produce a widebeam, a narrow beam, and a spot beam. A third dish produces nulls (for the uplink, to overcome jamming). There are also four wide-beam horns (one receiving, three transmitting). The 'wings' carry solar cells, which were expected to provide 1600 Watts of power when the satellite was new. The satellite's rated lifetime of seven years was set in part by the estimate that over that time damage due to space debris and cosmic rays would cut the cells' output to 1200 Watts. There are four SHF channels: (1) Earth Cover, (2) European Cover, (3) Wide Beam or Steerable Spot, (4) Central European Spot. The satellite can receive via its wide-beam Earth coverage horn, or via the steerable spot or the Central European spot beam. The UHF antenna serves two 25 kHz channels. SHF bandwidth is 60 to 125 MHz. The bandwidth makes it possible to use spread-spectrum waveforms to minimise the effect of jamming. (Matra Marconi)

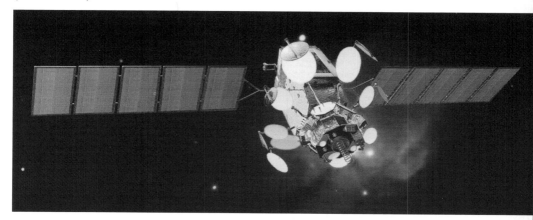

The Skynet 5 proposal offered by a consortium comprising Lockheed Martin, British Telecommunications PLC, and British Aerospace is shown. The satellite resembles that which Lockheed Martin plans to use commercially in its new Astrolink system (which is to become operational in 2001). Note the numerous separate dishes, for spot beams, compared to those of Skynet 4: Skynet 5 is intended to cover Europe, the Middle East, Africa, parts of Asia, the Atlantic Ocean, and the Eastern United States.(Lockheed Martin)

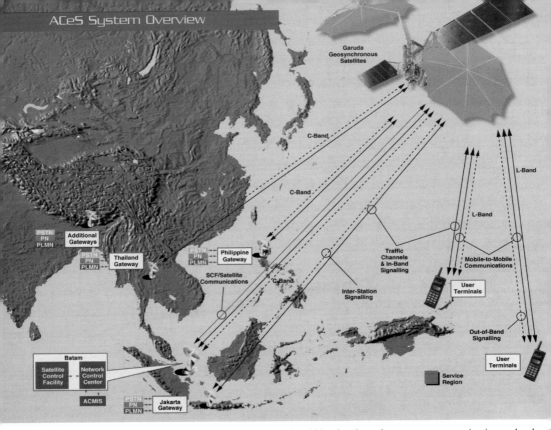

Above: The new Asia Cellular Satellite (ACeS), built by Lockheed Martin, shows how space communication technology is changing. With its pair of 12m dish antennas, a satellite in geosynchronous orbit can now communicate directly with hand sets similar in size to those used in standard cellular telephone systems. The system uses more conventional dishes to communicate with earth telephone services through the gateways shown, which have 13m antennas. Presumably the system's extraordinary capability can also be traced to the considerable processing power on board the satellite, which allows it to handle weaker signals (a user can improve signal quality by placing the handset in a special satellite cradle carrying a high-gain antenna). As of January 2000, the first of two Garuda satellites was scheduled for launch on 12 February 2000. ACeS service is to extend over Asia and into Europe. The high-gain technology clearly has implications for signals intelligence, at least at L-band and above. The satellite itself is marketed by Lockheed Martin as A2100 AX2; it was developed in 1995-7. Other versions of the same satellite bus are used as the U.S. SBIRS (Space-Based IR Satellite, the DSP successor), as the GE satellite communications satellites, and in a new Ka-band Astrolink satellite network. The same bus is the basis of Lockheed Martin's proposal for the Advanced EHF satellite, the Milstar successor. Smaller and less expensive than Milstar, it would also be faster and more powerful, with five or six times the throughput of current Milstar satellites. The Milstar LDR and MDR modulations would be merged into a single system running at 8.2 Mbps or more per channel, and offering 50 channels; total throughput would be about 500 Mbps per satellite. The full constellation would consist of four satellites with cross links (150 Mbps). The new system would be based on Lockheed Martin's commercial Astrolink. (ACeS)

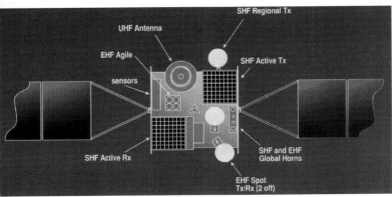

Left: A Matra Marconi sketch of a next-generation military satellite, presumably its concept for Skynet 5, shows how active arrays (SHF receiver and transmitter) can be combined with wide-area horns to provide the necessary spot beams without using numerous dishes. Global horns provide full Earth cover, to transmit to moving platforms such as ships and aircraft. Note that the satellite does use dishes for EHF spot coverage. (Matra Marconi)

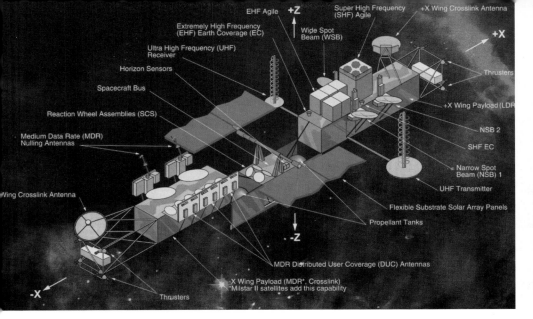

Above: Milstar is surely by far the most complex existing communications satellite, using digital coding for its messages and carrying out many functions formerly entrusted to ground control; it is sometimes called a 'switchboard in the sky.' Note that this drawing omits the solar panel 'wings' for clarity. (Lockheed Martin)

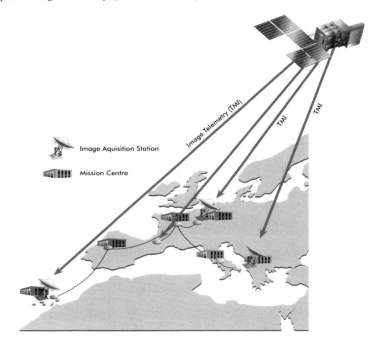

Right: Helios typifies current imaging satellites, military and commercial. It was funded by the French, Italian, and Spanish governments, of which France contributed the bulk of the support (Italy is responsible for about 14 per cent, Spain for 7 per cent). Multi-level encryption limits some images to only one of the three partners, France having the highest priority and the widest access. As shown here, the control centre is at Toulouse, France. Processing centres are in Creil, France, and in Madrid and Rome. The down-link antennas are in Colmar, France, in the Canary Islands, and in Lecce, Italy. There is also a Theatre Transportable Station. Not shown here is the laser cross-link to the French Syracuse communications satellite, which is described as permitting French surface ships to obtain satellite data. The first of two satellites was launched on 7 July 1995, the second on 3 December 1999. A Helios 2 is planned. Matra Marconi was prime contractor for the spacecraft. (Matra Marconi)

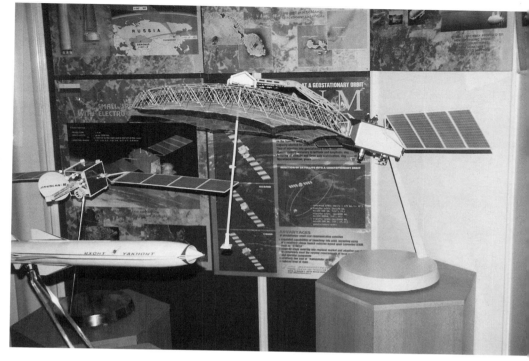

Above left: The beginning of the new kind of naval warfare: a submarine-fired 'Shaddock' (NATO designation) long-range cruise missile on display. Folding wings made compact stowage possible. The missile's range made it necessary for the launching submarine to obtain target information from an external sensor, via a data link. The canister in the background is for the new Yakhont missile, not yet in service. This particular missile seems not to have a nose radome (hence probably is not a P-6), but it does have a radio altimeter (the antenna is the rail alongside the nose), so it may be a P-5D.

Below left: The other main element of the Soviet missile system: a 'Bear-D' (Tu-95Ts) radar platform, showing its big belly radome (NATO 'Big Bulge'), for the Uspekh system which data linked radar video down to a submarine or surface ship. This particular 'Bear' was based in Cuba; it is being escorted away from the sea trials of the new carrier *Carl Vinson* by a U.S. F-4 fighter, 26 January 1982. (U.S. Navy)

Above: The ultimate remote sensor is space-based. This is a model of a small synthetic-aperture radar satellite, developed by Mashinostroyeniye, the lineal descendant of the Chelomey design bureau which produced both 'Shaddock' (and its successors) and the Cold War Soviet space-based naval reconnaissance satellites, US-A and US-P. It was displayed at the 1999 Chertsey (DSEi) military show. Mashinostroyeniye describes Almaz as both an earth-mapping and (with different processing) a ship-detecting or naval targeting satellite. In the latter role it forms part of a system with the same Bureau's new Yakhont missile (model in foreground). The Bureau may also see its Ruslan communications satellite as a potential part of the same global anti-ship system. Unlike the active radar satellite of the past, this one uses the recent technology of unfurling antennas in space to form a massive reflector, about as wide as its solar 'wings.' The reflector in turn offers a high enough gain to permit the satellite to fly high enough that it can use solar cells for power. As of early 2000, the satellite system had not been sold, at least publicly. Mashinostroyeniye was claiming that it had an overseas partner for Yakhont, presumably either India or China. (Author)

Above left: Existing submarines were modified to receive satellite data. This 'Echo II' (Project 675) has her Kasatka (NATO 'Punch Bowl') radome extended. Submarines so modernised could be recognised by bulges in their sails, to accommodate the new antenna. They fired Bazalt (SS-N-12) rather than 'Shaddock' cruise missiles. (MoD(UK))

Below left: 'Oscar' (Project 949) class cruise missile submarines were designed specifically to exploit space reconnaissance data, firing their big Granit (SS-N-19) missiles while submerged. The bulge on the sail of this 'Oscar,' between the open cockpit and the second periscope, covers a Kasatka ('Punch Bowl') radome, used to receive satellite transmissions. The submarine also has a towed array, which provides it with a secondary means of tracking distant targets (from 'under the horizon'). (Crown Copyright)

Above: Aircraft could also exploit the naval reconnaissance satellites, albeit apparently only indirectly. These Tu-22M bombers ('Backfires') carry Kh-22M 'Kitchen' (AS-4) missiles, which could be fired from below the aircraft's radar horizon, linking their radar video back to the bomber.

Above left: The aero-ballistic Kh-15S (AS-16) would have been the ultimate Soviet bomber-launched anti-carrier missile, fired from below the radar horizon on the basis of satellite data. The missiles shown are the nuclear version useable against fixed targets, on a rotary launcher on board a Tu-22M. The bomber also carries a pair of Kh-22s, underwing. Thus a single bomber might be able to deliver as many as eight missiles (six Kh-15s on the rotary launcher, two big Kh-22s underwing), overwhelming U.S. terminal defences. Hence U.S. concentration, in the 1980s, on systems to fight the Outer Air Battle, to deal with bombers before they could launch their weapons at all.

Below left: As in the case of anti-ship warfare, the U.S. Navy found that it could gain considerable Outer Air Battle leverage by exploiting existing space systems, which had been designed with quite different roles in mind. This DSP (Defense Support Program) IR early warning satellite is being prepared for a pre-launch test of its balance. Its deployed solar panel is visible below the satellite body, and the barrel pointing up houses the IR sensors which the orbited satellite will point at the earth. (TRW)

Above: The Global Positioning Satellite (GPS) system has radically changed warfare, making it relatively easy to guide weapons to map references. That change in turn makes it more important that targets be located correctly, since a weapon can be guided very accurately to the wrong place. This GPS satellite is being tested prior to flight.

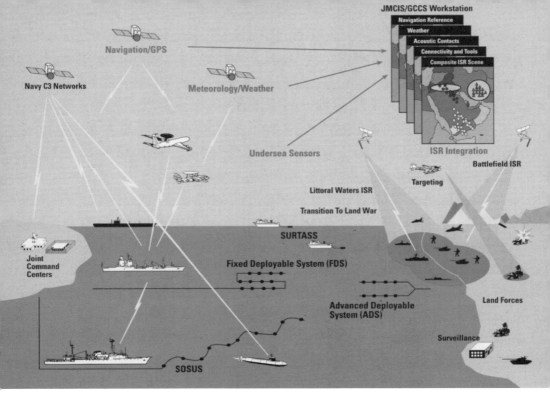

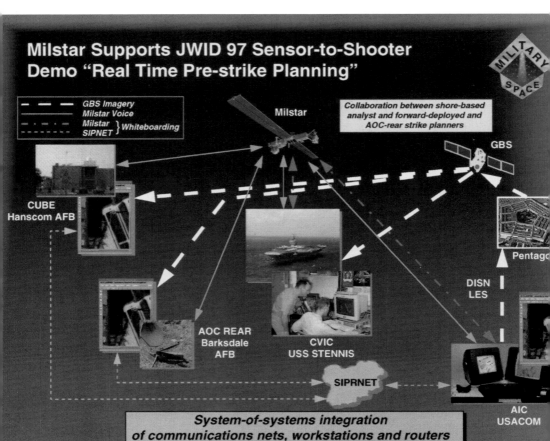

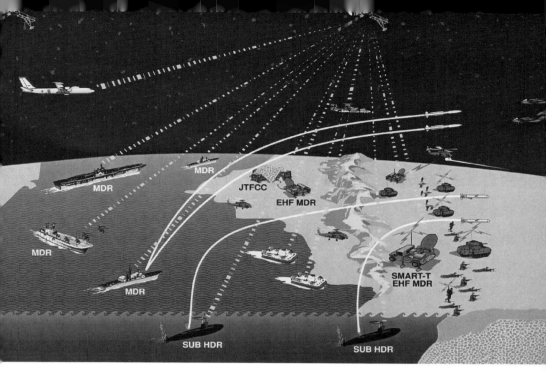

Above left: The U.S. view of the new style of warfare: numerous sensors, most of them outside the battle zone, build up a composite picture of the situation, which the fighting force uses. This TRW drawing emphasises fixed and semi-fixed underwater sensors: SOSUS (the Cold War Sound Surveillance System), the Fixed Distributed System developed at the end of the Cold War, and the new quickly-deployable Advanced Deployable System (ADS). Not shown are the various satellite sensors; the only explicit space contribution is that of communication satellites. (TRW)

Below left: The U.S. vision of over-the-horizon warfare: sensor-to-shooter connectivity, as demonstrated in a 1997 exercise. The sensors, which are not shown, are the space-based ones which feed the intelligence cell (AIC) in USACOM, the U.S. Command which ran the exercise. Its images are transmitted by high-speed line (DISN) to the Pentagon, and thence via Global Broadcasting Satellite (GBS) to users (shooters) such as the intelligence centre (CVIC) on board the carrier *Stennis*. However, this is not old-style command targeting. Rather, the users of the intelligence use satellite communication (in this case, Milstar), to query those holding the intelligence images (such as satellite photographs) they need. That is, they use the available data bases as though they are the outputs of their own tactical sensors. In an important sense the satellite-based sensors actually are the shooters' sensors (though in another sense they are not, since the shooters cannot command those satellites, at least not directly; in many cases, such as photo satellites, they must be satisfied with archived outputs). SIPRNET is a land-line computer net for classified data, which land users of the imagery use to query those holding it. (Lockheed Martin)

Above: The new concept of dispersed shooters concentrating their fire rather than their presence is illustrated in a littoral context, in which a Marine Expeditionary Unit (MEU) establishes itself ashore. The shooters are unified because they communicate via satellites (and also because they know their positions, thanks to GPS). Most have to fire well beyond their horizons, relying on remote sensors, many of which are necessarily space-borne (others may be aboard aircraft and UAVs). In this drawing, MDR and HDR are Milstar data rates (medium and high); submarines use the High Data Rate to minimise mast exposure, whereas surface warships can receive signals continuously. In this drawing, troops ashore use lightweight satellite communication antennas to communicate beyond their horizons, calling in fire support from the fleet offshore. Even the landing craft (LCACs) use satellite communications, since they operate beyond the horizons of the ships launching them (the alternative, HF radio, would be too easy to intercept and DF). Similarly, the E-2 Hawkeye radar aircraft reports via satellite, so that ships well beyond the horizon can fire anti-aircraft missiles at enemy aircraft menacing the troops ashore. The troops communicate ashore via VHF and standard radio nets; a SMART-T gateway on board a light vehicle connects them to the satellite. Via the gateway, the dispersed troops communicate with a distant Joint Tactical Force Command Centre – because it, too, has a satellite terminal. (Raytheon)

Above left: Tomahawk emerges from the sea, fired by the submarine HMS *Splendid*. Absolutely invisible is the massive, largely space-based, system which makes such a weapon effective: the remote sensors which detect targets and which enable a targeter to lay out the missile's course; the GPS system which guides some versions of the missile; the digital mapping satellites which provide the TERCOM reference material for others; the communication satellites which can provide missile guidance to the submarine itself. The Royal Navy uses Block III Tomahawks, which are GPS-guided (with TERCOM and DSMAC also available). Missiles are targeted by a Cruise Missile Support Authority (CMSA) ashore, whose flight plans are held aboard in the submarine's Advanced Tomahawk Weapons Control System (ATWCS). CMSA can provide updates via UHF communication satellite. The Royal Navy currently considers Tomahawk valuable mainly as a way of neutralising enemy command/control centres so as to make conventional air attack more effective: the limited number on board a submarine gain considerable leverage in this way. In future, however, the Royal Navy may arm major surface combatants with large numbers of surface-to-surface cruise missiles, as does the U.S. Navy, and they may become primary strike weapons. (MoD(UK))

Above right: The effects of satellites plus computers: a non-nuclear Tomahawk sent to destroy an aircraft inside a protective revetment. What made the test spectacular was that the Tomahawk had no onboard sensor with which to detect its target; it was merely directed to a designated point in space, which other sensors had determined was just above its target. That was very much an example of the new kind of space-based warfare. GPS was not yet in service at this time, so the Tomahawk was guided by terrain comparison (TERCOM) – which was based on digital mapping from space. Current Tomahawks use GPS instead. The test was somewhat poignant in that the target, an RA-5C Vigilante, was the last U.S. carrier-based strategic bomber and thus, in a sense, the Tomahawk's distant predecessor. (General Dynamics)

Under the terms of SALT I, a new treaty was due for signature in 1977. To get better terms, Nixon had to have new development programmes in hand, at least some of which might bear fruit by then, or otherwise the Soviets would have no incentive to come to terms. Fortunately the technology was already at hand to produce a U.S. cruise missile, in the form of work done for an abortive Air Force programme for an armed cruise decoy. Both air- and sea-launched versions of the new long-range cruise missile were ordered developed. Despite some hopes that a single weapon would be adopted by both the Air Force and the Navy, the two programmes diverged. The Navy version became Tomahawk, suitable for launch from a standard submarine torpedo tube. Should the Soviets reject SALT, Nixon could accelerate work on the new cruise missiles. Conversely, he could offer to trade them for some more attractive agreement on a future SALT II. Further in the future lay a new submarine-launched missile, Trident, and a new land-based ICBM, which became MX.

Thus Tomahawk was envisaged, at least at first, mainly as a bargaining chip. In part it was valuable as proof, to conservative Americans, that Nixon had not signed away U.S. interests in return for continuing detente. The U.S. Navy certainly did not press for the new weapon. However, by 1975 it was beginning to realise that Tomahawk could be quite valuable. Suddenly the U.S. ability to attack the Soviet Union could be distributed among numerous surface ships and submarines. If the Soviets were serious about tracking the U.S. naval threat, they would have to deal with hundreds of targets. To make sure the Soviets got the message, U.S. naval spokesmen began to emphasise the land-attack role of *Soviet* naval cruise missiles – to which Tomahawk was, in theory, equivalent.

Widely distributing nuclear Tomahawk was a direct attack on the centre directing Soviet naval strike forces. Armoured box launchers were mounted on board numerous cruisers and destroyers and, after the battleships were recommissioned, on board them as well. Tomahawk soon had a variety of non-nuclear versions; the Soviets could never be sure of whether a given ship was carrying the nuclear weapons they particularly feared. The situation was complicated further when ships were fitted with vertical launchers. During the design phase, the Mk 41 launcher was deliberately lengthened so that it could accommodate Tomahawk, as well as its original load, the Standard (anti-aircraft) Missile. Again, no one looking at a cruiser or destroyer equipped with vertical launchers could know whether nuclear Tomahawks were aboard.

Another response was the use of deception, which the centralised character of Soviet tactics made very attractive. The main Soviet search sensors, such as the satellites, were quite limited. Fairly simple devices could effectively simulate U.S. naval formations, as long as 'tattletales' could not close in to check that data visually. The U.S. Navy began to disclose the existence of devices such as acoustic simulators (to make a destroyer sound like a carrier), radar simulators, and air-deployed chaff bombs (which could produce the requisite radar returns).[33]

The motive was simple. Soviet tactics could work only as long as those directing the attacks retained their confidence in their sensors. As soon as they began to doubt them, as soon as they began to fear U.S. deception, they would hesitate to strike. They might well fail to mount attacks when necessary, and they might expend weapons needlessly. Deception mounted on a large-enough scale might, for example, attract valuable bombers into ambushes, or expose submarines as they surfaced to fire their long-range missiles. The deployed U.S. fleets formed special deception groups.

Deception on the desired scale required co-ordination through long-haul communications. HF communications could not possibly be satisfactory, as the Soviets might well intercept and DF them, picking out the fleet and sensing the deception in train. However, a satellite dish has a relatively narrow up-link. Its side lobe radiation is unlikely to travel beyond the horizon. Thus, except under extraordinary circumstances (a Soviet satellite in the up-beam), the link could not be intercepted. It seemed, at the beginning of the 1980s, that space communications would make deception work effectively. Space communications also made it possible for the fleet to adopt deceptive formations. As soon as the Soviets fielded their radar satellites, it was obvious that classical screens surrounding carriers would broadcast the identity and often the position of the carrier. Nothing but a high-value unit could possibly be surrounded by other ships. The fleet had already spread out in response to nuclear threats, but now it dispersed even more widely, out to the range limits of the HF Link 11. Moreover, there was still very reasonable fear that Link 11 transmitters would generate enough of a sky wave to be picked up by systems such as Krug. The obvious solution was to shift Link 11 to satellite transmission, which was applied during the 1980s. A satellite version of Link 11 could not be DF'd, and it could operate over a much wider area. Unfortunately the satellite link also imposes a considerable time delay, as the signal goes up and back from the satellite. That in turn limits the number of participating units. The slowest tolerable Link 11 speed, used to overcome noise, is 45 frames/sec (the normal speed is 75). The quarter-second delay imposed by transmission to and from a high satellite is equivalent to more than 10 frames. In the satellite version of Link 11 dummy pickets were inserted to free up the many slots between polling and response. That in turn drastically cut the number of participants which could be accommodated in the Link 11 net.

There was an alternative. An aircraft (*eg* UAV or airborne early warning) orbiting high above a battle group could keep all its dispersed units within its line of sight. Fitted with the appropriate transponders, these aircraft could become 'poor man's satellites', operating at UHF without imposing serious link delays.

The Soviet space systems inspired some other ideas. US-P could pick up radar signals. Carriers and the new Aegis ships had unique radars on board by which they could be identified, so the decision was taken to standardise on the SPS-49 air search radar. For example, it supplemented the Aegis radar (SPY-1) on board Aegis cruisers. Until an engagement began, the ships would avoid using any but their standard radars. All of this may seem unremarkable. However, before the advent of US-P, the U.S. Navy had followed a very different policy. The fear then had been, not detection by passive Soviet sensors, but rather successful jamming. The theory had been that the more variety in U.S. radars, the more difficult the Soviets' electronic attack. In fact SPS-49 itself had been conceived as part of a frequency-diversity programme.

At the very least, the fleet could be warned that a satellite was approaching. Given some knowledge of the satellite's search swath, the fleet could shut down characteristic emitters before a US-P could pick them up. US-A could not detect small ships, so if a carrier turned towards the satellite track, it might appear (to the satellite) as a very short ship (with length equal to its beam). Much therefore depended on the U.S. ability to track and identify Soviet satellites; information was provided by the Naval Space Surveillance System (NAVSPASUR).

Then there was ASAT. If the Soviets were coming to depend so heavily on their ocean reconnaissance satellites, perhaps destroying them would save the fleet from an early attack. By the late 1970s the U.S. Navy had already begun work on a one-man anti-satellite 'space cruiser' which could be lofted by a Poseidon (submarine-launched strategic) missile.[34] In 1978 the Joint Chiefs of Staff made Soviet ocean reconnaissance satellites the priority targets in a projected new ASAT programme, to be attacked upon the outbreak of war.[35]

Work on ASAT had continued despite the demise of the nuclear-armed Thor. Indeed, when he had ordered the Thor system terminated (May 1970), Deputy Secretary of Defense David Packard had called for both the Army and the Air Force to develop non-nuclear alternatives. The Army, which needed nuclear warheads to make its Spartan ballistic missile defence system work, was not interested, but in April 1971 the Air Force revived the earlier idea of an aircraft-launched missile. Air Defense Command proposed placing a modified Shrike (AGM-45) anti-radiation missile on its existing F-106 interceptor (Project Spike). Although nothing came of this proposal (analysis showed that it would not have been nearly as simple as imagined), the basic concept survived. Ultimately the Air Force proposed a direct-ascent 'miniature homing vehicle' (MHV) carried aloft by its F-15 fighter. It would fire the missile out of a steep near-sonic zoom beginning at about 40,000ft. Mid-course missile guidance would be inertial, based on calculated satellite position. The entire missile weighed only 1200kg (it was 5.2m long); its warhead, the manoeuvring kill vehicle, weighed only 16kg. It was expected to hit its target directly; it carried no explosives. The seeker was a set of eight IR telescopes. The fighter offered great flexibility, being able to attack satellites in a wide variety of orbits. However, its small missile could not hit satellites at high altitudes. The Soviet naval reconnaissance types were ideal targets, and the emergence of the U.S. ASAT system probably explains the Soviet attempt to build a higher-altitude radar satellite, at the end of their US-A programme. In 1978 it seemed that, to reach high-altitude satellites like Molniyas, the United States would eventually have to place an ASAT atop a long-range missile, such as Minuteman.

Unfortunately, it was by no means clear that attacks on a few satellites would solve the problem. It was obvious that the US-A and -P satellites were quite short-lived, so the Soviets had to have considerable stockpiles of replacements. Not only would they probably add satellites at the outbreak of war, but also, short of destroying the launch sites, the system could not be broken. Yet by the late 1970s the U.S. expectation was that any war would begin on a non-nuclear basis; clearly it would be very important to keep it that way (it is not clear that the Soviets agreed with this view). Only nuclear hits could destroy the key launch sites.

In 1985, for example, the CIA reported that just prior to the planned outbreak of war, the Soviets would expand the ocean reconnaissance net to four EORSAT (US-P) and seven RORSAT (US-A), compared to two of each in peacetime. This expansion would take no more than a few days, given the number of spare boosters (and, probably, spare satellites) at Tyuratam. Once the extra satellites were in place, the Soviets would be able to target ships anywhere in the critical area between latitudes 50 and 70 degrees North, data never being more than two hours old. Two years before the agency had reported that as many as seven US satellites were stored at Tyuratam, together with twenty-two Tsiklon-2 boosters (most of them for ASAT attacks on American satellites). It would take only an hour to fuel a Tsiklon-2, and the CIA estimated that the Soviets could launch six to ten satellites per day. A stored booster could be made ready for launch within seventy-two hours of a request.

Meanwhile the estimated cost of the ASAT programme kept climbing. Congress banned further tests in 1985, on the theory that ASAT was a barrier to further arms control (the missile had successfully destroyed a satellite). Early in 1988 the programme was cancelled altogether due to rising cost. Ironically, a Soviet naval officer later complained that the Soviets' ocean surveillance satellite system was unlikely to survive for long in wartime. He pointed out that any low-altitude satellite was vulnerable because it could hardly drastically change its orbit. Nor could satellites easily be hardened in any way. These statements may reflect Soviet expectations that their own ASATs would effectively destroy U.S. satellites soon after the outbreak of war.

Later it became evident that U.S. success against the Soviet surveillance system had been somewhat mixed. Certainly Tomahawk was an effective way of confusing the centralised Soviet targeting system. On the other hand, space communications were not quite as safe as had been imagined. As explained above, communication satellites added just enough Doppler to up-link signals that positions could be estimated (albeit not nearly as well as by earlier HF/DF systems). Furthermore, until the problem was discovered, satellite transmitters could be 'fingerprinted' (like naval radars), so a succession of space-Doppler 'cuts' could establish the approximate course and speed of a particular naval unit.

Milstar

The solution, which was not however conceived with this sort of interception in mind, was a new-generation satellite, Milstar, intended to replace DSCS. It operates at a much higher (EHF) frequency range; the U.S. Navy portion of the EHF programme is NESP (Navy EHF Satellite Communications Program).[36] Milstar treats the uplink signal as a series of digits rather than as an arbitrary waveform. The satellite encodes the digits and creates an entirely new digital signal which it sends down or across to another satellite. The result leaves no trace of the details of the uplinked signal.

Milstar began when Congress directed in 1981 that three ongoing programmes be consolidated: an Air Force strategic UHF satellite, a Navy tactical EHF satellite which was to useable despite nuclear explosions (TacSatCom II), and a classified Science Prime System (almost certainly a relay for reconnaissance satellite data). This combination insured that the system would employ heroic measures for survivability and wartime endurance (despite, for example, the destruction of ground stations) and that it would use stealthy anti-jam waveforms. A total of eight satellites in orbit (twenty would be bought) would provide both geosynchronous and polar coverage.

EHF (44 GHz for the uplink and 20 GHz for the downlink) makes for very narrow beams: the advantages are that less power is needed for transmission, smaller, more mobile antennas can be used, an enemy trying to jam the signal must be much closer to the beams and narrow beams in themselves make the signal more difficult to intercept. The waveform itself is a spread-spectrum frequency hopper (with a range of 2 GHz on the uplink and 1 GHz on the downlink). In addition, signals are interleaved (with three interleave lengths from which to choose), and they can be further encrypted. To do all of this, the uplinked signal must be reduced to its digital content before it is processed for retransmission. As a bonus, this reduction also cleans up the signal, so that no part of the downlinked signal is simply amplified noise. Spread-spectrum operation makes the downlink signal difficult to intercept. Indeed, the Milstar design emphasised stealthiness (LPI, low probability of intercept) over capacity. It actually provides less data/sec than its SHF predecessors. LPI is partly an automatic consequence of using a higher frequency, hence a narrower beam (given more or less fixed dish dimensions).

Security also demanded world-wide coverage without ground stations abroad. Much of the resource allocation usually carried out by a ground station is done automatically on board the spacecraft. The entire system needs only one primary and one alternate ground control side (Schreiver AFB, formerly Falcon AFB). Satellites not in view of the ground station are controlled via a cross-link between satellites. It operates at about 60 GHz, a frequency at which the atmosphere strongly absorbs signals; thus the cross-link cannot be intercepted from the ground. A related 'switchboard-in-the-sky' feature allows a user in any one beam (of a specified payload) to connect with a user in any other beam. It is also possible to cross-band (EHF to UHF). EHF was chosen for Milstar partly because of its advantages in the face of nuclear effects, which could black out satellites operating at lower frequencies. This was despite the greater propagation losses EHF signals were likely to suffer, *eg* in rain.

Work began in 1983. Milstar was considered important enough to receive overriding (Brick Bat) priority. It was sometimes described as a nuclear war-fighting tool. Its refinements, intended mainly to protect it against nuclear damage and interference, made it the most expensive communications satellite ever built, costing $1 billion for the original fully-equipped model. With the end of the Cold War, its high cost came into question. Congress ordered the programme restructured, the nuclear hardening reduced, and a medium data rate (MDR) added to supplement the existing low data rate (LDR) payload. Plans to include a classified payload (to transmit data from reconnaissance satellites) were dropped. It had already been installed on the first satellite, but the second had 800lbs of ballast instead. The Block II satellite, with MDR, costs $800 million.

The Low Data Rate is the 2400 bits/sec achievable in the Fleet Satellite, sufficient for encoded voice transmission. The Medium Data Rate (MDR) is 1.544 Mbits/sec, sufficient for a television channel (the T1 rate). Because the MDR channel is not spread as widely as the LDR channel, it is not as well protected. The satellite provides 100 LDR channels, equivalent to a total of 192 channels at 75 to 2400 bits/sec. The MDR modification adds thirty-two T1 channels (2 GHz bandwidth each), using eight steerable beams (they can be split into lower-rate channels). In addition, the satellite carries a UHF piggyback transponder carrying the Fleet Broadcast channel (25 kHz wide) and two AFSATCOM channels (5 kHz wide). One broad-beam (earth coverage) downlink antenna is supplemented by five agile spot beams. Downlinks are steerable and uplinks can be nulled to avoid jamming. The satellite weighs 4.67 tonnes.

Plans initially called for four geosynchronous satellites, plus three in Molniya orbits (to cover high latitudes) and three spares held in very high (above geosynchronous) orbits, providing them with vast footprints and with some protection against ASAT weapons which might be developed to destroy geosynchronous satellites. In the post Cold War Bottom-Up Review (1992) this plan was scaled down to four geosynchronous satellites and two spares, the first two of which are Milstar 1s (launched 7 February 1994 and 6 November 1995). The other four (the first launched in the spring of 1999) are Milstar 2s with MDR. Later five Advanced (but cheaper) EHF satellites are to be launched by Medium Launch Vehicles. The system may eventually include a Polar Adjunct. Transition to Milstar is to be completed in 2003, freeing both DSCS Channel 1 and the wideband UFO channel to tactical users.

Because the numbers were cut so drastically (as of early 1999 only the first two Milstars were in orbit), EHF transponders had to be piggybacked on the later UFO satellites (numbers 4 through 9) to maintain full coverage. UFO number 8 was used during the Gulf War. Compared to Milstar, these packages have fewer channels and fewer modes, and they are not cross-linked to the Milstar satellites. They have Earth coverage and 5 degree spot beams, but no high-gain beams.

With the cancellation of the Molniya-orbit Milstars, some alternative means of extending EHF communications to the polar area was needed. EHF packages (the 'polar adjunct') were placed aboard two new SDS satellites.[37] Also, Milstar had been intended to take over the reconnaissance satellite cross-link role, but the planned classified payload was cancelled. At least two more SDS satellites were, therefore, launched (2 December 1996 and 29 January 1998).

The Navy EHF terminal is USC-38, the 'U' indicating 'universal' or multi-service. The shipboard antenna is a 34.5in dish in a 56in radome; submarines use a 5.5in stabilised dish in a 7in radome. Each terminal can handle up to twelve channels. There is current interest in alternative antennas with lower radar cross sections, such as flat plate phased arrays. As of 1999, an Advanced EHF satellite was planned as a

Milstar follow-on; the first is to be launched in 2006. This concept came out of the same Bottom-Up Review which drastically cut the number of planned Milstars. It is to be far lighter than Milstar (about 4300 rather than 11,000lbs), but it is also to offer higher data rates and greater system capacity; the appetite for capacity seems endless. There would be no UHF package. There will be 192 to 256 LDR channels, with a new 4.8 kbits/sec data rate, and it will be possible to combine four such channels into a 19.2 kbits/sec uplink channel. Each MDR channel will be able to support up to 8 Mbps. Overall capacity will be five to ten times that of Milstar II, at 200-400 Mbits/sec per spacecraft. Parallel 18-month System Definition Contracts were awarded in August 1999 to Lockheed Martin Missiles and Space and to Hughes Space & Communications Co.

Satellite ASW

The Soviets must have been bemused by White Cloud and by the other major efforts to track their surface fleet. Since they considered their submarines the core of their fleet, it just did not make sense to them that the U.S. Navy would invest heavily simply to detect their surface ships (a U.S. fleet built around surface warships might think quite differently). Beginning in 1976, the Soviets began writing about supposed U.S. attempts to detect their submarines from space. Americans often read such accounts as evidence of *Soviet* attempts to develop space-based systems to detect U.S. ballistic-missile submarines.[38]

However, from a Soviet naval point of view it was perfectly rational to concentrate on surface ships, such as aircraft carriers, which really did represent a large fraction of Western naval striking power. Clearly Western ballistic missile submarines were important, but they were surpassingly important only if one took a very Western view, that strategic (not merely nuclear) strike systems were the essence of the balance of power. Many in the Soviet Union clearly did not make sharp distinctions between strategic and theatre and tactical weapon systems.

Westerners, particularly those concerned with strategic warfare, knew that their deterrent depended heavily on ballistic missile submarines. These craft became more important as the Soviets gradually developed what seemed to be the ability to destroy U.S. strategic missiles in their silos, while the Soviets' massive national air defence system seemed to threaten the viability of the U.S. strategic bomber force.

In the 1970s and early 1980s the U.S. Air Force showed particular interest in Soviet non-acoustic (space-based) submarine detection, just as its own future missile system (MX) encountered severe political problems. Anything which put the submarine deterrent in question would certainly help MX. This special Air Force interest undoubtedly coloured public interpretations of Soviet naval space systems, since the Air Force was the main source of leaked information about Soviet military space activities. The Air Force's main argument against the submarines was that the Soviets were developing the means to search their operating areas rapidly enough to find them. Once found, the submarines could be destroyed at the outbreak of war, possibly by a missile barrage. There were two wide-area search options. One was acoustic, like the Western SOSUS (Sound Surveillance System). Against any such option, the U.S. Navy could claim effective silencing as well as the consequences of geography: in most places systems emplaced in Soviet-controlled areas could not look out through the open ocean towards the areas in which Western strategic submarines were likely to operate.

The alternative was a space-based system based on some non-acoustic technique. By the late 1970s, there were rumours of Soviet attempts to detect Western submarines from space. The sheer potential for disaster, should the Soviets really have learned how to detect submarines from space, made non-acoustic detection a very hot (albeit highly classified) topic through the latter years of the Cold War. If the principle of the non-acoustic system was unknown to the West, moreover, Western navies would be unable to shield their submarines from it. On the other hand, the West was also vitally interested in non-acoustics. Conventional sonar systems were unreliable at best; if some alternative could be found, the massive Soviet submarine fleet might be neutralised.

Quite aside from any possible space applications, non-acoustics has a long history both in the West and in the Soviet Union. For example, during the Second World War the Allies relied exclusively on what would now be called high-frequency sonar, but in 1945 a German Type XXI U-boat could fire torpedoes from beyond the usual sonar range. The situation was so bleak that the formerly classified U.S. ASW papers of the late 1940s are full of non-acoustic techniques. The advent of lower-frequency sonar saved the day, and the exotic non-acoustic ideas were shelved. That is aside from the very commonly used non-acoustic techniques: ELINT (*eg* HF/DF and detection of periscope radar emissions), periscope or snorkel detection radar, and magnetic anomaly detection (MAD).

During the Cold War, the Soviets were often observed making what seemed to be ASW experiments which could not be explained in terms of Western acoustic techniques. Clearly they were interested in non-acoustics, possibly because their sonars were not particularly good. We now know, for example, that many Soviet ASW ships carried thermal wake sensors. Towards the end of the Cold War, many

Soviet submarines sprouted what seemed to be non-acoustic sensors, generally mounted on their sails. Given the location of these devices, they were presumably intended to detect and then to follow some sort of wake associated with a submarine. However, even this apparently obvious interpretation is open to question.[39]

The history of non-acoustics is littered with false breakthroughs. For example, in the early 1960s (and also much later) it seemed that blue-green lasers could easily penetrate to considerable depths. An airborne or even a spaceborne underwater laser radar seemed to be a real possibility.[40] Lasers certainly could detect static underwater features, including mines, down to a depth of about 150m (depending on water conditions). However, at least for the moment, it seems that they cannot reliably detect moving objects. To achieve a sufficient signal-to-noise ratio, the laser spot must dwell for too long at a given place. As a consequence, the net search rate – at least for the present – is apparently too slow to detect a moving object.

The great hope of non-acoustics has always been that a submarine moving through the sea leaves a subtle yet detectable signature. For example, because water is not compressible, any object buried in the sea pushes water aside – and up. The resulting 'Bernouilli hump' is not very massive, but in theory it might be detectable.[41] Some years ago radar satellite measurements of the height of the sea surface clearly showed underwater terrain features such as mountains and deeps. In theory, those features should have included submarines – except that the satellite measurements had to be averaged out over a considerable time to eliminate the effects of wave motion. Similarly, a submerged submarine pushing water away as she moves produces a wake at the surface of the sea. The shallower the submarine, the stronger the wake. In theory the wake should be detectable, because it is a regular disturbance of an otherwise random sea surface. Unfortunately many non-submarines create submarine-like wakes.

In May 1999 it was alleged that in 1997 an American physicist, Peter Lee, working at Lawrence Livermore (the nuclear lab) had betrayed a U.S. radar wake detection project to the Chinese. Supposedly the laboratory had been investigating the use of its supercomputers to distinguish faint submarine wakes from surrounding waves. As the system was described in the press, synthetic aperture techniques were used to image the surface of the sea. By declining to support prosecution, the U.S. Navy was able to avoid saying whether the programme had been particularly successful. Reportedly the scientist showed a seminar in Beijing a radar image of a *surface* ship wake (hardly a subtle phenomenon), which he claimed he then destroyed.[42]

Claims for this type of system actually go back about two decades. In 1978 the U.S. Navy briefly operated an experimental radar ocean imaging satellite, Seasat. Reportedly it sometimes saw unexplained wakes which were associated with submarines.⁴³ Some suggested at the time that the Soviet active radar satellites were also intended to detect wakes, but we now know that was not true. The great barrier to satellite wake detection seems to have been that many things other than submarines produced wake-like effects. It is by no means clear that this problem has been (or can be) solved, and in recent years research seems to have concentrated on a better understanding of the ocean surface. In theory that would allow natural phenomena to be subtracted out, leaving submarine wakes. In May 1999 the British press reported that Lee had been a prominent member of a joint US-British Radar Ocean Imaging Program (ROIP), presumably intended specifically to investigate the phenomena involved.[44]

The submarine also creates internal waves, *ie* waves in the layered structure of the sea. They cannot be detected directly, but in theory they should affect the surface wave spectrum, and they may be observable because they modulate ripples on the sea surface, thus changing its radar reflectivity. According to a recent Russian account, during the 1980s experiments were done using a laser to sample sea motion and to detect a submerged submarine by the changes it imposed on the energy spectrum of surface waves. The programme, pursued by the Kometa bureau developing space-based anti-ship reconnaissance, was called Spektr-RM. Apparently the Soviet Union (and research funding) collapsed before anything beyond interesting results could be obtained.[45]

The submarine leaves a thermal wake. The temperature of the sea varies with depth. A submarine passing through the sea presumably disturbs the structure of the sea, leaving a thermal wake at the surface – which is detectable. This is an application of the internal wave idea. In addition, the heat from the submarine's reactor is dumped into the water around it (not to mention the radiation from unshielded parts of the reactor). Such ideas were certainly valid, and they were enthusiastically pursued. The Soviets even tried a thermal detector aboard Be-12 seaplanes, and they convinced themselves that it should have worked. In fact effects like sun warming swamp the submarine's thermal wake; according to a Russian account of trials in the Mediterranean, the detector was useful mainly to distinguish land from water – and there were much simpler ways of doing that.[46] Even so, during the Cold War many Soviet surface warships were equipped with a thermal wake sensor called MI- 110.[47] This particular idea was publicised in the James Bond movie 'The Spy Who Loved Me', in which the criminal mastermind used a satellite-based sensor to find the strategic submarines he hijacked.[48] The Soviets seem to have suspected that the United States was using its DSP strategic warning satellites to detect submarine wakes. They certainly wrote publicly about the use of IR sensors.[49] A submarine also leaves a chemical wake, as particles of paint or other material flake from her hull. Her reactor irradiates the water around her hull. Although water itself cannot be made radioactive, some substances certainly can become radioactive in seawater, and in theory their presence can be detected, perhaps even remotely.

It also seems likely that the surface wave structure can be sampled using radar, and reportedly the internal waves affect the thermal reflectivity of the surface waves. Thus a microwave scatterometer might, in theory, detect them.[50] The faster (and shallower) the submarine, the stronger the effect.

There were some indications of Soviet experiments using space-based radars in the early 1980s. The U.S. press reported radar submarine-detection tests on board the Salyut-7 manned space station in 1983-4, but this identification was clearly speculative, and it could be associated with the U.S. Air Force.[51] Current accounts simply mention 'naval experiments' associated with Chelomey's Almaz-K radar spacecraft. They involved a big side-looking radar with a 10 to 15m resolution, which eventually flew in the late 1980s. This device was probably intended for surface ship identification, but its resolution is also roughly what Seasat had, hence might, in theory, have been intended for submarine detection.[52] Certainly it is arguable that the passive EORSAT would eventually have fallen victim to improvements in radar, particularly LPI techniques. In that case an imaging radar would have been a natural replacement, performing the target identification function.

Even more subtle effects can, in theory, be exploited. Sea water conducts electricity. Moving a wire (or any other conductor) through a magnetic field will induce a current in that wire. That is, after all, the basis of electric generators. Associated with that current will be a magnetic field. Surely, then, as a submarine creates a wake, the turbulence of the wake will create electric and magnetic fields. They would be in addition to the magnetic anomaly created by the submarine herself. Normally such effects are detectable only at short ranges, up to a few thousand yards. However, there are indirect alternatives. The magnetic anomaly detector on board a maritime patrol aeroplane measures the effect of a weak magnetic field on the energy levels of atoms. Atoms much closer to that field – in the sea – ought to feel stronger effects. The energy levels can, in theory, be probed from a considerable distance.[53] In the early 1990s, there was some talk that 'far-field magnetic anomaly detection' was extremely promising. Presumably that meant space-based mechanisms.

Yet another phenomenon is bioluminescence. In places the sea is thick with plankton, and they react to disturbance. Their luminescence is sometimes quite visible – but in many places it is not. Moreover, the deeper the submarine, the deeper in the water will be the layer of plankton affected by its passage. There will be more water above that layer, to absorb the light the plankton produce.

At least during the Cold War, the Western verdict on proposed space-based non-acoustic detectors seems to have been that all the phenomena were swamped by noise, or else (like bioluminescence) were effective only in a few places. On the other hand, an understanding of potential means of non-acoustic detection would make it possible for a submarine commander to use ocean conditions more effectively to remain undetected. On this basis the U.S. Navy conducted very aggressive non-acoustic research as part of its strategic submarine security programme. After the end of the Cold War, a semi-official Russian account indicated that the United States had worked 'rather successfully' on space-based submarine detection during the 1980s.[54] This statement cannot be taken at face value. During the Reagan Administration there was a concerted U.S. disinformation programme, intended to convince the Soviets that the United States was developing weapons which they absolutely had to counter. It was fairly clear that the Soviet leadership would credit U.S. claims of technological prowess, even if their scientists told it that the claims were groundless. That certainly applied to 'Star Wars', the Strategic Defense Initiative, and could well have applied to U.S. non-acoustic submarine detectors. The advantage of spreading claims of U.S. prowess in this field would have been twofold. First, the Soviets might well be induced to scale-down their very dangerous investment in a large submarine fleet. Second, the Soviet counter-programme, particularly if it had to involve super-computers, might well preclude investment in other more productive fields. Disinformation was generally in the form of apparent investment in highly classified ('black') programmes. To make matters more complicated, many black programmes, such as those for stealthy aircraft, were quite real.

The future of non-acoustics, and therefore of space-based submarine detection, is difficult to evaluate because the sheer number of potential submarine-detection phenomena is so enormous. The sea has a complex temperature, chemical, and even biological structure, and any large object moving through it inevitably creates many disturbances. Some must be visible from outside the sea. Yet the same complexity also generates many disturbances which might easily be mistaken for a submarine, or which may mask the subtle signature generated by a submarine. Continued heavy investment both in submarines and in sonars, both in the United States and abroad, suggests that no one has been able to achieve reliable non-acoustic detection from the air, let alone from space.

Space systems do, however, have more conventional roles in submarine warfare. The U.S. Navy is currently developing a 'net-centric' approach to ASW, in which ships automatically pool their data via UHF satellite links. This was first tried, in very elementary form, in Fleet Battle Experiment Echo (March-April 1999, off the California coast). A Web-Centric ASW Network (WeCAN) was set up by creating a site on the SIPRNET, analogous to a website on the Internet. Information posted included environmental data, search plans, commander's intentions – and the tactical picture. The site also supported a chat room, as in the IT-21 systems (described below), allowing on-line discussions of prosecutions, contact reports, and tactical decisions. Since submarines were connected to the SIPRNET via their UHF satellite systems, they could participate. Issues like waterspace management and blue-on-blue deconfliction could be resolved relatively easily. Because all communication was by text or by graphics, there was none of the confusion normally associated with voice links.

Space systems already provide valuable environmental data in support of ASW. During the 1980s the Soviets began to operate their new 'Typhoon' class ballistic missile submarines under the Arctic ice. The submarines would not have fired through the ice, because rising missiles would have been damaged trying to penetrate. Similarly, submarines could not fire through holes in the ice, because pieces of ice might drift over missiles as they approached the surface. Rather, the submarines would have surfaced in natural holes in the ice (polynyas) before firing. The 'Typhoons' seemed to have been designed specifically for this role, since they popped their missiles up from their tubes. Earlier Soviet strategic submarines floated their missiles up through the water before they ignited; they could not have fired while surfaced. It is possible that the Soviets chose under-ice operation on the theory that eventually open-water operation would be impossible, given projected advances in space-based ASW. The Arctic ice would have protected the 'Typhoons' from Western aircraft – and, it is tempting to imagine, from Western satellites. Like their Western rivals, the Soviets tended to mirror-image, supposing that the West would develop an equivalent to whatever they had. Whether or not that was the case, Soviet space assets were essential to the 'Typhoon' system. Earth observation satellites detected the polynyas through which the submarines would have surfaced to fire in wartime. Without that observation, the submarines could not have positioned themselves to fire in an emergency.

For that matter, oceanography conducted from space clearly has important ASW applications. Astronauts on early Shuttle missions, for example, could actually see eddy fields in the sea, in which a submarine might hide from hunters. By that time the U.S. Navy was already very much interested in satellite observation of the sea. In June 1978 NASA launched Seasat, the first satellite designed to remotely sense ocean conditions using a synthetic aperture radar (SAR). It was intended to define requirements for a follow-on operational system. Seasat orbited at 761 × 765km, at an inclination of 108 degrees. The satellite carried two radars, an L-band synthetic aperture radar (SAR) observing a 100km swath (with 25m vertical resolution), and a Ku-band radar altimeter (10cm vertical resolution, to measure wave height). A satellite scatterometer measured surface winds. A scanning microwave radiometer (600km swath) measured surface temperature, atmospheric water vapour content, rain rate, and ice coverage. Another (IR/visual: 0.25-0.73 and 10.5-12.5 microns, 1900km swath) radiometer identified cloud, land, and water features (to register data from the other sensors) and provided thermal images of the ocean. Unfortunately, Seasat's electric power system died on 10 October 1978, after only forty-two hours of data had been collected. Even so, Seasat seems to have inspired a Navy programme for a Navy Remote Ocean Sensing Satellite (N-ROSS, probably intentionally close to the acronym for White Cloud, the Naval Ocean Surveillance Satellite). Presumably, like Seasat, N-ROSS would have had both a SAR and a radar altimeter. It was cancelled in August 1984 due to rising cost.

Much of the Seasat/N-ROSS story was kept quite secret.[55] A U.S. capability to remotely determine ocean conditions, not only at the surface but below it, would have been extremely valuable not only in ASW but also in protecting U.S. submarines from exposure. Sonars are notoriously variable in their performance, because ocean conditions are so variable, not only from place to place, but also from day to day in any one place. Normally a ship or submarine takes daily measurements of the sound velocity profile (SVP), the way the speed of sound varies with depth, wherever it is located. However, these data may be almost useless if the ship or submarine is passing into an area of suddenly varying SVP, such as a major ocean current. Hence the great value of remote sensing of what submariners call 'underwater weather'.

Given the high cost of the full Seasat package, the U.S. Navy launched a much more limited satellite, Geosat (Geodetic Satellite), on 12 March 1985. It carried a precision radar altimeter to measure sea surface (and other) heights (with a horizontal resolution of 10-15km and a vertical resolution of about 3cm) and a wind sensor. The satellite's orbit roughly matched that of the earlier Seasat (775 × 779km, inclination 108.1 degrees). The satellite's Primary Mission lasted from March 1985 through September 1986, after which it began an Exact Repeat Mission to refine the earlier data. Data for the initial mission was classified, then released in the 1990s (all data south of 30 degrees South was released in 1992, and all remaining data was released in July 1995).

The data were important partly because they were essential to measurement of ocean density, hence SVP. Average sea height is determined by the shape of the ocean bottom, by the local strength of gravity, by tides – and by water density. The large radar footprint averages out waves to give the average height of the sea at any given place. Averaging over time, which is possible given frequent satellite revisits to every place on the ocean, makes it possible to estimate the effects of the different components. For example, when filtered Geosat data were released in 1995, they included a map of the ocean floor, as though the basins had been drained – because to some considerable extent the height of the surface reflects that of the bottom. While the Geosat mission ran, the U.S. Navy also invested very heavily in powerful computer oceanographic models, which it used to interpret the data involved.

Geosat data also provided a detailed map of the strength of gravity around the Earth – the variation in which affected inertial navigation systems, on board missiles, aeroplanes, ships, and submarines. This application alone may account for the decision to classify the Geosat data. This data became substantially less important with the rise of GPS systems in the 1990s. After the gravitational data were declassified, Honeywell used them to correct its aircraft inertial navigation systems. It found that uncorrected systems made substantial errors if aircraft flew along the major Pacific Ocean trenches.[56] Note, too, that U.S. submarines are beginning to navigate using passive gravimetrics.

The success of Geosat seems to have inspired an attempt to revive the more ambitious N-ROSS. About 1986 plans called for launching it on a Titan 2 rocket in September 1990. Then it died again (of high costs), and a 1987-8 attempt to revive the programme (for a 1991-2 launch) failed. Instead the original Geosat remained in service until both its tape recorders failed in January 1990. A Geosat Follow-On (GFO) contract was awarded in September 1992; the satellite was launched on 10 February 1998 (there are options for two more). The GFO orbit roughly follows that of GFO (785× 788km, inclination 108.1 degrees). GFO is to build on Geosat data. Given current computer programs, it is now possible to translate ocean surface heights into the SVPs ships and submarines need. A second satellite system, Coriolis/Windsat, is to measure surface winds. The Naval Earth Map Observer (NEMO) is a hyperspectral observation system tailored to littoral operations. These systems are needed because civilian weather satellites, which would seem to provide the same services, concentrate on the land, not on the open sea.

NINE

Enter Tomahawk: OTH Targeting

About 1970 the U.S. submarine community proposed a solution to the Soviet surface shooter problem: a fast missile-firing submarine which could work with a carrier battle group. The idea had been germinating for some time. First, the idea that a fast nuclear submarine should escort carriers had arisen about 1964, with the unpleasant discovery that the new Soviet nuclear submarines, fast enough to intercept a carrier, were operational. The sheer sustained speed of carriers, particularly nuclear-powered ones, was no longer enough to protect them. The U.S. submariners' solution was to build a submarine fast enough to escort the carrier, dealing with Soviet interlopers as they appeared: the *Los Angeles* class.

In 1968 this class was ready for production, but to many it seemed extremely expensive. To settle the issue, the Chief of Naval Operations convened an Ad Hoc panel of submariners. The committee is best known for having approved the *Los Angeles*, but it also produced a study of submarine weaponry looking towards the design of a new submarine, to be bought under the FY75 programme. The submariners pressed for a new dual-purpose weapon they called STAM, a ballistic rocket (like the existing SUBROC) to carry a dual-purpose torpedo about 30 miles. The emphasis on anti-surface as well as anti-submarine performance presumably reflected the new reality of a serious Soviet surface presence in the Mediterranean. The range was selected because it was within the capacity of current and projected submarine sensors, and because it would probably keep the attacking submarine beyond the range of Soviet counterattack. A sketch in the 1 November 1968 report of the Ad Hoc Panel showed a FY75 submarine armed with twelve vertical tubes (for STAM) abaft her sail, plus the usual torpedo tubes.

In September 1970 Admiral Elmo Zumwalt, Jr, the then CNO, reported to the Secretary of Defense that among planned counter-measures against increasingly aggressive Soviet surface forces was a projected anti-ship missile submarine, then in the conceptual study stage. Tests of tactics for nuclear submarines to escort U.S. carrier groups were planned, and to make them practicable improved nuclear submarine communications were being developed. Zumwalt did not mention satellites in this context. A submarine operating with the battle group would need a much longer reach, because 'Shaddock'-armed ships would be able to fire from much greater ranges. STAM soon evolved into ACM, the Advanced Cruise Missile, a weapon which could outrange 'Shaddock'. Although carrier aircraft could reach even further, every sortie spent on defeating Soviet surface forces would detract from the carrier's much more important strike role against targets ashore.

Thus ACM would be a natural extension of the escort role envisaged for the coming *Los Angeles* class. To carry out its dual role of anti-submarine and anti-surface protection, the submarine had to be able to fire while submerged, and it could not surface to provide its missiles with mid-course guidance, as in that case it would lose its sonar picture. Thus the outgoing missiles would have to rely on whatever data was fed in at launch time. The target would move while the missile was in flight. To minimise the area it would have to search, ACM had to get to the target in minimum time: it had to be supersonic. The combination of required performance and substantial warhead weight made for a missile much larger than a conventional torpedo; the projected cruise missile submarine would have been far larger than a *Los Angeles*. As it happened, by this time Admiral Rickover's Naval Reactors group was developing a big 60,000-horsepower plant, well-suited to exactly such a submarine. Cynics suspected that the whole point of the missile project was to justify a big new submarine to carry Rickover's big new reactor.

Whether or not that was the case, the missile project had far-reaching implications. The firing submarine could not detect its targets using its own sensors. On the other hand, by 1970 U.S. Naval Intelligence was tracking Soviet surface warships as a way of warning battle groups of impending danger; OSIS and the FOSIFs were being built. The OSIS/FOSIF combination was already expected to report the positions of key potential missile targets to the coming TFCCs on board carriers. The submariners envisaged an equivalent system for submarines, which would need their own satellite link (to be called SSIXS, the Submarine Information Exchange System). It would be fed by Submarine Targeting Terminals (STTs) at the existing Submarine Operating Authorities, which would be linked to the fusion centres. The STTs would also feed data into the old VLF Submarine Fleet Broadcast, but SSIXS offered a much higher data rate.[1] It would carry a stream of coded target locations. Periodically a submarine could extend a mast to receive a burst of data. The submarine's combat system would keep track of target course and speed data, projecting ahead target positions so that missiles could be fired effectively.

Zumwalt killed the big cruise missile submarine in 1971.[2] At this time the Navy was already beginning to develop the subsonic turbojet strategic cruise missile which became Tomahawk, which could fit into a standard torpedo tube. Zumwalt ordered a tactical version, using a Harpoon seeker, developed to replace the projected supersonic weapon. The existing *Los Angeles* class could carry the new missile. An encapsulated version of Harpoon would be developed as an interim submarine-launched anti-ship missile. Meanwhile, the SSIXS project proceeded. In 1974 it was demonstrated in a project called Outlaw Shark. Rota was able to provide a submarine with timely targeting data, in the form of target positions no more than a few minutes time-late.

Because OSIS was designed to project ahead target movement, its product was apparently exactly what the new missile needed. Anti-ship Tomahawk took about half an hour to cover the 250 miles to its target area. At 30kts, a target might move as much as 15nm from its position at launch time. A missile fired at the last known target position would most likely spend a considerable time weaving back and forth through the area of uncertainty around that last known position. The odds were excellent that it would be shot down passing its target, well before it ever locked on. With the position predicted, albeit only statistically, the odds that Tomahawk would find its target improved dramatically. The system was likely to work because tracking was passive and because it never concentrated on any particular potential target. Unaware that an attack was imminent, the target was likely to continue on the course and speed detected by the OSIS system. By this time the FCC/TFCC combination was being developed in earnest. In a 1974-5 Mediterranean experiment, Outlaw Hawk, the carrier *Kitty Hawk* received targeting data from Rota. Unlike the submarine, the carrier could receive messages on a continuous basis. Operation roughly paralleled that of the existing short-range Link 11 system. That is, the complete tactical picture was built up out of a series of updates. The carrier used a special computer to interpret and collate the messages received.

In both cases, the key was a satellite link, because only a satellite could convey enough information quickly and reliably enough. Because OTH data were based on electronic intelligence collected by national and theatre systems, they had to be carried by a special channel. In the developed version of the targeting system, the link from FCC to carrier TFCC was TADIXS (Tactical Data Information Exchange System), carried by a 2400 bits/sec (fast teletype) channel of the Fleet Satellite. It had two components: TADIXS-A for fused (processed) data, -B for urgent raw data, passed on a broadcast basis. Both used the same channel (TADIXS-A was interrupted for -B broadcasts), and -B used a dedicated processor, TRE (Tactical Receive Equipment). All messages were formatted, so that they could be inserted directly into a ship's OTH picture-compiling computer in her TFCC.[3] These broadcasts are still used, albeit in modified form.

To back TADIXS up, the satellite devotes one of its 25-kHz channels to a two-way TACINTEL (Tactical Intelligence) circuit which connects the shore OSIS elements with the intelligence centre on board the carrier. Using it, analysts can exchange data and analyses. Their messages are necessarily in text form, not computer-formatted. TACINTEL also carries urgent tactical messages: attack warnings (Indications and Warning). These messages are sent in special computer-readable formats, and take priority over other intelligence traffic. The system can carry them directly from ship to ship. This application was clearly developed as various forms of OTH targeting evolved. For example, the Indications and Warning messages included notice that particular Soviet missile bombers were taking off to attack the fleet. Like TADIXS, TACINTEL feeds teletypes and computers, normally operating at 2400 bits/sec.[4] The TACINTEL net can support up to twenty-three users, enough for the entire carrier, flagship, and large-deck amphibious ship force (currently it supports twelve Air Force, ten Navy, and one National Command Authority channels).

It was assumed that nothing short of a carrier or a major flagship could accommodate the personnel and equipment needed not merely to display the OTH picture, but also to process it for local use. The Tomahawk shooters associated with the carrier might be spread quite widely. To convey the carrier's assembled targeting picture to the shooters, a second Fleet Satellite circuit, OTCIXS (Officer-in-Tactical Command Information Exchange System), was provided. Because many Tomahawk-armed ships might lack the appropriate computers, OTCIXS carried both teletype messages and messages intended for direct insertion into a computer.[5] The system became operational in 1982 (FY83). The OTH-T satellite net within a battle group is now called BGIXS, the Battle Group Information Exchange System.

OTCIXS had another significance. In the 1970s many U.S. battle groups operated in classical ring formations, with the carrier in the centre. They were instantly recognisable from space. In the 1980s the U.S. Navy took the problem much more seriously. It adopted very widely dispersed pseudo-random formations ('4-W'). That was possible partly because the escorts' defensive weapons improved dramatically, with the advent of Aegis, and partly because satellite links (OTCIXS among them) could still tie together a force. As part of this change, a satellite (UHF) version of the tactical link (Link 11) was developed. The widely-dispersed formation was inherently deceptive, and its adoption coincided with a wider interest in Operational Deception (OpDec). The Fleet Satellite also carried SSIXS, a one-way broadcast at 2400 or 4800 bits/sec. The SSIXS broadcast included the content of the earlier submarine fleet broadcast, but it added OSIS-based targeting information.[6]

Tactical Exploitation of National Capabilities

By the mid-1970s the U.S. services all badly wanted access to the output of 'national' intelligence-gathering systems. They all created TENCAP (Tactical Exploitation of National Capabilities) programmes. In the Navy's case, the catalyst was the need for over-the-horizon targeting data. The Navy TENCAP programme probably began about 1977 (the key Naval War College study was completed about 1976); it was formalised in a 1978 Congressional directive, which was described at the time as a Congressional initiative. In 1980 President Carter approved PD/NSC-37, a National Space Policy which directed that the Air Force be directly involved in designs of future 'national' space systems. The Congressional report on the 1980 Defense Appropriations bill report specifically emphasised the need to take TENCAP into consideration in determining the capabilities of all 'national' systems. TENCAP soon became very important to the Navy: in 1986 it was claimed that Navy used over three-quarters of all the tactical data gathered by 'national' space systems. There is also an inverse process, National Exploitation of Tactical Capabilities (NETCAP).[7]

The agonised U.S. decision to create TENCAP reflected a point of view very different from that of the Soviets, in which national and local war-fighting seem not to have been nearly so clearly distinguished. From about 1969 onwards the Soviets enjoyed an invaluable source of intelligence in the form of broken U.S. codes, courtesy of the Walker spy ring.

All three services developed their own broadcasts of TENCAP-derived data. The Navy version is TADIXS, the Air Force equivalent is TRAP (Tactical Receive Equipment Applications), while the Army's is TRIXS (Tactical Reconnaissance Exchange System). Each operates at the high level of classification associated with signals intelligence. At a lower level of classification, the Air Force operates a Tactical Information Broadcast System (TIBS) to provide near real-time intelligence. All of these systems use encrypted (covered) UHF satellite channels.[8] The non-naval systems became more and more important to the fleet as it became more involved, in the 1980s and 1990s, in operations near and over the shore.

Navy interest in TENCAP was tied directly to the needs of the Tomahawk anti-ship missile. By 1975 the U.S. Navy was about to place Harpoon in service. Its range was at the very edge of what a ship could get out of her own sensors; for example, one concept was to use a destroyer or frigate sonar (SQS-26CX) as an *under-the-horizon sensor*.

Later low-frequency components of the standard shipboard ESM set, SLQ-32, would be extensively modified to exploit enemy emissions leaking over the horizon. None of this was likely to suffice for the longer-range Tomahawk, which was already in prospect. The most attractive possibility was to detect the surface-wave emissions of Soviet long-haul HF radios.[9] This technique was implemented in Classic Outboard, a combination HF/DF and code-breaking system.[10]

For the most part, then, Tomahawk would be fired at targets beyond a ship's horizon. Since the main point of the missile was to deal with anti-carrier forces, targets would generally have to be detected well beyond the horizon. The problem of Over-the-Horizon-Targeting (OTH-T), would form an extremely important theme of U.S. naval development through the mid-1980s. Because it entailed a merger of intelligence with operations, it became a dominate theme in Naval Intelligence, to the point where the new intelligence headquarters at Suitland, Maryland, was designed to include a large battle management space (which, in the event, became a cafeteria).

In 1975 anti-ship missile targeting was under study by, among others, the Advanced Studies Group at the Naval War College. As the only current users of anti-ship missiles, the Soviets were clearly worth studying. It was soon obvious that merely locating major U.S. warships was the least of the Soviets' problems. They were acutely aware that they had to distinguish the targets *within* the formations, because they believed that it would take several cruise missile hits to sink a carrier, and they had relatively few such missiles available. Thus they took enormous pains to find out which ship in a formation was the carrier. Their tactics in the Mediterranean were revealed as a case in point. The 'tattletale' destroyers shadowing the carriers generally carried four short-range ('Styx') anti-ship missiles, pointed aft. This unusual configuration was now easily explained. The role of the 'tattletale' was not to keep track of the carrier but rather to keep track of where the carrier was within her formation. When an attack was imminent, the 'tattletale' would turn away – pointing her short-range missiles back at the carrier. The turn-away was necessary, because without it Soviet long-range missiles might well be wasted on the 'tattletale'.

The implication was that much more detailed information would be needed. At the least, any group of Soviet ships surrounding a potential target would have to be identified. Because it had been conceived to find particular Soviet warships, White Cloud probably would not detect the other ships near the important targets. An approaching Tomahawk might well find itself seduced into hitting the wrong ship. Something more was needed. As an interim solution, Tomahawks were fitted with radar detectors called Passive Identification Devices (PID). They were not seekers; rather, they were a means of identifying the ship onto which the missile seeker had locked. That was hardly enough; for example, the particular radar the PID sought might not be operating.[11]

Initially it seemed that much more, most likely Clipper Bow (see the previous chapter) would be needed. Unfortunately, any such system would be quite expensive, and Tomahawk would be doomed if it required an expensive new class of sensors. It had been attractive because it was a relatively inexpensive missile. The Soviet surface fleet was an annoyance, not a primary U.S. naval target. With money very tight, there was barely enough for the vital combination of offensive firepower and defence against the main Soviet threats, which were still submarines and long-range naval bombers.

BRIGAND

It turned out that there was an alternative. One of the Newport analysts, Commander (later Captain) Hal Cauthen, observed that the necessary information was already being collected – and thrown away as unwanted. Commander Cauthen was well placed to understand the potential of TENCAP. He had previously been involved in a Vietnam War programme called Charger Horse, a sort of proto-TENCAP. By the time of the Vietnam War, signals intelligence detachments on board major fleet units, such as carriers, reported their product directly back to NSA, for strategic exploitation; NSA decided what was releasable. Under Charger Horse, detachments on these units provided tactical information directly. This was apparently a major departure from past practice, and it was extremely successful.[12]

Cauthen's key was probably a technique the Navy had long used, BRIGAND (Bistatic Radar Intelligence Generation and Analysis, New Development).[13] An intercept receiver picks up signals reflected by objetcs near the radar illuminating them. It can therefore form a version of the target's own radar picture. BRIGAND was developed by technicians of a Navy ELINT squadron (VQ-1) in 1960, using the highly directional antenna of an APS-20 radar on board an EC-121 electronic reconnaissance aircraft to feed a very sensitive radar receiver (an APR-9). Initially it was used against ground radars; by mapping nearby objects in relation to the radar, the radar itself could be located more precisely. However, BRIGAND also reveals nearby objects – such as ships in company with a ship whose radar is being detected, even if they do not radiate at all.

NSA's ELINT satellites apparently already collected signals in the radar frequency range (to pick up Soviet microwave signals). Radar signals themselves were of no interest to NSA: they were unwanted noise. Its mandate was to exploit message traffic. Moreover, NSA feared that any direct connection with military operations would compromise its security. For example, if the Soviets suspected that it was monitoring their internal microwave circuits, they might make the extra effort to lay buried cables (which would be impossible to monitor); they would at least try to scramble their messages. The shift from HF to microwave had certainly demonstrated Soviet alertness to threats to their communications security.

NSA's unwanted signals inevitably included BRIGAND data; they were just what the Navy needed. To read messages, NSA had to collect signals in exquisite detail. It would record not only the pulse of a radar, but also the smaller reflected pulses, coming a few microseconds later. A high-flying NSA satellite might not be able to locate the source of the main pulse very precisely, but BRIGAND analysis would depend only on the time differences between the main pulse and the reflected ones. Those differences would not be distorted by distance. They would tell the analysts how many ships surrounded the main radar, and where they were in relation to it (White Cloud tracking would tell roughly where the formation was). That BRIGAND data is typically supplied by electronic intelligence satellites is confirmed by an entry in an unclassified list of SpaWar projects. Radiant Crimson develops National Systems (*ie* NSA satellite) BRIGAND tracking capability and transmits the results over the special satellite data links. The combination of fully passive techniques – radar tracking by fingerprinting and BRIGAND, supporting a dynamic computer data base – had an enormous advantage. It could not alert the targets that they might be attacked. Best of all, that information would be obtained passively. There would be no U.S. equivalent to 'Bears in the morning, missiles in the afternoon'.

After a fight, NSA agreed to allow the Navy to use its unwanted – but tactically invaluable – data. The programme was called TENCAP, Tactical Exploitation of National Capabilities. In this case 'National' meant the sensors, mainly space-borne, which were used to gather intelligence, and which in the past had been considered far too valuable to risk for tactical purposes. The best historical analogy is with ULTRA during the Second World War, when the source of intelligence had to be protected so very carefully. In the Cold War case, national sensors fed into the U.S. capability to wage a nuclear war. Photo satellites and, to some extent, signals intelligence systems, detected the targets which might have to be hit and also provided a vital degree of war warning. This kind of war, which would have been directed from Washington, was distinguished, in the United States, from the sort of low-level war, like Korea or Vietnam, which deployed forces fought. Anything actually used in a war is likely to be compromised, so national systems were generally insulated from the forces which might have to fight local wars.

TFCC and JOTS/JMCIS

The system's great recognised drawback was the cost and complexity of the TFCC. Although the FCCs were quickly set up, TFCC moved ahead very slowly. A 1968-72 Grumman study (the '77 volumes') for the Naval Electronics Laboratory in San Diego identified the essential information a carrier group commander needed. On this basis, in 1972-3 the then Naval Electronic Systems Command produced a Request for Proposals. Meanwhile it conducted a TFCC test, Outlaw Hawk, in the Mediterranean, using the carrier *Kitty Hawk* (1974-5). An interim TFCC was installed on board the carrier *John F Kennedy* in 1975 and demonstrated successfully in 1975-6. In one test, Outlaw Shark, the flow of data between the FOSIF in Rota and the carrier was tapped by the Sixth Fleet headquarters in Naples and targeting messages were sent to a submarine. This validated the SSIXS idea the submariners had already developed. Meanwhile Martin won the TFCC contract and built a prototype.

TFCC required an extremely powerful computer merely to display the OSIS data. The carrier would also have to correlate that time-late data with the real-time data flowing in from battle group sensors. On the other hand, Navy funds for command and control, seemingly only distantly connected with war-fighting, were quite limited. Also, senior commanders could not agree on just how a TFCC should be arranged and used, echoing earlier hot disagreements in many navies on the arrangement of bridges and other command spaces. To make matters worse, early plans for a single terminal in the TFCC soon expanded to require four terminals. Probably that reflected the decision to make the TFCC the source of Tomahawk targeting data, rather than merely a source of data for the carrier commander. Martin's TFCC contract was cancelled in 1978 because its design was too costly. With only one experimental White Cloud in orbit, and with the Fleet Satellite not yet in place, TFCC would not yet have been particularly effective. Lockheed now won the contract with a new computer, USQ-81.

To make the system affordable, in August 1979 the Chief of Naval Operations approved a new plan. It was incremental, partly to spread out costs but also to limit the time a carrier would have to spend in a shipyard. Thus Increment 1 merely created the new space to be used for a TFCC: a 20 × 20ft space suitable for handling Special Intelligence (SI) data (*ie* TENCAP) with a central console, plots, and connections to the ship's combat system (*ie* to Link 11 data). Increment 2 added the heart of the TFCC, the Flag Data Display System. It provided a ship with the vital ability to receive, process, and display the ocean surveillance picture developed by OSIS. It would correlate offboard data from OSIS with battle group data, and it would provide multiple terminals. Increment 3 added further features, mainly in the form of new software.

By this time TFCC seemed more urgent, as Tomahawk was about to enter service, and development was ordered accelerated in March 1980. Carriers already had a Supplementary Plot (Supplot) which filtered intelligence data. It naturally fed into the TFCC. Similarly, the carrier's CIC filtered data from Link 11. OSIS data was also received and filtered. The TFCC displayed the combination of all this data for the carrier group commander. At this stage the TFCC did not yet read the local electronic warfare or ASW pictures. Typically it displayed the situation within a 1500-mile circle around the carrier (it could zoom in to show the situation more clearly on a smaller scale).

The main issue in designing the TFCC was to relate the time-late OTH data (supplied by remote sensors, such as White Cloud, via OSIS) with the picture created by the battle group's own organic sensors. In TFCC, that was done by an Afloat Correlation System (ACS) fed by an Ocean Surveillance Terminal (the initial version of which was POST, the Prototype OST). ACS was supposed to form a consistent tactical picture for the ship's combat direction system. It also sent battle-group track data back to OSIS to help ensure that the larger system carried the same tracks as the fleet at sea. As a measure of the scale of the problem, an *unclassified* description of ACS showed an ability to handle 4000 tracks simultaneously, compared to perhaps 256 for the carrier's local combat direction system.

Correlation was based on the same sort of statistical concepts which had been developed for OSIS itself: the computer could test whether a ship was likely to have moved from one reported position to another. Radar emissions (fingerprinting) were used for identification. For example, the same type of radar detected 30nm and 15 minutes apart would be taken as separate ships, whereas the system might conclude that two similar detections 30nm and 70 minutes apart came from the same ship.

Given a good picture of enemy ship positions and movements, the TFCC could use separate computers to evaluate tactical responses. For example, TFCC incorporated an EWCM (Electronic Warfare Coordination Module), which could evaluate alternative EW tactics on a large scale quickly enough for them to be implemented. Decision aids include selection of optimum power and frequency of communications equipment (to minimise the probability of intercept) and optimum decoy location. EWCM was clearly part of the deception strategy developed to counter the Soviet anti-ship targeting system (see the previous chapter). Once Tomahawk was in service, the Navy would finally have distributed firepower. The twelve existing carriers had been badly stretched to cover the many U.S. naval responsibilities. Non-carrier Surface Action Groups could help. They would enjoy even less organic OTH capability than the carriers, probably only Classic Outboard, so for them the new systems would be even more important. A cruiser version of TFCC would have to be developed.

Once more there were interim installations: in the carriers *Midway* (June 1980) and *America* (January 1981), and in the cruiser *Josephus Daniels* (February 1981). All of these proto-TFCCs used commercial-grade equipment. They were soon removed pending development of a military-grade (Mil-Spec) equivalent. Fleet experience had been so successful that in November 1981 the CNO ordered limited procurement. Two FDDS prototypes would be bought (one for the carrier *America* and one for the NOSC test site). Another six were bought in advance of approval for full production. Work on the first six carriers went ahead briskly: they were *America* (completed 15 May 1983), *Constellation* (15 July 1983), *Nimitz* (15 September 1983), *Midway* (15 October 1983), *Ranger* (15 February 1984), and *John F Kennedy* (1984). By about 1982 plans called for TFCCs aboard all fourteen carriers (the Reagan Administration was enlarging the carrier force), four *Belknap* class cruisers, the big nuclear-powered cruiser *Long Beach*, and, if possible, the two amphibious flagships (LCCs).

The TFCC programme was paralleled by developments in the Tomahawk weapon control system. When the system was introduced in 1983 (on board the *Iowa* and *Spruance* classes), a ship fired her Tomahawks at a single high-interest target (HIT), the position, course, and speed of which could be passed via the new satellite links. The fleet argued that targets would often appear in groups, and that the Tomahawk shooters needed the ability to maintain their own target data base. That was not too different from the argument that the carrier needed a sea surveillance picture, not merely a list of a few critical targets, and it demanded much the same response.

The solution, adopted in 1986, was to include the same computer, USQ-81, which formed the core of TFCC, in the Tomahawk Weapon Control System (TWCS). This track-control group supplemented the original core of the system, the launch-control group. It had two operator positions, one to form the track picture and the other to plan Tomahawk attacks.[14] A separate computer handled the OTCIXS link. Given the track-control computer, a Tomahawk platform could be designated Force OTH Track Coordinator (FOTC), supervising a group's sea surveillance picture.[15]

In 1983, with the first six carriers being outfitted, it was time to decide whether to fit TFCC to the other ships in the programme. The Navy had new standard computers (UYK-43, -44), and it might be attractive to redesign the TFCC around them, instead of the special

Lockheed computer. However, there was a much larger issue. The military was no longer leading computer technology; commercial machines were at least as powerful as standard military ones. That was why the interim TFCCs had been installed so quickly in 1980-1. The central feature of the TFCC was its computer, not the special space and the array of specialised displays.

As a carrier group commander in 1981, Rear-Admiral Jerry O Tuttle was partly responsible for assembling a tactical decision aid not too different in concept from TFCC, but infinitely less expensive. It was a package of programs running on a commercial desktop computer, providing a geographic plot, a data base, and an operator interface.[16] Given a more powerful desktop computer, Tuttle ordered the program package transferred to it. The result was demonstrated during a war game at the Naval War College in Newport and then placed aboard the carrier *America* for tests at sea. At this stage it was primarily a tactical decision aid, simplifying the tactician's job. Information had to be input by hand.[17] Tuttle's solution was initially called the 'Jerry O. Tuttle System;' its acronym, JOTS, soon came to stand for the Joint Operational Tactical System.

Within a few years it was clear that JOTS-type systems could provide the same service as the projected TFCC, but without requiring lengthy overhauls and expensive installations. The computer, not the system, was the key. The U.S. Navy was already buying standard Hewlett-Packard workstations (which it would soon designated Desk-Top Computers, or DTCs), which were at least as powerful as a USQ-81, though far less expensive. Because they did not meet military specifications for reliability and ruggedness, these computers could not be used in critical combat system roles. However, they certainly could run the software to provide exactly the picture the costly TFCC would maintain and display. All one needed was the satellite connection, via a standard modem.

The first version of JOTS to be integrated with satellite data was JOTS IV, on board USS *Forrestal* (Carrier Group Six, October 1985, under Rear-Admiral Duke Hernandez). A special CUDIXS-type satellite net, JOTSIXS, was used to provide the system with live tactical data.[18] Tuttle took JOTS IV with him to the Atlantic Fleet command centre in November 1985. An improved JOTSIXS became OTCIXS. Most importantly, the system was so inexpensive that it could easily be placed on board virtually all U.S. warships, even those without Link 11 capability.[19] TFCCs were redesigned around the new computers (the name survived).

On the other hand, the commercial computers could not replace the track-control groups aboard surface shooters. U.S. policy was that only Mil-Spec computers (which had been very thoroughly tested) could be allowed to launch weapons. That was not unreasonable.

Commercial hardware could easily harbour bugs which might have disastrous consequences.[20] TFCC was not, in theory, tactically crucial. If it sent an aircraft to the wrong target, the pilot could still decide not to shoot. But if it sent a Tomahawk to the wrong target, the missile would have no such potential.

JOTS software reflected its connection to distant, often space-borne, sensors. The OTH-T data base, essentially a map of world shipping, was keyed to geography, not to the current position of the ship carrying it. As the system evolved, other data bases keyed to geographic position could be added.[21] Overlaying them, the terminal could build up its complex tactical picture. This was very different from the plot of shipping within a 1500-mile circle initially envisaged in TFCC, supplied mainly by a nearby FOSIF. Operators generally did not want the full global map, so JOTS was equipped with geographical filters based on current ship position. There were also security filters. Parts of the JOTS picture were based on very sensitive sources of information, such as NSA satellite data. In some cases they had to be withheld. That also applied to Allied navies using JOTS. More generally, multi-level security measures became an important feature of JOTS and its successors. The key to JOTS was that the picture compilation function was not, at least in theory, mission-critical: it did not directly involve firing a weapon. Navy regulations therefore made it possible to use something other than a standard militarised machine.

JOTS II, using a follow-on DTC II workstation, was operationally evaluated on board the *Independence* and *Jouett* in June 1990. Not long afterwards, the Iraqis invaded Kuwait, and the UN approved an embargo. Suddenly the U.S. ship-tracking system was extremely important. There were never many warships available to cover a vast area of ocean funnelling into Iraq. They had to be used economically. To do that, they had to be kept aware of the movements of merchant ships headed towards the Gulf. Moreover, it was vital that effort be concentrated on the right ships. This was not too different, in principle, from the problem of tracking and attacking the Soviet surface fleet.

JOTS was installed on board British flagships, including the three carriers as part of the British TFCC equivalent, PFSS (the Pilot Flag Support System). A version of the U.S. TADIXS broadcast was supplied via Saclant in Norfolk to the British naval headquarters (OPCON). There it was multiplexed with the Opcon broadcast for transmission to PFSS via Link R.[22]

Admiral Tuttle took JOTS and the accompanying philosophy with him when he went to the Atlantic Fleet FCC. It was pointless to spend heavily on Mil-Spec computer hardware when commercial machines could do much more at much lower prices. Better to standardise on commercial machines and to concentrate on the special software the Navy needed. Moreover, computers were changing rapidly. The Navy had to keep up. It could not do so as long as it spent heavily on specialised machines. In Tuttle's view, which prevailed, it would be much better to standardise on a particular upgradeable commercial-grade machine, then replace it wholesale with a new standard every four years or so. Ideally, software would be written in a modular fashion so that it could be transported from generation to generation.

Tuttle sold his idea. He ended up as commander of the ideal place from which to do so, the U.S. Navy's Space and Electronic Warfare Command (SpaWar), formerly the Naval Electronic Systems Command – which had been responsible for TFCC. At the Atlantic Fleet FCC, Tuttle replaced the Mil-Spec terminals with the standard off-the-shelf computers used in JOTS. As commander of NavSpaWar, he enforced the idea of regular computer upgrades. JOTS and its immediate successor, JOTS II, used Desk Top Computers I and II (DTC I and II). By the time JOTS II was being developed, the Navy had redesignated these machines as Tactical Computers (Tacs), even though they were only ruggedized versions of civilian workstations. What would have been JOTS III (using a Tac 3 computer, which might otherwise have been called a DTC 3) became instead JMCIS, the Joint Maritime Command Information System. Because the idea of continued upgrades was so central to it, the S was soon changed to mean Strategy, meaning the JMCIS development strategy.

JMCIS (1992-3) was designed specifically to support further applications, so that previously-separate systems could be unified. It provided the necessary Common Operating Environment (COE) running on TAC-series computers, within which the necessary software could run. The tactical data base manager was extended to buffer incoming information, correlating it with existing information, and then sending the result automatically to the appropriate section of the software. The idea, born in JOTS, of keying everything to geographic position, continued to be extremely fruitful (like JOTS, JMCIS is built around a combination of data base and chart). The JOTS/JMCIS concept was so attractive that a derivative was later adopted by the U.S. national military command authority and by the CINCs.

The computers bought for JOTS II could do far more than compile and display the tactical picture being developed by OSIS or by the FCC. A commander could use that picture to ask 'what-if' questions needed for tactical planning. Tactical Decision Aids (TDAs) were added to JOTS when the DTC II computer, which can do multi-tasking, was adopted. The results of TDA calculations can be displayed as windows over parts of the basic data base. This capability was actually already present in the Tomahawk anti-ship fire control system. That is, given some indication of where the target was in the midst of other ships, the Tomahawk targeter could try various approaches to make sure that the missile's seeker locked onto the right ship. There were far too many variables for such estimates to be done by hand. Thus the Tomahawk anti-ship system had to incorporate not only a picture-keeping capacity (using Lockheed's USQ-81, the same machine bought for TFCC), but also a mission-planning capacity. That should not be confused with mission-planning for strategic or land-attack Tomahawks, which was a very different proposition.

Clearly 'what-if' questions were important on a much wider scale.[23] What is the best way to deploy escorts, given local water conditions? The best counter to a particular kind of air attack? The best plan for underway replenishment? TDA programs were useful because, thanks to JOTS and the satellite links, both FCCs and deployed ships already had before them a reasonably accurate computer picture of where U.S. and foreign ships were. It was natural to add such other factors as weather conditions into exactly this data base.

By the early 1990s, JOTS and the associated satellite terminals had spread to virtually all U.S. surface combatants; the system was much more widely used than Link 11. Because all the JOTS stations could receive the OTCIXS satellite broadcast, they amounted to a fleet-wide command and control system, which was called NTCS-A (the Navy Tactical Command System – Afloat Element). The headquarters (FCC) end of the system was a series of computers and displays running a software package called the Operations Support System (OSS).[24]

The system moved away from the original FCC-TFCC model, in which the TFCC had been little more than a passive display of the picture compiled by OSIS. The system's distributed computers, linked by the satellite-borne OTCIXS, came to co-operatively develop the global shipping picture. It was similar in concept to the tactical Link 11 net, but on a much larger scale. All users hold the same database (the world shipping picture). As a user updates the database, the updated data are transferred to the other users. In addition to OTCIXS, the users communicate via a High-Interest Target (HIT) radio teletype net. The latter sends only the positions of targets, not the more general updates. Within a battle group, one ship is designated FOTC, the Force OTH coordinator (track manager). It serves as an agreed basis for Tomahawk anti-ship targeting, and it provides the force with consistent target tracks, based on the available OTH data. The others interpret the data independently, but accept the FOTC choices (they provide backups if the FOTC goes out of action).

OTCIXS supported communication *within* a dispersed battle group. It led to a new concept, the Battle Group Information Exchange System

(BGIXS), in which a battle group flagship could send messages in the format used for SSIXS traffic. A submarine equipped to receive the SSIXS satellite broadcast was automatically equipped to participate in the BGIXS net. In the past, submarines generally received information. They passed very little back, either to ships or to their distant force commanders, because it was very dangerous to transmit HF messages. UHF satellites seemed to offer safety. The uplinks were extremely difficult to intercept, and the high data rate would drastically reduce transmission time. The submarine did have to come to periscope depth to use the satellite, but that was also the case with most other forms of radio (she could receive, but not transmit, ELF messages at depth, but they would have only very limited content).[25] A submarine using BGIXS routinely transmits periscope pictures, graphics, and even video back to the flagship.

NATO adopted a version of the OSS/NTCS-A combination as NACCIS, the NATO Command and Control Information System. It became operational 1 January 1995, at SACLANT (Norfolk, Virginia), CINCIBERLAND (Lisbon), CINCEASTLANT (Northwood, UK), and at CINCNAVSOUTH (Naples). As in the British PFFS system, SACLANT distributes a sanitised NTCS-A picture to shore headquarters, where national information is added for retransmission (by NATO SHF satellite or by HF radio) to ships at sea. The system was adopted at national naval command centres in the Benelux countries, Canada, Denmark, and Norway, and in France alongside AIDCOMER at Toulon (and is being considered for Brest). The Royal Australian Navy has the compatible OSS system.

Concentrating on software while using standard hardware can make for very rapid development and deployment. Once a software package has been demonstrated, it can be reproduced extremely quickly. There was, therefore, a considerable incentive to replace the mass of specialised computers in U.S. Navy command systems with more powerful units running the new software. As that happened, many systems which in the past had clearly been separate no longer had to be; they could be reduced to different software running on the same machines. For example, OSS absorbed the formerly separate submarine Shore Targeting Terminal (STT) and the Force High Level Terminal (FHLT) which controlled maritime patrol aircraft. NTCS-A software expanded to absorb the functions of the formerly separate ACS, EWCM, and POST, among others.

Thus one might say that the combination of the Soviet surface-shooter threat and the advent of anti-ship Tomahawk brought forth dramatic changes in the U.S. Navy. The threat caused the Navy to create an ocean-surveillance system and its ocean terminal, TFCC. The advent of TFCC in turn inspired the shift to commercial computers and to an emphasis on software, which in turn made for very rapid development of new or evolutionary systems. Meanwhile the appearance of Tomahawk itself led to a change in basic U.S. naval policy, TENCAP. Space assets made all this possible. Space systems, first White Cloud and then also those provided by TENCAP, made sea surveillance affordable. Satellite links, provided by the Fleet Satellite, tied the system together and provided its products to its sea-based users. For that matter, the existence of a satellite link made it easy to tie the entire fleet together, once Admiral Tuttle made the extremely perceptive observation that the right software could provide every ship with something close to a TFCC at very low cost. The same ships had never been tied together by Link 11 partly because it entailed fitting a very expensive computer – but also partly because it required a massive HF radio installation, including a big antenna. Moreover, the systems built up for OTH combat against Soviet missile-shooting surface warships proved extremely valuable in very different contexts. The OTH system made the blockade of Iraq possible in 1990. Its extensions proved essential to the new naval strategy the United States adopted in the 1980s.

There was, to be sure, a major irony. The Soviet surface shooter threat amounted to very little, at least until when it touched off the U.S. programmes in the 1970s. Throughout the Cold War, Soviet missile-bearing bombers and submarines were a much more serious threat, and the OTH-T systems were unlikely to do much to stop either. The Soviets built eight 'Shaddock'-firing light cruisers (four 'Kynda' [Project 58] and four 'Kresta I' [Project 1134A]). About 1970 it seemed that the Soviets were going into mass production with two new classes, 'Kresta II' (Project 1134B) and 'Krivak' (Project 1135). They were also converting destroyers into Styx-armed 'tattletales.' The Soviets designated their new ships (and the 'Kresta Is') as ASW units, but that could be discounted as deception. There was no evidence that the Soviets had the slightest idea of how to hunt quiet U.S. submarines. Surely the strike carriers, the core of the U.S. Navy, were still their main targets. The new ships were armed with a new missile. As in previous systems, the weapon was carried in a closed tube. With very limited electronic coverage of the Soviet Union, there was almost certainly no intercepted telemetry to reveal the new weapon's performance. It was assigned the NATO designation SS-N-10; it was tentatively identified as a short-range anti-ship missile. The systematic study of Soviet naval tactics conducted as part of the Harpoon/Tomahawk targeting project suggested that in general the Soviets tried to attack their targets from the greatest possible range, specifically to limit any counterattacks. SS-N-10 seemed to be a major exception, possibly marking a drastic change in thinking. By this time there was already a minority opinion within

U.S. naval intelligence: the Soviet designations were accurate, and SS-N-10 was actually an anti-submarine missile like the U.S. ASROC or the French Malafon or the Australian Ikara. For a time it was supposed that such a weapon (SS-N-14) could be carried as an alternative to the anti-carrier SS-N-10; then it became clear that SS-N-10 did not exist at all.[26] The surface anti-carrier threat suddenly collapsed to all of eight ships. The surface threat did revive to some extent, as the Soviets built the *Kuznetzov* and the *Kiev* class carriers (from 1975 on), two classes of cruisers (*Kirov* [beginning in 1980] and *Slava* [beginning in 1982]), and destroyers (*Sovremennyy* class [beginning in 1980]) armed with powerful anti-ship weapons. Thus the U.S. anti-surface shooter programme matured (with systems such as White Cloud and the Fleet Satellite) just in time for the true flowering of the Soviet anti-ship force.

The new U.S. OTH-T systems depended entirely on whatever signals potential targets and non-targets produced. Those using the system were very reluctant, in practice, to fire weapons based entirely on passive data. They suspected that non-radiating ships were also present; the system would miss them altogether. Thus they strongly preferred to use the OTH-T system to cue a battle group's own airborne active reconnaissance sensors, which would be flown out before any missiles could be fired. That of course largely negated the most important virtue of the OTH-T system, that it made non-carrier surface action groups effective counters to other surface ships. This problem persisted through the end of the Cold War. It could, of course, be argued that virtually no ships were non-emitters; large ships generally have to keep their navigational radars running at all times. All of those radars, moreover, can be detected from space. One major question was whether radar 'fingerprinting,' which certainly made sense in the context of major naval radars, could be extended to mass-produced navigational sets. In the 1990s this question became more urgent because the U.S. Navy badly wanted to be able to track particular merchant ships, carrying embargoed weapons (*eg* ballistic missiles). The new version of fingerprinting was called Specific Emitter Identification (SEI); it included a 1996 feasibility study, in which the Naval Research Laboratory used a specially precise ESM set to make the necessary measurements. The ultimate promise of SEI is a catalogue of all the world's radars – which can then be used for space-based tracking via passive satellites.[27]

From about 1970 on, the U.S. Navy's dependence on space systems had changed drastically. In the early 1970s space systems were used almost exclusively for strategic support: the reconnaissance programme, for strategic targeting; Transit, for navigation; DSCS, for strategic connectivity; the Defense Meteorological Support Programme to provide weather prediction for strategic forces (including ice reporting for strategic submarines); and the Defense Support Programme (for strategic warning). In virtually all cases the Navy's strategic requirements coincided more or less precisely with those of the U.S. Air Force, which bought, launched, and controlled the spacecraft. A decade later the Navy was far more interested in tactical applications, such as OTH-T and the Fleet Satellite. It was becoming obvious that these applications were essential to naval operations. Perhaps more significantly, the Navy's tactical needs were very different from those of the Air Force.

The U.S. Navy, then, needed some way to make its voice heard in the formulation of space policy. Its tactical users needed leverage. On 1 October 1983 the Naval Space Command was established to provide a fleet focus for the growing specifically naval applications of space technology. It took over naval TENCAP as well as naval space communications and the naval space surveillance system (SPASUR). However, the U.S. Air Force continued to be the lead service in space operations. It was (and remains) responsible for launching and controlling satellites. After 1983, as before, most naval needs were met by levying requirements on satellites most of whose cost was borne by the Air Force or by the Defense Department.

The Gulf Embargo

As it turned out, the great test of the OTH-T system was not Cold War combat, but rather the embargo enforced before and during the Gulf War, in 1990-1. Like an OTH war against Soviet surface shooters, the blockade required the quick interception of ships spread out over a very wide area. Because the embargo fleet was small, errors could have been very costly. Identification mattered. Some readers, for example, may recall the 'baby milk ship', crewed largely by Iraqi women. The Iraqis hoped that they could capture on film American Marines attacking those women when they boarded the ship. Boarders would have been deterred, but the system had tracked the ship from the point at which she had been loaded with contraband – ammunition – under the layer of baby food. The Marines therefore had the confidence to search fully, and they found the ammunition. Overall, the system worked extremely well, overcoming Iraqi attempts at deception.

A picture of surface ship traffic approaching the Gulf was distributed by satellite to JOTS terminals on board the embargo-enforcing ships. Because JOTS was software running on a standard computer, it was very easy to deploy. The computers could be taken from warehouses, and software could be reproduced very quickly. More than 200 systems were hurriedly produced to support Desert Shield. In fact JOTS was soon spread among not only the U.S. ships in the Gulf, but also among Australian, Canadian, and Dutch ships helping enforce the embargo.

Clearly many of the sensors were different, but the central element, the ship tracker, was the same as the one developed initially to track down Soviet missile-firing surface warships. It turned out to be very important that it could accept a very wide variety of inputs, by no means limited to satellites. For example, a P-3 spotting and identifying a merchant ship could insert that data into the track files used for the embargo – if the position of the merchant ship was precisely known. Otherwise the P-3 might merely be adding a spurious extra track. The embargo command might then have assigned ships to run down both contacts. This was much the same problem that faced an anti-aircraft commander whose tactical picture was fed by Link 11. If one of the ships was out of the grid, her radars would generate spurious contacts. Space systems provided an answer in the form of GPS, the Global Positioning System (see Chapter 13).

When the embargo was ordered in September 1990 a P-3C (from VP-9) had already been modified to support Tomahawk targeting under the code-name Outlaw Hunter (Orion being the hunter of mythology). This quick demonstration conversion was completed in August 1989 after nine months, at a cost of $700,000, including development; further conversions were expected to cost $250,000. The usual P-3C search radar (APS-115) was replaced by an ISAR type (APS-137) which would give an imaging capability for ship identification, and the position (via GPS) and identification data were correlated with the JOTS data base (received via OTCIXS) and then transmitted to other platforms via the OTCIXS link (OTH-T Gold message format).[28] GPS was essential because the aeroplane's usual inertial system was not accurate for Tomahawk targeting (the missile cannot be allowed to spend much time searching for its target at low speed within the target's self-defence zone; position errors translate into search time).

In 1989-90 tests, it took about three minutes for the aeroplane to transmit a target track or an Opnote to a ship or a shore site (*eg* for further transmission to a submarine). Routing data for a submarine through a Shore Targeting Terminal and then through SSIXS entailed a total delay of five to eight minutes. In the 1989 exercises, Outlaw Hunter successfully cued the submarines *Salt Lake City* and *Omaha*. The system also provided targeting and prestrike surface pictures to a carrier, both to the combined strike warfare commander (Alfa Sierra) and to the aircraft themselves (via a secure voice system, SURPICS). The P-3C could also use its Link 11 to transmit target data. Again, although one might think of Outlaw Hunter as anything but a space system, it was practicable only because of space systems, in the forms of GPS and the Fleet Satellite (carrying the

OTCIXS channel). A few months later, Outlaw Hunter proved its value during the embargo against Iraq (Desert Shield).

Given the success of the conversion, an improved version with better electronics and software, OASIS (OTH-T Airborne Sensor Information System), was ordered; OASIS I, an Atlantic Fleet P-3C, was converted; it was just too late for the Gulf War. OASIS II later became part of the P-3C surface warfare upgrade programme now in progress. Similar systems for the E-2C (Outlaw Hawkeye) and the SH-60 (Outlaw Seahawk) were fielded. OASIS for the S-3B (Outlaw Viking) was developed, but it did not go into production. Outlaw Seahawk differs from Outlaw Hunter/Viking in that data is down-linked to the mother ship for onward transmission via OTCIXS.

TEN

Defending the Fleet: The Outer Air Battle

U.S. naval strategy shifted dramatically with the accession of the Reagan Administration in January 1981. Secretary of the Navy John Lehman articulated the Navy's new Maritime Strategy. One of its facets was a new mission, which exploited space assets, the Outer Air Battle against Soviet missile-armed bombers. The Maritime Strategy reversed the strategy proclaimed in 1970 by the then-new CNO, Admiral Elmo Zumwalt, Jr.

Zumwalt had announced that the Navy would concentrate on sea control, which meant mainly insuring that in wartime NATO would be able to use the sea relatively freely to transport vital supplies to support a war in Europe. From a naval point of view, this was a defensive posture. Admiral Zumwalt sometimes said that he feared that the rising Soviet Navy might actually be able to defeat a U.S. fleet shrinking in the aftermath of Vietnam. To Zumwalt, sea control meant very largely guaranteeing the security of the North Atlantic shipping on which NATO would depend in wartime.

Prominent among the threats Zumwalt's fleet faced were Soviet landbased 'Badger' (Tu-16) bombers carrying long-range stand-off missiles. Although U.S. shipboard missiles were far better than they had been a decade earlier, there was still real scepticism that they could react quickly enough to deal with attacks by regiments of aircraft armed with new supersonic rocket-powered weapons. Typical convoy escorts (frigates and destroyers) certainly could not do so. Probably the main burden would fall on carrier-based fighters, and even they might not be effective against incoming anti-ship missiles.

In Zumwalt's view, in wartime the U.S. carrier force would be split up, single-carrier battle groups covering either vital convoys or straits through which Soviet anti-shipping forces would have to come.[1] The carriers faced not only the bombers but also a rising Soviet nuclear submarine threat, which the carriers could no longer outrun. In this the Soviet ocean surveillance system was crucially important. It seemed shocking when a Soviet 'November' class (Project 627A) submarine intercepted the carrier USS *Enterprise* en route to Vietnam in February 1968. It was assumed that interception had made possible by the Soviet Ocean Surveillance System, which at that time did not yet include space components. In the late 1960s it had become customary to operate specialised ASW carriers in support of attack carrier groups. Now the ASW ships, which dated from the Second World War, were wearing out, and no replacements were in sight.[2] Zumwalt ordered the big carriers equipped with anti-submarine aircraft, their designations changing from CVA (attack carrier) to CV (multi-purpose carrier). As a consequence, the surviving carriers could operate in a specialised ASW role.

Zumwalt could not eliminate the key strike role of carriers in the Mediterranean and the North Pacific, but he did drastically reduce the likely number of ships assigned to that role in wartime. One reason he did so was that the anti-carrier threat in the Mediterranean seemed to be rising very rapidly. As a surface warrior, Zumwalt was very impressed by the new Soviet surface ships so aggressively shadowing U.S. carriers in the Mediterranean. At the least, they would make the Soviet missile-carrying bombers far more effective than in the past. It seemed entirely possible that any carriers on station in the eastern Mediterranean at the outset of a war would soon be sunk, in the Soviets' 'battle for the first salvo'. Americans began to talk about fighting a campaign to regain control of the Mediterranean before sending in carriers – but the carriers would probably be most valuable for their ability to strike at Soviet and allied forces coming south as the war began. One outcome of this thinking was the OTH-T systems and techniques described in the previous chapter.

Ultimately, the eastern Mediterranean seemed an almost hopeless proposition because the fleet there could not easily withdraw out of missile-bomber range, nor could it evade the 'tattletales' directing a missile strike. Elsewhere, for example in the Atlantic, the different Soviet threats could be somewhat separated. Convoys running far enough south would not encounter Soviet bombers until they reached European ports – which would, presumably, be covered by local fighter forces. Thus in the most important theatre of a future naval war sea control might reduce largely to a struggle against the Soviet force which *could* attack anywhere on the world ocean, the submarine force. It would be prosecuted by submarines in forward areas and by frigates covering convoys.[3]

Zumwalt deliberately downplayed another naval mission, projection of power. Since the Second World War this had generally meant carrier strikes into the Third World; such operations seemed far less acceptable in a post-Vietnam era. Power projection near the Soviet Union, for example by Sixth Fleet carriers, seemed suicidal – largely because of the Soviet shore-based bomber fleet and the new missile-firing surface ships and submarines, all directed by the Soviet Ocean Surveillance System. In the early 1970s, when Zumwalt was making his key decisions, the effects of the build-up begun under Brezhnev, in the late 1960s, were first being seen. The Soviet ground forces seemed more and more powerful, while U.S. ground and ground-based air forces assigned to NATO were still feeling the effects of drawdowns for the Vietnam War. Within the U.S. government, the obvious solution to this problem was to spend more of a limited defence budget on forces assigned directly to the Central Front in Europe. It may well have seemed to Zumwalt that only naval forces explicitly required to support the Central Front could be justified in so difficult a time. That meant mainly sea control forces, although Zumwalt continued work on OTH-T and on countering tattle-tales, both of which would be essential if the carriers were kept in the Mediterranean. The overall war plans apparently required their contribution.

Zumwalt's detractors within the Navy charged that he had emphasised sea control because it seemed to be a safe naval sanctuary: none of the other services could achieve it. If that ever changed, the Navy would be doomed. Moreover, Zumwalt's division of naval tasks into sea control and power projection was hardly as natural as it seemed. It was, rather, an extension of an ongoing attempt within the Defense Department to split up military functions in order to encourage rational budgeting. Zumwalt had served in the Pentagon when this practice had begun, under Secretary of Defense Robert S McNamara.

In fact, in classical naval thinking sea control and power projection were inseparable sides of the same coin. For example, sea control could be seized by going after the enemy fleet (power projection). The threat of power projection might tie down enemy forces, which would otherwise attack vital NATO shipping. More importantly, a naval threat might tie down Soviet ground forces, which otherwise would help overwhelm NATO. This concept was later called 'virtual attrition.' It turned up in the Gulf War, where the threat of an amphibious operation tied down several Iraqi divisions. Indeed, until Zumwalt entered office the U.S. Navy had planned exactly this sort of integrated wartime strategy. An early offensive (by the NATO Strike Fleet Atlantic) into the Norwegian Sea would have paralleled strikes by Sixth and Seventh Fleet carriers. These attacks would destroy the Soviet bomber threat. Soviet submarines trying to deal with the carriers would be destroyed by their escorts, drastically reducing the threat faced by convoy escorts in the open ocean.

In a European war, moreover, power projection might be the single most important contribution to the ground war that the Navy could offer. Its carriers were the most powerful NATO strike assets which could operate on the strategic flanks (Norway and Southern Europe) of a Soviet advance through the heart of Europe. Merely threatening strikes and amphibious landings would probably force the Soviets to keep strong forces on the flanks; those forces could not move quickly enough to contribute to the main attack against the outnumbered NATO armies. The sheer mobility of the naval forces would tie down disproportionate Soviet forces ashore. Moreover, attacks from the flanks would probably cut into the lines of communication on which a Soviet attack would depend. The Navy had made exactly this argument during the bitter post-1945 fight over which forces to maintain as the U.S. defence budget was squeezed. It had lost, because the Army and the new Air Force had said that the Navy's contribution duplicated theirs; the only unique naval mission was protection of vital shipping. In framing his own approach to naval forces, Admiral Zumwalt probably remembered that earlier disaster, during which he had been a student at the Naval War College, surely well aware of what was happening.

After 1976 the Carter Administration pressed Zumwalt's logic further. It had no real interest in power projection into the Third World (the President sometimes said that the central problem was the conflict between the developed North and the undeveloped South, not that between East and West). The Administration reviewed its military programme and, not surprisingly, concluded that the expensive carrier navy was not really needed. The Navy's reply was a study called Seaplan 2000.

The Emerging 'Backfire' Threat

By this time the Soviets were introducing a new supersonic naval missile-bomber, the Tu-22M (NATO 'Backfire'), which had twice the range of the existing subsonic 'Badger', and could carry a much larger payload. It also had a more flexible flight profile. Now the bombers were, if anything, an even greater threat. The new Navy study concentrated on the potential offered by some new technology, the F-14 fighter with its very long range Phoenix missile, the shipboard Aegis anti-aircraft missile system, the new towed arrays (and other advanced acoustic sensors), and the projected follow-ons to the Mk 46 and Mk 48 anti-submarine torpedoes.

Given the new acoustic systems, a fleet could probably beat off Soviet submarine attacks. U.S. submarines might even detect and kill submarines before they could get close enough to fire their anti-ship missiles. The fleet could then concentrate on a different task: winning what was now called maritime air superiority by destroying the Soviet shore-based missile bombers, keeping them out of the fight against NATO-bound shipping. Since the bombers were the only major Soviet naval weapon which could overwhelm convoy escorts, such an offensive might well be decisive. It was a perfect example of winning sea control by projecting naval power, a forerunner of what, a few years later, would be called the U.S. Maritime Strategy.

The F-14/Phoenix combination had been conceived specifically to break up large air raids directed against the fleet. For that to be practical, the fighters had to engage the bombers before they could launch their missiles. In the late 1960s, when the Phoenix weapon system was conceived, that seemed entirely practical: fighters would orbit about 150nm from battle group centre, and the Soviets would have to fire weapons such as K-10S (AS-2) and KSR-2 (AS-5) at about that range. In the mid-1970s, when F-14s were entering service, the situation worsened significantly, because the Soviets were deploying longer-range AS-4 (Kh-22) missiles (in fact it now seems that the Soviets considered the early version of Kh-22 inadequate in the face of an F-14 defence). U.S. practice now shifted. Since the bombers might well launch before they came within fighter range, the fighters would concentrate on the missiles as soon as the bombers fired them. That was despite the fact that their own air-to-air missiles were far less effective against missiles than against bombers.

The fleet just could not rely on its existing surface-fired defensive missiles as a backstop. The net effect of the new tactics would be to allow the bombers to escape after firing, to reload and return to attack again. There was very little point in sending the fleet north into Soviet-dominated waters, if all it could do was survive volleys of AS-4s. If the fleet's fighters were occupied dealing with the incoming missiles, the fleet would have to destroy the bombers in their bases. That might well be impossible, since a President probably would forbid any attacks on Soviet soil at the outset of a war – yet it was essential that the bombers be eliminated as early in a war as possible.

The new Aegis system promised such high performance that it could serve as the badly-needed backstop. Given enough Aegis ships, the F-14s could concentrate on the bombers, both before and after they launched their weapons. This practice was later called concentrating on 'the archer, not the arrow'.

The U.S. Maritime Strategy

When he became Secretary of the Navy under President Ronald Reagan in 1981, John Lehman asked his naval staff to articulate the Navy's strategy for a major war. He felt that without such an articulation, he could not convince Congress to finance the much larger fleet the United States badly needed. As secretary, Lehman was unusually sophisticated. He and his CNO, Admiral Thomas Hayward, had both worked on Seaplan 2000. Moreover, Lehman had helped convince President-elect Reagan that the future defence emphasis should be on non-nuclear warfare, *ie* on exactly the sort of protracted conflict for which the Navy was well suited. This achievement was all the greater since during his campaign Reagan had concentrated on the Soviet strategic nuclear threat.

Lehman's staff almost literally stitched together existing war plans to produce a global concept of future naval war. The result embodied the classic offensive concept – with which Lehman was already somewhat familiar thanks to his experience in the 1970s. Lehman's great contribution was to embrace the classical ideas (which seemed quite radical in the wake of Zumwalt's work). He articulated the concept as the U.S. Maritime Strategy, and it shaped the programmes the U.S. Navy followed during the 1980s. Later it would become clear that the strategy was strikingly like that another great sea power, Great Britain, had followed in confronting a land power, Russia, more than a century earlier. Indeed, it turned out to be much like that which all major sea powers had followed in fights against land opponents. Such wars inevitably involved coalitions, the sea power contributing mobility which made the war global and thus tied down much of the opponent's army.[4] They were also sequential. First sea control had to be guaranteed, after which the Navy could support the coalition on land by attacking on the enemy's strategic flanks.

For example, in the usual view a war between NATO and the Soviet Union meant a war in Central Europe. It was not even clear to what extent the direct defence of flanking areas such as Norway mattered. The Far East was clearly irrelevant. To a naval strategist, potential naval pressure in the north and in the Far East was extremely valuable, because it would tie down important resources which would otherwise be fed into battle in Central Europe. As Pacific fleet commander, in the late 1970s Admiral Hayward had argued exactly this point. Pentagon planners proposed that, in the event of a European war, the Pacific Fleet should 'swing' to the Atlantic, to reinforce the relatively weak force already in place. Hayward argued the opposite: his fleet should be kept in the Pacific specifically to tie down Soviet naval bombers and submarines there. That was another way in which power projection (in this case, mainly its threat) and sea control went together. Hayward's ideas were developed further by a Strategic Studies Group at the Naval War College at Newport.

The growing Soviet 'Backfire' force would have to be dealt with at the outset. Lehman made a point of advertising that, early in any big war, a carrier strike force would run north into the Norwegian Sea to attack Soviet naval bases in the Kola Peninsula. To make his point, Secretary Lehman sponsored the sort of multi-carrier exercises needed to develop the necessary tactics. He knew that the area was extremely sensitive, since it supported the forces protecting the 'bastion' in which Soviet strategic submarines would operate in wartime. Moreover, he made it clear that exactly those submarines would be pursued by various U.S. naval forces from the outbreak of war.

The hope was that the Soviets would take Lehman seriously. Since they could not afford to lose their bombers on the ground, as the fleet headed north they would stage the most massive air strikes they could, exposing their valuable bombers to the fleet's massed F-14s. The great fear was that the Soviets might mount co-ordinated attacks from several directions. To prevent this, they had to be induced to attack the carriers at the greatest possible ranges. The further the bombers had to fly, the looser their co-ordination. The U.S. Navy thus had to prove to the Soviets that it could mount strikes from very great distances; therefore it flew thousand-mile A-6 missions, using tankers. The fleet's nuclear-armed Tomahawks offered the Soviets a similar incentive to deal with the U.S. fleet as soon as possible.

For their part, the F-14s had to begin their attacks on the bombers as far away as possible. That would help guarantee against the possibility that the Soviets would field even longer-range anti-ship missiles. It was not yet realised to what extent 'Backfire' could fire its Kh-22M (AS-4) missile in a pop-up trajectory based on ocean surveillance data; indeed, U.S. intelligence did not differentiate between Kh-22 and Kh-22M. Through the 1980s it was expected that the Soviets would soon field a pop-up missile which would lock onto its target after launch. Another reason to concentrate on the bombers was the possibility that a future Soviet anti-ship missile would be significantly smaller than a Kh-22, and therefore would be carried in greater numbers. If the bombers managed to launch such weapons, they would have a much better chance of saturating the fleet's anti-aircraft defences. This threat would have materialised in the form of the Kh-15 aero-ballistic missile, six of which (twice the Kh-22M load) a 'Backfire' could carry internally.

In pre-1980 tactics, the fleet's horizon was defined by its longest-range sensor, the big radar on the E-2C airborne early warning aeroplane. It could detect targets 250nm away (in 1983 a modified version, with a range of 350nm, was ordered). If the fleet knew that the bombers were coming from a well-defined direction, then the E-2C could loiter a few hundred miles out, in that direction, to push the horizon out to 400 or 500 miles. The battle group's horizon could be pushed even further out if the signals from the Soviet bomber radars could be exploited. Radar detectors (ESM) on board specialised U.S. aircraft might pick them up well before they detected any ships. This concept, of a Battle Group Passive Horizon Extension System (BGPHES), was eventually embodied in the ES-3A Shadow.[5] Much would depend on how accurately the bombers could be coached into position by the Soviet Ocean Surveillance System, *ie* mainly by the satellites. The better the coaching, the less the bombers would have to rely on their own radars. In the worst possible case, a pop-up launch from low altitude, the bombers might not use their radars at all. On the other hand, even the satellite system provided time-late data, and it might not reveal the carrier's position within a dispersed formation. In that case the bombers might well have to use their own radars, giving themselves away, and they would be unable to launch at maximum range.

The Outer Air Battle

In the early 1980s it was assumed that the usual attacking force would be one or more regiments (eighteen to twenty-four 'Backfires' or 'Badgers' each) supported by reconnaissance and jamming aircraft, so that the defence would have to be mounted under intense jamming conditions. The regiment would probably try to launch missiles at a range of about 200nm. The Navy called the long-range attack on the bombers, from about 100 to 200nm from battle group centre, the Outer Air Battle. The Inner Air Battle was the multi-layer fight against Soviet missiles, using area- and then point-defence missiles, as well as various kinds of jammers. If the direction of the raid could be estimated in advance, one or more missile ships could be placed to attack the incoming bombers before they could fire. Such ships might even maintain radar silence, waiting for cues from an E-2C.

The ideal was to destroy the bombers before they could fire their missiles – which meant beyond the carrier's sensor horizon. This was a new kind of OTH combat: the bombers would have to be destroyed at or beyond the edge of the battle group's sensor horizon. Space assets made the Outer Air Battle possible. The earliest warning would probably be provided by the radio chatter of the bomber pilots as they prepared for take-off. If the Russian accounts of the NSA satellites are correct, they had already detected exactly such chatter during the Soviet build-up in the Far East. Chatter was probably unavoidable if safety during take-off was to be ensured.[6] The fleet would receive its warning via its satellite links.

Given a take-off warning, it was possible to estimate, to within a few hours, when the bombers would appear near the fleet. It was also possible to guess that they would have to approach within a given sector. These guesses became more difficult as the Soviets switched from slow 'Badgers' to supersonic 'Backfires,' which might fly out to approach the fleet from its flanks. The further out the defence was mounted, the narrower the sector a carrier could cover with her F-14s. The narrower that sector, the worse the consequences of a wrong guess. Moreover, as the F-14 flew out, it would soon be beyond the usual line-of-sight radio range. It had no satellite link (nor did any other tactical aeroplane), so continued communication required some sort of link back to the carrier. One possibility was to orbit S-3s between the F-14s and the carrier(s). Because they were fitted to join the high-frequency (beyond-the-horizon) Link 11 net, they could remain within radio range of the carrier.

This point suggests just how elaborate Outer Air Battle tactics could be. Not only would a carrier have to assemble multiple lines of F-14s, but to keep the defence alive for long would require numerous tankers, not to mention radio relay aircraft and perhaps jammers. Just about all the carrier's aircraft, other than strike aircraft, would be involved. In 1982 it was estimated that simply to reconfigure a carrier's aircraft and her flight deck for Outer Air Battle operation would take about twelve hours. It was assumed, in 1982, that a suitably-armed F-14 could destroy two or three bombers. To deal with a raid by twenty-four aircraft, then, the carrier would have to deploy a full squadron of F-14s. An F-14 squadron could mount a defence for about six hours, so the carrier needed enough warning to estimate the attack window within four to six hours.[7]

Before the 1980s, fleet air defence tactics had envisaged a ring of CAP stations distributed around the carrier at a range of about 150nm. As bombers crossed the radar horizon of the E-2 over the fleet, the fighters could be vectored in to attack them. The Outer Air Battle demanded interceptions considerably further out. An F-14 could fly, untanked, about 500nm from the carrier, loiter briefly, then return. It could act as an armed Outer Air Battle sensor. The screen of distant F-14s could be maintained by launching a replacement aeroplane as the first aeroplane reached its maximum range. If the long-range F-14 spotted any bombers as it flew out, it could report their approach, then attack them as it flew back in. If the carrier only had to maintain a screen over a limited sector (thanks to the estimate of bomber attack course), then the CAP could be doubled up, a long-range layer added to the usual layer at about 150nm.

It might well be impossible to predict the attack axis, so carriers would have to work together, much more closely than was usual in the 1970s, when they so often worked alone. To make that possible, Outer Air Battle tactics were standardised, a practice often expressed as the adoption of 'grid logic'. The grid was a set of co-ordinates centred on the battle group centre (which might lie between two or more carriers), in which the CAP stations were placed. Typically a defence in depth was envisaged, in which the aircraft on the outer stations would engage the bombers first. Because it was so important to kill the bombers, aircraft on the inner CAP stations would move to the outer ones as the latter were uncovered by F-14s attacking bombers. That was the opposite of what would have been done had the incoming missiles been the primary targets: the *inner* stations would have had priority, as backstops to protect the carriers from the missiles.

By 1982 it was possible to envisage Outer Air Battle tactics using the F-14s, but it was clear that they would be difficult to carry out. It would take hours to configure a carrier's aircraft to support the Outer Air Battle; for example, the carrier would not be able to conduct air strikes (among other things, all her tankers would be occupied supporting F-14s on distant CAP stations). Even so, the Outer Air Battle stations could not be maintained for very long. Moreover, much depended on estimating the direction from which the raid would approach.

Sensors

For Outer Air Battle tactics to succeed, the bombers had to be tracked after they took off. Spaceborne radars were an obvious possibility. Such a system would correspond in principle to the Soviet RORSAT (US-A), but with important differences. On the positive side, radar detection of aircraft might be easier than detecting ships. Although they are much smaller than ships, aircraft at high altitude are not surrounded by reflecting clutter. Also, Doppler techniques might be used to distinguish a fast aircraft. On the negative side, the revisit times, typically two hours, entirely adequate for ship detection and tracking, would be unacceptable for aircraft. 'Backfires' would take an hour or less to reach their target areas; the satellites would have to detect them well before they got there. They would have to visit everywhere in the North every half hour or so. A satellite at high enough altitude might scan so vast an area that it could keep a bomber in sight for several hundred miles, measuring its course and speed.

The satellites could not fly too high, so they would not spend a very long time over any one area. An anti-'Backfire' system would, therefore, require multiple satellites. In one study made in the early 1980s, alternatives included twelve satellites at about 2000nm altitude, or twice as many at about 1000nm. Either system would have been very expensive, yet anything less would not be particularly effective. Also, in the early 1980s stealth technology was clearly on the horizon. It would be particularly effective at the short wavelengths mandatory for spaceborne radars. Their range would be drastically reduced. Satellites would have to fly much lower, and the swaths they searched would be much narrower. Many more satellites would be needed to achieve the sort of solid coverage the Navy needed to fight the Outer Air Battle.

The Outer Air Battle was important, but the Navy's budget was quite finite. Instead of the satellites, Secretary of the Navy John Lehman chose the ground-based ROTHR – the Relocatable Over-the-Horizon Radar.[8] Staring up into the Norwegian Sea, a ROTHR in Scotland would provide exactly the sort of continuous coverage the Navy needed. Another, in the Aleutians, could look south over the potential bomber routes used by the Soviet Pacific Fleet. Certainly a few massive OTH radars would not cover the world the way satellites could, but then again the 'Backfire' threat was limited to a few key areas. At first blush, a massive OTH radar might seem to be the antithesis of naval operations, since it is anything but mobile. However, ROTHR could be moved (in theory in about two weeks) to meet changing requirements. Moreover, it operated at lower frequencies (longer wavelengths) which would probably be much less affected by stealth techniques. The prototype, for Amchitka in the Aleutians, became operational in 1985.[9]

Moreover, in the early 1980s there was no unified Outer Air Battle policy. Despite the build-up of U.S. defences, money was tight. Competition was triggered by the surface force's requests for a new surface-to-air missile to attack high-altitude targets at very long range; called SM-3 or ASAM (Advanced Surface to Air Missile), it enjoyed important Congressional support. Although an Aegis ship might not be able to detect an incoming bomber early enough, it might be able to hand over control to an aeroplane or even an unmanned air vehicle, a concept then called Forward Pass.[10] Within the Navy's R&D budget ASAM competed with a new long-range air-launched missile (the Advanced Air-to-Air Missile [AAAM]). There were continued calls for a radar satellite programme, and proposals to use unmanned air vehicles to detect the incoming bombers.

The surface community argued that given the kind of command and control demonstrated in their big exercises, the Soviets might well be able to mount co-ordinated multi-regiment attacks from different directions. For example, a battle group running north up the Norwegian Sea might face simultaneous strikes from the north (*ie* from the Kola Peninsula) and from the east (from the Baltic Fleet). It seemed unlikely that a few F-14s on four or eight CAP stations could handle this sort of onslaught. Moreover, a great deal depended on early warning. If the carrier had to maintain air defence around the clock, she would soon exhaust her few radar aircraft.

In 1983 Secretary Lehman ordered a study of all possible approaches. Industry was asked for total systems approaches, from indications and warning (take-off warning) to final engagement. That was very different from the usual requests for replacements for specific weapons. The programme stalled for two years because of Congressional and Defense Department pressures to concentrate on a proposed improved SM-2.[11]

Lehman's approach extended well beyond the issue of the next missile or missiles. Key to fighting any Outer Air Battle would be the extent to which the fleet could co-ordinate its assets. It needed much faster combat system computers and the ability to control many more aircraft. Thus the Outer Air Battle helped push the Navy to buy a longer-range ship-to-air link, JTIDS, and also a new generation of ship combat direction systems.[12]

Companies submitted their approaches in January 1986. One revived the concept of radar satellites. For years, Martin-Marietta had been developing a rocket-ramjet engine, which it had offered the Air Force as the powerplant of a next-generation supersonic cruise missile. The company proposed to the Navy that a long-range supersonic missile powered by its engine supplement the Standard Missile (SM-2) which armed Aegis ships. The new missile would be cued by radar satellites with down-links to the ships. Like SM-2, the new missile would have an autopilot controlled from the ship, so it could be directed into position to deal with an air target located by the satellite (it use its own active seeker to home on the target, well beyond the ship's horizon). One virtue of the proposal was that it would make maximum use of existing ships and their weapon systems. As the targets came closer to the ship, the satellite would hand over to the ship's own radar, and SM-2s would be fired. The idea was elegant but, as in the earlier analysis, the satellites would have been horribly expensive.

Another proposal was to use technology then under development for a manoeuvring re-entry vehicle for ballistic missiles. ROTHR defined a series of resolution cells (each quite large); it could only say whether a bomber was or was not in each cell. Normally a cell would be far too large for a missile's active seeker to search, which was why ROTHR was considered a surveillance tool rather than a fire control sensor. However, a sensor on board a missile approaching the bomber stream from *above*, *ie* from space, might view the entire resolution cell. It turned out that a ballistic missile, LORAINE, could be fired from a standard vertical launcher cell, and plans called for equipping its manoeuvring re-entry vehicle with a millimetre-wave radar. The weapon would dive so fast that the bomber would have little chance of escaping. The space aspect of the system would have been the high-capacity link between ROTHR and the firing ship. LORAINE was not ready for tests until 1990, by which time the Cold War, and the Outer Air Battle, were both effectively over.[13]

As it was, the only viable Outer Air Battle weapon was the F-14. Given its very powerful on-board radar and the associated computer system (AWG-9), it could operate even when not under direct control of the carrier, *ie* beyond the carrier's horizon. Outer Air Battle funds paid for the design of a new longer-range missile, AAAM (Advanced Air-to-Air Missile) and associated sensors, which would extend the aeroplane's effective range even further. However, the F-14 would be useless in the Outer Air Battle without space assets. Only a space link could ensure that the carrier group received either the warning of an impending strike (as the Soviet bombers assembled for take-off) or the strike course and speed supplied by ROTHR. These data had to be fused with whatever other data other U.S. systems could provide, then supplied very quickly to the fleet, so that F-14s could be launched in the appropriate direction. Special techniques of fighter control had to be developed to support the very long range use of F-14s. Within-horizon interceptions were ordered on the basis of the carrier's own radar, or else by the E-2C controlling the fighter.[14] In either case, what counted was the range and bearing of the target relative to the controlling ship or aeroplane. Now the carrier had to send the fighter off in a direction defined geographically, by a distant sensor or sensors.

Because they were working beyond the carrier's horizon, the F-14s would have had to navigate independently. The existing Transit system could not supply their needs, because it required considerable time in which to fix a platform's position – time during which the F-14 would hardly remain in one place. By the early 1980s the Navy's solution, both for the F-14 and for very long range strikes by its A-6 bombers, was inertial navigation, adjusted at take-off time by the carrier's own inertial system, CAINS. At this time GPS was not yet operational. F-14s are now being fitted with GPS because it is superseding all other forms of carrier navigation.

Slow Walker

As the Outer Air Battle studies continued, it turned out that, as in OTH-T, a 'national' sensor was already providing some of what the Navy needed. A series of geostationary DSP (Defense Support Program) satellites had been in place since the early 1970s to provide early warning of Soviet missile firings. They used arrays of infra-red detectors, which periodically scanned across the cool Earth.[15] The hot exhaust of a rising rocket would automatically register. As it turned out, so did afterburning jet engines running at high altitude. The great threat was a Backfire cruising at supersonic speed – which meant flying at high altitude on afterburner, hence detectable by DSP.[16] Given the vast area each DSP covered, a jet seemed to move slowly across the screen. Jets were called 'Slow Walkers', in contrast to satellites (Fast Walkers, visible due to the glint of sunlight reflected off them). To reach as far as they had to, to attack distant U.S. formations, Soviet bombers had to fly high enough to be detected. If the Slow Walkers were reported promptly enough, the Outer Air Battle could be fought and won.

In fact, the idea of using a space-based sensor to detect afterburning jet engines dated back to 1956. Lockheed's original MIDAS proposal, written that year, envisaged detection of both missiles and high-flying bombers. Early in the DSP programme, in 1974, Kenneth Horn, then head of the advanced mission requirements section of the Aerospace Corporation (which was responsible for

DSP design), showed that this was a reality, by matching DSP data with intelligence on 'Backfire' bomber flights. After eight years of fruitless attempts to interest the Air Force in this capability, Aerojet-General (which built the satellite) turned to the Navy in about 1982. Almost at once the Navy sent a contingent to the Nurrungar (Australia) DSP ground station, to see whether DSPs could pick up either a bomber takeoff or the launch of a bomber's AS-4 missile. The Navy briefed the commander of the Air Force Space Command on this new capability, renamed Project Slow Walker, on 3 May 1983.

The first formal U.S. Navy Operational Requirement was drafted in 1984. Navy personnel at Nurrungar would actually see the Slow Walkers crossing the screens, and would enter their data into communications processors for transmission to the Fleet via an AFSATCOM channel on the Fleet Satellite. This operation was called the Slow Walker Reporting System (SWRS), and it was effective because it used the techniques already developed for reporting ship movements for OTH-T. When the Slow Walker early warning technique was demonstrated in a fleet exercise (Ocean Venture), the fleet commander commented that he had never before received such timely data. SWRS became the Tactical Event Reporting System (TERS) in 1989. Slow Walker was so important that it was mechanised; in 1990 (FY91) IBM received a contract for a specialised Slow Walker ground station, which became operational in 1992 (FY93). A later project, Radiant Ivory, provides a down-link receiver by means of which a command ship can receive DSP data. It is fused with other national sensor data. New DSP satellites (5R and 6R) were fitted with sensors intended specifically to detect Slow Walkers. Their sensitivity could be increased for a designated area of interest (AOI) on the Earth.

The success of the DSP programme apparently led to installation of an IR package, Heritage, on SDS satellites in Molniya orbits. This was an intelligence-collection project intended to obtain signature data on fast-burning high-acceleration missiles comparable to the U.S. Sprint anti-missile missile.[17] The SDS Heritage programme was eventually renamed DSP-A (for Augmentation); there were allegations that the warning role was, in effect, added to save it from budget cuts. Whatever the truth, Heritage had definite naval value, because SDS satellites in Molniya-type orbits could see bombers taking off from bases high in the Arctic. It is also possible that the staring sensor could pick up cooler IR sources, such as the thermal trails left by subsonic bombers. The Navy set up a new system, Jogger, to convey bomber flight data from the DSP-A/Heritage IR sensor.

As in the case of OTH-T, the Outer Air Battle had an interesting echo during the Gulf War. Slow Walkers opened up the possibility of using

DSP data tactically. There were no important Slow Walkers coming out of Iraq, but the DSP satellites did detect Scuds being launched. Their stereo data was precise enough to predict roughly where missiles would fly (to Saudi Arabia or to Israel). Warnings to take cover could be issued, and the Patriot batteries alerted. The interception footprint of a Patriot was such that without some cueing it probably would not have been able to hit an incoming Scud. In effect the satellites were supporting the Patriots' OTH battle.

At the time, the Navy lacked any suitable anti-Scud missile, but the SPY-1 phased-array (electronically-scanned) radar of the Aegis system could clearly detect and track missiles. However, a SPY-1 simply scanning the sky spends so little time looking in any one direction that it has only a very limited chance of detecting a very fast incoming missile. If it is cued by another sensor, its probability of detection increases enormously. As in the Outer Air Battle, TENCAP cueing makes it possible to concentrate the defence (in this case the radar) in the appropriate direction. During the Gulf War, DSP data were used to cue SPY-1s (as well as the land-based Patriot batteries which actually engaged the incoming missiles). It followed that, given sufficient cueing and the appropriate missile, Aegis could engage ballistic missiles like Scud. The Outer Air Battle lives on, albeit in drastically altered form.

A major Gulf War issue was whether data from the DSP satellites should be centrally controlled and issued to those who seemed to need it, or whether it should go directly to users in the field. The Air Force, which operated the system, took the first approach; the Army and the Navy preferred the second. They argued that any attempt to process DSP data and then distribute it overseas would make for fatal delays, and would also be vulnerable to failures in long-haul communication. Eventually they prevailed, to the extent that they developed a portable (by C-141) JTAGS (Joint Tactical Ground Station). JTAGS receives raw DSP satellite data; it also receives battlefield sensor data, such as that from radars. The system gives the location of the launcher, the time of firing, trajectory data (needed for interception), and the estimated impact point. Its calculations are based on stereo comparison of the data from two or more DSP (or successor) satellites. JTAGS can track multiple incoming missiles simultaneously. In theory JTAGS can both cue defensive missiles and provide targeting data for missiles aimed at mobile launchers; it can also warn targets that they are about to be hit.[18] There are currently five JTAGS ground stations, manned by Army-Navy crews. They support the forward-deployed CINCs, providing early warning of ballistic missile launches by directly receiving DSP data.

JTAGS is part of the larger Tactical Event System (TES) which had grown out of Slow Walker.[19] The first prototype was delivered in 1993.

The first production version was installed in Korea in 1997, replacing a prototype. Another is in Germany. JTAGS may also be integrated into the Navy's own emerging missile defence system (see Chapter 13).

Space-Based Wide Area Surveillance

The DSP system is now deteriorating. It is to be replaced by a two-tier system, SBIRS (space-based IR), designed to support the planned U.S. National Missile Defense: four satellites in geosynchronous orbit, two in highly elliptical orbits, and a low-altitude constellation (SBIRS-LEO or SBIRS-Low). Current (1999) plans call for a total of eighteen to twenty-four satellites capable of tracking missile re-entry vehicles in mid-course. The first are to be launched in 2004.

In the early 1990s the Air Force and the Navy jointly proposed a new Space-Based Wide Area Surveillance System (SB-WASS) which would have replaced, among other things, White Cloud. The Air Force hoped that new technology would finally make space-based air search radar affordable. If that succeeded, it could move its AWACS (airborne early warning) operation to space and achieve continuous global coverage. It might also move J-STARS (detection of moving ground targets) into space. The Navy was more interested in the infra-red technology pioneered by Slow Walker. For example, it expected stealth (anti-radar) technology to spread rapidly, drastically reducing the value of costly space-based radar. Reportedly a Navy experiment, conducted about 1992, showed that a space-based IR system could detect even stealthy aircraft.[20] Reportedly the Navy's preference for IR over radar represented a reversal, the Air Force initially taking that position, then reversing it.[21]

Reportedly the Navy's current view is that IR is quite sufficient for fleet air defence. The Air Force sees a space-based radar as a potential ship detector and as a means of detecting strategic cruise missiles which might have very low IR signatures. Ship detection would aid in the war against drugs. From the Navy's point of view, passive systems such as White Cloud (and any successors) already provide much of the desired ship detection capability. In June 1997 the Defense Department conducted a study of whether SBIRS-Low might have important applications beyond national (*ie* very long range) missile defence.[22] They were theatre missile defence, gathering technical intelligence, and replacing the current radars used for missile attack warning. The missile defence part of the study examined not only cueing but targeting (using multiple satellites to locate a target in space, as an extension of current DSP stereo practices).

The study concluded that SBIRS-Low would be a natural complement to the Navy theatre ballistic missile defence system now under development, because it provides global data well beyond a ship's horizon. There are problems. The very long range targets of national missile defence move at high altitudes, hence are relatively easy to detect and to track. A shorter-range missile spends more of its time below the horizon. SBIRS-Low operates in IR bands which can detect a missile against the glow of the earth, but it may find resolution (of a missile against debris or decoys) more difficult.

SBIRS-Low would feed into CEC and the emerging Joint Composite Tracking Network (JCTN). These systems were all designed to operate at a very high data rate, mainly using radar data. The SBIRS downlink might have to be modified to match a radar data rate, or the other systems would have to be modified to accept occasional SBIRS data at a rate much lower than the usual radar update rate. There is also some question as to how well SBIRS can discriminate between a real missile and debris. In the national missile defence role it has far more time, perhaps as much as twenty minutes, to do so; but decisions must be much faster in theatre missile defence. Discrimination is particularly important because there will be relatively few defensive weapons, all of them non-nuclear. Waste may be intolerable. Even if decoys fly relatively close to a real warhead, a defending weapon attacking them probably will not damage the warhead.

SBIRS is attractive as an intelligence-gatherer because it is to operate in several different IR bands. Depending on its resolution, it makes an interesting complement to conventional intelligence-gathering satellites. With many low-altitude satellites in orbit, one will almost always be overhead. For example, it will be difficult or impossible for any Third World country to test a new ballistic missile without a satellite observing it, and collecting its signature for later possible defensive use.

Finally there is the key Integrated Tactical Warning/Attack Assessment role in ballistic missile defence, the warning so clear that it can be used as the basis for a counterattack. Current doctrine demands 'dual phenomenology', *ie* detection by two systems so different that it is very unlikely that both are wrong. This requirement was adopted as a result of numerous false alarms – the famous cases of birds or even the Moon fooling radars come to mind. In this case the combination of two SBIRS components operating in different IR bands might suffice. The big ground radars, which have their own problems, might be retired.

Discoverer II

The Air Force apparently sees a space-based radar not as a complement to existing surface systems (such as naval radars) but more as a replacement for airborne radars such as those aboard AWACS aircraft.[23] The aircraft need bases near combat zones, but such facilities may well not be available in future wars (a point often made by the Navy). Furthermore, AWACS is a key element in the new style of information-based warfare. The Chinese in particular have said that they cannot hope to match U.S. investments. Instead they will employ asymmetric measures. In their case that apparently means a combination of deception and direct attacks on key sensor systems such as AWACS. For example, the Chinese have announced a surface-to-air missile, FT2000, intended to attack AWACS aircraft. Whether or not this particular weapon is effective (it seems to lack the requisite range), the threat is certainly apparent. Moving the AWACS function to space would solve the basing problem, and it would also make attacks against the system far more difficult. Against these advantages, space systems are extremely expensive, perhaps altogether unaffordable.

The joint programme was tentatively titled the Space Based Wide Area Surveillance System. In some documents the Air Force concept is called AWACS-B.[24] An infrared system would require eight to ten satellites for full global coverage; a radar system would require eight to twenty-four. The Air Force's June 1998 estimate of future space launcher requirements shows a series of launchings ('Payload B') which seem to correspond to AWACS-B, scheduled for FY04-FY12 (six payloads), with replacements beginning in FY16.[25]

J-STARS (Joint Surveillance Targeting Attack Radar System), the big airborne system which detects moving ground targets, has a roughly analogous role with respect to attacks against enemy vehicles. It is a key targeting system for ship-launched missiles supporting ground combat. At present J-STARS requires a very large aircraft (a converted airliner), which must fly from a large airfield near the battle zone.

Although the Air Force wants to move both AWACS and J-STARS into space, apparently J-STARS is by far the easier of the two. In 1997 Lincoln Lab, which does much of the Air Force's electronics research, reportedly found that it would be nearly impossible to duplicate AWACS functions in space.[26] A 40 × 60m antenna would be needed; the lab estimated that the necessary spacecraft would weigh nearly 40,000lbs. To maintain constant coverage, twenty-five to thirty satellites would be needed. By way of contrast, a J-STARS equivalent would need a 3 × 16m antenna, and the satellite might weigh as little as 7500lbs. Because the targets move slowly, coverage need not be as complete, with a revisit time of 100 seconds vs. 10 seconds for the space AWACS. Only fifteen to twenty satellites would be needed for effective continuous coverage (critics pointed out that J-STARS provides battle management as well as the sensors).

Discoverer II/Starlite is a current proposed space-based equivalent to J-STARS. It would provide a combination of synthetic-aperture radar (SAR) imagery and high range resolution (HRR) ground moving target indication (GMTI).[27] Unlike typical reconnaissance satellites, but very much like J-STARS, this data would be down-linked directly to the tactical user. Coverage would be nearly continuous; how nearly would depend on how many satellites were placed in orbit. With the twenty-four satellites currently envisaged, GMTI coverage would be very nearly continuous.

Discoverer II is a prototype demonstrator. Full capability would involve a total of twenty-four Starlite satellites, to be in place by 2008 (the two prototypes are to be launched in 2003-4). In October 1998 the U.S. Defense Science Board supported Discoverer II, calling for a follow-on system which would incorporate a radar capable of penetrating foliage. Whether the programme is affordable is still open to question. In 1998 the House of Representatives tried to cancel Discoverer II altogether, on the ground that it was too expensive, but the Senate saved it.

ELEVEN

Copernicus

Both OTH-T and the Outer Air Battle made long-haul communication with a deployed battle group much more extensive and much more urgent. In 1989 Admiral Tuttle went from the Atlantic Fleet FCC to head the Naval Space and Electronics Systems Command (SpaWar), which was responsible for naval communications. Most naval officers probably took communications for granted. SpaWar's predecessor command, NavElex, had always seemed relatively insignificant compared to the main platform and weapon commands, NavSea and NavAir. Tuttle pointed out that communications, his command's responsibility, became crucial as the shooters came to depend more and more on off-board sensors. He declared it his responsibility to insure better 'sensor to shooter connectivity'.

Clearly the fleet's need for communications would continue to rise, particularly as it came to rely more and more on off-board sensors such as those on board satellites. With HF radio capacity very limited, most communication had to be via satellites. Yet satellites were very expensive, and capacity would always be limited. Tuttle's best solution would be more efficient use of the existing channels. As it was, when the Admiral took command at SpaWar, each separate long-range national sensor system reported through its own communication system, using its own kind of formatted messages. It had its own dedicated satellite channel. The same was true of other communication systems. This connection between function and physical channel was called 'stovepiping'. No matter how little of channel capacity a message system (*eg* TADIXS) took up, no other channel could take advantage of that under-use to put through the additional information it was trying to transmit. Moreover, new functions would be difficult to accommodate within the very limited capacity built into existing and planned satellites.

Tuttle realised that the stovepiped channels had been created to connect distinctive kinds of hardware. However, he was unifying the hardware, using a single type of computer for all the different command/control/communications functions, as he had learned to do in the FCC. Now it was no longer so clear that the divisions between the channels were natural ones. Clearly the Fleet Satellite in its high orbit could not be modified to change the number of radio channels it carried, each 25 kHz wide. However, those channels could be used much more efficiently.

Multiplexing

There was already some effort in this direction. Single satellite channels, such as that devoted to the Fleet Broadcast, were already time-division multiplexed, each user being assigned a time slot; the messages were transmitted in sequence. Often a message took up less than its allotted slot, but the slot lengths were inherent in the multiplexing hardware, so they could not be shortened to match actual message lengths; the longest standard message block determined the length of the standard slot.

It had long been realised that this was inefficient. When Tuttle joined SpaWar, a long step towards greater efficiency had already been taken by developing DAMAs, Demand-Assigned Multiple Access devices. By the early 1970s ARPA, the Defense Department advanced development agency, was deeply interested in improving the efficiency of communication (largely in connection with its new ARPANet, which later became the Internet). Just as in the Navy's satellites, setting slots long enough to accommodate full messages was clearly wasteful. Ideally, time should be chopped into slots so short that none would be wasted. A powerful computer could chop up each incoming message into slot-length blocks, numbering them so that a computer at the receiving end would know how to reassemble them into the full message. This packet-switching technique was central to the success of the Internet. Users downloading material generally see it arrive in sections, which are the chopped-up blocks.

ARPA had another requirement in mind. Its computer net was intended as a prototype system for a national command system which could continue to function even if parts of the country were destroyed by a nuclear attack. The system would automatically switch around areas which had been destroyed, or which were clogged by traffic. To do that, it would send its short message blocks independently, each along whatever channel was best at the moment it was sent. Because they were numbered and headed by their intended recipients, they did not have to arrive in the order sent. For that matter, blocks could be sent along different channels at the same time, for much faster overall transmission.

These ideas were incorporated in the Navy's DAMA. A DAMA accepts standard streams of data at the usual high-speed teletype or voice rate, 2400 bits/sec. It can send blocks in parallel along different satellite channels to achieve effective rates of 9600, 19,200, or 32,000 bits/sec.[1] The Navy released an initial paper describing DAMA in July 1972, and the Joint Operational Requirement followed in January 1975. At this stage DAMAs were planned for surface ships and for the shore communications stations serving them, both of which could accommodate the necessary large computers. By 1982, DAMA was far enough advanced that the Navy communications handbook promised that it would quadruple service on standard 25 kHz channels (in terms of numbers of users or nets).

A DAMA provides bandwidth on demand; more accurately, it denies bandwidth when there is no demand, providing a user resources only when a signal is present. Overall, it increases effective channel capacity and it can support the transmission of several types of data on the same channel. To make it work, however, terminals have to become much more complex, providing timing, burst rates, and buffers. Network control is more complex. Significantly longer delays affect voice communications. Furthermore, a DAMA system can deny service when it is needed. DAMA techniques are, therefore, unacceptable for tactical data links.

DAMAs are controlled by the NCTAMS. There are three alternative levels of control. The least efficient is distributed, configuration information being entered manually. In semi-automatic operation, a channel controller (a computer operated by a human) allocates time slots based on user requests. In automatic operation, channel control is autonomous, based on user requests.

During the 1990s, DAMAs were applied to the UHF satellite channels. The capacity of each 25 kHz channel was increased four-fold, to four 2400 bit/sec sub-channels. Each would suffice to carry the standard secure (encrypted) voice channel, the Satellite Voice Variant of the Advanced Narrowband Digital Voice Terminal (ANDVT).[2] This increased capacity came entirely from equipment on Earth; once in orbit, a geosynchronous communications satellite cannot be modified. One shore station in each satellite footprint has four four-channel DAMAs. Each typically handles four kinds of message: secure voice (SECVOX), fast teletype, TACINTEL, and CUDIXS/NAVMACS. DAMA capacity was increased fourfold when these switches were converted to automatic (Auto-DAMA) control, assigning slots on the basis of measured message length (an intermediate semi-automatic system entered service in 1993).[3] The shipboard end of the system requires computers to reassemble messages from the blocks sent down the different channels.

Plans called for more ambitious applications of these ideas to Channel 1, which is used for the Fleet Broadcast. DAMA would increase its capacity to 9600 bits/sec. Its software would create twelve variable-capacity sub-channels, one of which (4800 bits/sec) was to be used for an intelligence broadcast (INTELCAST). In fact funds were limited, and by the mid-1990s the Navy's communications system was being integrated with those of the other services. The need for military broadcasts like the Fleet Broadcast was growing. The Fleet Broadcast will ultimately move to the joint-service Global Broadcasting System, along with the intelligence broadcasts.

As OTH targeting became more important, submarines and airborne early warning and patrol aircraft had to join the satellite net.[4] They were fitted with Mini-DAMAs (USC-42s), which were practicable because computers had become so much more compact. Moreover, they were now so powerful that a Mini-DAMA offered twice as many channels as a DAMA as well as many encryption and data transfer functions which had previously required separate equipment. The first pre-production Mini-DAMAs were delivered during FY93; integration began into P-3 (ASUW), E-2C, and EP-3E aircraft. Work on a secure voice/data encryption version began in 1993.

Meeting Increasing Demand

Tuttle found that, in a typical eighteen-hour period, a carrier battle group in the Mediterranean might receive 500 tactical reports (HITs), each amounting to 5200 bits (ten lines, sixty-five characters each): a total of 2.6 Mbits.[5] There was certainly enough current message capacity in the Mediterranean, spread over all available media, including all the satellite and HF radio systems (a total of 155.5 Mbits in eighteen hours). However, much of that capacity was clearly not in the Fleet Satellite, and much of the high capacity, DSCS, was not dedicated to naval purposes.

The problem continued to grow through the 1990s, as the needs of OTH targeting and the Outer Air Battle were joined by those of strikes against land targets. Also, the sheer number of platforms, each of which requires high data rates, has increased as surface combatants now carry and independently fire Tomahawk. In 1992 (FY93) the U.S. Navy required that each flag-capable ship be able to handle 1.2 Mbits each second, almost half of what the carrier a few years earlier had needed in eighteen hours. As of 1992, that was expected to increase to 2.0 Mbits per second in FY96 (1995), and to 6.0 Mbits/sec in FY2000 (1999). At this time DAMA applied to SHF satellites was expected to meet the FY93 requirement – about 1996. A single commercial T1 channel (television channel) carries about 1.5 Mbits/sec, so the expected FY2000 demand was for the equivalent of four such channels to serve each flag-capable ship. This was quite aside from the growing needs of other ships.

Tuttle's solution was threefold. First, messages would have to be spread across the available media, extending the DAMA idea. That alone would greatly increase effective capacity. Available DSCS channels would no longer be considered separately from the Fleet Satellite or, for that matter, from HF radio. At the ship end, all radio signals would have to pass through a single automated switching system, which eventually materialised as ADNS.

Second, Tuttle proposed to impose discipline on the message traffic. The Fleet Broadcast was a mixture of urgent tactical messages (the HITs) and administrative or logistic traffic. Messages are now tagged in such a way that the computers at naval messaging centres can save the administrative ones for periods in which little of the system capacity was being used for the urgent tactical traffic.

Third, he would make the best possible use of the bandwidth he had by pushing for the fastest possible computers. In the past, military components had typically run about a generation behind their commercial counterparts, and there had been few opportunities for block upgrades. The military programming cycle was just too slow. Tuttle extended the idea he had developed into JOTS. The navy would buy standard commercial-grade computers, and periodically it would replace all of them.

Since all computer systems would operate on the same type of computer, software could be standardised. The concept is much the same as in the civilian world, in which numerous computer programs operate on the same machines because they all fit a very few standardised operating systems (environments), the best-known being Microsoft's Windows series. The naval equivalent of Windows is the Common Operating Environment (COE), which is now part of a Defense Department-wide COE.[6] This apparently recondite development has huge implications.

For example, as the Navy shifts its emphasis from blue-water to littoral operations, it needs access to other services' systems, such as the JSTARS radar. JSTARS uses moving target indication techniques to detect vehicles moving along roads; it was extremely successful in the Gulf War. Originally it reported its data to a stand-alone console on the ground. If missiles like Tomahawk are to be used effectively against the vehicles ashore, JSTARS data must be feedable directly into the Tomahawk mission planning system. It cannot be enough to have a JSTARS console alongside the Tomahawk consoles, so that an operator can try to match data. The key is that the only really special thing about the JSTARS console is its software. That software can be modified so that it works with the COE system, just as so many stand-alone software packages now exist mainly in forms compatible with Windows. Just as in Windows, once the software is compatible, data from one package can be moved over to be used in another. Obviously modification is not a trivial proposition, but this sort of commonality and interaction is the essence of the Copernicus system.

From the user's point of view, under Copernicus channel capacity would no longer be set nearly so tightly by satellite design features settled years earlier. The new system would adapt channels to the *user's* needs. Tuttle called the new system Copernicus, to emphasise that it drastically simplified communications, just as Copernicus himself had drastically simplified man's view of the universe (if Earth is taken as the centre of the universe, the planets' motions are exceedingly complex). He also had in mind the revolutionary impact that

Copernicus had had in his own time. Like JOTS, Copernicus was possible only because computers were now so powerful. The crucial role of computers was reflected in references to Copernicus as a C^{4}I architecture (the usual C^3 – command/control/communications – plus computers and intelligence). The Copernicus programme replaced existing special-purpose terminals with more powerful workstations derived from the JOTS/JMCIS programme.

Copernicus

Copernicus was apparently derived from a Unified Networking Technology (UNT) concept announced in 1986 by the Naval Research Laboratory, the Naval Ocean Science Centre (NOSC), and SpaWar. It exploited ARPA's new multi-media packet-switched radio net (Aloha) to connect multiple users aboard a ship with multiple shore terminals. For the first time, the channel through which a shipboard user communicated was defined by software rather than by a particular radio. The object of UNT was to make the usual combination of HF and UHF (line-of-sight) communications more robust and to increase the capacity available to any one user. The interconnected net was called an Internet.

The Copernicus project office was established on 1 October 1990. A six-year procurement plan was begun in 1991. It was overtaken by the move towards greater jointness between the U.S. services. Copernicus ideas were adopted as the basis for a new Global Command and Control System (GCCS). Ultimately all satellite and other media capacity would be available to all military users, with priorities based on current needs. Ultimately, then, all of them need DAMAs and DAMA-based communication systems.

Copernicus envisaged the creation of sets of software-defined channels: TADIXS, NAVIXS, and GLOBIXS. One effect of drastically increasing available communications capacity was that the ashore data fusion developed for the OTH mission could be replaced by 'distributive fusion,' in which each node in the network contributes to the overall tactical picture. Thus there could no longer be a clear distinction between OTCIXS and TADIXS; the new TADIXS would replace both, as well as the Fleet Broadcast. The new TADIXS would distribute the full tactical data to all subscribers (rather than merely to a carrier) on the theory that it was far simpler to do so than to pre-edit data; the computers on board each ship could filter out what they did not need.[7]

The new TADIXS would replace, among other things, a series of separate formatted OTH broadcasts, each associated with a particular type of sensor. Examples included special Slow Walker and land-based HF/DF. Using so many separate nets was not only inefficient; it also discouraged the sort of sensor fusion which the fleet increasingly needed. By creating functional nets instead of the earlier stovepiped ones, Copernicus would automatically encourage attempts to combine the outputs of the many available OTH sensors to form a single tactical picture. NAVIXS would carry administrative traffic, mainly during periods of low TADIXS activity, while GLOBIXS was conceived as the net feeding the shore intelligence and fusion centres; its data would go over DSCS and the ashore Defense Data Network (DDN).[8]

These nets would have to take over *all* available long-haul communication links, because the philosophy of Copernicus was that the fleet's growing demands could not otherwise be met. Hence the need to deal with administrative traffic, which could not be given any sort of parallel link.

In fact the Copernicus vision was not quite realised, for two reasons. First, although it was relatively easy to unify UHF media, DSCS turned out to be a very different proposition. Second, as Copernicus developed, so did the vision of jointness. Copernicus merged with the larger project to develop a new Defense Department-wide global command and control system, which became the current GCCS. One consequence was that many 'stovepiped' systems and broadcasts survived much longer than Tuttle and his successors had imagined.

It turned out that DSCS was ill-suited to DAMA techniques. The satellite system had been designed to handle a steady data load, such as might be expected between fixed headquarters on land. Its transponders shared the same power supply, and they were ill-suited to fluctuating tactical loads. Although there was some flexibility in establishing and breaking circuits, there was much less than in the UHF systems. Ships have access to channels which they can use in alternative ways; for example, telephone lines can give way to a video teleconference circuit. However, one ship cannot set up telephone lines at the expense of another, because she cannot pick up the other's capacity.

GCCS (Global Command and Control System)

GCCS replaced an earlier World-Wide Military Command and Control System (WWMCCS). WWMCCS was very much a top-down system by means of which Washington could control nuclear forces to avoid untoward incidents. Its structure was much influenced by the Kennedy Administration's perception that only its own intervention had kept the Cuban Missile Crisis under control. There was little interest in, or provision for, deployed forces to obtain information (rather than orders) from the United States. WWMCCS fitted the top-down command style favoured by the Air Force and by the Army, rather than the decentralisation favoured by the Navy.

In the post-Cold War era, however, the United States found itself deploying forces from all the services in a style closer to the Navy's than anyone else's. The deployed shooters (or peacekeepers) needed information much more than they needed detailed orders from Washington. In a world far less prone to accidental catastrophic war, the Navy's vision of a decentralised information system was far more attractive than it had been in the past. Copernicus became the basis of a broader system. That broadening in turn delayed measures like the widespread use of DAMAs, because new satellite waveforms could not be introduced until all of the services had refitted their radios (all the services are now mandated to apply DAMAs to both 5 and 25 kHz UHF satellite channels). Non-DAMA stovepiped systems operate under waivers, and as of early 1999 it appeared that some would survive until 2003.

Conversely, the DAMAs applied to the UHF satellites now have to deal not only with the standard Navy channels (25 kHz wide), but also with the narrow-band channels (5 kHz). As an interim measure, the standard modems of the WSC-3 receivers are being modified to tune in 5 kHz steps. For the longer term all the services are buying digital modular radios (DMRs) with full DAMA capacity across all the UHF channels. At the same time, to add UHF capacity, the old 500 kHz strategic channel is being broken up (see above), and more efficient waveforms were planned, to get more bits/sec out of each channel.

In GCCS a combined command centre (CCC rather than FCC) communicates with deployed commanders using TCCs (rather than TFCCs). The communication concept is called Copernicus Forward; the TADIXS replacement is BCIXS (Battle Cube Information Exchange System). GLOBIXS is replaced by a series of ashore nets, such as JWICS (Joint Worldwide Intelligence Communications System, handling compartmented data), SIPRNET (Secret Internet Protocol Router Network), and NIPRNET (Non-classified Internet Protocol Router Network: unclassified but sensitive information). Under the current IT-21 programme (see below), ships' computer nets are connected to the shore nets via satellite modems.

JMCOMS, the Joint Maritime Communications System, is the Copernicus communications element.[9] In the spirit of Copernicus, JMCOMS is built around a multi-medium switching system (automated digital network system, ADNS, for a four-fold increase in throughput). The same programme deals with the central problem of satellite communication, that there is no growth in shipboard real estate to allow for the proliferation of satellites. Like any other antenna, a satellite dish has to be able to point at its (electronic) target; otherwise the ship cannot communicate. The more separate antennas a ship must carry, the smaller their clear arcs. For example, a typical Aegis cruiser with one INMARSAT dish often has to steer a specific course to remain in contact with the satellite. Dual dishes are far better (the ship has blind arcs limited to 10-12 degrees).

The projected solution is a combination of broader-band antennas and broadband terminals. Hopefully as many functions as possible will be combined into a single aperture, preferably on a stealthy basis. This idea is being applied both to ships and to the new JSF strike fighter. A new Multifunctional Multi-beam Broadband Antenna (MMBA) is to be able to receive both EHF and GBS signals. It may be part of the projected Low-Observable Multi-function Stack. All terminals handling signals above 2 GHz are to be replaced by a single type of terminal under an Integrated Terminal Program (ITP).[10]

A single lightweight low-radar cross section antenna is being developed to combine UHF line-of-sight communications (the current WSC-3 ship-to-ship role), JTIDS (Link 16), Combat Direction Finding (SRS-1), and IFF roles. Of these, the UHF role is equivalent to that of the UHF satellite communication antenna, except for orientation. Clearly the antenna must have a very wide frequency range, since JTIDS and IFF operate at L-band (the high end of UHF), whereas SRS-1 operates at HF frequencies. A new UHF satellite antenna will handle both UHF and INMARSAT signals. UHF terminals are to be replaced by broadband software-controlled wideband radios of the Digital Modular Radio System (DNR: 'slice' radio). It is necessary in order to take advantage of channels previously used by the other services: under GCCS, all users ultimately gain access to all channels.[11] On average, JMCOMS is to offer a communication rate of 9600 bits/sec.[12] It will connect to the Army's Enterprise and to the Air Force's Horizon systems.

There is also revived interest in a satellite version of the standard data link, in this case Link 16 (TADIL-J, or JTIDS). Link 16 normally operates at L-band, within line of sight. Now it is being tested using the WSC-3 UHF up-link. The Link 16 architecture may be better suited to satellite use than that of Link 11. In a Link 16 net, each participant is assigned a time slot. Presumably time slots can be juggled to take account of transmission time to and from a satellite (that was impossible for Link 11, which operates on a query-and-response basis, the response necessarily following the query before any further query can be made). The satellite version is associated with a new generation of Link 16 terminals based on commercial micro-processors, which presumably have the computing power to take the satellite delays into account. As of late 1999, satellite Link 16 had been installed on board the ships of the *Abraham Lincoln* and *Harry S. Truman* battle groups, with full tests scheduled for March 2000. Installation is apparently quite simple, since as of late 1999 the *Enterprise* and *Constellation* battle groups were scheduled to test satellite capability in May 2000. It seems likely that the commercial processor will be installed on board numerous ships not already fitted for Link 16, such as *Spruance* class destroyers and *Perry* class frigates. These ships already have satellite modems and antennas, so they will then gain satellite Link 16 capability.

The one great question which arises is the central issue in all satellite communication: capacity. Satellite Link 16 would, in effect, dedicate a portion of the UHF capacity to a single purpose, the data link. That is directly opposed to all the efforts to tear down stovepiped systems. UHF, moreover, has much less capacity than, say, SHF or EHF, so such stovepiping should have more serious consequences for it. On the other hand, satellite Link 16 would offer enormous tactical advantages, particularly if it then became the standard not only for warships but also for ground units communicating with ships, *eg* to request fire support.

IT-21 and INMARSAT

Even with Copernicus/JMCOMS in place, the Navy's communication needs continue to outpace military satellite capacity. The main commercial supplier of satellite communications for ships is an international non-profit consortium, INMARSAT (International Maritime Satellite Organization), formed in 1979 by representatives of sixty-three countries, including the United States (23.37 per cent share at the outset) and the Soviet Union (14.09 per cent); the headquarters is in London. Service began on 1 February 1982, and aviation services were added in 1988-90. Until the advent of Iridium, INMARSAT provided the only commercial satellite telephones (the Planet One service, using the Mini-M terminal).

INMARSAT provides most navies with their satellite communications service. Terminals are relatively inexpensive, and they are widely available. For example, during the 1982 Falklands War, the Royal Navy provided many ships with satellite capacity simply by installing INMARSAT terminals. That was far easier than providing them with full long-haul communications suites. INMARSAT began by taking over the U.S. commercial maritime satellites (Marisats) on which the Navy had leased its Gapfiller transponders. Then it launched its own MARECS satellites.[13] The system currently employs eight satellites and about thirty Coastal Earth Stations (CES). All traffic goes via satellite to a CES, whence it can be rerouted to another ship (via satellite) or to a shore telephone.[14] The U.S. Navy is currently the single largest INMARSAT user, accounting for nearly 3 million minutes of telephone time in 1995.

The Inmarsat A analog terminal offers what amounts to on-demand telephone service (initially 9600 bits/sec, but 16,000 bits/sec is now available) with a 1m dish, at a high cost, about $5.50 per minute. A 64 kbit/sec package is available at $12 per minute. The follow-on digital Inmarsat B terminal offers similar performance but at a fraction of the price, because it is much more efficient; the cost per minute may be as low as $1. Inmarsat M uses a much smaller dish and is suitable for mine countermeasures craft and for the Special

Forces' *Cyclone* class. Because gain is lower, so is data rate, typically 4800 bits/sec for voice, 2400 bits/sec for facsimile.

In 1997 the U.S. Navy began to lease full-time access to the new 64 kbit/sec INMARSAT service (the Navy was the launch customer) as part of its IT-21 (Information Technology for the 21st Century) program. The single INMARSAT channel provides multiplexed voice (four telephone lines) and data (32 kbits/sec). A ship can gang two data channels together to get a net rate of 128 kbits/sec, and some current techniques will compress images well enough to transmit useful quantities at that rate. Ships are outfitted so that all radio signals can feed into the IT-21 local area net (LAN) via the ADNS automated switching system. Thus, at least in theory, beside the leased INMARSAT capacity, a ship can feed her LAN with SHF and EHF satellite links, with the DAMA links from UHF satellites, and with the narrowband UHF signals which the navy can now use (thanks to jointness, no service has exclusive rights to any part of the satellite system). Even so, INMARSAT is the key IT-21 link.

At the land end, INMARSAT data go into public telephone lines, and thence to the NCTAM communication system. Data are bulk-encrypted; the land lines are not specially secured. Until the smaller ships get the capability to receive the current imaging broadcasts (*eg* using GBS), INMARSAT may well be their only important high-data-rate satellite system. It is, moreover, received not only by combatants but also by important auxiliaries such as combat support ships (AOEs).

IT-21 was first tried by the Third Fleet, in the eastern Pacific. As of early 1999, two battle groups (*Enterprise* and *Kitty Hawk*) were regularly using it (60 per cent of ships had been fitted). The *Enterprise* battle group was the first to deploy completely outfitted for IT-21; it returned from deployment in May 1999. The *John F Kennedy* battle group was scheduled for completion with IT-21/INMARSAT in the summer of 1999, and the others will soon follow. Probably the most important aspect of IT-21 is that it gives ships at sea the equivalent of a connection to the Internet, albeit a slow one (each ship has the equivalent of a 32k modem, which is shared among several computers).

The satellite-mediated net changed the way IT-21-equipped battle groups operated. Under IT-21, local nets of standardised commercial computers were installed aboard each ship, handling both tactical and administrative tasks.[15] Using standard commercial equipment made for a very fast programme, less than two years from inception to fielding. Current (1999) plans call for the Navy-Wide Intranet (the shore system plus IT-21) to be complete by the end of FY01 (October 2001).

IT-21 offers all of the internet services so familiar to the public ashore, including both E-Mail and chat-rooms. Situational awareness in the IT-21 equipped *Enterprise* battle group was dramatically better than in more conventional groups, because so much more information could be exchanged between ships. For example, the group commander could post intentions, optasks, etc, on a web page, available to all ship commanders. He no longer had to refer back to previous text messages, as had been customary. E-mail greatly simplified the hand-over from the *Vinson* to the *Enterprise* battle groups: it was relatively easy to pass massive operational files back and forth without physical contact (or even close proximity).

E-Mail had an additional, unexpected, advantage. When maintenance difficulties arose, equipment could be digitally photographed and the results sent back instantly to the United States to be examined by experts there – and their advice could be sent back to the ships via E-Mail. The satellite links also provided previously unavailable telephone service.

Using a chat-room, those aboard all the ships of a battle group could easily consult with each other; for the first time officers on board lesser units could easily discuss battle group policy with those on board the flagship. In the case of the *Enterprise* battle group, the frigate *Klakring*, running a maritime interception mission (blockade) in the Northern Arabian Sea, could easily remain in contact with the flagship hundreds of miles away. Using a chat room, her officers could easily discuss their operations with the group staff on board the flagship. These net meetings were particularly valuable when voice satellite communication failed. It was no longer necessary for officers to fly to the flagship for regular conferences, yet those same officers were far more likely to influence decisions made on board the flagship. In effect the battle group became far less centralised, at least partly because ships other than the flagship or the major units were no longer information-poor. Another standard internet service amounts to providing a user with a virtual library – which, for the navy, could mean providing ships with what they needed on an as-needed basis. IT-21 has other, administrative, implications. Clearly manning will be increasingly difficult. Some shipboard administrative functions, such as personnel, can be shifted ashore – if the ship has a very reliable communications link to the shore centre. Clearly a dedicated point-to-point link would be pointlessly expensive, but the new Internet-style connection is quite adequate.

One implication of IT-21 is that ships will transmit nearly continuously – as they already do on their standard digital links. As of 1999, the operational security implications of such operation were still being debated. Systems must sometimes be turned off, but it is not yet clear how often, or under what circumstances.

The IT-21 E-Mail function has had another, unexpected, consequence. As any Internet user knows, E-Mail is extremely easy to send. As a consequence, there is a great deal of it, far more than the volume of earlier kinds of message traffic. There is a real danger that tactical users may find themselves buried in E-Mail. This situation is very different from that of formatted systems which feed into computers maintaining various kinds of plots. In their case, the computer can digest information very rapidly. Text, which has to be read by people, is another matter. The problem may become particularly acute if automation drastically reduces staff numbers, as it promises to do.

It does not take the full IT-21 system to provide a ship with an E-Mail capacity. For example, a few years ago the Royal Australian Navy equipped its ships with small computers and data buses connected to radio (and, later, satellite) terminals – and made it possible for them to send and receive E-Mail in a burst mode. Quite aside from the tactical or administrative value of such E-Mail, crew members could stay in touch with their families at home. The U.S. Navy, and presumably also the RAN, considers such contact extremely valuable for morale (it is called Quality of Life, or QOL, communication).

There is a submarine equivalent of IT-21. As of early 1999 the Submarine Force was trying a new concept, Asymmetric Communications, in which the submarine would transmit at UHF rates to request data from the Navy SIPRNET, which is sent at a much higher rate (typically 33 kbits/sec). The low-rate side of the system is called reachback; it is used to pass data and to notify the shore system of transmission errors. As of mid-1999 this was still an experimental programme.

Iridium and Other Telephone Systems

The U.S. Defense Department has already bought heavily into the new Iridium system, whose sixty-six low-altitude satellites offer telephone service over the Earth's land and coastal areas.[16] In contrast to a geosynchronous system, a low-altitude system can re-use its bandwidth many times over, because only a few customers see each satellite at any one time. Low altitude reduces the power demanded of any one transmitter on the Earth's surface (which is why it is attractive for telephone service) and reduces transmission delays (so that voice communication is easier). However, to make such a system work, satellites must be able to hand over communication as they pass beyond the horizon. The whole constellation must be in place before it offers any service at all.

Iridium sea and polar coverage is limited because the satellites use their time over the sea (where there are few potential commercial customers) to recharge their batteries. In contrast to other systems, Iridium uses its customer's signals to fix the customer's position (for billing purposes). To avoid this potential breach of security, the U.S. government bought a special secure 'gateway' in Hawaii, through which its Iridium calls will be handled. The gateway strips off the location information and calculates billing for the Iridium consortium.

This is not a specially naval system. Secure handsets are distributed to joint commanders (CINCs) on a bulk basis. They in turn will be billed for the calls. As of early 1999 the military end of Iridium had just become operational. The civilian market failed to materialise; by the end of April 1999, there were only 10,000 customers, rather than the 100,000 predicted. The cellular telephone system had expanded much faster than the Iridium backers had expected, and their own telephones were poorly designed and expensive to use. Their main competitor, INMARSAT, was already providing much of the service Iridium promised. In August 1999, Iridium filed for bankruptcy. Possibly it will end up as a U.S. government system. The failure of Iridium has raised questions about the viability of any space-based mobile telephone system, but as of mid-1999 Globalstar (with thirty-six satellites already in place) was ready to begin service in October 1999. Overall, however, there are real question about the projected market of 30 to 40 million customers worldwide.

The apparent failure of Iridium opens important questions for future U.S. military space policy. Iridium was backed by a major U.S. electronics firm, Motorola, and was generally expected to succeed. That expectation in turn made it reasonable for the U.S. military to finance the Hawaii 'gateway', and to rely on Iridium for a large fraction of its communications needs. The Iridium system is expensive to maintain, even if the cost of its satellites is written off. If it cannot find a civilian buyer, it may be allowed to collapse. What happens to the defence users then? After all, they need more than simple access to long-distance telephone lines. They need security measures such as those provided by the gateway. How wise, then, is it for the U.S. defence establishment to bet on the viability of particular commercial systems in a very competitive communications market?

Follow-On Satellite Capacity

By the spring of 1995, new satellites were badly needed in all three bands used by the U.S. Navy (EHF, UHF, SHF). Several cost-cutting options were proposed. One was to combine EHF and UHF in five satellites, rather than the existing combination of Milstar and UFOs. Another was to simplify Milstar further by eliminating its crosslinks and its agile beams. The SHF satellites (mainly DSCS) could be simplified by eliminating their anti-jamming features (nulling to avoid jammers operating against their up-links, agile antennas to limit ground coverage and thus to increase effective power to overcome down-link jamming, and hard-limit transponders which shut down when the up-link signal is too badly swamped by a jammer). Without the special features, it could be argued that the military geosynchronous satellites were not too much better than transponders on board existing commercial satellites (Milstar was still an exception, given its digital character).

Under emerging ideas of jointness, all the services will now share all satellite media available to the Defense Department. There is now a DoD Space Architect, in charge of developing a uniform satellite program for the coming century. In line with Copernicus, the satellite system is conceived as a connection between a unified Defense Information Support Network (DISN, son of GLOBIXS) and users; the system is to be driven by information services rather than stovepiped. Because the Army and the Air Force need even more portable terminals than the Navy, a major goal is to reduce the footprints of these devices. That translates into pressure to use higher frequencies.

At the same time, the commercial world is using space communications far more heavily, and its up- and down-links compete with the U.S. military for frequency allocations. In the mid- and late-1990s the Clinton Administration tended to favour civilian applications over military arguments that blocks of frequencies should be retained against possible future needs.

In the late 1990s, then, follow-ons to the existing EHF (mainly Milstar), X/Ka-band (SHF: DSCS), UHF, and Polar systems were all considered. As the Navy showed, more could be extracted from the existing satellites; it seemed likely that the Milstar medium-data-rate waveform could provide 6-8 Mbits/sec, given better ground equipment (such as DAMAs and larger dishes). A new system could provide tens of megabits/sec in a format compatible with follow-on SHF satellites, and thus with plans for a single integrated EHF/SHF terminal, the ITP (Integrated Terminal Program).[17]

For UHF, the great question was whether to continue at all. Certainly UHF requires a separate terminal (UHF technology is not compatible with EHF/SHF), and as ships get stealthier the shipboard real estate available for dishes of any kind is likely to shrink. In the 1990s, the UHF system was unique in its ability to net together mobile users. As the other services follow the navy in deploying regularly on a mobile basis, the net appetite for UHF capacity keeps booming; the system is chronically over-subscribed. However, systems like Iridium offer comparable services, with which the Defense Department could experiment before the UHF systems needed replacement.

As of 1999, the Defense Department was defining a Mobile User Objectives System (MUOS) as the ultimate replacement for UFO and other UHF satellites. Because it is so dependent on satellite communications, the U.S. Navy is the lead service in this effort. MUOS will include a Gapfiller-type UFO satellite (Number 11) to bridge the gap to next-generation systems.

As for Polar (north of 65 degrees North) communication, the choice fell on EHF transponders on board SDS satellites (see above) in

Molniya-type orbits, two of which were needed for full coverage. By piggybacking the EHF function on another satellite, the U.S. Navy (the primary user) was able to obtain full communications coverage above 65 degrees North at an affordable price. An earlier project for a communications-only Articsat had been dropped at the end of the Cold War. Presumably a future SDS satellite will have a similar dual mission. The nominal designation for a future SDS system replacement is the National Space Communications Program (NSCP). It may handle communication with long-endurance UAVs as well as with satellites, on the theory that UAVs are, in effect, low-altitude reconnaissance satellites, with the same need for high-capacity data extraction. Given current interest in coalition operations, a Universal Modem is being developed by the United States and the United Kingdom to allow U.S. and Allied forces to pool their satellite communications systems.

There is still intense interest in getting more out of existing satellite systems by using more efficient forms of multiplexing. One possibility now under investigation for EHF (Ku-band) down-links is CDMA, Code Digital Multiple Access. Each message is separately coded, and multiple messages are transmitted together. A receiver strips off the appropriately-coded message. In this way a single channel, however defined, can carry several messages at the same time. The associated modem is being developed under an HDR (High Data Rate) programme.

Until recently civilian communication systems roughly paralleled the military ones, using geosynchronous satellites. As noted above, the U.S. military has leased transponders on commercial satellites, and has gone so far as to buy telephone time on INMARSATs. During the Cold War, the single great argument against depending on commercial satellites was their vulnerability to nuclear effects (see Chapter 5). Now, however, the main argument probably ought to be that anything other than Milstar or something like it carries a real potential that Doppler effects may betray the positions of the ships transmitting via the satellite.

Overall, shipboard satellite capacity is growing, but not as fast as demand. The current validated goal (PR-99) is that each major combatant ship – which includes Tomahawk-firing missile destroyers – have a capacity of 2.3 Mbits/sec. That would provide ships with the sort of imagery they need. Much of it will be provided by GBS, but installation of ship terminals for GBS has been deferred to FY02 (October 2001 through September 2002). Current rated capacity, on board missile destroyers, is 64 kbits/sec of UHF and 128 kbits/sec of each of EHF and SHF, a total of 320 kbits/sec, about an eighth of what is desired.

The U.S. Navy leases commercial Intelsat 8-series satellites to provide a fleet television service, Television Direct To Sailors (TV-DTS), in cooperation with the Armed Forces Radio and Television Service (AFRTS). The system was conceived mainly as to give ships better situational awareness by providing current news programmes. It was promoted by former Chief of Naval Operations Admiral Boorda, and it was pressed forward by Secretary of Defense Perry (its approval was one of Perry's last acts as Secretary). This one-way service was activated in the fall of 1997. Receiving dishes were placed aboard the carriers. The large-deck amphibious ships had already been fitted with 3m dishes for what was called TV-RO (receive only); the receivers had been bought by the Type Commanders. In 1999 smaller ships were being fitted with 1.5m dishes. The C-band down-link carries 3 Mbits/sec, providing two full-time television channels as well as audio bandwidth for AFRTS radio and a 128 kbits/sec data channel. This is not a tactical system, although in the system's concept stage the data channel had been considered for tactical purposes. Now it is used for administrative files.

Lightsats

There is one other way to increase communications capacity: adding more satellites, preferably cheap ones, easily manufactured and launched. One obvious goal was to maintain command and control in the aftermath of a nuclear strike, when primary launch facilities had all been destroyed, and when many satellites already in orbit had been disabled, *eg* by EMP. In that case very small satellites on mobile platforms might survive to be launched into service. Another possible lightweight satellite role was reconnaissance tied not to organisations like NRO, but to tactical commanders. Beginning about 1980, DARPA tried to develop the necessary light satellite bus. The first example was the Global Low Orbiting Message Relay (GLOMR). The programme was designed to show that a digital communications satellite could be built in less than a year. Because it flew at low altitude, GLOMR operated in a store-and-forward mode; for example, it could read the output of unattended sensors, such as ice-penetrating sonobuoys in the Arctic. It could also locate surface emitters by measuring the Doppler shift of their signals. GLOMR was designed to fit into a Shuttle Getaway Special canister, and it was successfully orbited (at 176-mile altitude) by the Shuttle in October 1985. When it decayed out of orbit in December 1986, it was still operating. A follow-on (and slightly heavier) GLOMR satellite was launched by Pegasus in April 1990.

In May 1990 two 68kg Multiple Access Communication Satellites (MACSATS) were orbited by a Scout rocket. They were placed in the same orbit, but with velocities different enough to place them 180 degrees apart within two and a half years. They provided logistics communications in support of a Marine air wing during the Gulf War; they were also used to support an Army special forces exercise in Thailand, and to communicate with Navy forces in the Antarctic. Like the big fleet satellites, the MACSATs operate in UHF.[18] Both satellites failed in 1994.

TWELVE

The Global Positioning System

OTH and the Outer Air Battle both required precise global navigation: those units using data from distant or even space-borne systems had to know where they were relative to those systems. Transit did not quite solve the problem, because a fix took several minutes. The ideal would be a system giving instant (or near-instant) and extremely accurate navigational fixes. It emerged in the 1980s as the Global Positioning System (GPS), using Navstar satellites. Once GPS had been accepted, it became the least expensive way of guiding missiles to their fixed targets, and it ushered in (at least for the United States) an era of guided weapons to replace earlier unguided ones. That was quite aside from such more traditional navigational feats as helping the Allied armies find their way through the trackless Iraqi desert in 1991 to catch Sadaam Hussein's force in the flank, thus winning the Gulf War far more quickly than might have been expected.

GPS can be traced back to the Naval Research Laboratory's idea for a navigational system based on the precise time standards it had developed: Timation.[1] Two were launched (31 May 1967 and 30 September 1969). Given that radio signals travel at the fixed speed of light, measuring a time interval is equivalent to measuring the distance a radio signal travels in that time. Both the GPS receiver and each GPS satellite contain a clock (the receiver clock is less precise than that in the satellite). The satellite bleeps its radio in time to its clock; the receiver notes the time at which it receives a bleep. It converts that into a range to the satellite. The satellite is in a precisely-defined orbit, data on which is carried by the GPS receiver. The range defines a circle on the surface of the earth. The range to a second satellite defines a second circle, intersecting with the first in two places; a third range circle (from a third satellite) selects one of the two intersections. By using four satellites, the GPS receiver can also measure its height above the Earth.[2]

Work on GPS began in 1973, Rockwell receiving the system contract the following year. Two Navigational Test Satellites were launched to test GPS hardware, one on 14 July 1974 and one on 23 June 1977. The first Block I GPS satellite, Navstar 1 (Navigational Development Satellite 1), followed on 22 February 1978. Block II is the operational version. Although GPS was originally intended for military use, in 1983 President Reagan ordered an open-use policy, in the wake of the loss of Korean Air Flight KAL-007 due to a navigational error. The entire system was to have been operational by 1988, but launches were delayed by the *Challenger* disaster and the consequent suspension of Shuttle launch service. GPS was declared operational in May 1990; that August there were six Block I and eight Block II satellites in service. By the eve of Operation Desert Storm (January 1991) another two Block II satellites were operating (they had been launched deliberately into orbits giving maximum Gulf coverage).[3] However, a full twenty-four were required to give full 24-hour three-dimensional service. The satellites available in 1991 provided 22.5 hours of two-dimensional GPS service and 16.75 hours of three-dimensional GPS service daily.

The twenty-four satellites (three spares) of the full system lie in six orbital planes in 10,898nm orbits (12-hour period) at 55 degrees inclination.[4] Satellite subsystems lifetime is 7.5 years (average lifetime should be 6 years). Each satellite carries fuel, since it has to correct its orbit periodically on command from the ground station.[5] Each carries four atomic clocks, to ensure that its pulses are sent at precise time intervals. It emits a coarse (public) signal at 1575.42 MHz (accuracy 90 to 100ft) and a coded precision signal at 1227.6 MHz (30-52ft).[6] Signals at these frequencies easily penetrate clouds, rain, and snow, although really dense vegetation (as in a triple-canopy jungle) can block or weaken them. Because not enough military receivers were available, the precision signal was transmitted uncoded during the Gulf War.

Rated accuracy, for military users, is 16m (users in the Gulf got accuracies of 7.5 to 13m). Speeds can be measured to within 0.1m/sec. In fact signal processing (such as Differential GPS) can make even the civilian signal good enough for very precise location.[7] Reportedly the civilian signal can be intentionally degraded in wartime, to deny it to enemy users. Because it is a public system, unless it is intentionally degraded GPS is also available to enemy forces (reportedly the Iraqis used it to help target mobile Scud missiles during the Gulf War).[8]

The Soviets developed their own equivalent to GPS, GLONASS; under an international agreement the two are interoperable. This programme began in the 1970s, reportedly with the particular support of the Soviet Navy. Launches began in 1982.[9] As of 1999, the European Union countries are considering developing their own GPS equivalent. They are currently developing a European Geostationary Overlay Service which is to improve the precision of GPS and GLONASS from 2002 on. A European Global Navigation Satellite System (Galileo, or GNSS-2) is being discussed, with parallel eighteen-month studies (by the EU and by the European Space Agency) of potential architectures planned.

Tomahawk Guidance

Quite aside from making it much easier for the fleet to exploit distant sensors like satellites and ROTHR, GPS had an enormous impact on missile guidance. The land-attack version of Tomahawk was the first really prominent example of this. At the outset, Tomahawk was conceived as either a nuclear attack weapon (BGM-109A) or an anti-ship weapon (BGM-109B). However, as it became obvious just how accurately it could strike, a third version emerged, a long-range conventional weapon (BGM-109C). In a hot war arising out of the Cold War, its great advantage might be that it could be used before the war entered a nuclear phase. Another version (BGM-109D) carried bomblets, specifically to strike enemy airfields. The long-range conventional version, the one for which GPS was to become very important, was first used in combat during the Gulf War, when it proved extremely successful. Compared to a ballistic missile, Tomahawk presented a defender with much less of a signature. Since it would generally fly quite low, a defender might not even detect the missile until it was nearly upon its target. By way of contrast, a good enough radar would certainly see the ballistic missile at very long range. On the other hand, a low-flying cruise missile required much more targeting effort.

From the beginning, the missile's developers were interested in high accuracy at an affordable cost. The relatively simple inertial mechanism they had in mind would certainly drift during the missile's nearly three-hour flight, just as SINS would inevitably drift as a submarine cruised underwater. They needed a means of updating the system, just as SINS used the space-based Transit. When Tomahawk was conceived in 1973, there was no space-based navigation system which could provide instant fixes. With its need for lengthy reception, Transit clearly was not suitable.

The solution adopted was a technique, TERCOM (Terrain Comparison), which already existed, having been invented during the 1960s.[10] It turned out, remarkably, that if the missile periodically checked the profile (*ie* a series of heights) of the terrain along a straight line over which it flew, it could tell whether it was on course by comparing that profile with a numerical map in its memory. TERCOM would even work in apparently featureless terrain like the Soviet tundra. The problem was that before any missiles could be programmed, let alone fired, someone had to map that territory with sufficient precision. Existing maps were not nearly accurate enough. A second problem was tactical. It would take several terrain comparisons to fix the missile's course accurately enough. A target on or very close to a coast would not provide enough such fixes if the missile flew directly from the sea. It would, then, have to take a circuitous course simply to be sure of hitting accurately.

The non-nuclear versions needed even better accuracy. For them, TERCOM was supplemented by another technique, DSMAC (Digital Scene-Matching Correlation), which was used near the target. To take a DSMAC fix, the missile compares its image of the ground nearby with a digital scene stored in its memory. The match is based not on a particular detail of the observed scene, but on the overall match. Thus a minor change, such as damage to a building or cars moved along a street, will have little effect, as long as the overall scene registers strongly enough. This technique is *not* the same as homing on a target building; all DSMAC does is set the missile's precise position so that its onboard inertial system is more likely to function accurately during the last few moments of flight. During the Gulf War, Tomahawks were sometimes observed using strobe lights. They were illuminating the scenes they were using for DSMAC fixes. Tomahawk is stealthy partly because it has no ahead-looking seeker, whose radiation an enemy can detect. That also means it cannot detect obstacles in its path, including walls, church steeples, or even rising hills. Tomahawk is compact because it is powered by a very small engine – which offers a very limited rate of climb, so it cannot negotiate a steep hill or mountain; it will simply crash.

Targeting the missile is therefore much more complicated than targeting a ballistic missile like, say, Poseidon or Trident. In the case of the ballistic missile, all that is really needed is knowledge of the position and nature of the target (which may include knowledge of anti-missile defences around the target). In theory, once the position of the firing submarine is known, the inertial guidance unit in the missile can be adjusted so that the missile will hit the target as required. In fact the missile is given correction coefficients its computer is to apply as it senses its precise upward motion, as affected not only by its motor but also by, for example, wind. The coefficients in turn depend on exactly where the submarine is when it fires.

When the United States first deployed Polaris, computers sufficient to develop the necessary coefficients were quite massive. First-generation submarines went to sea carrying decks of computer cards produced at the Naval Weapons Center at Dahlgren, Virginia. The next step was to place the computer aboard the submarine. The advantage was that the submarine could be effective despite having been re-routed once at sea. Calculations were still elaborate; the necessary cards were produced while the submarine ran out to her patrol station. With Poseidon (1968) came the next step, real-time calculation, the coefficients automatically being fed into the missiles as needed.[11]

A Tomahawk targeter had to lay out the missile's entire flight plan, from launch to target, using very precise maps of the area over which it would fly. Considerations included avoiding obstacles (which was why the map had to be so complete) and also avoiding enemy radars and air defences. Also, the missile had to be routed over areas including the necessary TERCOM maps. The process was lengthy because it was repetitive. That is, a targeter would start the missile down a promising path. As the path continued, it might well run into an obstacle – so the targeter would have to start again. A full flight plan would take several days to complete. In addition to laying out the missile's course towards the target, the targeter had to select the appropriate scene or scenes for DSMAC. Targeting Tomahawk was so complex that the work had to be done ashore, at special Theater Mission Planning Centres, deployed at nine locations, but the two main TMPCs are at the two Cruise Missile Support Activities (CMSAs), at Norfolk and at Camp Smith, Hawaii. Each CMSA had three sections, a TLAM Planning System (TPS), an imagery analysis section, and a section which produced the DTDs. The necessary software for the Tomahawk Mission Planning System was delivered in 1978.

Sets of flight plans were compiled onto massive (80lb) hard disks called Data Transfer Devices (DTDs), which were placed on board flagships (mainly carriers) for off-loading to Tomahawk shooters (surface ships and submarines). Each plan involved so much data that it was impractical to transfer by radio; a ship was limited to the plans physically placed aboard (in the case of a carrier, perhaps by cargo plane). For each mission, the DTD also contained a data package (TCI, Tomahawk Command Information) that described the mission and could be read into a computer display (a mission display system, MDS) or used to prepare a paper mission folder. The ships had no means of developing their own Tomahawk flight plans. To hit specific targets, they could vary the end point on a flight plan.

The Fleet Satellite considerably improved the situation. Its special secure channels (OTCIXS/TADIXS) linked the mission planning centres with the new mission display systems (MDS) on board the Tomahawk shooters, just as the same channels linked the FCCs with the fire control systems on board the Tomahawk anti-ship shooters. The first engineering-development model MDS were installed on board several ships and at shore stations beginning in May 1987; the improved XMDS version (which could support strikes by the bomblet version, BGM-109D) was installed on board the carrier *America* in May 1989. Soon the MDS software was rewritten so that it could run on the DTC II computer used by the JOTS II system.

Tuttle's vision of unified hardware linked by the satellites was coming closer to reality.

Only a shore centre could re-write a Tomahawk mission, but the links made it possible to transmit that revision (a mission upgrade) directly to the shooters, via an MDS. Because each MDS could communicate with others, it could review, not only missions held by one ship, but those held by the others. Missions were far too complex to be sorted or reviewed by hand. An MDS could search other data bases to decide whether a given possible flight plan would or would not be feasible, given constraints such as enemy force dispositions and permissable ship launch areas. Given the review, the MDS could be used to coordinate air strikes. As the number of versions of Tomahawk multiplied, it became important that the MDS could also review those on board the ships to check which were suitable. It could display the target area and its air defences (a facility which air strike planners could use). It was designed to produce displays and charts showing Tomahawk and aircraft routes overlaid, so that strikes could be properly tailored.

For Tomahawk, then, one key sensor was the 'national' imaging satellite which provided the targeter with the photographs on which each flight plan was based (ELINT satellites helped indicate where enemy defences might be located). Both would naturally contribute to planning for nuclear Tomahawk strikes. However, as Tomahawk evolved into a non-nuclear strike weapon, access to such data became an example of TENCAP. Another key to using Tomahawk was a radar mapping satellite capable of building up the maps used for TERCOM. During the 1980s Tomahawk attacks against the Soviet Union were fully planned, and the Soviet Union was fully mapped for TERCOM. The system did have a minor problem, in that it was too rigid to take quick changes in Soviet defences into account. For example, it probably could not have changed flight plans quickly enough to deal with the quick placement of mobile surface-to-air missiles during a crisis.

Then the Cold War ended, but the world did not become more stable. On the contrary, crises erupted in quite unexpected places. The Gulf War was the first prominent example. Iraq had not figured on Cold War target lists, so it had not been mapped with Tomahawk attacks in mind. For about six months after the Iraqis invaded Kuwait, the Defense Mapping Agency feverishly developed the necessary maps, and the two Tomahawk planning centres developed flight plans. They were ready just in time. Even so, there were some bizarre incidents. Missiles fired from the Mediterranean flew very circuitous routes simply to overfly TERCOM-mapped areas. Also, the packages of flight plans were quite limited, because those who designed them could not be sure of what the tacticians in the Gulf would want. For example, there were very limited numbers of routes into Baghdad. Eventually the Iraqis discovered what they were, and they set up anti-aircraft weapons to match. Although the first missile generally got through, in one case several of a string of six were shot down.

Even so, Tomahawk was a very striking success during the Gulf War. It demonstrated remarkable accuracy under all weather conditions, often flying when cloud cover prevented aircraft from hitting targets in Baghdad.[12] Clearly it would be used again, whenever the United States had to deal with a crisis in a Third World country. Equally clearly, such crises were likely to be unpredictable. Surely the United States would not have six months in which to prepare for strikes.

GPS could solve part of the problem, since it could replace TERCOM. Moreover, a Tomahawk using GPS rather than TERCOM updates could approach its target directly from the sea, since there was no longer any need for detours to check guidance accuracy. Tomahawk-armed ships could now attack from greater ranges, the detours having consumed fuel (this was quite apart from the use of indirect approaches to confuse target defences). As long as the position of the target was known, the missile could be flown there. A GPS-equipped Tomahawk could be used even in a surprise crisis. That actually happened in Bosnia, where GPS Tomahawks (Block III) were used for the first time.[13] As an incidental advantage, the missile became stealthier. There was always some fear that the radiation from the altimeter required for the TERCOM guidance system was detectable. Because it is a passive system, GPS should provide no warning at all.[14]

Meanwhile the larger Tomahawk weapon system became significantly more flexible. By the late 1980s computers capable of targeting Tomahawk were far more compact than those in the planning centres; they were even small enough to go to sea. They were now so fast that the day or two for planning a single mission could be reduced to an hour or less; Tomahawk could be fired in near-real-time. New software was designed for two parallel programmes, a land-based Theater Mission Planning Center Upgrade (TMPCU), and a shipboard Afloat Planning System (APS), the latter intended for the new TFCC on board carriers. The system was still too massive to go on board Tomahawk-firing ships, but it did provide more flexibility. Instead of producing a DTD (or, better, a much more compact DTD successor), the APS on a carrier would produce mission tapes, which the carrier's helicopters could deliver on board the shooters. Both programmes ran more slowly than expected; the TMPCU came fully into service in 1993 (FY94), and the first production APS went aboard the carrier *Nimitz* in 1995. APS includes a Tomahawk Strike Coordination Module, intended to integrate Tomahawk and tactical air strike planning.

Admiral Tuttle's key perception, that very soon compact standard Navy computers would be able to take over all the tasks assigned to specialised ones, was still quite valid. As APS entered service, the Navy was planning the necessary further step. Tomahawk shooters already had computer systems designed to support the anti-ship version of the missile. They used the antiquated computers of the late 1970s, and by the early 1990s they were due for replacement. It was now time to integrate the APS function into a next-generation Advanced Tomahawk Weapons Control System (ATWCS) using software running on a Tac-3 (successor to DTC II) computer. The existing MDS became an important element of the new system.

Until this point ships had separate systems to control land- and sea-attack Tomahawks. ATWCS would replace both; it would also control the other major new long-range land-attack missile, SLAM. For Admiral Tuttle's philosophy of using commercial computers, this was a drastic advance: for the first time, a Tac-series machine was to be used for a mission-critical role, replacing the military computers used in earlier Tomahawk fire control systems. To speed mission design, ATWCS was intended to include an auto-router. The computer would automatically develop the vertical track associated with a ground track it generated, then modify the track as necessary to make it practicable. No human flight planner would be needed, only an analyst to decide exactly where the missile had to go. Prototypes began shipboard tests in 1994, and the first version of the system entered service in 1996.[15]

The new shipboard mission-planning computers still needed a mass of data, mostly in image form. The bulk of such images came from satellites, which dumped their data down to the U.S. National Reconnaissance Office (NRO) or its successors. The data had somehow to get to the targeters afloat. Space assets offered the only possible answer, this time in the form of a new class of high-capacity communication satellites. No other communications medium could provide anything like as much data as quickly, nor could any other guarantee prompt delivery almost anywhere in the world.

The current land-attack Tomahawk with its supporting targeting system, then, exemplifies the way that space assets have changed naval warfare. It is pilotless because it can be fired accurately at a target hundreds of miles away; and that accuracy depends entirely on space assets, GPS and satellite-acquired and -delivered images. Without the precision due to space assets, Tomahawk would have needed some sort of human intervention to guide it to the target. At the least, it would have needed a large thermonuclear warhead, to make up for guidance errors – and thus would be unusable in the sort of low-level conflicts which are now the rule.

GPS Missile Guidance

By the late 1980s, with GPS soon to become operational, there was considerable interest in using it to guide a wide range of tactical weapons. It was by far the least expensive guidance mechanism in sight. If a target's location could be specified in the geographical terms used by GPS, then a missile could be sent to it; in effect, GPS became the missile seeker of choice for the U.S. military. It superseded an aborted generation of missiles and guided bombs which would have used cheap inertial guidance devices.

For the U.S. Navy, the major new tactical weapon of the 1980s was the air-launched AIWS, the Advanced Interdiction Weapon System, which would replace homing missiles like Maverick.[16] It was conceived in 1986 in the wake of the U.S. air raid against Libya. The main requirements were very low cost and a stand-off range (initially 5nm) which would protect the attacking aeroplane against enemy close-in anti-aircraft missiles. It was essential that the attacker not have to pop-up and stay up to designate the target, which had to be done when laser-guided bombs or missiles were used. Nor could the attacker maintain contact with the missile via a data link.[17]

AIWS soon superseded an alternative Air Force weapon to become the current JSOW (Joint Stand-Off Weapon, AGM-154). Given severe limits on its cost, it could not possibly accommodate a conventional seeker. At first the solution seemed to be a cheap inertial system, but in the end GPS was chosen. A future version may have a seeker. Even then GPS will drastically cut its cost, by bringing the missile very close to the target; the seeker will see far less non-target (noise) than will a conventional seeker looking out over a wider area.

Another important new Navy strike weapon, SLAM, also depended heavily on GPS. SLAM (AGM-84E) achieved enormous precision by transmitting its optical seeker picture back to a controlling aeroplane; the controller selected the target and locked the missile on. This was not a new idea; during the Vietnam War the Navy had pioneered the idea of optical lock-on guidance with its Walleye glide bomb, and it had developed a powered equivalent, Condor (AGM-53). Condor had died because it was far too expensive and far too difficult to use. The controlling aeroplane had to command it all the way to the target, maintaining contact via a data link. The missile had to be controlled from beyond anti-aircraft range. It was impossible to provide a jam-resistant link at such a distance at any affordable price.[18]

GPS changed the situation. One aeroplane could launch a missile programmed to fly to a selected place, and to point its seeker in a selected direction. A second aeroplane could take over control at the specified time. Since the seeker would already be pointing at about the right place, the control link would not have to be maintained for very long. Control could, moreover, be exercised from fairly close to the missile – in fact, SLAM was controlled via the relatively simple link developed for Walleye. SLAM was first used, quite successfully, in combat during the Gulf War.

Harpoon itself, on which SLAM was based, may soon benefit from GPS insertion. As naval operations move closer to the shore, precise navigation towards the target area becomes more critical. For example, an attempt to hit one ship in a harbour is quite difficult. The more precisely the missile's flight can be planned, the less the chance that its seeker will lock on to the wrong target. The currently projected Block II version of Harpoon incorporates GPS and other SLAM guidance features. Because the missile has a GPS receiver, it can be commanded to hit a target ashore, even one without any radar signature (without GPS, Harpoon relies entirely on a radar seeker). Thus GPS makes Block II into a dual-purpose missile.

During the 1980s there was also considerable interest in cheap guidance for conventional bombs. In theory, very large bomb tonnages are needed because most bombs miss. During the Gulf War, laser-guided bombs accounted for about 10 per cent of the weapons used, but they made most of the crucial hits. One possibility, investigated at length during the 1980s, was to attach a simple inertial guidance package to an iron bomb. Unlike a laser-guided bomb, such a weapon would not require the launch aeroplane to risk itself by lingering near the target. Also, it could be salvoed, since the debris thrown up by one bomb would not interfere with the others' guidance. Again, in the end GPS proved simpler and less expensive, and it was adopted, in what is now JDAMS (the Joint Direct Attack Munition System). Later there was considerable interest in naval fire support for the battle ashore. Again, GPS seemed to be the solution, in the guided 5in shell and in a land-attack version of the Standard Missile (LASM).

Overall, GPS helped inspire a new U.S. Navy Precision Strike concept of operations, which might be considered the latest stage in the space-led military revolution. As important as missile guidance itself is the use of GPS on board surveillance platforms, such as aeroplanes and UAVs, to define their own positions and thus to define target positions more precisely.[19]

Aircraft Guidance

For the Gulf War, the U.S. Air Force tried to coordinate the movements of very large numbers of Allied aircraft, using a computer-generated Air Tasking Order. To avoid conflicts, aircraft had to navigate very precisely, both in space and in time. Distances were vast, and there were few conventional navigational aids. GPS helped enormously. The main heavy strike aircraft, the Navy's A-6 Intruder and the Air Force's F-111, were both fitted with GPS receivers. In the case of the Intruder, GPS was integrated with an existing inertial system keyed to the carrier's SINS. The Air Force installed GPS receivers on seventy-two F-16Cs and on its F-15E strike fighters. In the case of the F-16s, GPS was used to cue the LANTIRN infra-red strike sensor. The F-16s also acted as Fast Forward Air Control aircraft, using their GPS sensors to determine the locations of targets; that data could then be passed to other strike aircraft. The F-16 case exemplified the new type of warfare, in that targets had to be handed off quickly from sensor to shooter. The longer the shooter had to search, the more danger it would be in, or even the better the chance that a mobile target would escape. At the very least, GPS made for a precise target location, which could be used by other aircraft, or by other sensor systems.

GPS also makes low-altitude map-based air navigation far more practical, since the aeroplane need not pop up to detect landmarks or to receive data from surface beacons (as in Loran or tactical beacons). Since its position is accurately enough known that it can rely on the terrain as shown on a map. In effect the map is information from remote sensors (distant in time as well as in space), which can be exploited because position can be determined so precisely. Without GPS, aircraft are subject to sufficient cumulative errors that they must update their navigational systems periodically. For example, at the outbreak of the Gulf War GPS-equipped Special Forces helicopters (MH-53Js) led Apaches (which did not have GPS) to attack two key Iraqi early warning radars from very low altitude. Each MH-53J dropped a bundle of 'chemlite' flares about 14km from the radar targets to provide the Apaches with a visual reference. That eliminated virtually all of the cumulative navigation error (about 8 to 11km) built up during the lengthy flight to the target.

There were also some unusual applications. GPS data were used to stabilise the APY-3 phased-array radar on board J-STARS ground surveillance aircraft. In effect they were replacing classical gyro or inertial sensors. Clearly GPS on board the J-STARS registered their radar data in coordinates which strike aircraft or ground-based weapons (such as the Army's ATACMS missile) could use.

Moreover, unlike most theatres of land combat, most of the Gulf War theatre was trackless desert. Indeed, Sadaam Hussein seems to have been surprised by the 'right hook' attack because he doubted that Western armies could make long-range runs through the desert. That was mainly a navigational issue. An army has a very limited striking range. It must periodically stop to refuel and to replenish. In the Gulf, it was possible to set up dumps along the expected line of advance, and to specify their location in geographical terms against which troops could set their GPS receivers. That had never before been possible. The alternative would have been a series of radiating beacons, subject to destruction or, even worse, to spoofing.

GPS also made it relatively easy for aircraft to support Special Operations Forces deep in the Iraqi or Kuwaiti desert, since they could fly to particular coordinates, rather than depend on beacons – which would have given away the Special Forces' locations. That may be an important pointer to the future. It may well be that the United States will find itself relying on very small ground formations, not too different from Special Forces. They in turn will have to be able to call in air and missile support, to survive the attentions of substantial enemy ground forces. To avoid destruction, they will have to minimise any radio transmissions of their own. In the past, however, any attempt to call in fire would have required some kind of beacon, or at least considerable discussion to determine coordinates correctly. At least in theory, with GPS all units involved are well aware of the correct coordinates, so a very brief message giving them suffices.

Mine Countermeasures and Amphibious Warfare

Mine countermeasures demands particularly precise navigation, since a minehunter must be able to specify exactly where it has detected and neutralised mines. Before the advent of GPS, that was generally done using a shore-based precision system such as Hyperfix, in effect a variant of Loran. That made combat mine countermeasures, off a hostile shore, particularly difficult, since no enemy was likely to permit the erection or use of any such system. Similarly, deep water countermeasures operations (*eg* against rising mines) might have to be conducted so far from any shore that no shore-based system would be effective. In the case of the Gulf War, the Iraqis were required to hand over maps of the minefields they had laid. Even though the maps were considered generally accurate, the situation was complicated in that the Iraqis had destroyed the Hyperfix system in the area. The mine clearance force relied instead on GPS.

In a wider sense, GPS supports new kinds of mine countermeasures. For example, there is now considerable interest in the use of anti-mine torpedoes, which can be delivered by helicopter. A helicopter can tow a minehunting sonar, or it can carry the torpedoes; generally it cannot do both. The past objection to a two-helicopter combination was that the killer helicopter probably could not come precisely to the mine position detected by the hunter. With GPS, however, a helicopter really can be flown to a set position. In fact the hunter can transmit its sonar data back to a mothership for analysis, in the knowledge that the data are all keyed to fixed points. Similar considerations apply to the semi-submersible currently being developed to support surface forces. As it runs ahead of the force, the semi-submersible uses its sonars to detect mines – and keys them to its GPS-derived position.[20]

Navigational issues also arise in amphibious operations. With the advent of air-cushion landing craft (LCACs), the U.S. Marine Corps gained the ability to assault a shore from beyond the horizon. At the same time it lost the ability to track (hence control) the craft as they approached the beach. Yet such control is crucial. For example, it is generally impractical to clear mines from the entire area off a beach. At best lanes into and out of the beach are likely to be cleared. The LCACs need some means of guidance. Although they were conceived well before the advent of GPS, this form of precise navigation is likely to be quite useful. GPS positions can easily be keyed to a computer navigation system incorporating digital maps – including the positions of mines and underwater obstacles.[21]

More generally, the Marines would like to be able to bypass beach defences entirely, in a form of infiltration tactics. To do this, they would disperse their attacking forces, foregoing entirely the sort of logistical build-up formerly conducted at a beachhead. Instead, small individual units would move ahead, converging only at the objective inshore. The concept is called Ship to Objective Maneuver (STOM). It is practicable, if at all, only thanks to space-based assets. Thanks to GPS, small dumps can be established in fixed positions along the routes of advance of all the sub-units, and the sub-units can find their ways to the dumps. Thanks to reliable OTH communications – which almost inevitably means satellite communication – the small sub-units can call in assistance, such as long-range naval fire support (which uses GPS guidance) as required.

Situational Awareness Beacon and Response

GPS turned out to have some other interesting applications. One was a new means of positive identification, Situational Awareness Beacon and Response (SABER), intended to provide a better picture of friendly forces while reducing the possibility of friendly-fire casualties. Each user has a UHF beacon capable of sending its GPS location via a satellite channel or a line-of-sight radio channel. System terminals, which can be aboard ships or aircraft or in ground facilities, act as network controllers; they can query the beacons or order them to report automatically. Given beacon reports, each terminal develops and maintains a picture of the location of friendly forces. The key to the system is that all beacons worldwide share the same 25 kHz UHF satellite up-link channel; their reports are automatically interleaved, and are available to every net controller.

Thus the system automatically operates far beyond the horizon. The satellite channel can carry 60,000 GPS positions per hour. Point-to-point (line of sight) local UHF radio can carry up to sixteen positions per second per channel (48,600 per hour). The satellite channel data can be received by any suitably-equipped platform, so the friendly-position data is easily available.

This positional data can also be used as a kind of IFF. SABER can operate in an 'intent to shoot/friendly identification' mode. The shooter broadcasts the estimated target location, together with a radius of concern (generally the damage radius around the target). Upon receiving the intent message, each beacon automatically checks whether it is within the radius of concern. If it is, it responds with a 'don't shoot' message.

Another application is to combat air control. In the past, carriers have provided their aircraft with position information using Tacan, a beacon which produces very characteristic signals. Using GPS, an aeroplane can always orient itself; if the carrier periodically announces its position, the aeroplane can return home as required. Aegis cruisers have used SABER when returning to a force (without using conventional IFF) and when controlling helicopters (without having to use Tacan beacons).

Clearly any transmission by numerous beacons entails some vulnerability. However, the transmissions are very short (less than 0.063 sec), and in the satellite version they are more or less vertical, with small side lobes. The system is being modified, moreover, to use a spread-spectrum signal (COBRA waveform).[22]

SABER was first tried in the field in March 1995. It then performed very successfully in the 1995 All-Services Combat Identification Evaluation Team (ASCIET) exercise: in 8500 beacon-hours of operation there were no friendly-fire casualties. The system was so simple that it could quickly be placed in service. The 22nd Marine Expeditionary Unit (MEU) took twenty-four beacons with it when it deployed to Bosnia in January 1996; the command terminal was on board USS *Guam*, acting as amphibious flagship. When Secretary of Commerce Ron Brown died in an air crash, SABER beacons helped the staff on board *Guam* maintain a clear picture of the search operation. The 22nd MEU later used SABER in an exercise off the coast of Sardinia, and then during the crisis in Liberia in April 1996. There the system tracked helicopters beyond the ship's horizon, as they picked up and evacuated Americans trapped in the Liberian civil war. Shortly afterwards SABER was tested in another ASCIET, this time in combination with systems used by the other services, such as the F-16's Improved Data Modem (Situational Awareness Data Link) and the Army's Enhanced Position Location Reporting System (E-PLRS). SABER was also very successful in Marine Corps exercises.

In the ASCIET exercises, SABER data transmitted automatically through the JMCIS satellite net was translated into messages transmitted via TENCAP (OTCIXS Gold), so that users from all the services could receive and distribute it. JMCIS in turn translates all SABER tracks into standard combat direction system formats (Links 11 and 16). Improvements beyond the original version include encryption of GPS data, the use of a stealthy waveform (COBRA), and the use of a more accurate GPS receiver.

SpaWar tried a more exotic application under a project called Radiant White. 'National' (often space-based) systems often cannot locate an emitter very precisely, since the satellite's orientation cannot be perfectly controlled. However, the satellite's accuracy can be dramatically improved if it locates the emitter relative to a known reference point. It occurred to SpaWar that the reference point could be a coded beacon incorporating a GPS receiver (so its location was known). It could be emplaced by Special Operations Forces. Emitter location accuracy might then be as good as 100m.

Vulnerability

There is, of course, a potential downside to all of this. The more the U.S. military came to depend on GPS, the more nervously it contemplated anti-GPS measures an enemy might take. Navigation was now so important that there was talk of a new kind of electronic attack, 'navigation warfare', the object of which would be to deny an enemy navigational capability. There was a general perception that in wartime the U.S. government might choose to turn off at least the public GPS signal altogether. On this basis in the 1990s the French decided not to use GPS to navigate their new Apache cruise missile.

The GPS signals received by a missile or a ship or aeroplane are quite weak. Although the receiver looks up, like any other receiver it has sidelobes, through which an unwanted signal can come.[23] There are actually two kinds of GPS countermeasure. The simplest is jamming: a noise signal covers the GPS signal and thus causes the receiver to break track. The other is spoofing: a pseudo-GPS signal confuses the receiver, taking it over. A GPS receiver can tell when it has lost track of the satellite, in which case it generally shuts down. It might well, however, be unable to distinguish a spoofing signal from a real one.[24] One major objection to shifting from earlier radio navigation systems such as Loran-C to GPS was that these systems are far more difficult to jam; the jammer would have to be far more powerful. Moreover, because the systems operate at much lower frequencies (shorter wavelengths) any jamming antenna would have to be quite large. On the other hand, GPS was twenty times more accurate than the next best system, Loran-C, and Loran-C was not available worldwide.

That countermeasures are a serious issue is suggested by the fact that later Tomahawks have anti-jam GPS receivers rather than the original ones. At least some U.S. Air Force (and almost certainly U.S. Navy) anti-jam receivers lock onto eight rather than four satellites, averaging some of their data. Also, many missiles have integrated inertial-GPS navigation systems, on the theory that they can use the inertial element if GPS is somehow disabled. The U.S. Navy is developing a low-cost inertial unit specifically to provide its cheaper weapons (such as JSOW and JDAM) with this sort of insurance. The Russians have now advertised a GPS jammer, which can be placed near a target, and which can disrupt GPS-based navigation over a radius of about thirty miles. Reportedly a jammer in a balloon might be able to disrupt GPS operations over hundreds of miles, a consideration raised when the U.S. Federal Aviation Administration proposed an air traffic control system based heavily on GPS. The HARM anti-radar missile is currently being modified to attack GPS jammers.[25] The hope is that the threat of such attack will keep an enemy from turning on jammers in the first place, and thus will open the way for attacks by GPS-guided weapons. For that matter, tests showed that if chaff, the common radar countermeasure, was placed between satellites and receiver, the receiver might find it difficult to maintain its lock on all four satellites. Given possible enemy countermeasures, some asked whether the GPS miracle was also a new kind of potential disaster.[26]

There is current interest (1999-2000) in a new generation of GPS satellites producing much stronger signals (perhaps 100 times the strength of existing ones) and also in using UAVs as surrogate GPS transmitters. The latter would amplify and re-transmit GPS signals so as to overcome jamming. There must, however, surely be some question of the extent to which a UAV, whose flight path is strongly affected by winds aloft, would distort signals originating from a satellite in a well-determined orbit.

THIRTEEN

The Navy and the Battle Ashore

In the wake of the Cold War, the U.S. Navy changed focus. Although it no longer faced a major naval threat, the United States clearly still had rivals, and the Navy was best equipped to deal with them. A policy document, '... From the Sea', expressed the new orientation in its title. By the mid-1990s the Navy's view was that its main future role would be to dominate the battle ashore, not merely at or near the water's edge, but well inland. To do that, it would arm its ships with a variety of guided land-attack weapons: short-range guns and medium- and long-range missiles. Carrier air attacks would be integrated with missile strikes and with long-range gunfire.[1]

The new kind of war was called 'littoral', to distinguish it from the blue-water warfare of the past, and also to distinguish it from coastal warfare, which involved only a limited area to either side of the shoreline. By way of contrast, littoral warfare included overcoming enemy coast defence forces, landing Marines, and attacking enemy positions well inland. The land-attack versions of Tomahawk were clearly well adapted to littoral operations, as were the new guided munitions developed using that miraculous positioning device, GPS. So, for that matter, were strike aircraft delivering their guided weapons. Clearly littoral warfare was nothing altogether new. The U.S. Navy had been attacking enemies ashore for many decades. However, the shift towards missiles changed tactics. So did the 'jointness' which was increasingly demanded of the U.S. services. Jointness demanded that the Navy adapt itself to some of the practices of the services with which it had to cooperate. The problems such jointness raised first became evident during the Gulf War.

The Air Tasking Order (ATO)

In this war, for the first time, virtually all air activity over the enemy area was tightly co-ordinated, in this case by a Joint Air Commander in Riyadh, who produced a voluminous Air Tasking Order (ATO). The ATO is the basis of an automated attack planning system. Typically it takes seventy-two hours to complete an ATO, but (at least since the Gulf War) there are also contingency strike plans, which can be executed in four to eight hours. Typically the ATO is accompanied by special instructions (SPINS).

The Air Force creators of the ATO argued that strict co-ordination is absolutely essential if large numbers of aircraft are to be used effectively. Both in the Gulf War and in Kosovo, rigid Air Tasking Orders were used. In neither case were there mid-air collisions nor, it would seem, were there any losses of aircraft to friendly fire. These successes tend to support the Air Force's case. On the other hand, rigidity is ill-adapted to hitting moving tactical targets.

During the Gulf War, the mass of Allied aeroplanes had to be routed through the defended airspace of Iraq without unduly concentrating (in which case they would have been attractive targets) except at their objectives. Collisions had to be avoided. Radios had to be tuned to avoid interference, and identification codes had to be set. The entire process demanded very careful and precisely timed flights. For example, to overwhelm the defences of a target, aircraft had to arrive together from many different directions. To avoid tipping off those same defences, the aircraft could not fly out in a stream, separating just before attacking; the Iraqis did, after all, have numerous surveillance radars. Thus the co-ordinated attack had to be timed so that aircraft from widely separated fields would arrive close enough together to saturate defences. In the Gulf the ATO provided both the overall squadron assignment and other crucial information such as allowable routes to the target and radio frequency assignments.

It gradually became clear that the ATO imposed rigidity on air operations over much of the battle area. It typified classic air attack practice, in which targeters take the products of reconnaissance systems, nominate targets, and help plan attacks. These steps were followed by the commanders' decisions as to which targets to strike, and then by the attacks themselves. The cycle is necessarily slow and cumbersome; it cannot easily respond to a dynamic enemy. For example, late in the Gulf War, the Iraqis began moving aircraft around city streets on a 24-hour cycle, *ie* much faster than the 72-hour ATO cycle. This timing generally protected the moving aircraft from being attacked.

There was, to be sure, an important exception. The area near the Saudi border was not controlled under the ATO. Army attack helicopters and Marine Corps tactical aircraft, which operated as long-range artillery, did not come under the ATO. That was one reason why air strikes could be made so effectively against an Iraqi force which attacked Khafji prior to the main Gulf War offensive. There were also serious efforts after the Gulf War to refine the ATO process so as to make it less rigid, but it appears that for the Kosovo air war ATO rigidity was restored to its pre-1991 level, perhaps to make political control of the operation more effective.

In a larger sense, the ATO was designed to hit fixed targets. That had been the rule for both aircraft and surface fires in earlier wars. For example, the Marines estimate that typically 70 per cent of all supporting fire is against relatively fixed or suspected targets; the figure in Desert Storm may have been about 64 per cent. That applies to air attacks as well as to gun fire support missions. In a future war many more targets may be mobile. For example, the Marines currently plan to move inland as fast as possible, preferably without stopping to develop a beachhead. At least at the outset, they will have to rely on weapons delivered by the fleet to fend off enemy counterattacks, *eg* by armoured forces, simply because they will not be able to take very much with them as they move.[2]

The ATO concept encountered one very real disaster in Iraq. It was very important to the coalition to deal with the mobile transporter-erector-launchers (TELs) which the Iraqis were using to launch Scud ballistic missiles against Saudi Arabia and Israel. Given the ATO structure, it was quite expensive (in terms of operational capacity lost) to assign aircraft to quick-reaction stations above presumed TEL areas. Unless an aeroplane happened to be nearly overhead, it had to be vectored to a presumed TEL location – from which the TEL would run as soon as the missile was away. Time was inevitably lost in translating DSP data (which was considered a form of TENCAP) into TEL location data at distant ground stations, then transmitting it to air control in Saudi Arabia. Iraqi decoys proved very successful. The main lesson learned at the time was that forward commanders should be able to receive DSP data more directly; hence the later development of JTAGS and the TES system. The effect of the ATO was probably to make the unsuccessful 'Scud hunt' disproportionately expensive in terms of the loss of other target coverage.

The ATO had to be distributed to all participating units in time for them to work out their roles in the overall attack. Distribution within Saudi Arabia was no great problem, but the carriers were offshore, in the Red Sea and in the Gulf. Ideally the ATO would have gone to them by radio – but it turned out that the Fleet UHF satellite systems lacked the necessary capacity. Instead, special S-3 flights to and from Riyadh had to be organised, simply to distribute the ATO. Ideally, the ATO data would have fed directly into the carriers' mission planning process via planning computers. That was impossible for the carriers in the Gulf. Postwar, simply in order to be able to receive ATOs in digital form, required carrier data-receiving capacity was increased more than tenfold, from 64 kbits/sec to 2 Mbit/sec on a joint mission. The Fleet Satellite and its UHF successors had nothing like the required capacity. Instead, SHF channels like that provided by DSCS would have to be used much more completely. A fleet-wide SHF upgrade began, including an interim DAMA capability (Quick-Sat) on board selected flagships and carriers. DSCS earth stations were upgraded.

Higher antenna gain was a key to this kind of improvement. Existing satellites could not be modified (although new ones were). They merely received a waveform, amplified it, and sent it down at a different frequency. Information capacity was set by how easily a receiver at the surface could distinguish information in the waveform from the noise mixed with it. The higher the gain of the satellite receiver, the more signal (and the less noise) it would pick up. The higher the signal-to-noise ratio, the more the receiver could extract from what it received. Thus, as part of the SHF upgrade, ships received a larger-diameter, hence higher-gain, antenna (for a modified WSC-6 terminal).[3] Given the new antenna and a new DAMA modem, ships can receive data at the rate of 256-640 kbits/sec, depending on antenna size (*ie* gain).

Using the new capacity, current practice is to provide the carrier strike planning team with the ATO and SPINS twelve to eighteen hours prior to a strike, via the Contingency Theater Automated Planning System (CTAPS). Strike teams include representatives of all the squadrons – which on a carrier means aircraft types – likely to be involved in the strike.[4] The product of the strike team is a rough battle plan, including projected routes to and from the target and defence suppression techniques. Once approved by the carrier air group commander (CAG), the plan is developed in greater detail. Squadrons work out their roles, and each pilot then uses an automated planning system, TAMPS (Tactical Aircraft Mission Planning System) to develop his own flight plan.

TAMPS exemplifies a long-standing trend, both in the Air Force and in the Navy, towards automated air planning. Its basis is not too different from that of Tomahawk targeting. In the past, planning an air attack against a heavily defended area was quite as laborious as planning a Tomahawk strike. The only simplifying factors were that aircraft flew somewhat higher, and that their pilots could take evasive action when they encountered enemies. On the other hand, a pilot planning a mission needed (apparently) complete information to know when to make a long detour to avoid an enemy early-warning radar station or a long-range missile battery. Just as in the case of a Tomahawk planner, the pilot had to be sure that the detour would not send him into greater danger, or exhaust too much fuel before he reached the target area. Like a Tomahawk flight planner, TAMPS can test alternative flight paths, all the way to the target. This kind of system is attractive because it automates away much of the detail work normally associated with strike planning. The planner can try alternative plans (equivalent to the 'what-if' questions which can be asked using high-capacity tactical computers). Given really comprehensive photography and maps – and a very powerful computer – a path can be planned automatically, a process called auto-routing. In either case, the output of the mission planner is a digital flight plan which is inserted into the aeroplane's mission computer before take-off. As the aircraft flies along the planned path, orders to turn at way-points are displayed on the pilot's head-up display (HUD). The aircraft computer (cued by TAMPS) also controls settings for radio channels and for IFF codes. TAMPS was actually derived from the standard Tomahawk planner. Under the procedure developed for the Gulf War, each squadron receives its orders under the ATO. Each pilot in a squadron uses TAMPS to plan his own mission to a designated target. The computer produces images (visual, radar, FLIR) of what the pilot is likely to see as he approaches the designated target.[5]

TAMPS was first deployed on board the carrier *Carl Vinson* in August 1986. By the early 1990s it was on board eleven carriers and at more than thirty land sites; it was standard for both Navy and Marine Corps. For the Gulf War, twenty-six TAMPS terminals were on board six carriers, a battleship, and the major amphibious ships. TAMPS still used pre-JMCIS computers, and it had some major limits. It provided no advice on the choice of weapons for a strike (weaponeering) nor was it adaptable to anti-ship strikes (the 'war at sea' role). Nor was it directly connected to the other major naval air planning system, TEAMS (the Tactical EA-6B Mission Support system). The pilot can use the computerised TAMPS-generated mission plan to rehearse the planned mission, using a system called TOPSCENE. The TOPSCENE console shows roughly what the pilot should expect to see along the mission route, allowing for manoeuvres and for possible detours. The images fed into it come mainly from photo satellites; very powerful computers adjust them to the pilot's likely point of view.[6]

The process from ATO through strike planning team to TAMPS is described as deliberate planning, designed to deal with pre-chosen targets. The process becomes more complex as aircraft increasingly share the same airspace with Tomahawks and other missiles, such as SLAM, JASSM, and ATACMS. Strike planning thus requires careful de-conflicting. Any last-minute changes become more and more difficult. For example, diverting a Tactical Tomahawk from its planned path (or sending it into action from a loiter orbit) may affect the aircraft strikes being carried out at the same time. The resulting rigidity is a real problem. For example, in June 1995 a mission was mounted to rescue Air Force Captain Scott O'Grady, whose F-16 had been shot down by a Serbian missile. The crucial condition for the actual rescue was that no Serbian mechanised force be within thirty minutes of O'Grady. Given the inertia inherent in the strike planning system, it would have been impossible to mount a sufficiently powerful air attack in time to have prevented the Serbians from overrunning the small force which actually rescued O'Grady.[7]

The goal is to make strike planning responsive, so that tactical aircraft can efficiently attack transient or pop-up targets. As of mid-1999, responsive planning was to be available by the end of FY00 (October 2000), and merged planning (responsiveness at the force and unit levels) in FY04. The key to mission flexibility is increased automation. Ideally, that makes it possible for the system as a whole to take into account any change at the individual aircraft target level. One key is auto-routing. Carriers are being provided with an automated Tactical Strike Coordination Module (TSCM) fed by the ATO.[8] It in turn feeds into the individual planning systems, APS (Afloat Planning System) for Tomahawk and TAMPS, TEAMS, MOMS (Map Operator and Maintenance), and PTPS for aircraft. Also feeding into the planning process are weaponeering systems used to choose aircraft load-outs: ATACS (Automated Tactical Supplement), NSPW (Naval Strike Planning Workstation), TARGET, WINJAMS.

These planning levels merge in a carrier's Naval Strike Warfare Planning Center (NSWPC), the first of which was installed on board USS *Nimitz*. For the near term the connection between the three is an electronic Strike Planning Folder (SPF) which all of the computers involved use. This admittedly inadequate technique will be in service in FY00. A Tactical Strike Coordinating Manager (TSCOM) is being combined with TAMPS as a force-level planning module. It adds a multi-element dimension to TAMPS. TAMPS itself is being replaced by JMPS, the Joint Mission Planning System (JMPS) being co-developed (with different plug-in elements) with the U.S. Air Force, the Marines, and the Special Operations Forces. It is to be fielded in 2002. JMPS in turn is to be part of the force-level Naval Warfare Planning System (NWPS), which embraces all warfare areas.

Rapid Targeting in the Cockpit

The demand, then, is for real-time targeting well beyond the fleet's horizon, and also for extremely quick responses. Quite aside from the problem of co-ordination with other plans, a quick strike requires a great deal of information. The pilot must be provided with a route which, at the least, minimises exposure to enemy fire. The pilot will not want to be guided by rigid commands. Instead, he will want sufficient visual or radar cues to enable him to reach (and to find) the target, and with information about enemy anti-aircraft firepower near the target. If the target is mobile (*eg* a Scud launcher), the aeroplane flying towards it needs constant updates. The new process of quickly routing individual aircraft to strike at transient or pop-up targets is called Rapid Targeting in the Cockpit or Real-Time in the Cockpit (RTIC).

In RTIC, reconnaissance data (including satellite data) is used directly to designate mobile targets for attack. The initial Air Force demonstration programme (1993-4) was Talon Sword. In one demonstration an EA-6B launched a HARM missile at a patrol boat on the basis of satellite OTH-T information. In another, an EA-6B combined satellite data with its own sensor data to guide an F-16 firing a HARM at a radar site beyond its own horizon.[9] In yet another demonstration, the satellite data was correlated at an Air Force Constant Source ground station, then transmitted to an F-16. In July 1994, satellite data cued a long-range electro-optical sensor on board a P-3, which used it to confirm and refine attack data. The satellite data and the P-3 image were transmitted to an F-16. Further satellite data were used for battle damage assessment, to plan a follow-up strike.

RTIC clearly equates to the use of space resources, both to see the targets and to transmit data reliably. Satellites may often offer the clearest pictures of a target area, prior to combat. The fleet must be able to receive them as needed, often as a baseline against which to estimate the positions of the targets picked up by tactical sensors such as reconnaissance aircraft and UAVs (whose pictures will be distorted by, for example, their manoeuvres). To aim a weapon at GPS coordinate points, moreover, requires that the target be located extremely precisely, in GPS terms.[10]

More generally, the products of national sensors – signals as well as image intelligence – may have to be used for targeting, and even provided in some form to the attacking pilot. It becomes vital to be able to sanitise such information to limit the cost if an aeroplane is shot down and its contents recovered. In the past, many documents carried markings forbidding their carriage on board combat aircraft. Now that the documents are so often electronic, the strictures have to be handled on a more automated basis. Hence the current interest in Radiant Mercury, the automatic multi-level security sanitise.

In effect, RTIC is an extension of earlier close air support practice. When aircraft were all fitted with mission computers, forward air controller practice changed. Instead of voiced instructions, aircraft now typically can receive a stereotyped 'nine-line targeting message' via a digital link from a forward controller (pilots tend to like the combination of voice and text). The text message is automatically displayed on the aeroplane's digital display screen, and necessary headings pop up on the pilot's HUD.

In RTIC, the mission planning centre rather than the forward air controller sends the nine-line message. Because the screen is digitally controlled, it can also display an image of the target area, with the target itself highlighted. In RTIC, that image is sent either by Link 16 or via a special Link 16-compatible digital link (the Improved Data Modem of F-16s), or via a pod normally used to receive the digital images by means of which a pilot guides electro-optical weapons (the Navy AWW-13 for SLAM, or the Air Force AXQ-14 for GBU-15/AGM-130). There is, in 1999, some question as to whether the image of the target area itself suffices, since the pilot must guide the aeroplane into the area. Experiments run by the Naval Air Weapons Center at China Lake suggest that it is best to provide a pilot with three images in sequence, to lead him in: 10×10nm, 3×3nm, and 1×1nm. Only the last need show the target itself. In the others, it is far more important to provide landmarks visible to the pilot and also to show where defences are. The larger-area images are provided by satellite or even by stand-off electro-optics (*eg* the P-3's Cluster Ranger). Many details will have been supplied by signals intelligence systems.

RTIC requires that strike aircraft be able to receive the images a pilot would need to recognise the new targets. The pilot needs imagery because he is generally moving too fast to scan and interpret what he sees; he has to be cued, at least to the point where he can see and recognise the target itself. Without cues, a pilot may well be unable to acquire the target.[11] It is unlikely that each tactical aircraft can be provided with a new wideband link adapted to receive images (Link 16 can transmit images, if its whole bandwidth is used for that purpose). However, images can be compressed drastically. Radiant Tin is a Navy/Marine project to develop the necessary technology; a typical compression ratio is 175:1. Beside tactical aircraft, it is to be used on board submarines.[12]

Quite aside from targeting, the concept of supplying external data directly to the pilot has other advantages. Enemy missile systems may be hidden by terrain; given an RTIC link, the pilot can be given their locations and characteristics, to keep him from blundering over them. When NATO air forces attacked Serbian troops in Bosnia, they found that they had to deal with pop-up targets. Without troops on the ground, there were no forward air controllers. Instead, there were UAVs observing the Serbs and relaying back images of the targets which had to be hit. Suddenly RTIC became extremely important. It became even more important in Kosovo in 1999. Typically a targeting centre on the ground (and well to the rear) receives information from a source such as a UAV over the battle area. It processes that information, and transmits orders up to an assigned aeroplane via a surface-to-air link. The Link 16 information automatically goes into the aeroplane's mission computer to be fused with other data, and it is displayed on the computer-controlled panels in the cockpit.

Satellite data is at the least indirectly involved. The UAV or other source imposes some distortion on the photograph showing the pop-up target. It has to be registered against a data base of ground photographs – provided by satellite – carefully adjusted to cancel out distortion. Other satellite-derived data indicates threats which aircraft must evade, such as mobile missile batteries.

Given the combination of data, a mission planning system such as TAMPS can automatically plan routes into and out of the target area. Given the use of TAMPS (or a comparable system) to plan all routes, this type of re-routing becomes essential. Not least is the need to avoid interference with other aircraft nearby, all of which are following other computer-generated flight plans. Thus the future of RTIC is tied to the future of aircraft avionics and of the ways in which data are fed into them.[13]

The great unresolved question is how to integrate RTIC with the ATO process. Different approaches are currently being employed by different C-in-Cs and by different services. Current attempts to do so use the planning and targeting computers to select aircraft already in the air for RTIC missions, and to adjust other aircraft routing to allow for their new routes. Ideally all the aircraft will eventually be connected by Link 16 (once that is in widespread service) to the planning (*ie* control) centres. Changes are automatically entered into their mission computers. For example, re-routing is accomplished partly by displaying new waypoints and courses in the pilots' head-up displays. The pilots' Link 16 displays will show their new orders, as well as any updates concerning threats they may encounter. This type of operation demands a degree of artificial intelligence at the mission planning level.[14]

The problem is somewhat simplified if aircraft can generally fly at medium to high altitude, unaffected by terrain. That was certainly the case in Iraq in 1998-9 and in the former Yugoslavia, because enemy long-range missile and other air defences which had not largely been destroyed at the outset were holding their fire.[15] The situation would be considerably more complex if the strikes had to be mounted against an enemy with powerful surviving long-range air defences, because in that case mission planning would have to take into account the details of terrain.

Getting More Bandwidth

Both Tomahawk targeting and air planning require vast amounts of detailed, largely graphic, information about not merely the target area but also the area over which the missile or aeroplane would fly. Moreover, the information must be current, since it would not do for a missile to fly into a building erected in the last few weeks, or for a pilot suddenly to encounter a newly-sited missile battery.

Prior to the advent of new satellite links, carriers already had an ability to obtain the necessary photographs, using their own aircraft carrying special TARPS pods. Clearly, Tomahawk-firing ships lacked any equivalent – except what they could obtain from libraries of satellite photos held in the United States. It could also be argued that even carriers could not obtain pre-strike intelligence before a surprise attack, such as the air strikes on Libya in 1986 or on Iraq in 1991. Aircraft overflights would inevitably cause problems, whereas satellites could take their photographs quite freely.

Ships had long been taking files of satellite photographs to sea, to support their planned wartime strikes against the Soviet Union. That had been practicable partly because there were relatively few likely targets. Everyone in the Seventh Fleet, for example, could guess that places like Vladivostok harbour and the Komsomolsk-na-Amur shipyard were on the target list. In a post-Cold War world (as foreshadowed by the 1986 strike against Libya), long-suppressed tensions cause conflicts in what almost seem random places. To the extent that the U.S. government wants to intervene, it has to be able to assemble strike forces very quickly. Once assembled, they must be able to plan and execute their attacks on places not even under consideration when they left port months earlier.

The fleet must therefore be able to gain instant access to libraries of intelligence data which, due to their sheer size – and to the way they are updated – cannot be duplicated aboard ship. Access must, moreover, be secure, otherwise the potential victim of the strike might suddenly discover that the fleet was receiving satellite photos of the targets. Only satellites offer the combination of long-haul capacity and security that this sort of situation demands. The rub is that each image equates to an enormous amount of data.

In 1991, when the Gulf War was fought, commercial satellites already offered television-quality (T1) channels, which could transmit images in the required quantities. In 1992 GTE funded a six-month demonstration of such capacity aboard the carrier *George Washington*.[16] It was so successful that SpaWar leased a satellite to support the fleet flagship *Mount Whitney* for the 'Ocean Venture 1993' exercise (it began operation in May 1993). The project was called Challenge Athena.[17] The satellite was one of nine GTE Intelsats, with a capacity of up to 2.048 Mbits/sec (T1 is 1.554 Mbits/sec). The uplink is SHF (C-band) and the down-link is SHF (Ku-band, to minimise the size of the receiving dish).

In the *George Washington* experiment, the satellite linked the ship to a CCC (formerly FCC) ashore, typically operating at 768,000 bits/sec. This is enough bandwidth for simultaneous transmission and reception of voice, full-motion video, and compressed data and imagery. Probably most importantly, it could provide the images necessary to plan air and Tomahawk strikes against land targets. In this sense it was the littoral warfare equivalent of TADIXS. As it happened, the enormous capacity of the link could also be used for numerous simultaneous telephone calls back to the United States, a feature which proved quite popular among deployed sailors. Using the satellite, the carrier was able to receive an Air Tasking Order (ATO) from Shaw Air Force Base. Later, when IT-21 was implemented, an ATO could be passed from ship to ship (or from shore to ship) in the form of an E-Mail file.

The wideband link was also able to support a video teleconference between the Joint Intelligence Center (Norfolk), the National Military Joint Intelligence Center (Washington), and the carrier's Strike Planning Center (16 October 1992), something not previously possible. Such live conferences were considered exceptionally valuable, since they could tie together widely dispersed ships. During the 1996 Taiwan Straits crisis, for example, daily plans were distributed in the form of charts and other graphics shown on satellite displays.

A second system, Challenge Athena II, was funded by the Congressional Intelligence for Targeting Initiatives programme. It employed the Intelsat V satellite (COMSAT experimental Wideband Mobile Service). It used a 2.4m shipboard dish to receive the 36 MHz-wide C-band global satellite beam; data rates were T-1 shore to ship and 1.152 Mbit/sec ship to shore (*Mount Whitney* used a Ku-band spot beam steered to follow the ship). This system was used during a deployment from the Atlantic to the Mediterranean and then to the Gulf during May through November 1994. It was reactivated, at CNO request, as Challenge Athena IIA in the fall of 1995.

By this time the new Defense Information Supply Agency (DISA) had announced a Commercial Communications Satellite Initiative (CSCI). A fleet-wide Challenge Athena III, broadcasting to several ships simultaneously, was the first project.[18] In 1996 wideband terminals (WSC-8) were approved for all command ships.[19] By 2003 there should be up to twelve space transponders serving between forty and sixty-five ships.

If the problem of receiving images might hamstring surface Tomahawk shooters, submarines had far worse problems. An image equates to a million bits or more, yet submarines typically receive at 2400 bits/sec, the capacity of a Fleet Satellite channel. It might take a half hour or more to receive a single picture, and for all that time the submarine would have to keep her mast exposed. To do better, the submarine needs a larger antenna connecting them to higher-frequency satellites with higher data rates. In a May 1994 experiment, for example, the submarine USS *Albany* set up a 64 bits/sec link. It received an image in 4.6 minutes, using a 12in dish on a mast. Current plans for next-generation submarines include much larger dishes, which may be integrated with the submarine's sail or opened umbrella-style from a mast.

The Global Broadcasting System

Although video teleconferencing was clearly valuable, in fact most Tomahawk shooters needed high capacity on a one-way basis. If they could contact the libraries of images held in the United States, they could state their requirements relatively simply. Fulfilling their needs would require T1 channels or better, since each image might equate to as much as 100 Mbits. To transmit them in the great quantities required, the Defense Department developed a new Global Broadcasting System (GBS), modelled on existing direct television broadcast technology (which gives consumers 100 channels, using an 18in satellite dish). The same technology has been offered to Internet users: they use their relatively slow modems to request data, but the data is blasted back to them via television relays (the disadvantage is often limited capacity at the transmitter end, so not every customer can be served in quick enough sequence).[20]

GBS has another face as well. It offers a natural expansion of the Fleet Broadcast concept. Indeed, current plans call for the Fleet Broadcast to migrate to GBS so as to free up valuable UHF band-width. Similarly, plans call for GBS to carry other one-way broadcasts (multicasts) currently carried by UHF channels, such as the TRAP and TIBS intelligence broadcasts and the missile warning (DSP) broadcast. A planned channel, for the Consistent Operational Picture, would be an expanded equivalent to TADIXS.

GBS uses Ka-band (SHF) transponders on the last three UFO satellites. The frequency was chosen over X-band to avoid terrestrial interference; however, wide-area coverage is difficult due to the need for power margins against rain blockage. Capacity is 23.6 Mbits/sec (over fifteen T-1s) spread over multiple channels. GBS can send images at 12 Mbit/sec (one image in 8.4 seconds). That compares to 30 seconds for the enhanced version of DSCS (3 Mbit/sec) or 1 minute for Challenge Athena (T-1) or 3.3 minutes for the 7ft antenna of the enlarged WSC-6 (512 kbits/sec) – or 3 *hours* for a commercial Inmarsat, operating at 9600 bits/sec.

This Phase II system replaced an initial version using commercial Ku-band transponders, and became operational in the Continental United States and in Bosnia, carrying intelligence, weather, and entertainment broadcasts. Phase III will be part of the X/Ka-band Gapfiller; it has been downsized to match the current capacity. Phase IV may be part of a projected Advanced Wideband System. GBS began operation in 1998; by 2005 every surface combatant will have a GBS dish.[21]

GBS exemplifies a point which goes well beyond Tomahawk or air targeting. Ships go to sea with limited amounts of data on board. They are sent to unexpected places, to carry out missions not envisaged when they leave port. To do those things, they need access to information which may not be on board when they steam out of their bases. Generally that information is available at home, and virtually all of it is in the form of computer data. Equipped with a satellite link, a ship at sea can, at least in theory, call up whatever it needs. Examples include not only satellite images but also tactical decision aids (TDAs) to be used by the ship's computers.

OTH Fire Support

The Navy's new concept of dominating a battle ashore required more than accurate imagery. In the past, naval fire support had generally meant firing at a target within sight. Forward observers could easily designate it, perhaps in relation to a known landmark or reference point. However, by the 1980s there was general agreement that ships supporting even an amphibious operation would have to remain well off shore, probably beyond the horizon. There they would be relatively immune from coast defence missiles, and they would probably be outside any minefields. Moreover, an amphibious force attacking from beyond the horizon could assault beaches over a very long stretch of coastline. An enemy would find it difficult to predict where troops could come ashore, hence find it almost impossible to concentrate forces to oppose them. The great 1980s symbol of this new assault technique was the fast air-cushion landing craft (LCAC), which could reach a distant beach in about the same time that its slow predecessor, the LCU, would need to cover the much shorter distances of earlier assaults.

The Marines were still interested in gun fire support, however. Through the 1980s there were numerous proposals for new support systems, ranging from guns firing rocket-boosted shells to an adapted version of the Army's MLRS. As in the case of the Outer Air Battle, there was little real problem in lobbing explosives over the required distance. The real problem was in locating targets and guiding relatively cheap munitions accurately enough to hit them. Proposed solutions to the targeting problem, such as unmanned aircraft (UAVs), were all rejected as too expensive.

GPS solved the problem. Suddenly a forward observer equipped with a GPS receiver could accurately report a target's position. An ERGM shell could be guided to that place by its own cheap GPS system. Satellite links could reliably connect observer and distant shooter, at the same time limiting any danger the observer might feel from exposure due to using his radio.

In fact a somewhat more systematic approach is being adopted. In its own attempt to automate command and control, the U.S. Army devised a family of command systems based on an electronic map of the battlefield, which includes both terrain features and the positions of friendly and enemy troops. One of them is AFATDS, the Army (actually Army/Marine) Field Artillery Tactical Direction System. Given the order to destroy a target, it assigns the weapons to do so. AFATDS was successfully installed on board the large amphibious ship *Saipan*. Ultimately the AFATDS function is to incorporated in a new naval fire control system (NFCS), which will control guns and missiles and will communicate with air planning systems (TAMPS).

Coordination matters partly because the new long-range shells share the same air space with aircraft and missiles; it would not do to fire a shell through a friendly aeroplane. The combination of precise navigation (via GPS), reliable communication (via space), and very fast computers make it possible to coordinate on the necessary scale.

Mine Countermeasures

Littoral warfare naturally involved other kinds of naval operations, such as mine countermeasures. It is more and more difficult to detect and clear mines once they are in the water. However, few countries posses very large mine stockpiles. In many cases a naval force can elect to operate anywhere along a considerable stretch of beach. Much of that area is likely to be free of mines. By 1990 there was considerable interest, at least in the U.S. Navy, in an alternative to minehunting or sweeping: mine reconnaissance. If the minelaying operation itself could be observed, then the minefield could be mapped, at least approximately. The entire mined area could be avoided. This strategy was tried during the Gulf War. It was not entirely successful, apparently because Allied intelligence did not realise that many sorties by small Iraqi craft were for minelaying, rather than simply for transporting loot back from Kuwait to Iraq (many time the boats did carry loot).[22] Even so, the idea was enormously more attractive than any more conventional kind of mine countermeasure.

Space assets could help, because reconnaissance satellites could capture images of the areas an enemy might mine. To be effective, a minefield had to be fairly dense, so minelaying would surely be protracted. A satellite visiting an area every day would almost certainly capture images of some activity. However, because the satellites could not dwell over the areas which might be mined, they could not be relied upon to map out the minefield. Even so, activity would indicate the approximate area mined.

It seems likely that in 1990 the Iraqi mining activity was observed mainly or entirely by fleet aircraft. In a future war, satellites would offer an important advantage: the minelayers would be unaware that they were under surveillance. That would preserve an important element of surprise. In 1994 (FY95) the U.S. Navy ran an advanced concept technology demonstration programme in mine countermeasures, Hamlet's Cove, an attempt to improve mine reconnaissance by using all-source information, including that from satellites. Probably Radiant Clear, an ongoing Joint Countermine advanced technology demonstrator conducted by the Navy's Space and Electronic Warfare Systems Command, was a related programme. It sought minefield and barrier detection by using satellite imagery. The reference to barriers and the Marine Corps' stated interest suggest that Radiant Clear is concerned mainly with shallow water, where barriers might be detectable by wave patterns which they distort. In such areas mines themselves would have little visible effect on the surrounding sea. However, their presence might be reflected in non-acoustic phenomena such as bioluminscence.

Theatre Ballistic Missile Defence

As Third World countries acquire medium-range ballistic missiles, the question naturally arises of whether they can deny entry to U.S. forces by using them against natural bottlenecks such as ports or airfields or the beaches amphibious forces must cross. The threat of such attacks may also be used to deter friendly countries in an area from supporting a U.S. operation – even one in their own defence. This is not a new idea. After Iraq invaded Kuwait, U.S. carrier fighters provided initial cover for Saudi Arabia. Since they could do so whether or not the Saudis officially asked for such cover, the Saudis were unaffected by Iraqi threats against inviting unbelievers into their kingdom. They were able, then, to invite in the Western armies which successfully liberated Kuwait the following year.

Warships are natural platforms for mobile wide-area missile defence, and the U.S. Navy is actively pursuing this mission in two phases, Area Defense and Navy Theater Wide (NTW). To fuse data from an entire theatre of operations, cruisers are being modernised with AADCs (anti-air defence centres). The current 'smart ship' programme reduces a ship's complement to provide the necessary space. Unlike the usual flat combat system displays, that of the AADC provides a three-dimensional image of the air space, so that the path of an incoming ballistic missile is clearly distinguishable from an aircraft track.

The Area programme, using the SM-2 Block IVA missile, is to become operational in 2003. NTW, using the SM-3 missile, is to become operational in 2005-07, depending on funding. The Area programme is intended to protect specific targets, such as ports or troops coming ashore. It deals with short- and medium-range missiles, such as Scuds. NTW can protect much larger areas; for example, a Navy map shows a single cruiser off South Korea protecting most of that country and Japan. This system defends against medium and long-range missiles, and it can attack a missile during its ascent phase. Technically, it is far more demanding than the Area system, which seeks a hit as the missile warhead descends, using a high-explosive warhead which need only come close enough to damage the incoming missile badly enough to cause it to break up. NTW, on the other hand, uses a guided hit-to-kill weapon, LEAP, which can destroy a missile anywhere in its trajectory (*ie* ascent, midcourse, or descent). In a defence in depth, Aegis NWT ships would coordinate their attempts at interception with one or more Aegis Area ships closer to the missile's likely target.

The first step towards NTW is Aegis LEAP Intercept (ALI), LEAP being a powered third stage with an IR seeker. This version of the SM-3 missile has a single-colour IR seeker and a dual-pulse third stage. It is intended only for tests, so the Aegis system supporting the missile has minimal capability and depends heavily on external cues. However, it can detect and track targets at Area defence ranges, 400-500km. The first operational version, NTW Block I, is designed to defeat missiles like the extended range Scud-C, the North Korean No Dong, and the Chinese M-9. It has to be able to distinguish different missiles at a range of about 1000km, and to discriminate between the missile warhead and debris (such as a separated booster). To do that, the missile is provided with a combination radar and IR seeker. The SPY-1 radar is modified to discriminate between debris and the missile, so that it can command the missile effectively. This version relies on theatre cueing, *ie* on JTAGS. Current plans call for deployment on board four cruisers (with eighty missiles) no later than 2007.

NTW Block II responds to a predicted future threat, such as the Chinese CSS-5 (DF-21) or the North Korean Taepo Dong. It must separate threats at ranges beyond 1500km, and it must deal with natural debris such as body fragments, as well as with decoys. The associated version of the SM-3 missile is to have improved discrimination and guidance and improved propulsion, and the Aegis weapon system version will have a high-power discrimination radar (in addition to the usual SPY-1) and will be associated with a battle management combat direction system. The radar (HPD) currently envisaged is a Raytheon unit originally developed to support the Army's Theater High Altitude Area Defense (THAAD) interceptor missile.

Area system detection and tracking were successfully tested by the cruisers *Lake Erie* and *Port Royal* on 18 and 20 November 1998 in a 'Linebacker' exercise, using both a Terrier Missile Target (TMT, simulating a Scud-like weapon) and an Aegis LEAP Intercept Target Test Vehicle (TTV, simulating a longer-range missile). They successfully passed cueing and tracking data to each other, as well as to other theatre defence systems (such as the Army Patriot). USS *Shiloh* (CG 67) is the current ALI test cruiser, firing SM-3 test missiles. Current plans call for developmental and operational tests at White Sands between August 1999 and June 2001, followed by a Linebacker test off Point Mugu in September 2001, and then full-up developmental and operational tests off Point Mugu between May and November 2002. The first operational Area Defense ship should be ready in 2003.

The space aspect of missile defence is cueing and communication. Without cueing, a ship's SPY-1 radar probably would not pick up an incoming missile, or at least would not pick it up in time for successful interception. At present DSP (in future, SBIRS) is likely to be the best way of spotting and tracking an incoming missile early enough. That makes JTAGS a crucial element of naval ballistic missile defence. Ideally, a JTAGS capability would be built into the ships. However, ship real estate is very limited. JTAGS uses large dish antennas to receive the DSP downlink. The solution, if it comes, will presumably be to use the new multi-function broadband satellite antennas (see below) for this purpose as well as for communication. JTAGS software functions will have to be integrated with other combat direction software. JTAGS is already a theatre-wide system. It communicates with distant missile batteries and command centres via UHF satellite (TRAP broadcast). Thus the interim solution to the cueing problem is simply for the ship to receive cueing data from a theatre JTAGS via her existing satellite communication system.

The Navy already plans to use netted radar data, transmitted via its CEC (Cooperative Engagement Capability) system, to help its ships detect and track incoming ballistic missiles. As conceived, CEC used a line-of-sight link, so it was limited to a few ships. Then it turned out that CEC could link ships with distant land sites. In 1998 the Navy gained permission to use an EHF satellite link. The satellite link is probably easier to implement for CEC than it was for Link 11, because CEC does not use a roll-call; instead, it uses a series of independent one-to-one connections, which need not be synchronised.

Theatre ballistic missile defence is to be a joint venture, uniting weapons and systems developed by the different services. Thus the Navy's versions of Aegis will work together with the Army's ground-based THAAD missile and with the Air Force's airborne laser (ABL). To the extent that these very varied weapons are to form a unified system for defence in depth, they must be linked by a high-speed net – which automatically means by satellite communications. They are much too far apart to rely on line-of-sight radio, and HF cannot carry the necessary traffic at the necessary rate. The time line to intercept a theatre or short-range ballistic missile (like a Scud) is just too short. Current U.S. plans are to create a Joint Composite Tracking Network (JCTN), which will fuse data from all available sensors to provide each defensive weapon with adequate information for fire control. This is the CEC concept on a grand scale, and the need for JCTN explains why CEC has received so much investment in recent years – and why it is being loaded with additional functions the presence of which has apparently slowed development.

Certainly the combination of the Aegis system with an evolved version of the Standard Missile offers a remarkably robust and mobile defensive system, in a very easily deployable package. Equally, there is no other way of providing an expeditionary force with such defence, at least until it is established ashore. Moreover, the current Marine Corps thrust towards an operation based at sea, rather than ashore, precludes any transition to a shore-based system, such as THAAD, early in an expeditionary campaign. In this sense the theatre ballistic missile defence ships perform much as carriers now do, in supporting an amphibious operation: they remain on station until sufficient resources can be built up ashore. That is a natural Navy function.

Beyond this function is a real danger, however. The main advantage the Navy offers the country is its mobility. If defensive systems on board cruisers are the best U.S. theatre defence assets, then local commanders may well want to keep those assets permanently in place offshore. Just as the Navy has avoided spending its limited budget on the space assets it uses, these theatre commanders are unlikely to agree to spend other limited budgets to buy more cruisers specifically to be tied to the theatre defence mission. Instead, sea power designed for mobility will be used statically. To the extent that non-naval commanders control missile defence ships, they are likely to misuse them, to reduce the sea power upon which the United States has come to depend. The situation will be much worse if the Aegis-based system evolves into some kind of national missile defence.

This is not an altogether new idea. Until the mid-1960s the U.S. Navy contributed ships and aircraft to the Continental Defense System managed by Norad, the predecessor to the current Space Command. Although the picket ships were specialised, and could not have contributed effectively to open-ocean operations, the manpower certainly could have been used elsewhere. Budgets were limited; money spent on national air defence could not be spent on, say, carrier-based striking power, or on amphibious forces. The situation now is more difficult, in that the ships which would be devoted to missile defence have other important alternative roles.

It is very important to keep in mind that the question is not what the U.S. Navy might prefer for itself, but the way in which the finite resources offered by the Navy can best support the United States. Naval forces are innately flexible, in a way that, say, a THAAD/Patriot battery or a laser aircraft is not. Those very familiar with missile defence, and uninterested in other types of warfare, may not appreciate the wider potential of naval forces, and may see the cruiser as little more than a mobile THAAD battery. That would be unfortunate, not for the Navy, but for the country it serves.[23]

FOURTEEN

A New Kind of War?

At the end of the Cold War the United States was faced with several new realities. One was that the world was becoming far less predictable. Crises could erupt suddenly in widely-separated places and forces had to be agile enough to deal with them, yet the forces would be far weaker than those raised during the Cold War. A second reality was that far less money would be available for defence. A third was that nuclear weapons would be very nearly unusable in these crises. A few very precisely-delivered weapons would often have to suffice.

Joint Vision 2010

Out of these considerations the U.S. Joint Chiefs of Staff fashioned a new military doctrine, Joint Vision 2010. The key perception is that the United States can no longer afford to oppose mass enemy forces with comparable masses. Instead, it must always seek to attack an enemy's centre of gravity, something so vital that its destruction will cause the enemy to collapse. The hope is that by co-ordinating enormous amounts of information decision-makers can identify the centre of gravity, and long-range weapons can destroy it. They offer a degree of agility which is vital in a confused post-Cold War world, in which limited U.S. forces cannot possibly be in place wherever crises erupt. With his centre of gravity destroyed, an enemy will collapse even though most of his forces have not even entered combat.

Conversely, it is vital to shield U.S. forces from the enemy's reconnaissance assets, and to keep the enemy from corrupting the key information. This aspect of the new thinking supports further efforts to make naval forces stealthier. That raises some interesting questions. In theory, the use of off-board sensors might dramatically reduce a ship's visibility. In some cases, for example, she need not radiate at all. On the other hand, the dishes required to receive all that satellite-borne information (or, for that matter, information from UAVs, or from early warning aircraft) are likely to be quite massive. The higher the required data rate (the better the information), the bigger the dishes – and the larger their radar cross-sections. Thus the demand for stealth can impede any attempt to spread satellite capacity through the fleet, a major issue largely unresolved early in 1999.

Satellites dominate this sort of war. They collect most of the data about the enemy, although other intelligence sources are also important. Almost no matter what data are collected, they have to be transmitted (generally by satellite) to some central point for analysis. If several analysis centres are involved, they must be able to transmit enormous amounts of data back and forth – again, most likely by satellite. Satellites provide the vital navigational information needed by the precision weapons. Satellites are even valued because they are likely to be more secure (against eavesdroppers) than are Earth-based systems.

This new doctrine is the new naval warfare writ large. Indeed, naval forces will probably be its chief elements. Uniquely they combine mobility (to get to the scene of a crisis) with endurance (so they can remain on station, and need not escalate the crisis by shooting out of turn) and with the ability to remain whether or not the local powers want them. Supporters of the new kind of warfare proclaim, moreover, that, although their weapons capacity is limited, they can achieve decisive results with those weapons – as long as the other elements of the overall system, the intelligence-gathering and -assessment, the communications, the navigation – work as advertised.

Unlike land warfare, naval warfare has always focused on a few key targets – such as major warships. Because the ocean is so vast, reconnaissance or targeting has always been a key problem. Long-haul communications connects widely-dispersed sources of information and their users. Quite naturally, then, the new style of space-oriented warfare began with navies.

The Transformation of Surface Combatants

In the U.S. Navy, space systems transformed the status of surface warships, from subsidiaries supporting strike carriers, to something far more like the capital ships of the past. Space systems have given missile-armed surface ships the reach (though not the weight of fire) of aircraft carriers. Such warships had been eclipsed during and after the Second World War not because they lacked punch, but because they were badly outranged. Carriers could strike at ranges of hundreds of miles; surface ships, at less than twenty.[1] Surface ships remained valuable after the war mainly for their ability to shield carriers (and other ships) from threats such as submarine and air attack. Even when anti-ship missiles were added, their ranges were inherently limited by the ranges of the ships' own sensors. Off-board sensors completely changed the situation. With the demise of the A-6 Intruder attack bomber, the Tomahawk missiles fired by U.S.

surface warships outrange the aircraft on board U.S. carriers. That does not mean that the carrier is finished, rather that the surface ship is back as a complementary capital ship.[2]

It costs much less to build and operate a missile-armed surface ship than an aircraft carrier. Indeed, it can be argued that without such ships the modern U.S. Navy would be unable to meet the country's obligations at an affordable cost. To some extent this impression is illusory. Tomahawk is useable and flexible because a space-based GPS system can guide it wherever it is wanted. The surface ships are effective because they have access to space resources, particularly to the fruits of national sensors. Those sensors are not – and cannot be – part of the Navy budget. However, many of them are ageing. The future efficacy of the surface fleet will depend on just how their replacements are chosen.

The New Role of Intelligence Agencies

Whether the new style of warfare is likely to be altogether effective remains to be seen. At the least, it moves intelligence agencies into unaccustomed operational roles as tactical targeters. In the past, to the extent that they had a targeting role, it was in the planning for massive strategic nuclear attacks, in which mistakes presumably would not have been noticed. Very limited strikes on point targets are a different proposition, and errors can be extremely embarrassing. During the August 1998 strikes, Tomahawks were fired at a pharmaceutical plant in the Sudan, the claim being that it was producing precursor chemicals for nerve gas. Some weeks after the attack the U.S. government found itself retracting the claim.

During the war in Kosovo, U.S. GPS-guided JDAMS bombs targeted in error hit the Chinese Embassy in Belgrade, causing serious political problems. It did not help that the weapons had been characterised as extraordinarily accurate; the Chinese refused to believe that the targeters had simply erred. This error was comparable to Tomahawk errors because the weapons were guided automatically to GPS coordinates chosen, not in the theatre of operations, but back in the United States. The bombs were dropped by a long-range B-2 bomber during a 31-hour round trip mission from a base in Missouri. The pilot simply places the bomb in a 'basket' from which it can guide itself to the coordinates inserted before the aircraft takes off.

The key difficulty is in the meaning of accuracy. Normally it is quite simple: an attacker sees a target and tries to hit it. The more accurate the weapon, the closer it comes to the desired target. However, for JDAMS (and, incidentally, for most strategic weapons) accuracy merely means coming close to coordinates fed into the weapon. Whether those coordinates coincide with the desired target, which is seen by someone far from the war zone, is an entirely different question. The targeter looks at a photograph which he tries to register against GPS coordinates, a process called mensuration. Without someone on the ground carrying a GPS receiver, the targeters cannot really be certain of the coordinates of any given building (in this case it was the Yugoslav arms development agency); they can only estimate the coordinates to which the bomb is sent, probably based on an offset from a nearby building. Much clearly depends on just how precisely the orbit of the satellite which took the key pictures is known; a very small error can equate to a very large one in mensuration. It does seem to be the case that the underlying data base, which should have warned the targeters that they were calling for a strike near the Chinese Embassy, was outdated (the Embassy had moved four years previously). In the rush to digitise the U.S. military mapping archive, current maps may not have been obtained.

It does seem clear that for their own political reasons (a desire to whip up anti-American sentiment, partly to discredit the democracy movement whose ten-year anniversary was approaching), the Chinese apparently refused to realise that what imaging satellites saw (the embassy looks very different from the arms agency) would not automatically translate into GPS coordinates. However, the larger political fall-out of the Chinese bomb scandal is that politicians, who have little interest in (or understanding of) the details of weapon systems, will be far less confident in the precision of GPS-guided 'surgical' weapons. They might have done well to remember the old computer mantra: 'garbage in, garbage out'.

Surely a key question is whether an enemy's centre of gravity is always physical. Even if it is, it may not always be open to attack and destruction. For example, when the Tomahawks hit the terrorist training camps in Afghanistan, they did not destroy the terrorist movement based there. The movement's centre of gravity, Osama bin Laden, probably moves far too often to be killed by a long-range missile. It may be that his main vulnerability is financial: if his money can be diverted or seized, his threat may be neutralised. Or, it may be that his only real vulnerability is psychological or moral; if he is somehow shown to be stupid or corrupt, his followers will desert. Neither is really a task for cruise missiles. In the aftermath of the raid, it seemed likely, moreover, that the failed attack had made bin Laden far more of a hero than he might otherwise have been.

The moral of this story is that the new style of warfare demands far more of intelligence analysts than mere identification of physical targets. Centres of gravity exist – if at all – because of the way in which a country, or a terrorist organisation, or a society is organised. They have to be chosen with a view towards understanding the likely effect of their destruction, which may not be at all obvious at the targeting level. This is a very different proposition from the photo analysis which apparently dominated intelligence during the Cold War.

The claims that an enemy's centre of gravity can always be identified and destroyed recall earlier claims for strategic bombing, which were never quite proven. It may be that, just as strategic bombing was a reaction to the unacceptable human cost of the First World War, the 'revolution in military affairs' is a reaction to the high economic cost of fielding mass forces to fight the Cold War. It is also a reaction to the political cost of protracted war in Vietnam. In each case, technology promises a cheaper and hence more acceptable way of war.

The Impact of the OODA Cycle

Advocates of Vision 2010 suggest an alternative, Boyd's idea of accelerating the OODA cycle (see Chapter 7). His explanation of the German success in the Battle of France in 1940 fascinated Americans who could not imagine building up a powerful enough force in Western Europe to match the Soviets. By the late 1970s, the main NATO equaliser in Europe was a powerful tactical air force. In traditional air operations, reconnaissance forces found targets, which were then passed to planners. There might well be a considerable delay between the discovery of a target and the despatch of an aircraft to attack it. Given a powerful enough computer, however, reconnaissance data could very quickly be turned over to strike aircraft. The cycle of target discovery (Observation), of target allocation (Orientation and Decision), and of strike (Action) could be made extremely short – shorter, say, than the usual Soviet decision cycle. The Soviets understood the U.S. concept, and in the late 1970s announced that the creation of 'reconnaissance-strike complexes' (*ie* systems) would made for a new military-technical revolution, comparable to the revolution caused by nuclear warheads and guided missiles. When the United States began to rearm in the early 1980s under Ronald Reagan, quite naturally it bought the new generation of computer-oriented weapon systems.

From a Soviet perspective, this was a disastrous development. If Boyd was right – and the Soviets apparently thought he was – then a NATO force equipped with the new reconnaissance-strike complexes could win a conventional war in Europe (the Soviets seem to have accepted, by the early 1980s, that nuclear weapons might be unusable). Marshal Nikolai Ogarkov, chief of the Soviet General Staff, demanded that the Soviet Union produce its own 'reconnaissance-strike complexes'. Without them, its land forces would lose the OODA battle and collapse in combat. The Soviets already had naval 'reconnaissance-strike complexes', but land warfare was a different proposition. A land battle involves far more separate units operating much closer together. Much more information has to be processed. Without realising it, Ogarkov was really demanding a computer revolution for the Soviet Union, since only computers (in vast numbers) could make a land-warfare reconnaissance-strike system practicable.

As it happened, Ogarkov's perception that the OODA battle was crucial was also deadly to the Soviet system. By this time the Soviet economy had been almost completely militarised; it had very little slack which could have been taken up in the creation of a Western-style computer industry. No Soviet ruler after Khrushchev was willing to dismantle existing entrenched industries to free the necessary capacity. Such a step would have been politically suicidal (those running the existing industries had considerable power). Mikhail Gorbachev tried an alternative, expanding the Soviet economy. To do so he had to liberalise, and the brittle Soviet system fell apart. Ironically, some of the systems Ogarkov so feared materialised only after the Soviet Union itself collapsed.

Clearly space-based systems make it possible to build a very agile force, which can operate inside its opponent's OODA cycle. However, surely it is also important that that agile force do something to its more sluggish opponent which has such effect that the cycle matters. Ogarkov had to worry about agile Western systems because NATO had sufficient firepower to destroy an oncoming Soviet army. In the wake of the Cold War, Western forces have been drawn down to the point where it is no longer clear that they can do enough damage to force an opponent to react to them, *ie* to fall victim to the OODA cycle. That is one way to interpret the Serbian non-reaction to much of the NATO bombing campaign in Kosovo. Eventually the Serbs did withdraw, but hardly in a state of collapse. For that matter, the nature of the coalition operation precluded the sort of agility which is envisaged; target lists had to be approved at too many levels. Unfortunately, the future seems to belong to coalition (not even alliance) operations. Technology does not always dominate events.

Both Boyd's perception and the broader language of Vision 2010 demand enormous agility and responsiveness. Because forces will generally be very thin, weapons will have to be used very economically. That is taken to require the ability to retarget weapons dynamically. In the past a strike was planned and weapons were assigned. Once the strike aircraft took off or the missiles were launched, the target choices could not be changed; the only exception, in the case of aircraft, was that there were secondary targets in the event the primaries could not be hit. After the wave of strikes, target damage would be assessed, and a second series of strikes planned and then executed. In Boyd's terms, the length of the OODA cycle was set by the time taken to plan, launch, execute, and evaluate.

Now that cycle is considered too slow. Enter dynamic retargeting. As the initial strike weapons hit, their effects are assessed. Normally a targeter allocates several weapons to each target, so that as the first weapons strike, others are already in the air. The capacity to retarget dynamically means that those other weapons can be reallocated based on the observed effects of the first ones to hit. Dynamic (or responsive) weapon allocation demands two things. One is much quicker evaluation of battle damage, most of which occurs well beyond a commander's horizon. The other is a reliable high-capacity link to aircraft and to guided weapons. The new Tactical Tomahawk is intended specifically to fill these requirements. It carries a strike assessment camera, and it has a high-capacity two-way link back to a controller, presumably via a UHF satellite. The missile can be launched, then ordered to loiter awaiting targeting instructions.

Replacing the Satellites

At the end of the century, the United States finds itself with two related space problems. The Cold War satellite sensor systems are ageing. Replacements will be expensive. Can cheap UAVs do much the same job? A UAV can obtain outstanding images, and if it flies high enough it can maintain a line-of-sight link back to its control station. It can certainly provide both tactical intelligence and enough photographs to target any missile. However, the UAV has one great drawback. It is a local rather than a global system. That is true even of long-endurance systems, with endurances measured in thousands of miles: they must spend much of that endurance loitering over an operational area. Most UAVs arrive in the theatre of operations only after bases have been set up, or after large warships have arrived. Thus they do not really fit the current fluid world situation. Crises generally surprise policy-makers; they are unlikely to have set up systems, such as UAVs, in the expectation of a need. The system of fixed U.S. bases is much more likely to contract than to expand or even to remain as it is. That is, after all, why the fleet is valuable to the United States: it can go where it is needed. The surveillance which supports it must quickly be shifted to match unexpected developments. It must be good enough to support early missile and air strikes.

The Future of the Sea Sanctuary

The other problem is that civilian space enterprises are providing some of the communication and even the sensing services which the

United States uses to support its new style of warfare. To what extent can other countries use those same civilian assets to gain the same sort of capabilities the U.S. Navy now enjoys?

There is a widespread view, at least in the U.S. military establishment, that somehow the growth of commercial space systems has doomed surface navies. Surveillance will soon be global and nearly continuous, it seems. Perhaps, within a few decades, the only viable fleet will be an underwater one. This argument recalls similar statements made early in the nuclear age: after the target fleet was wiped out at Bikini, who could really support surface warships? Yet the surface fleets have survived, not out of some outworn sense of tradition, but because the logic of the situation is much more complex than has been imagined. In the case of the Bomb, the attacker had to find a ship before he could vaporise it. That turned out to be quite difficult, as both the Soviet and U.S. Navies discovered – and as previous chapters explain.

For the future, the great barrier to universal sea surveillance is its sheer cost – which is much more than the cost of the satellites and the computers. No likely enemy seems to have the resources to spare for such a project. That leaves the commercial sector. Yet there seems to be no real incentive for any system of world-wide ship tracking. There is not even serious interest in a system of world air traffic control, which would probably be far more attractive commercially. Nor has there been any attempt to get part way to world ship tracking by demanding that all merchant ships carry satellite transponders – as is the case with cargo containers – a step which, incidentally, would not solve the warship detection and tracking problem, but which would completely fill the commercial demand for ship position information.

Imaging satellites, which certainly are proliferating on both military and commercial levels, will not end the sea sanctuary. Current resolutions seem to be as good as 1m. The problem is not quality (resolution) as much as quantity. Any such satellite uses a downlink to its ground station. It transmits images while within sight of the station; that time window and the bandwidth of the downlink together determine how many images of what quality the satellite can provide at any one time. The only way to overcome the down-link limit is to cross-link to another satellite (or satellites), so that transmission time is greatly extended. The United States does that with the SDS satellites, but the commercial satellites are unlikely to enjoy similar facilities (commercial communication satellites generally lack cross-link receivers and transmitters). This limit probably also affects foreign military imaging satellites, except perhaps for the Russian ones.

Moreover, any imaging satellites covers only a limited swath, a very small percentage of the ocean, as it orbits.[3] That is inescapable, given the high resolution it generally enjoys. That having been said, it is true that ships leave very detectable (and persistent) wakes not only on the sea, but also in the clouds above the sea. Thus a medium-resolution image of an area through which ships have passed may yield valuable data. It cannot, however, suffice to maintain any sort of open-ocean picture of naval activity. Nor, probably, does the cloud trail identify the ship which makes it.

Thus it seems likely that ELINT will continue to be a prerequisite for any sort of open-ocean surveillance. It is most unlikely that any commercial system will provide a viable alternative. While there is real profit to be made in communications, and while commercial imaging is apparently profitable, there does not seem to be any commercial point in electronic intelligence gathering. The satellite end of an ELINT system, moreover, is only the beginning of what is needed. Somehow that information must be translated into tactical intelligence. Emissions which the satellite picks up must be interpreted and collated.

The main inadvertent emitters which satellites are likely to detect are shipboard radars. A ship-detection/tracking system must somehow identify a particular emission with a particular ship; it must fingerprint the ship's radars. To do that requires not only satellites but also platforms which can pick up emissions and identify the emitters. That is a global task. The satellite end is almost the least complex; the really difficult task is data fusion on a vast scale. This task was difficult enough for the Cold War United States, which had to track only a limited number of Soviet and Bloc warships. It probably strains current U.S. capabilities to track all the world's warships, but that is necessary given the ambiguities of current politics. An enemy would face similar problems.

When Hughes sold China a satellite-based telephone system, it was suggested that it was in fact partly an eavesdropping system, which could help the government suppress dissidents. That is, each telephone call made through the satellite could be routed through a central government office. That in turn raises another question. To what extent is a communications satellite an ELINT satellite, over its normal operating band?[4] The success of Grab suggests that there was little difference. Modern communications satellites operate at radar frequencies. What distinguishes signals from radars from the signals generated by, say, satellite telephones? Presumably the satellite normally rejects any signal which does not fit the desired format. However, it can probably be commanded from the ground to change formats. Analysis would be done on the ground (as in the Grab system of 1960).

However, even if commercial geosynchronous communications satellites can be used (or adapted) to gather electronic intelligence, it is unlikely that they can pinpoint emitters accurately enough for tracking and targeting. They are just too far away, their beams too broad at the earth's surface. Thus any country interested in denying the United States the sea sanctuary would have to invest in a specialised low-altitude (for precision location) ELINT system for that purpose. Moreover, emitters would have to be fingerprinted on a global scale, and the library of fingerprints kept up to date as radars were repaired or replaced. No truly commercial system would offer this capability. It seems unlikely that the new low-altitude communications systems, like Iridium, can be adjusted for this purpose. A flood of ELINT data would, moreover, be only the beginning; there would still be the vast data fusion problem. Moreover, the task of collection is probably becoming more difficult. Many navies are buying stealthy radars like the Signaal SCOUT. Others will likely follow. Quite possibly next-generation search radars will use stealthy or LPI techniques. In that case passive surveillance from space will become increasingly difficult.

True, the Soviets apparently demonstrated that even without satellites of their own, they could exploit satellite downlinks for approximate force location. However, their techniques are probably far less effective against current frequency-hopping satellite communication systems. Without fingerprinting of the uplink systems, the Soviet technique loses much of its value; and fingerprinting requires a massive force of ELINT ships and/or aircraft. Furthermore, the Soviet system requires intercept stations within the footprint of the satellite downlink, and that may become much less likely as the fleet shifts more of its traffic to higher-frequency systems (SHF, EHF) with narrower downlink beams.

Certainly it will still be possible to do active radar surveillance. Current radar-imaging satellites, which are commercial, certainly can detect ships at sea, but they have the same the limited search swath widths that afflict other imagers. A satellite designed for ship detection needs a very wide swath, and thus captures only very limited data. It probably has no commercial application at all.

All of this suggests that navies will continue to enjoy a significant sea sanctuary when they pass over the horizon. Indeed, the rise of commercial imaging satellites will widen the gap between naturally stealthy naval forces and relatively immobile (hence not at all stealthy) forces ashore. If that is the case, then forces ashore will feel an increasing pressure to reduce their footprints and to disperse. The U.S. Navy's current interest in distant fire support to small groups of Marines or Army troops ashore will seem more and more valid. For that matter, current Marine interest in using offshore basing and manoeuvre (as opposed to a build-up on a beach, with manoeuvre inshore) is also consistent with continued faith in a sea sanctuary beginning at the horizon.

Imaging satellites are likely to make surprise attacks on the ground very difficult, though hardly impossible (if advantage is taken of a known satellite ground path). Even then, the satellite must be targeted, and it may well fail to cover a desired area at all, or at least during daylight. Full coverage of a large ground area probably takes days or even weeks, using a space camera capable of being aimed away from the ground directly under the satellite. Furthermore, there is a real question of the satellite's ability to see through tree cover. Camouflage is likely to remain effective, although advocates of the new kind of warfare claim that it can be defeated by viewing the same scene using several different spectra (IR, visual, radar).[5]

Almost certainly those foreign powers which have invested in military imaging satellites (Russia, China, and France) see them mainly as aids to targeting weapons and sometimes as a means of monitoring their enemies' forces to preclude a surprise attack.[6] A few photos of this type suffice for ballistic missiles. Something like Tomahawk requires far more, probably more than these systems can provide on a timely basis. Probably a U.S.-style cross-linked system is necessary for cruise missile or aircraft attack planning.

The experience of the last forty years, as reflected in this book, and as elaborated in concepts like 'Net-Centric Warfare' and 'Vision 2010' is that information gathered by multiple remote sensors, fused and overlaid, and distributed reliably by satellite is the key to future victory at sea and, increasingly, on land. It follows that eventually war will be fought in space. After all, air warfare began with reconnaissance. The first fighters were designed to deny enemies that information. Both major space powers developed (and one deployed) ASAT systems during the Cold War. Does the U.S. desire for information superiority demand a new generation of ASATs? For that matter, will other countries develop ASAT systems as an asymmetrical response to the new U.S. style of warfare? For example, there are rumours of Chinese interest in ASATs.

Yet it is not clear that key foreign space-based capabilities are really vulnerable to attack. The new style of warfare depends more on communications links than on space-based sensors, and any enemy can make use of a growing number of commercial communications (not to mention imaging) satellites. They cannot be attacked, at least not in peacetime. The only hostile capability which would probably depend uniquely on an enemy country's satellites would probably be electronic surveillance. Even in that case, an enemy might be able to obtain the necessary assistance from an ally – which might not be a formal enemy of the United States. Moreover, the United States has no monopoly on either the design or the launch of the commercial satellites, so it cannot enforce legal sanctions on enemies using these capabilities. For example, the United States cannot prohibit a foreign imaging company from selling photographs of its most sensitive installations.

All of this suggests that the only really effective way to conduct warfare against space assets would be to attack their ground-based control and downlink facilities. If the attacks were covert enough, they might be made even in neutral countries. Overt destruction of commercial satellites would be a different proposition, equivalent, perhaps, to the unrestricted submarine warfare of the two World Wars.

Even so, U.S. interest in ASAT systems continues. There are two current U.S. alternatives. One is the Army's kinetic-energy (KE) device, a manoeuvring kill vehicle which would deploy (in effect) a fly-swatter with which it could hit and disable a satellite in space. Another possibility is an anti-satellite variant of the Air Force's Airborne Laser (ABL), which is intended mainly to destroy theatre ballistic missiles in their boost phase.[7] There is also current interest in a space-based laser, mainly for missile defence, but clearly applicable to anti-satellite missions.

The U.S. Navy's reaction to satellite surveillance during the Cold War was to devise electronic countermeasures which might deceive the Soviet ocean surveillance systems. With the demise of the Soviet satellites, naval interest has moved elsewhere. However, the Navy's concepts certainly apply to future foreign use of satellites for sea surveillance. Unlike the Air Force's ideas, the Navy's can be used whether or not the surveillance is being done by a military satellite, and whether or not rules of engagement permit the satellite's destruction.

The U.S. Space Command

Within the U.S. military establishment, there is a very real policy question, potentially no less decisive than the question of whether naval air arms should have been folded into national air forces. The U.S. Navy is the single greatest user of American space systems, yet they are largely operated by the Air Force, which wants full control of U.S. space systems under its Space Command. There is a widespread feeling that, should that happen, Space Command will be far more interested in space combat than in the much less glamorous operation of communication links and space-based sensors to support sea and land operations.

Certainly the U.S. Air Force was the first service to create a Space Command, on 1 September 1982. This organisation was the lineal descendant of the North American Air Defense Command (NORAD), which was responsible for ballistic missile warning. By 1982 the Air Force was deeply involved in managing space systems not only for itself but also for the National Reconnaissance Office (NRO), which worked for the independent intelligence agencies (CIA and NSA). The Navy and then the Army created their own space commands (the Army command is now the Army Space and Strategic Defense Command). In 1985 a unified U.S. Space Command, whose commander is the commander of the Air Force Space Command, was formed. Its components are the Army Space Command (Forward), the Naval Space Command, the 14th Air Force, the Cheyenne Mountain Operations Centre (the old NORAD command centre), the Defense Department's Manned Space Flight Support Office, and NORAD/US Space Army command (ballistic missile defence). U.S. Space Command operates satellites once they are in orbit; it is also responsible for space surveillance and for ballistic missile warning. Air Force Space Command, which launches satellites, is a separate entity. Actual combat against incoming missiles is a separate responsibility. The Air Force is currently pressing for unification of all Defense space activities under a single command.

Air Force control of U.S. Space Command is somewhat diluted in that a Vice-Admiral (*not* commander of Naval Space Command, though a former commander of that organisation) is now deputy commander of Space Command.[8] By the late 1990s the Air Force hoped that Space Command would take over many of the space-oriented responsibilities of the regional commands. The fight over JTAGS (see above) is symptomatic of the struggle between centralisation and regional control of important space systems. In the January 1998 version of the Unified Command Plan, CINC Space Command was made the sole point of contact for military space operational matters.

There is also a National Security Space Systems Architect in the Defense Department, responsible for the actual choice of new systems. In 1998 the Air Force tried to take over this role within (the Space Architect would have had a formal relationship only with Space Command and with the National Reconnaissance Office, responsible for intelligence satellites). The Army, Navy, and the Marines all objected, and the idea was dropped. In return, they agreed to increased centralisation of operational control in the Space Command. Not only will the Space Architect not have any special relationship with the Air Force; his staff will be manned by representatives of all the services (rather than predominantly by the Air Force and NRO, augmented 'as required' by representatives of the other services). All the services will be members of the Senior Steering Group (replacing an earlier Joint Space Management Board) advising the Space Architect; each will have a veto on space policy.

The first Space Architect is Air Force Brigadier-General Howard Mitchell, who is also director of space systems for the National

Reconnaissance Office. Soon after entering office in September 1998, he called for better dissemination of data from space sensors. Existing stovepiped systems, such as TADIXS and TRAP, would be merged with other data sources such as GPS, to feed a common ground processor (a super-TRE). This concept would eliminate direct feeds from satellites to warfighters, such as JTAGS.

From a naval point of view, perhaps the main significance of the Air Force's increased interest in space warfare is as a potential threat to U.S. Navy space requirements. In the past, the cost of naval space operations has not figured in Navy budgets, mainly because the most expensive items involved have come out of the intelligence and Air Force budgets. Well after the end of the Cold War, expensive intelligence-gathering systems built to wage that war may well not be replaced. The U.S. Navy may well now be called upon to pay a much larger share of the cost of the space systems it uses. There is a real danger that 'black' intelligence systems, whose TENCAP products have proven so valuable, will die because, due to security measures, most users of those products are entirely unaware of their sources. Budgets are tight, and are likely to grow tighter, particularly if the United States pursues a national missile defence programme. It is easy to see, say, DD 21 (the new land-attack surface combatant) as an economical way to deal with enemy forces ashore. It is much more difficult to explain that funding for DD 21 should include money for the satellite systems which make it effective.

Yet those systems are as essential to the success of the ship as are its Tomahawk missiles, or its radar. DD 21 is an expression of the way in which space systems have transformed the U.S. Navy. Without the space capability, the current reach of U.S. seapower would necessarily disappear; only the carriers would be really effective in the deep strike role. It would be very ironic if the Navy's general ignorance of its space assets caused it to allow those assets to disappear while at the same time the Navy pursued programmes which, like DD 21 really rely heavily on space systems to make them viable.

Other Potential Space Powers

During the Cold War, only the United States and the Soviet Union fully integrated space systems into their military forces. The United States went much further than the Soviets in using space-oriented communications and navigation, partly because its forces were generally more dispersed. Since the end of the Cold War, much of the Soviet military space system has collapsed, and it seems unlikely that the Russians and other successor states can develop anything equivalent to the emerging U.S. military system – at least not in the short term.

Nor have the European allies developed their own equivalents of 'net-centric warfare' and Vision 2010. The great barrier is cost. No single European country can buy the necessary combination of surveillance and high-volume communication systems. Problems arise even for groups of countries, as witness the failure of the French-led Helios II and Horus reconnaissance programmes. There has been some talk of a European space initiative (*eg* involving the Galileo navigation system) based on the emergence of the Western European Union (WEU) or of the European Union (EU) as a military power in its own right. It is possible that nothing short of such a combination would have the financial and industrial power to build up the necessary space-based systems. There must be some question, however, as to whether even such combinations would be interested in doing so. At present Europe as a whole spends about as much as the United States on defence, but it gets far less hardware for the money; far more goes in salaries and in manpower, some of it clearly duplicative. There seems to be little sense in European defence circles that the sort of sophistication represented by the U.S. net-centric or digital battlefield or Vision 2010 initiatives is really necessary; there seems to be a hope that somehow a simplified version would be enough for Europe. One observation from the war in Kosovo was that the NATO air arms were not really equipped to fight the same sort of war as the U.S. Air Force and the U.S. Navy. That applied to all-weather GPS weapons and also to air retargeting. It seems possible that few in European governments or think tanks appreciate the extent to which the space-led revolution in warfare is seen in the United States as not merely attractive but inescapable.

Obviously adopting a space-oriented form of warfare means accepting vulnerabilities. In recent months the Chinese military has published books and articles advocating 'asymmetric' warfare. Admitting that China cannot possibly match U.S. capabilities, at least in the short term, the authors have suggested that high technology carries its own vulnerabilities, which an intelligent Chinese military might exploit. They make much of the potential for hacker attacks on U.S. civil computer systems, for example. However, it is also true that distant sensing systems can be fooled by carefully-constructed decoys. For instance, in Kosovo the Serbs built wooden MiG-29s, which NATO aircraft dutifully bombed. They also claimed success for decoy tanks. These devices worked well enough against aircraft marking their targets with lasers; they would work even better if the targets were designated from photographs to be hit by GPS-guided bombs. It is possible that the European allies' reluctance to join the space-led revolution derives in part from a certain scepticism. Some of them have been burned before, *eg* by excessive claims for strategic air power.

Another factor is a very natural reluctance, in Europe, to depend entirely on U.S.-supplied space services. For example, the French Apache cruise missile, which is not too different from Tomahawk, was deliberately designed without GPS guidance on the grounds that in a crisis the United States might exercise a veto on its use by turning off the GPS system, at least over the combat zone (the advent of Galileo would presumably change matters).

There is some evidence that the situation is changing. The Royal Navy is reportedly planning to arm its Future Surface Combatant largely with cruise missiles, making it a sort of anglicised DD 21. It has been suggested, moreover, that the weapon of choice will be a surface-launched version of the Anglo-French Storm Shadow/Apache, which may also be launched from the vertical tubes of the projected Type 45 frigate. If the British follow the U.S. pattern, they will soon find that to use such weapons effectively they need OTH-T capability. They may feel that it can be provided by organic aircraft, for example by a next-generation radar helicopter with a dual-mode (AEW and ground moving target indicator/ASTOR) radar. They may then discover that the best over-the-horizon link between helicopter and ship – and headquarters – is a satellite. If the missiles are envisaged as army support weapons, the fleet needs a link with an army well over the horizon, which is also likely to be a satellite link. Perhaps this logic will drive the British along the path already trodden by the United States. It is not clear whether other European countries, less interested in power projection from the sea, will follow suit.

The Coalition Issue

One much larger question overlies these technical issues. In its post Cold War operations – in the Gulf War, in former Yugoslavia, even in Haiti – the United States tends to seek coalition partners. Yet the United States is developing a form of warfare in which few if any of our prospective partners, perhaps not even those in NATO, seem interested in participating or adopting. In the past, for example, cooperation between armies was generally assured by exchanging staff officers. In the new vision of a very agile army, computers displace many traditional staff billets; there literally may not be staff officers available for cooperation. In the NATO navies, only fairly recently has Link 11, the means of exchanging a tactical picture, become virtually standard. To many of our partners, the ceaseless shifts between different computer systems, satellites, even message formats are bewildering. Partnership seems to demand open-ended purchases of U.S. hardware and software, on a scale few navies are willing to match. This is not altogether a new problem (the Royal Navy was complaining about the cost of U.S.-capacity communications in the 1950s) but it is still a very serious matter. It becomes far more serious when the United States tries to cooperate with our former enemies. Military co-operation is often the best way to cement friendship. Yet the navies and armies involved do not follow our tactical style; our command and control systems are unlikely to be very relevant to them. To the extent that the new style of warfare we are adopting places greater and greater emphasis on space systems – and therefore on more and more powerful computers to receive and interpret their products – this is a growing problem.

The situation may seem less serious than it is because forces still can co-ordinate under benign conditions. Thus navies with fundamentally incompatible systems and tactical ideas can achieve the appearance of co-operation, which often seems to suffice for political reasons. This is the case, for example, with recent exercises joining Western and Russian warships, or ships from different Baltic navies. There is no question but that, whatever changes we see due to computers or satellite systems, ships can still manoeuvre together – when there is no real threat. Once they are under the threat of air attack, however, the situation changes dramatically. Without a shared tactical picture, which means without computer data exchanges, ships cannot co-ordinate defence against aircraft. They probably cannot agree about the identity of incoming aircraft, so the fleet has the terrible choice of either risking the destruction of friendly aircraft or opening itself up to attack. The situation for a force operating off a hostile coast is even worse.

Another problem, paradoxically, is that some navies have gone much further than they realise in adopting U.S.-style Net Centric Warfare systems, such as JMSCOMS/JOTS and local equivalents to IT-21. As a consequence, their estimates of how much further they would have to go for a useful degree of commonality may be grossly excessive, and that error may well be a barrier to further modernisation. Here the fault may lie partly with advocates of the new style of warfare, who have failed to distinguish essential from optional elements.

What does seem clear is that the United States cannot easily retreat from the new style of warfare. It is being developed, not because of some pressure to utilise shiny new technology, but to meet overwhelming economic pressures. It seems to be the only way in which the forces we can afford – in terms of numbers, of troops and of platforms – can deal with the military problems we face.

Notes

Chapter 1: Introduction

1. For an early view of this new kind of warfare, see Friedman, Norman, 'C3 War at Sea', in the May 1977 (*Naval Review*) issue of U.S. Naval Institute *Proceedings*. See also the author's chapter ('U.S. vs. Soviet Style in Fleet Design') in Murphy, Paul J (ed) *Naval Power in Soviet Policy* (Washington 1978, Vol 2 of 'Studies in Communist Affairs').

Chapter 2: Satellites and their Mechanics

1. Latitudes of some other important current launch sites are: Chinese: Jiuquan, 40.7 degrees North; Xichang, 28.2 degrees North; Japanese: Tanegashima, 30.2 degrees North; U.S.: Vandenberg Air Force Base, 34.7 degrees North; Wallops Island, 37.9 degrees North; and Russia: Plesetsk, 62.8 degrees North.
2. Example given by Dutton *et al*, *Military Space*, p15.
3. In 1961 the U.S. Navy seriously considered converting four seaplane tenders into satellite-launch ships (AGSL), at a cost of $33 million each. They would have been equipped to fire navalised versions of the lightweight solid-fuel Scout booster, and ultimately to fire rockets weighing up to 60,000lbs. Among the arguments for the programme (which ultimately was not carried out) were that such launches could be covert (because they were conducted at sea), and that they could be made in any direction, without fear of having the rocket fall onto populated land. Technology developed for the AGSL could later be extended to really dangerous rockets; at this time the U.S. government was seriously considering a nuclear-powered heavy rocket. The system's proponents also argued that a sea base would be an ideal venue for launching other countries' satellites, thus improving relations within the Free World. They did not point out that, unlike Cape Canaveral, a ship would not be subject to the sort of press attention that made the early U.S. Vanguard launch failures so humiliating (the Soviets suffered no corresponding problems when their rockets failed). Vanguard had been a Navy project.
4. To make a simple plane change (no change in orbital shape), the satellite requires a thrust applied at right angles to its motion along its orbit. The faster it is moving, the greater the required thrust, so that it pays to make the change at minimum velocity. The larger the change in angle, the greater the thrust needed. For a 60-degree change in orbital plane, the added velocity (at an angle to the original direction of motion) is equal to the satellite's velocity along its orbit (the initial and final velocities are vectors, and the change in velocity is also a vector; in this case all are legs in an equilateral triangle).
5. This is *not* the maximum distance from the centre of the Earth, because the Earth is not at the centre of the elliptical orbit (it is at one focus). It is half the sum of the apogee and the perigee.
6. The situation can be difficult, since the low orbit may not be equatorial. In the U.S. case, it is typically inclined at 28.5 degrees. A kick at perigee places the vehicle in a transfer orbit, also inclined at 28.5 degrees, the apogee of which is at geosynchronous altitude – and over the equator. Once at apogee, the satellite receives a second kick (thrust) which simultaneously changes the orbital plane (to equatorial) and changes the orbit to circular. The plane change is done at apogee because there the satellite is moving most slowly, hence requires the smallest kick to change orbital plane and shape.
7. Generally, satellites use rockets. The alternative, which has been under development for decades, is electric propulsion, in which ions are ejected at very high speed (their velocity, about ten times that of a rocket exhaust, makes up for their very low mass). This technique was first used in 1997 by Hughes Space and Communications Co., in the PAS-5 (PanAmSat) communications satellite. This Xenon Ion Propulsion System (XIPS) uses only about 10 per cent as much propellant as a conventional rocket, since its energy comes from electricity generated by the satellite's solar panels.
8. EMP effects, which seem to have been unexpected, were first demonstrated after a 9 July 1962 high-altitude nuclear test, Starfish Prime (1.4 MT, altitude 248 miles). Exploding above Johnston Island, this burst blacked out Pacific communications and knocked out lighting and even burglar alarms in Hawaii, 800 miles away. Less publicised was unexpected damage to the solar cells of the British Ariel, U.S. TRAAC, and U.S. Transit IV-B satellites nearby. The great Cold War fear was that a single massive EMP burst high above the United States would destroy virtually all key American electronics; solid-state devices were considered far more vulnerable than the vacuum tubes the Soviets still used. The fact that the Soviets still did use archaic vacuum tubes was often cited as evidence that they planned to use EMP in war. At short ranges, effective EMP can be created by a non-nuclear explosion, properly harnessed. For example, during the Cold War non-nuclear EMP was used to test U.S. military aircraft. Readers may recall that in a recent James

Bond movie, 'Goldeneye', EMP was to have been used to destroy computers in the City of London.

Chapter 3: Getting into Space: Boosters

1. The only important U.S. solid-fuel booster was the Scout (Solid Controlled Orbital Utility Test) rocket, capable of placing a 385lb satellite in a 500-mile orbit. Design work began in 1957, and Vought (later Loral Vought and then Lockheed Martin Vought) became prime contractor in 1959. Motors were taken from existing weapons: the first-stage motor was based on that of Polaris, second stage on that of the Army's Sergeant, and the third and fourth stages were based on motors developed for the U.S. Navy's Vanguard satellite launcher. Scout was last launched in 1994; the follow-on is Pegasus.
2. Initially there were national British and French programmes using, respectively, Black Arrow (based on the abortive British Blue Streak IRBM) and the French Diamant (based on a series of sounding rockets). The French Diamant successor, L-III-S (third-generation launcher to replace the failed Europa, which had a Blue Streak first stage, a French second stage, and a German third stage), became Ariane, the first of a series of European Space Agency launch vehicles developed by Aerospatiale (an Arianespace corporation was established in 1980). The French contributed 62.5 per cent of the financing. The Ariane series was conceived from the start for direct ascent to a transfer orbit, the theory being that virtually all commercial satellites would go into geo-synchronous orbits. It was also conceived as a very conservative evolutionary design, Europa having failed due to the unreliability of its third stage. Ariane 1, first launched on 24 December 1979, used three liquid-fuel stages, the third being cryogenic (liquid oxygen and hydrogen). It could place 1750kg in a transfer orbit, or 2400kg in a sun-synchronous orbit, or 4900kg in a low earth orbit. The rocket motors were Vikings (four for the first stage, one for the second); a version was license-produced in India. First fired in 1984, Ariane 2/3 had a stretched third stage (Ariane 3 added two solid-fuel boosters strapped to the first stage) and more powerful engines (Viking 4B/5B rather than Viking 4/5). Ariane 2 could place 2000kg in a transfer orbit (2500kg for Ariane 3). Ariane 4 (1989) has stretched first and third stages. It has to be launched partly fuelled if it is fired without strap-on boosters. There are different versions with alternative strap-ons: 42L (two liquid-fuel strap-ons), 44L (four liquid-fuel strap-ons), 42P (two solid-fuel strap-ons), 44P (four solid-fuel strap-ons), 44LP (two liquid-fuel and two solid-fuel strap-ons). The most powerful version, 44L, can place 4200kg in a transfer orbit. The completely new Ariane 5 (6800kg into transfer orbit, 18,000kg in a low orbit) was conceived as the first European manned space launcher (however, no such programme emerged). It is in the Titan IV category, and has much the same configuration, comprising a core flanked by two very large solid-fuel boosters. The core has a new single Vulcain cryogenic rocket engine, and the second stage has a more conventional engine. Ariane 5 was first launched successfully on 30 October 1997 (a 4 June 1996 launch failed). The second lot, ordered in 1999, is to consist of several different versions (AR 5 Plus, AR 5 Versatile, AR 5 cryogenic), at a unit launch price 40 per cent less than that of the first series (lot P1). The transition to Ariane 5 is to be complete during 2002.
3. The Soviets also tried amine and kerosene as fuels with nitric acid, and liquid oxygen as oxidiser with UDMH.
4. The typical measure of engine performance for comparative purposes is specific impulse, in seconds: lbs thrust for duration of burn divided by pounds of propellant (this figure is independent of the size of the engine). To give an idea of relative propellant effectiveness, the Space Shuttle main motor (cryogenic) has a specific impulse of 455 sec (Centaur is 444). The liquid oxygen/kerosene first-stage engine in Saturn V has a specific impulse of 260 sec (in Atlas it was 258 sec for the sustainer and 220 sec for the main engine). The storable liquid propellants in a Titan motor (LR-87-11) has a specific impulse of 302 sec at sea level. The Russian Energiya core engine, RD-170, has a specific impulse of 308 sec at sea level (336 sec in vacuum). Castor 4A, a modern solid-fuel motor used as a strap-on booster, has a specific impulse of 237.6 sec at sea level. Although this figure (typical of solid fuels) is comparable to that for liquid-fuel engines, for many years it was very difficult to build a large solid-fuel rocket motor, hence to develop the sort of power offered by conventional liquid-fuelled motors.
5. The basis of the first stage of the Navy's Vanguard satellite-launcher, which first flew successfully in 1958, was the Viking sounding rocket. Viking was conceived in 1946 (and first flown in 1949) as a V-2 replacement (the supply of captured V-2s was limited, and there was no interest in making new ones). It was the first large U.S. liquid-fuelled rocket. A 1951 proposal to develop it into a tactical weapon had been rejected. The Vanguard second stage was based on a smaller liquid-fuelled sounding rocket, Aerobee. The third stage was a solid-fuelled rocket (Vanguard III used a new Altair third stage, which would have other space applications). In theory, Vanguard should have been quite reliable, since it used two well-tested rockets. In fact the first stage was considerably modified, with a new engine. The first Vanguard test vehicle with all three planned stages, TV-3, exploded on take-off, 6 December 1957, well after the Soviets had successfully launched two satellites. The U.S. Army had already pressed for permission to use an upgraded version of its tested Redstone missile (Jupiter C) to launch a satellite, which it did on 31 January 1958.
6. A great deal obviously depends on the launch site and on the orbit. Fired from Cape Canaveral into the most efficient possible orbit (28.5 degree inclination), a Titan 405 (with no upper stage – NUS) can place 39,100lbs in a low earth orbit. With an Inertial Upper Stage (IUS) a Titan 402 can place 38,780lbs in a low earth orbit. Titan 401 (Centaur upper stage) can place 10,000lbs in a geosynchronous orbit. Fired from Vandenberg into a polar orbit, however, a Titan 403 (the West Coast equivalent to the Canaveral-launched 405, ie without an upper powered stage) can place 31,100lbs in a low earth orbit. The upgrad-

ed solid-fuel motor offered, respectively, payloads of 47,800, 47,000, 12,700 (13,560 with Centaur-G), and 38,800lbs. The first Titan 4B with upgraded solid-fuel motors, K-24, was launched on 23 February 1997. A rocket of this type failed on 9 April 1999.
7. If Thor Able, Thor Ablestar, and Thor Agena are considered the first three satellite launcher versions of Thor, then the new version bought by NASA was clearly the fourth – hence the name, Delta, which was applied to the upper stages when the contract for their fabrication was signed. Note that Delta was conceived primarily as a civilian satellite launcher. By way of contrast, Thor derivatives retaining the Thor name (with a suffix indicating the upper stage) were essentially military.
8. Saenger's proposed spaceplane was named 'Silverbird'. Later he collaborated with (and married) Irene Bredt, and his concept is often called a Saenger-Bredt aeroplane. During the Second World War Saenger proposed a bomber version, which could have reached the United States. It was called the 'antipodal bomber' because it would have skipped halfway around the world on one flight (it was, however, beyond the level of existing technology). After the war both the Americans and the Soviets considered developing Saenger's ideas (Stalin nearly ordered Saenger and Bredt kidnapped for this purpose). The Germans did (unsuccessfully) fly a winged version of their V-2. In 1952 Bell Aircraft proposed a Bomber-Missile (BoMi) based on the Saenger idea, but using a very different kind of lifting body. One version would have attained orbit; the other would have had a range of 3000 miles, attaining Mach 4 at 100,000ft. Bell later received a development contract for a higher-performance BoMi, and in 1956 it received a contract for a spaceplane reconnaissance system, Brass Bell (System 459L) boosted by an Atlas ICBM; the air force was also financing several studies of rocket bombers (RoBo). These ideas in turn led to an air force programme to develop the necessary spaceplane, which became the X-20.
9. NASA had planned a vehicle with stubby straight wings, to enter the atmosphere at a steep angle and thus to take most of the heat of re-entry on its belly. This would have had excellent subsonic gliding characteristics; the design actually chosen was optimised more for high-speed flight, and has to land at a steeper angle and at higher speed. Compared to the NASA concept, the design actually chosen endures far greater structural and thermal loads during re-entry. The shallow re-entry angle chosen recalls Saenger's sub-orbital atmosphere-skipping bomber concepts.
10. As the projected cost of launches at Vandenberg increased, it was pointed out that Kennedy (Cape Canaveral) could place satellites in polar orbits, but that in any direct flight solid boosters and the ET might well fall over populated land, quite possibly in the Soviet Union. Instead, the Shuttle would have to execute a dog-leg manoeuvre in space, at the cost of considerable fuel – and weight. Payload in a sun-synchronous orbit might be cut from 32,000 to 22,000lbs. That would rule out a KH-12 mission. Moreover, unless the Shuttle could fly from Vandenberg it seemed to be impossible to launch and retrieve satellites in the same mission (Mission 4).
11. The bay is 4.6 × 18m and can house up to four satellites. It is designed to fly at altitudes up to 950km; satellites intended for higher orbits require their own boosters.
12. This account is based largely on Andrew J Page's article in *SpaceViews*. The China Lake web site includes a briefer 'Notsnik' account, which does not claim that the satellite reached orbit.
13. Yangel had previously headed a Special Research Institute (NII) of the 7th Main Administration of the Ministry of the Defense Industry, charged with missile research (NII-88).
14. The version of the ICBM actually fielded at Baikonur and Plesetsk was R-7A (8K74).
15. There was also an 11A514 version, designed to launch the Soyuz R (reconnaissance) and Soyuz P (interceptor: ASAT) manned satellites, work on which began in January 1964 (they were cancelled in 1966). It would have placed 6700kg in a 65 degree orbit. This project was proposed in December 1962 by Korolyev. It included a Soyuz A for a manned orbit of the Moon. Soyuz P was intended to rendezvous with the target; the cosmonaut would then exit and inspect it. This concept was abandoned because Soviet satellites were all equipped with automatic self-destruct devices, and it was feared that the U.S. satellites were similarly equipped. Therefore Soyuz P was redesigned to be armed with eight small rockets, of 1km range. The ASAT project was abandoned when Soyuz development proved too slow. However, manned military spacecraft (Soyuz R) development continued, and a design was ready in the first quarter of 1967. The existing 11A511 could place 6300kg in the desired 65 degree orbit, which would have limited the craft (Soyuz 7K-VI) to one cosmonaut. The Soviet military wanted a second pilot, who would add 400kg. Development of an improved 11A511M was therefore authorised by the Soviet Central Committee (21 July 1967), for first flight in 1968. Eventually the manned spacecraft (by then called Soyuz 7K-S) was cancelled (1974), but eight 11A511M had been built; one launched a Zenit-4MT reconnaissance satellite (the others were launched by 11A511Us).
16. Russian figures published in 1993 for weights in low earth orbit are: Soyuz, 7250kg; Vostok, 5000kg; UR-500, 12,200kg; UR-500K, 20,000kg; UR-200, 3000-4000kg; UR-700, 150,000-230,000kg; Energiya, 100,000kg; Energiya-M (ie, developed), 200,000kg; Molniya, 2000kg; Zenit-2, 13,800kg; Tsyklon, 3000-4000kg. The figures in the text are taken from James Wade's *Spacepedia*. They are significantly lower than those current during the Cold War. It is not clear whether the U.S. figures took into account the high inclination of the orbit the Soviets were forced to use. Now that their rockets can be launched from more favourable sites, presumably in-orbit performance will improve.
17. SL-7 (Kosmos 63S1) was developed to meet a 1960 requirement for a space booster for payloads not requiring the power of an R-7. A new second stage using a unique combination of liquid oxygen and UDMH was developed specifically for this rocket, and its first successful flight (after two failures) was on 16 March 1962. The production version was 11K63. Kosmos 65S3 (production version 11K65M) was an analogous launcher based on R-14.
18. Two ICBMs were developed as successors to R-7: Korolyev's cryogenic-fuelled R-9 (8K77; NATO

SS-8) and Yangel's R-16. The Soviet rocket forces rejected R-9 because, like R-7, it was too unwieldy. R-16 could be kept on alert for six months and, unlike R-7 and R-9, it used all-inertial guidance. The Soviets found R-16 difficult to build in quantity. Hearing at a meeting in February 1962 that this programme was unlikely to match U.S. missile production, Khrushchev decided to place shorter-range missiles (which were available in quantity) in Cuba, within range of American targets. The same problem offered Chelomey his great opportunity. He told Khrushchev that the problem was that the missiles were being made by the artillery industry, which had little ability to mass-produce airframes (the missile programme had gone to artillery makers because of Stalin's early post-war purge of the aviation production ministry). By way of contrast, Chelomey was involved in the aviation industry thanks to his work on cruise missiles; he could offer the mass production Khrushchev wanted. Hence his victory with UR-100. Korolyev was well aware that the Soviet rocket force was unhappy with his liquid oxygen weapons, and he seems to have begun work on solid-fuel rockets in parallel with the design of R-9, producing RT-1 (8K95), a massive IRBM. In 1962 Chelomey absorbed part of the Lavochkin design bureau.

19. Khrushchev apparently initially favoured Korolyev because he thought nuclear-armed missiles, rather than aircraft and conventional weapons, were the wave of the future; if the Soviets could dominate the new technology, they could overcome the industrial advantages of the West. According to Barry, by 1959 it had become apparent to Khrushchev that Korolyev was too interested in space exploration at the expense of weapons development. Chelomey offered no such problems. Korolyev was brushed aside partly by eliminating his allies in the governmental and Party organs nominally overseeing the new missile industry. Among the victims were Leonid Brezhnev and Dmitri F Ustinov (later Brezhnev's Minister of Defense). Chelomey had had reason to seek political cover. Lacking it, he had had his own design bureau (OKB-51) disbanded in 1953 (see below). His OKB-52 had opened in 1955. Given Khrushchev's influence, he was able to absorb into it another major design bureau, that of Myasischev (which had developed the 'Bison' heavy bomber). Myasischev's engineers could design ballistic missiles, and his bureau was associated with a big factory capable of building the necessary big airframes (which ultimately became Khrunichev). At this time Korolyev absorbed an ordnance design bureau, Grabin, in which he was interested as a source of solid propellant talent.

20. There was considerable talk in the 1960s and 1970s about an attempt by the Strategic Rocket Forces to replace Soviet warships with shore-based anti-ship ballistic missiles, presumably UR-100s. In the event, the only anti-ship ballistic missile fielded by the Soviets was the short-range R-27K/4K18 (SS-N-13) based on the R-27 (SS-N-6) ballistic missile. In 1974 it became operational (unknown to Western intelligence, which considered it experimental only, as SS-NX-13) on board one Project 605 (modified 'Golf' class) submarine.

21. However, the follow-on UR-100N (NATO SS-19) was energetic enough to be marketed, after the fall of the Soviet Union, as the Rokot commercial space booster.

22. Khrushchev's fall did not quite destroy Chelomey's position. When it was time to develop a UR-100 successor, the fight between Chelomey's supporters and Yangel's supporters was so intense that missiles developed by both organisations were built, as SS-17 (MR-UR-100, by Yangel) and SS-19 (UR-100N, by Chelomey). Unaware of the political background, U.S. analysts thought the emergence of two parallel programmes proved that the Soviets were particularly determined to gain missile superiority. They were paralleled by the expected R-36 successor (SS-18, R-36M) and by what turned out to be an unsuccessful solid-fuel ICBM (SS-16).

23. Just why is not clear, since recent accounts of UR-200 credit it with the ability to launch only a 2000kg satellite; however, accounts of the Soviet ocean reconnaissance programme all claim that the system had to be split up because Tsyklon could not launch a satellite designed for UR-200. It may be that simple weight growth in the planned satellite was to blame.

24. For example, the rocket's last launch, on 15 December 1997, was of a Yantar-4K2 surveillance craft.

25. Development of the Raketoplan was authorised under a 23 June 1960 Council of Ministers Decree 'on the production of various launch vehicles, satellites, spacecraft for the Military Space Forces in 1960-67'. Raketoplan would weigh 10 to 12 tonnes and gliding range during recovery would be 2500 to 3000km. An unmanned prototype (1960-1) would be followed by a two-man piloted version (1963-5) with 24-hour endurance, capable of intercepting U.S. satellites at altitudes up to 290km. Designs were developed by Chelomey, Myasischev (whose bureau Chelomey absorbed in October 1960), and Tsybin; Chelomey won the competition in November 1960. An unpiloted small-scale version of his Chelomey R was tested on 21 March 1963. In 1964 Chelomey proposed a single-seat Raketoplan, to be launched atop a Soyuz or a UR-500. At about the same time the U.S. Air Force was developing its own manned space plane, the abortive single-seat Dynasoar/SAINT II (X-20), which would have been launched atop a Titan III. Both the Soviet and the U.S. programmes continued into the 1970s, the follow-on to Chelomey R being the MiG Spiral, and the follow-ons to Dynasoar being the Prime, Asset, X-23, and X-24 programmes, plus a reported black programme.

26. The masses of the stages were increased by 71, 30, and 27 tonnes, the latter almost doubling. Low earth orbit payload increased from about 12,000kg to nearly 20,000kg. Compared to Soyuz, this rocket was 2.22 times as heavy but could launch payloads 2.78 times as heavy.

27. R-56 was designed to launch 12,000kg into a lunar orbit, or a heavy satellite into a geosynchronous orbit. These figures considerably exceeded what Chelomey claimed for UR-200. It was conceived as a drastically scaled-up R-36 (rather than as a cluster of R-36 stages). At least one version of the R-56 design seems to have used four clustered R-46s as its first stage.

28. GSR was to have been powered by four hybrid turbo-ramjet engines. A first-phase version using kerosene fuel would have accelerated to Mach 4 and

risen to 22-24km before launching the RB rocket. A second-phase version would have burned liquid hydrogen fuel, to reach Mach 6 and 28-30km altitude (about 140,000ft). No prototype was ever built. The flat lifting-body OS was reportedly based on a Tsybin design, presumably for the Raketoplan requirement (MiG absorbed the Tsybin bureau).

29. The objective was to be able to place 30 tonnes in a 200km northeast orbit and to be able to return a 20-tonne payload. A formal development plan was issued on 21 November 1977, and the technical project (design) was completed in May 1978. Reportedly the main engines were intended specifically to match the main engines of the U.S. Space Shuttle, with the same thrust rating and specific impulse. They were the first fully throttleable Soviet rocket engines. The RD-170 booster engines were reusable (they were rated for ten missions).

30. The U.S. Defense Department issued drawings of Uragan atop its Zenit booster in 1986; it was to be armed with a recoilless rifle and with space-to-space missiles. According to Mark Wade's *Spacepedia*, Russian sources deny that Uragan ever existed. The key question is whether hypersonic re-entry tests of scaled-down OS models (1982-4) were related to Uragan, or were merely tests of heat shield materials planned for the Buran space shuttle.

31. Alternative upper stages are Icare, Fregat, and the cryogenic H-10.

32. Start is a converted SS-25 (RT-2PM: 600kg or 1000kg payload); Rokot/Strela is a converted UR-100N (SS-19 with a new Breeze upper stage: 1800kg); Dnepr is a converted SS-18 (R-36MU: 4000kg); and Shtil-1 is a converted SS-N-23 (R-29RM: 270kg).

33. The current target seems to be a 40 per cent cut in launch cost. See Lardier, Christian, 'Les Lanceurs du XXIe Siecle' [The Launchers of the 21st Century] *Air & Cosmos* 13 (November 1998). Lardier quotes Steve Dorfman, Vice President of Hughes Electronics. Typical current U.S. figures are about $22,600 to $29,900 per kg in orbit. Chinese and Russian launchers cost about $11,200 to $14,800 per kg in orbit. The European Ariane consortium tries to match the high end of the Chinese-Russian price range. The U.S. Air Force hopes to launch a medium EELV for $50 to $80 million, rather than the current $80-$110 million, and a heavy EELV for $100 to $150 million rather than the current $350 million for a Titan IV. Under the EELV contract, each medium EELV is to cost $72 million, which is about 35 per cent less than current costs (*eg* Delta II). Against the low prices available in China, Russia, and Ukraine must be set relatively low commercial launch success rates: 86.5 per cent for Proton, 83 per cent for Zenit-2, 71 per cent for LM-2E (75 per cent for LM-3). Delta-1/2 achieves a 94.3 per cent success rate, Titan IV achieves 92 per cent, and Ariane 1/4 achieves 93.7 per cent.

Chapter 4: Polaris and Precise Navigation

1. Early sketches of Polaris submarines show radar or radio sextants, with which star sights could have been taken even in bad weather.

2. A purely ballistic guidance system cannot compensate for the effect of wind as the warhead comes down, unpowered, through the atmosphere. The most prominent exception was Pershing II, an intermediate-range ballistic missile with a radar correlation terminal guidance system. Because early Polaris re-entry vehicles were subsonic in the atmosphere, they were probably heavily affected by windage (not to mention being easier anti-missile targets). Later vehicles re-enter at high supersonic speed, and some of them spin (like a rifle bullet) for better accuracy.

3. Loran C was an Air Force development. It was first used by the U.S. Navy on board the Polaris test ship USS *Compass Island*, and then aboard Polaris survey ships. By 1962 networks were in place in the North and West Atlantic, in the Mediterranean, around Hawaii, and in the North Pacific. Initially three stations were needed for a fix. Under an early-1970s improvement programme, however, a new 'phase shift' technique was introduced, in which very accurate onboard clocks made it possible to deduce ranges from two stations and thus take a fix from them. That greatly enlarged the area in which Loran-C was effective. By September 1965 the Navy planned to phase out Loran-C in favour of the Transit satellite.

4. SOFAR may have been a cover for a classified Air Force acoustic system designed to pick up the sound waves from Soviet nuclear tests. This system picked up the sound produced when the U.S. nuclear submarine *Scorpion* sank in the Atlantic in 1968. Unlike SOSUS, SOFAR did not form underwater beams. Instead, it located objects by measuring the time of arrival (at different stations) of a well-defined sharp sound. RAFOS would have been used to provide a submarine with a reference point; she could then survey the sea area around that point, so that she could navigate by bathymetry. This concept was proposed for ASW submarines (SSKs) which would have to operate in waters near the Soviet coast.

5. These points were forcefully made by the British in connection with their own Polaris programme. ADM 1/27851 (Public Record Office) describes the navigation system on board USS *Patrick Henry* as of mid-1961 (ADM 1/28987 is the report of a mission to the United States to discuss Polaris navigation). The U.S. accuracy requirement was $1/2$nm in range and $2-2^{1}/_{2}$ minutes of azimuth. The British suspected that azimuth was the weak point; at 1500nm range, an error of 1 minute would throw the aim point off by half a mile. U.S. submarines used every available method: Loran C, Transit, bottom contour matching, and even star sights taken automatically through a special Type 11 periscope. Estimated Loran C error was 1ft per nm from the Loran station. Transit was probably jammable (the British did not plan to fit it aboard their submarines, although they would use it on survey ships). As for bottom mapping, as of June 1961, only nine areas had been surveyed for bottom contour matching, and each was only 8nm square. Moreover, the surface survey ships depended on Loran C fixes, so any subsequent submarine fix was limited to Loran accuracy. Nor could just any place be used; the bottom had to shelve gradually; peaks and valleys were unsuitable. The optical star tracker was intended as a direct check on the submarine's

inertial navigational system, but its accuracy was probably no more than 1 or 2 minutes, which was not good enough to decide between a Loran C and a SINS fix. Worse, the periscope took its vertical direction from SINS – which it was intended to check.

At least as of 1963, the British preferred to have their submarines or survey ships plant underwater beacons in prearranged and pre-surveyed spots. A Staff Requirement for the necessary coded sonar beacon originated in February 1960, and was approved that May.

According to a Secret (now declassified) 20 September 1965 memorandum in the CNO file (OP-09A2/rle Ser 511P09), as of that date high-resolution bottom contour charts were being prepared for bottom topography navigation, with special emphasis on Pacific SSBN patrol areas. Some areas in the Atlantic and in the Mediterranean had already been surveyed and were in use by SSBNs. According to the memo, it would take 10-15 years to complete charts covering the 70 per cent of the ocean judged useful for bottom-contour navigation, after which ships and submarines with precision depth recorders would be able to locate themselves with accuracies of 0.25nm or better.

6. A special Transit commemorative issue of the *Johns Hopkins APL Technical Digest* (Vol 19, Number 1, for January-March 1998) reproduces part of the key memorandum, written on 18 March 1958 by Frank T McClure (the system's inventor) to the director of the Applied Physics Laboratory: after spending an hour discussing work on Doppler satellite tracking, he realised that it could become the basis for a 'relatively simple and perhaps quite accurate navigational system'. McClure had been spending much of his time with the Polaris (Special Projects) office, and was well aware of its concern with the problem of precise navigation. On 4 April the director, Ralph E Gibson, proposed Transit to the U.S. Navy's Bureau of Ordnance, noting its simplicity (it needed only one antenna and no base line) and its relative immunity to jamming (it would be affected only by line-of-sight radiation). The same issue of the *APL Technical Digest* includes William H Guier and George C Weiffenbach's article 'Genesis of Satellite Navigation', an account of early APL work tracking Sputnik. It turned out that, quite by accident, APL had only Doppler data on Sputnik; all other organisations used dishes to track the Soviet satellite (*ie* angular data). APL was also already using the maximum slope of the Doppler shift to estimate the distance of closest approach of a missile to its target, so it was natural to use Doppler data to try to estimate the shape of Sputnik's orbit.

7. There is also an error associated with the receiver's altitude, since the system measures range to the satellite, not latitude and longitude. Its computer assumes that the Earth is spherical (it is not), and due to the error in the Earth's shape a receiver is often at some height above or below the assumed spherical shape.

8. The Navy rejected Transit because of its cost and its high risk; the programme was initially financed by the Defense Advanced Research Projects Agency (DARPA). The Navy took the programme over in the early 1960s.

9. Prototype satellites weighed 300lbs each, but APL decided to use the small solid-fuel Scout rocket, which had the lowest cost to orbit, to launch operational satellites. Because it lacked power, satellites had to be redesigned to cut their weight to 120lbs. Later satellites were launched piggyback fashion (multiple satellites per launch) by liquid-fuelled Thor-Able-Stars and then on improved Scouts (two per booster).

10. As originally stated, the ultimate requirement was that each ballistic missile submarine (SSBN) be able to get a fix accurate within 0.1nm (200yds) several times a day (the initial requirement was 0.5nm). Later the Navy Space Command specified an accuracy of 0.042nm (about 80yds) in each direction (0.06nm on a radial). The waiting time for a navigational satellite pass (between 15 and 75 degrees North or South) should not be more than 4 hours; the percentage of time it was more than 8 hours had to be less than 5 per cent, and the longest interval not more than 24 hours. System reliability was to be 98 per cent.

11. As it happened, some of the accuracy of Loran C was attributable to Transit, which made it possible to measure the location of Loran C sites much more precisely.

12. The British view, as reflected in 1961-3 papers now released to the Public Records Office, was that the U.S. system was far more complex than it had to be, partly because the U.S. Navy demanded that Polaris be fired within 15 minutes of the order (the British suspected that this time limit had been set to overcome Air Force objections to the system). Since the entire 15-minute period would be occupied by system checks, there would be no time for navigational corrections. Instead, the system would have to be accurate at all times, which would require frequent updates to the submarine's SINS.

13. Unlike Loran C, Omega was widely associated with the U.S. ballistic missile submarine force. As a consequence, there was local opposition to the construction of stations in Australia, New Zealand, and Norway, demonstrating the vulnerability of any key strategic support system not based in space (hence operating mainly from U.S. soil). Omega was recommended by a 1965 JCS study, 'World-Wide Navigation Requirements'. The Navy formally proposed Omega development in a 19 May 1965 Program Change Proposal (money was needed for aircraft receivers), and an Omega project office was set up that June. The Interim Operational Capability was established using sites in Trinidad, Norway, and Hawaii. As of mid-1965 plans called for completing the necessary eight transmitter sites in FY66 (2), FY67, FY68, and FY69 (4); the first 100 shipboard receivers would be delivered in FY67, followed by 405 in FY66 and 432 in FY69. There would also be integrated ship-aircraft units (532 in FY68 and 692 in FY69). Omega replaced Loran A on board U.S. Navy ships and aircraft.

14. Stellar-inertial navigation became possible only after the missile system had evolved considerably. Originally Polaris used Q-guidance ('fly the wire'), in which the elements of the trajectory from a known launch position were computed before the missile was fired. That greatly simplified onboard guidance, but it did not allow for any correction to the initial

position fed in by the submarine's navigational computer. The computation load was so great that Polaris submarines initially went to sea carrying decks of punch cards produced ashore, for different possible launch points (if the submarine was between launch points, the necessary numbers could be calculated by interpolation). Q-guidance was needed to shrink the size of the computer on board the missile (it was digital, but wired for only a specific type of calculation, solving the differential equations involved in Q-guidance). After the first ten Polaris submarines, a new fire control system (Mk 84) was introduced, incorporating a digital computer which could calculate the necessary data as the submarine moved out to her patrol area, and which could retarget missiles as required. As computer technology improved, however, it was possible to substitute explicit guidance, a scheme in which the missile continuously calculated its position, and adjusted its planned trajectory accordingly. The system the U.S. Navy adopted in 1964 for the multi-warhead Poseidon was based on that used in the Apollo programme and the analysis favouring it was based on examination of the guidance system of the MIRV version of the Air Force's Minuteman missile. In this case the advantage of explicit guidance was that a single bus, properly programmed, could launch all of the separate warheads. The alternative to this 'Mailman' concept was to provide each warhead with its own Q-guidance package ('Blue Angels'). As an incidental advantage, an explicit-guidance system could accept and use a gross position correction provided by a star sight. Unlike Q-guidance, explicit guidance required a general-purpose digital computer (in this case, with 12k of memory). It also required a digital model of the Earth's gravitational field (in Q-guidance variations in gravity were fed in with the precalculated parameters). That model was based on Transit measurements. About 1966, as Poseidon was being developed, stellar-inertial navigation was proposed for the missile: information from star sightings would supplement that the missile assembled using its own inertial measuring devices. The idea was initially offered to an uninterested Air Force. Test flights using surplus Polaris missiles were successful, and the proponents of the system, who worked for the Kearfott Division of the General Precision Corporation, turned to the Navy. They soon found that they could make do with a single star sighting (the 'unistar' system), and that greatly simplified the new kind of guidance. Since the star sight could overcome much of the inaccuracy in submarine location, a stellar-inertial missile could be at least as accurate as one launched from a fixed silo. This technique was never used in the Poseidon missile, because Congressional opponents argued that if it became really accurate it might seem to be a first strike weapon, and hence destabilising. Stellar-inertial guidance was, however, introduced in Trident. It was pointed out that in order to be more survivable (against improved Soviet ASW) the submarine would need more sea room in which to operate, and that she might have to remain submerged for much longer between SINS updates. These concerns were quite compatible with the vision of Trident as an 'assured-destruction' deterrent missile rather than a hard-target killer. The requirement for greater sea room translated into a demand that accuracy at 4000nm range be equal to that of Poseidon at 2000nm.

Chapter 5: Passing the Word: Reliable Communications

1. To preclude attacks made in error, bombers would execute their missions only if they received the 'go-code'. This precaution was not publicised for fear that the Soviets would realise that they could neutralise the U.S. deterrent simply by destroying command centres authorised to issue that signal; there was a widespread misapprehension that the bombers would strike unless they were recalled. That was the basis, for example, of the movie 'Doctor Strangelove'. This explains why SAC was so interested in reliable long-haul communications.
2. Initially it seemed that hardened missile silos for Minuteman missiles would also be able to ride out a Soviet missile strike, because Soviet missile accuracy was limited. Later the survivability of the land-based missiles became very controversial.
3. As a corollary, the Administration feared that any ally with an independent nuclear deterrent could trigger a nuclear war. It therefore tried to replace the British and French forces with a NATO-wide 'multilateral force' which would be subject to U.S. veto. The British and French were, understandably, less than enthusiastic.
4. This apparently theoretical argument had a very practical side. Under Eisenhower, the U.S. Army was gutted, on the theory that it was unlikely to be needed. The deployment of tactical nuclear weapons in Europe underlined, for the Soviets, the reality that any invasion of Western Europe would have gruesome consequences. The Army argued that it was essential to have some non-nuclear ability to fight in Europe, since otherwise the Soviets might subject Western Europe to nuclear blackmail (most Europeans found Eisenhower's views perfectly reasonable). President Kennedy bought the Army's argument, and after the 1961 Berlin crisis rebuilt the U.S. Army – providing the force needed for intervention in Vietnam.
5. It is difficult to say precisely when the Soviets gained access to U.S. naval communications. Walker began selling key lists for the standard KW-7 cryptographic device in December 1967, and after the electronic surveillance ship USS *Pueblo* was seized the following January the Soviets had the corresponding device itself. In December 1968 it became known that the crew had not had time to destroy their coding equipment, but the NSA decided that the system had not been compromised because the keys were still secure (Walker would not be exposed for about 15 years). Therefore the KW-7 system, which was very widely used, was not withdrawn from service.
6. According to Hezlet, *The Electron and Sea Power*, p157, after the First World War radio amateurs first discovered that HF signals could skip to great ranges by bouncing off the ionosphere (there was a general movement to higher frequencies to obtain

more channels). In 1921 amateurs in Britain and in the United States managed to send low-powered signals across the Atlantic at 1.3 MHz, and by 1923 HF was being used across the United States and in Europe. It was soon clear that world-wide transmission was possible at moderate power, but reception was intermittent and the signal could not be heard at all in the skip zone between bounces. The British felt that due to its skips, HF was unsatisfactory for communication with ships. Instead, they used it to connect fixed land sites, which could themselves communicate with ships on MF/LF frequencies (ships were fitted with HF equipment, but it was considered secondary). The Admiralty was suddenly able to communicate with distant stations without recourse to fixed cables. According to Gebhard, *Evolution of Naval Radio-Electronics*, p43, the U.S. Navy was loath to adopt HF because performance was considered too erratic; existing radio theory did not explain the skip effect or its variation. Meanwhile, with the growth of commercial radio, the Navy had to abandon use of the 550 to 1500 kHz band. By 1925 the U.S. Naval Research Laboratory had developed the basis for the necessary theory; in 1926 it managed to send signals all the way around the world. Meanwhile, in 1925 the U.S. Fleet had steamed to Australia, and it had used HF radio to maintain contact with the United States. On the basis of that success, HF systems were widely installed on board U.S. warships.

7. LF (10–300 kHz) has a ground wave which travels about a thousand miles over water (and a shorter distance over land). The associated sky wave is trapped in a kind of waveguide between the troposphere and the low ionosphere and the surface; a sufficiently powerful transmitter can be heard at any distance. On this basis in 1921 Marconi proposed that a chain of Empire stations be connected directly with the United Kingdom by LF transmission. MF (300 kHz – 3 MHz) also has ground and sky waves. During the day, the sky wave does not refract at all, so transmission is limited to the ground wave. At night the sky wave refracts from a low layer of the ionosphere (the D layer), hence reaches shorter ranges than HF. The existence of a refracting ionosphere was guessed after Marconi successfully sent signals across the Atlantic in 1901 (previously it had been supposed that signals transmitted only in straight lines), but in fact his LF transmissions followed the Earth's surface in the waveguide described above.

8. In pre-Second World War radio DF systems, a receiving beam is slowly rotated. The operator finds the direction to the target by seeking a null (zero signal), and clearly that must be done while the signal is still being received. If the signal is short enough, the operator cannot check all directions during it. Second World War HF/DF solved the problem in two alternative ways. One was to set up two beams at right angles, continuously displaying the sum of the signals they received on an electronic screen (the sum indicated the direction to the transmitter). Since there was no time lag during direction-finding, this beam-forming technique could deal with very short signals. The alternative was to spin the receiving beam at high speed, again displaying the signal on a screen. In this case, the target is DF'ed as long as the spin time is shorter than signal duration.

The second solution offers greater sensitivity, hence longer range. The German land-based Wullenweber technology, which was adopted post-war by the major navies, uses a fixed array of antennas to form a number of beams, the outputs of which can be continuously compared, as in the first wartime method described here. Because there are more, narrower, beams, such a receiver is more sensitive.

9. Pearl Harbor was a case in point. On the morning of 7 December 1941, the U.S. government had just decoded Japanese messages which implied that Japan was about to attack. Senior officials, including President Roosevelt, debated whether to warn distant outposts, including Pearl Harbor. The fear was that the Japanese might intercept the warning and realise that their own codes were insecure. This was not a matter of possible wire-tapping. Since there was no undersea telephone cable to Hawaii (technology did not yet permit it), communication had to be by HF radio – which was essentially a public channel. Despite this fear, eventually they decided to send a coded message. Then they discovered just how unreliable HF could be. First the Army was unable to raise its station in Hawaii. Then the Navy failed. Then the message had to be entrusted to a commercial carrier – and the radiogram was carried to the base a few hours after the Japanese struck.

As an indication of just how difficult it could be to find optimum frequencies (if they exist at all, at times), as late as the early 1980s (as described in NWP 4, the handbook on naval tactical communication, 1982) the U.S. Navy was just beginning to install a joint-service device, Chirpsounder (TRQ-35). Its TCS-4B transmitter periodically emitted a chirp, an HF signal whose frequency increased at a constant rate. Ships carried the corresponding special receivers (RCS-4B), which tuned at the same rate, and which were timed to synchronise with the transmitter (each receiver could be synchronised with up to three transmitters). A special display showed which frequencies gave the best propagation for particular stations, in 6 kHz increments. It also indicated the degree of multipath distortion. Plans called for installations on board all carriers, flagships (LCC), large-deck amphibious ships (LHA/LHD), and large dock landing ships (LPD). The ionosphere could change so rapidly that NWP 4 recommended making a sweep every 15 minutes; each sweep (2-16 or 2-30 MHz) lasted 4 minutes 40 seconds.

10. That is, 45 bits per word, or 9 per character, including a one-bit spacer between characters and one check bit, leaving 7 bits, enough for up to 128 distinct characters.

11. At the lower end of HF, 3 MHz, the wavelength is 100m; at the high end, 30 MHz, it is 10m (32.8ft). These figures explain why whip antennas about 35ft long can operate effectively at HF frequencies.

12. Ironically, HF was initially welcomed because its antennas were smaller than those of the earlier LF and MF radios, hence could be placed more conveniently and would not interfere. The mutual interference problem apparently first appeared on board amphibious flagships (AGCs), which needed more command circuits than other ships.

13. Two principal approaches were used. 'Sleeve' antennas (cylinders consisting of two concentric

tubes) offered a 3:1 frequency range. They were first widely used on board the fleet flagship *Northampton* (1953), ten such units (five receiving, five transmitting) replacing fifty conventional antennas. The other technique, important at the lower end of HF frequencies, was to use the superstructure itself as part of the antenna. This concept was first applied to a 'Guppy' submarine (*Dogfish*) in 1948, then to a *Sumner* class destroyer, and then to *Northampton*. The latter technique was probably the beginning of the need for brass modelling of ship structures, since the behaviour of integrated antennas would have been extremely difficult to predict (the model technique was first used in 1948). In some cases the lower part of a sleeve antenna would have been so massive as to restrict visibility needed to dock the ship. NRL therefore developed a 'conical monocone' antenna using four wires supported (and fed from) a central mast, with further pyramids of wire further down, for higher frequencies. Such antennas were first used on board the test ship *Observation Island* in 1959. This type of antenna can support a frequency range of 5:1, so that three stacked sections can handle the entire HF band, 2-30 MHz. The number of antennas was also limited by development of techniques of multi-coupling, whereby several transmitters could use the same antenna simultaneously. A four-transmitter system developed soon after the war went into service in 1950, followed by an eight-transmitter multicoupler developed in 1959 (in service in 1964). Multiplexing receivers was simpler; typically eight receivers could use the same receiving antenna during the Second World War. By 1966 twenty receivers, their frequencies more closely spaced, could use the same antenna.

14. Navies dependent on MF radio used shore stations connected to home by cable, which did provide world-wide reach. The loss of these shore stations (or of the cables) had crippling consequences. The importance of the radio stations for controlling naval forces explains why it was so important to the Royal Navy that the Royal Australian Navy roll up the German island possessions in the Pacific in 1914. It also explains why the British considered it so important to control the world's undersea cables, and why it was so important to cut key cables at the outbreak of a war. With HF radio, a shore station could receive by radio from home, then retransmit. Cables became less important.

15. The land elements of the U.S. naval communications net are illustrated in successive editions of NWP 16, *Basic Tactical Communications Doctrine*. The 1953 edition (page 19-4) shows Port Lyautey, in what was then French Morocco, as a primary teletype relay station, communicating not only with the fleet but also with three major relay stations: London, Naples, and Tripoli. Tripoli in turn communicated with the two lesser stations, Athens and Asmara. There was no African coverage beyond Asmara. In the Atlantic, Balboa (in the Canal Zone) served as the primary relay station, transmitting to the fleet; but Rio de Janiero was a secondary relay (connected directly to Washington rather than to Balboa), and San Juan, Puerto Rico, was a major relay station. The primary Pacific relay stations were Pearl Harbor and Guam, both on American territory; Tokyo was a major secondary relay, and there was a minor relay station at Sangley Point in the Philippines (Subic Bay had not yet been completed). Overseas relays were all connected to the United States (mainly to Washington) by HF radio links – which were themselves vulnerable to environmental problems. A chart of Fleet Broadcast areas shows Sangley Point responsible for the entire Indian Ocean and Port Lyautey responsible for the East Central Atlantic, the Mediterranean, and the Red Sea. Washington was responsible for the entire Atlantic, and Honolulu for most of the Pacific (Kodiak for the North Pacific). Balboa was responsible for the West Coast of South America and the South-Eastern Pacific.

About a decade later (December 1962) areas of responsibility had not changed very much, except that Asmara had been upgraded to serve the Western Indian Ocean and East Africa. With the construction of the major base at Subic Bay, the Philippines had been upgraded to a major station, and it now communicated with the British base at Singapore. Honolulu, Guam, and San Francisco were still the primary radio stations in the Pacific. In the Atlantic, the primary overseas stations was still Port Lyautey. A Turkish station, Karamursel, had been added, and Tripoli was already gone (even though the friendly Libyan monarchy would not be overthrown for almost another seven years). By this time Spain was a U.S. ally, so a major station there supplemented the earlier ones in Greece and at Asmara. Probably the most dramatic improvement was the construction of a land line, to insure reliable service, connecting Washington to the European naval headquarters (a primary station) in London, via Goose Bay, Resolution Island, Cape Dyer, Sondestrom (Greenland), Kulusik (Greenland), Iceland, and Thurso. Another land line connected London to the major radio station in Londonderry. Except for this land-line, all overseas communication with fixed stations was still by radio. At this time Balboa was still a major station, and there was still a minor station in Rio.

By 1970 the situation had changed again. The major communication centres (NAVCAMS) connected to Washington were Norfolk, Londonderry, Morocco, and Asmara in the Atlantic, Puerto Rico in the Caribbean, and San Francisco, Honolulu, and the Philippines in the Pacific. They were interconnected. For example, Asmara also had connections with Morocco and with Puerto Rico. Now there were many more major outstations, such as Cam Ranh Bay in Vietnam and Harold E. Holt in Australia; the main overseas Pacific Naval Communications Centres were Japan, the Philippines, and Cam Ranh Bay. Major Atlantic stations included Iceland, Thurso, and Rota. The station at Rio was gone, perhaps due to Brazilian disagreements with the United States over, among other things, attitudes towards Cuba.

By 1982, Port Lyautey was gone. Major responsibility for Mediterranean communications had shifted to Naples, with connections to Spain (Rota) and to Greece (Souda Bay). Major stations remained in London, Thurso, and Iceland, and there was a new station at Edzell, Scotland. In the Pacific, the major stations were in Japan, the Philippines, Australia (Harold E. Holt), and in newly-leased

Diego Garcia, in the Indian Ocean. Cam Ranh Bay was gone with Vietnam, and there was no nearby replacement. Asmara was gone, due to the Ethiopian coup and that country's shift towards the Soviets. Apart from Balboa and Puerto Rico (Roosevelt Roads), there was no coverage of the South Atlantic. No later communications NWP has been declassified, but the Philippines centre probably went when Subic Bay and Clark Air Force Base shut down.

16. For example, in 1962 Port Lyautey, Morocco, which serviced the Sixth Fleet, and which was considered inadequate, provided five separate shore-to-ship broadcasts (each using four or five high-powered transmitters), twelve two-way ship-to-shore circuits (four 2-channel multiplex, three duplex for teletypes, five for CW [Morse Code], each requiring a medium-power transmitter and a receiver), and four long-haul point-to-point 16-channel circuits; plus a facility to receive facsimiles and a variety of special-purpose circuits such as those used by the Naval Security Group (SIGINT) and the HICOM high-command voice circuits connecting the Sixth Fleet commander with task force commanders.

17. Official sketches of the AGMR and AGMT, prepared about 1960, give some idea of their layout. Each has a single big broad-band mast carrying HF antennas (2-30 MHz) and higher-frequency units (VHF, UHF, SHF) at its head. The AGMR had a single 100ft pole mast, a 2-30 MHz mast, forward of its main receiving mast. The AGMT, the transmitter, had a directional HF (11-23 MHz) antenna protruding from its principal mast. Presumably the masts were for omni-directional transmission to the deployed fleet, which might be widely distributed, whereas the directional antenna was for point-to-point communication back to the United States or to a designated shore station.

18. In fact only one escort carrier, *Gilbert Islands*, was converted, becoming the communications relay ship (AGMR) *Annapolis*. The other AGMR was a converted light fleet carrier, *Arlington* (ex-*Saipan*), which had been earmarked as a command ship. Both ships spent the bulk of their operational careers supporting the U.S. carriers on the 'Yankee Station' off Vietnam, 1965-70. *Annapolis* was laid up in December 1969, *Arlington* in 1970. As an AGMR, *Annapolis* offered twenty-two long-range transmitters (twenty MF/HF and two LF), forty-eight receivers (forty-four MF/HF, three LF, one VLF), as well as eight UHF circuits, nine multiplexers (sixteen channels each), and crypto facilities. Using the multiplexers, the ship could create eighty two-way teletype channels (100 words per minute, *ie* about 50 bits/sec). Most ships still lacked 16-channel multiplexers (they operated on either a single channel or had a 4-channel multiplexer), so on each frequency the ship could send as much to shore as five or six ship channels. As a relay to shore, she could use her two directional antennas and her ten powerful 10 kW transmitters, where ships generally had 500 W or 1 kW omni-directional transmitters. Relay facilities would be particularly valuable about 200nm from a ship, since MF/HF signals could be received accurately at that range (where they were ground waves), whereas at 250-500nm they would be mixed sky and ground waves, subject to fading.

19. In fact these signals are subject to ducting; sometimes they carry remarkable distances.

20. In contrast to the U.S. Navy, the Royal Navy of the 1970s considered the risk of interception unacceptable. It fitted its ships with a computer link (Link 10), but apparently rarely used it. In the Falklands, however, information provided via the link proved so valuable that British practice changed. All major British warships were then fitted with Link 11 (previously the Royal Navy had fitted only key units with Link 11, using them as 'gateways' to its Link 10-equipped fleet).

21. NWP 4, the U.S. Navy handbook on tactical communications (July 1982) describes an HF Limited Range Intercept (LRI) concept for use by a dispersed force. It was based on exercise experience and tests; the handbook referenced a 1979 Second Fleet TACMEMO (tactical memorandum). The technique exploited the fact that HF signals would not bounce off the ionosphere above or below certain frequencies: those above the Maximum Useable Frequency (MUF) between sunset and sunrise and below the Lowest Useable Frequency (LUF) between sunrise and sunset. Thus effective HF range might be limited to the 300nm of the ground wave. The Naval Oceanographic Science Center (NOSC) at San Diego developed HF propagation prediction programmes to run on the Wang 2000 computer then on board several carriers; it could also use the Tektronics 4052 graphic system which ships carried under the Naval Science Advisory Programme. As of 1982 NOSC was developing a Spectrum Utilization Management System (SUMS) and a predictor for the programmeable hand-held HP-67 calculator (Pocket Prophet III). The latter would calculate times to begin and end LRI broadcasts during the day. At the same time, HF transmitters were being field-modified to cut their power incrementally to the lowest LRI level (special kits were required for the URT-23; all others could be modified by ships' forces). Priorities for modification were (i) battle force, submarines; (ii) service force supporting the battle force; (iii) amphibious forces; (iv) all others. In effect the range to ships in the net would be estimated and power limited in hopes of limiting propagation beyond them. This was not the first U.S. Navy effort to limit HF interception. For example, a 1963 paper in the declassified CNO records mentions HARE, a recently-deployed anti-intercept technique, which was to be used very sparingly to avoid compromise (presumably, to Soviet HF nets). HARE was probably a new burst communication system. This type of countermeasure would be ineffective for Link 11, which had to be used more or less continuously, and which needed all available message time.

22. Sometimes even surface wave HF signals propagate too far. It now seems that the *Vincennes* incident in 1988, in which an Iranian Airbus was shot down by accident, is traceable to the inadvertent merger of two Link 11 nets whose radio coverage overlapped. As it happened, the other net assigned a nearby A-6 the track number *Vincennes* had assigned to the Airbus. Because *Vincennes* was the second ship in her group to have detected the Airbus, that aeroplane's track number was automatically changed (to avoid confusing the joint tactical picture) – but no one in

the CIC on board the cruiser noticed. Thus the computer never reported the double use of the same track number. When the cruiser's captain asked about the behaviour of the Airbus, he naturally asked by track number – so in fact he was asking his ship's computer what the aeroplane which still carried the track number – the A-6 – was doing. As it happened, it was landing on a carrier (*ie* diving). To the ship's captain, that dive was clearly a hostile act. That is, the ship's computer easily kept tally of changes in track number assignment, but those changes were much less obvious to those using the computer system.

23. The Link 11 Kineplex technique (different frequencies [tones] are sent simultaneously) works because radio is linear: the sum of signals of two frequencies can be broken down again into signals at each frequency. The simpler Link 10, which did operate at 75 baud, carried far less information. It was characterised as operating as a serial, rather than a parallel, link. It was neither encrypted nor did it use the stacked frequencies of Link 11. Compared to Link 11, it transmitted only position reports, never vectors or identifiers. Ships using Link 10 added their own identifiers and estimated their own vectors. Similarly, Link 10 did not transmit targeting data or firing tell-backs.

24. According to Breitler and Huang, the current expected HF rate is 2400 bits/sec, which matches the old U.S. Navy 25 kHz satellite channel. A new HFRG (High Frequency Radio Group) programme will double or, in some circumstances, quadruple this rate, to 9600 bits/sec. HFRG also offers the kind of automated controls needed to merge HF with the various satellite and current line of sight radio systems.

25. After the war the Allies discovered that the Germans had been able to communicate by LF radio with submerged U-boats in the Indian Ocean; the Soviets captured their big LF transmitter, Goliath. For the Atlantic, where U-boats were centrally controlled, the Germans preferred HF radio, which was effectively two-way, and which could carry much more information. For both the U.S. and the Soviet fleets, the standard LF receiving antenna was a DF loop (enclosed, in the case of U.S. submarines). The Soviet LF submarine antenna was code-named Park Loop by NATO. The associated long-range radio system (for all types of warships) was Pobeda (equipment for which appeared in 1946-50); the first Soviet VLF transmitter was the captured 100 kW Goliath, which was placed in service in 1952.

26. The six main U.S. VLF transmitters, all operating at megawatt power levels, were built in the late 1950s at Annapolis; Cutler, Maine; Oso, Washington; Wahiawa, Hawaii; Yosami, Japan; and North West Cape, Australia. There are also twenty-one LF transmitters. VLF penetrates sea water to a depth of 9m, compared to 5m for LF. The standard U.S. LF receivers are in streamlined bodies atop periscoping masts; they can also be placed in towed buoys. The standard VLF receivers carry a towed buoy carrying a loop larger than that used for LF (allowing the submarine to submerge to about 45m; she is then limited to a speed of about 4kts) and a buoyant wire (550-640m long, limiting the submarine to a depth of 15-18m and to a speed of 10kts or less). The buoyant wire must be towed in the appropriate direction, further limiting the submarine. Early in the Polaris programme, moreover, it was discovered that the wire could be a source of vulnerability. For example, sea birds tired of swimming had a disconcerting habit of riding it. The Soviets introduced a towed VLF antenna, Paravan, in 1967.

27. ELF research began in 1958 and feasibility was demonstrated in 1962, but development was slow because, given its very long wavelengths, ELF required a massive transmitting antenna. That is, at 30 kHz VLF has a wavelength of about 1km, but at (say) 30 Hz ELF has a wavelength of 10,000km. Thus the first serious U.S. ELF proposal was for a 10,000km buried antenna (which would have covered much of the state of Wisconsin) fed by an 800 MW transmitter. Eventually a much smaller ELF system became operational in 1989.

28. Originally the aircraft were modified C-130s. During the 1970s, as it seemed that the deployment of an ELF system was imminent, the TACAMO force was drawn down prematurely. The current E-6 is a modified Boeing 707; the E-6B version also incorporates the Air Forces strategic Airborne Command Post (ABNCP) function. In the 1980s a Soviet article claimed that TACAMO operates at 21-26 kHz at 200 kW, using an active antenna 3km long and a reflector (also a wire) 10km long; other sources place the lengths at 5000ft and at about 35,000ft (of which 16-20,000ft are typically used).

A classified summary of SSBN communications (June 1969, in the declassified CNO files) describes the system as it then existed: four dedicated VLF stations, five dedicated LF stations, plus six VLF and thirteen LF/MF stations for general service, backed by designated ships and aircraft. The VLF net included three NATO stations available in an emergency (in Scotland, Italy, and Norway). This system could deliver a 'go' message within 7 minutes of transmission by the National Command Authority, with an estimated reliability of 97.7 per cent (according to the 1968 Polaris evaluation report). Backing it up were the twelve EC-130Q TACAMO aircraft. According to the 1969 paper, it would take the co-ordinated effort of thirteen VLF/LF jammers in the Atlantic, eight VLF/LF jammers in the Pacific, and at least fifty HF jammers world-wide to disrupt the fixed communication system. Because this was within Soviet capabilities, the Navy was buying an anti-jam VLF/LF system, VERDIN. As for physical destruction, a world-wide strike would have to be executed within 7 minutes after the National Command Authority had ordered a U.S. strike. Again, that was difficult but possible, so the Navy wanted more back-ups: an ELF system (Sanguine), which could not be jammed except at extreme cost, which could be hardened against nuclear effects, and which could be received by a submarine at depth [eliminating the possible vulnerability inherent in using a floating wire antenna]). It also wanted a shipboard VLF system (SMVLF), which could communicate with commands outside the United States, and which might therefore survive a Soviet nuclear strike. Finally, it wanted an improved TACAMO system (TACAMO IV). In fact ELF development was very protracted, largely for political reasons. It is not clear whether shipboard VLF ever really materialised. TACAMO was improved.

29. A transcontinental teletype link was demonstrated on 29 November 1955, using four linked SK-2 parabolic radar antennas at the receiver. With eight SK-2 antennas, a message was successfully sent from Washington to Hawaii, 4350nm away, on 23 January 1956. This system was called CMR (Communications Moon Relay). CMR became operational in 1959, using trainable 84ft radio telescope dishes at Annapolis, Maryland, and at Opana, Oahu, in Hawaii, operating at UHF frequencies (445.1 and 435.1 MHz). CMR was considered a necessary back-up to the existing HF link between Washington and Pearl Harbor; it could accommodate one multiplexed channel (at that time, four teletype channels, although 16-channel multiplexers [60 words/min each] were in prospect). No shipboard UHF system was possible because the antennas involved were so large. However, it was clear that the same gain could be achieved in a much smaller antenna at much higher frequency. Using a larger antenna at the shore end, moreover, could compensate for using a smaller one at the ship. In February 1957 NRL bounced a 2860 MHz (SHF) signal off the Moon, using its 50ft steerable antenna. In December 1961 the research ship USS *Oxford* demonstrated shore-to-ship CMR operation using a 16ft antenna (the shore end was a 60ft antenna at Stump Neck, Maryland). In March 1962 the ship was able to transmit a message back while steaming in the South Atlantic. A CMR system was established in 1964, with six ship and four shore stations (Cheltenham, Maryland; Wahiawa, Hawaii; Okinawa; and Oakhanger, England). The ships were *Oxford*, (1964), *Georgetown* (1965), *Jamestown* (1966), *Liberty* (1967), *Belmont* (1968), and *Valdez* (1969). This system was placed in reserve status in August-November 1969.

30. According to Brannegan, *Space Radio Handbook*, p218, by 1960 amateurs were regularly using Moon bounce at frequencies of 28 MHz and above. Initially the technique required a large antenna array, but by the 1980s electronics had improved to the point where a backyard was quite large enough. Brannegan reports some multipath problems due to irregularities in the lunar surface and to wobble in the Moon's motion (but note that NRL suspected that the Moon had a useable ionosphere of its own, from which signals at the appropriate frequencies could bounce without much distortion).

31. This particular scheme aroused the fury of radio astronomers, because the needles would also have bounced most of the signals they studied back into space. On 9 May 1963 the needles were placed in a 2000-mile orbit. As it turned out, the cloud of needles broke up quickly, and was gone altogether within three years.

32. Clarke seems first to have proposed his idea in letters to the British radio magazine *Wireless World* in February and October 1945, and in a 25 May 1945 memorandum to the British Interplanetary Society. He envisaged a large manned space station, which even now cannot be sustained at geosynchronous altitude.

33. Above the HF range, the atmosphere is fairly transparent to radio signals at frequencies up to above 10 GHz. Much above that, absorption increases as frequency rises, except for several windows: one at about 30-40 GHz and others at 94, 150, and 240 GHz. These frequencies are important for both radar and communication. Absorption is particularly strong at 22.5, 60, 117, and 186 GHz.

34. The Army's need to use very small dishes to communicate on the move (as opposed to communicating at the halt, using a deployable antenna) greatly complicates UHF usage. Because the small dishes are virtually omni-directional, they send their signals to every UHF satellite above the horizon (and, similarly, they receive from each such satellite). Each channel can be used only once per hemisphere, regardless of how many satellites there are. All UHF users, not just U.S. ones, must co-ordinate in setting up a frequency system (currently four frequency plans, 21×5 kHz and 18×25 kHz channels, totalling 2.2 MHz out of a total bandwidth of 175 MHz [225-400 MHz]).

35. Given a signal spread over many frequencies, it may be difficult for the interceptor to be certain of the base frequencies, hence of the magnitudes of the Doppler shifts.

36. Traffic analysis plots the gross features of messages without any attempt at decoding them. For example, before an attack the sheer level of traffic should increase. Plotting the number of messages to particular addressees can also be useful. These techniques are quite old, and the countermeasures are very well known. For example, dummy messages maintain a constant level of traffic, and addresses in messages can be encoded.

37. The 10 per cent figure is set by the requirement that the imposed signal not so distort the carrier as to make it unrecognisable.

38. That is why 'squash' (compression) transmitters appeared only after high- (as opposed to medium-) frequency radios entered service. Such transmitters cut message time in hopes of precluding direction-finding by conventional rotating DF systems, since a 'cut' took a finite time.

39. Another way to see why is to observe that any waveform – such as a square wave representing a single bit of information – is the sum of pure waves of different frequencies, at different strengths. It is a mathematical fact that the product of the bandwidth – the frequency range required to produce such a square wave – and its duration is at least one. It follows that at least 5 kHz - 5000 Hertz – of bandwidth is needed to form square waves $1/5000$ of a second long, ie to send data at the rate of 5000 bits/sec. The effect of noise is to add random amounts to each wave making up each character, distorting it to make it difficult to tell where one begins and another ends. The cure is to spread out the characters, ie to accept a lower data rate.

40. Representing numbers through sixteen requires four binary digits (bits); representing those through four requires two. Thus a scheme involving sixteen levels gives twice the length of binary number as a scheme involving four levels.

41. For example, a common specification for information transmission is R3/4QPSK, meaning a rate of 3/4 QPSK (Quad Phase Shift Keying – 4 bit coding). The ratio is the ratio of information bits to data bits. If the user wants an effective data rate of 2400 bits/sec, 4/3 as many (3200) must be transmitted each second; that means adding 800 check (error-

correction) bits. Typical ratios are 1/2 (very robust), 3/4, 5/6, 7/8 (less robust); uncoded would be 1/1.

42. The U.S. Army Signal Research and Development Laboratory at Ft. Monmouth was ordered to produce a communications package (maximum weight 150lbs) late in June 1958, and the missile was orbited on 18 December. The orbit was low (100 × 800nm, 32 degrees), and the satellite re-entered the atmosphere on 21 January 1959 (its batteries had failed after 12 days). Besides the President's tape-recorded greeting, SCORE carried both a real-time and a store-and-forward communications package which transmitted seventy-eight messages (both real-time and delayed) between stations in Arizona, California, Georgia, and Texas. Its up-link operated at 150 MHz, its down-link at 132 MHz (8 Watts). The recorder (300 to 5000 Hz band) had a capacity of 4 minutes. Apparently inspired by the SCORE idea, in September 1958 the Army Signal Corps proposed a more sophisticated store-and-forward satellite, Courier, which was launched into a low orbit (525 × 624nm, 28 degrees) on 4 October 1960. It operated at 1.8-1.9 GHz up-link/1.7-1.8 GHz down-link (two 2-Watt transmitters with two back-ups) and it carried four digital (each 4 minutes at 55 kbits/sec, 13.2 Mbit total) and one analog recorder (4 minutes with a bandwidth of 300 to 50,000 Hz). It could handle a single half-duplex real-time voice channel. Courier communicated successfully between ground stations in New Jersey and Puerto Rico. The satellite failed after 17 days, apparently because its internal clock fell out of step with that on the ground. The access codes were clock-based, so once that happened the satellite would no longer respond to commands.

43. Advent was to have been a geosynchronous satellite with four repeaters, each with the capacity for twelve one-way voice links or one spread-spectrum voice link; there would also have been a secure command system.

44. This roughly paralleled the decision to develop a Single Integrated Operational Plan (SIOP) for nuclear attacks by U.S. forces.

45. The choice of SHF seems to have been based on Lincoln Laboratory's experience with SHF communications in Project West Ford. The Laboratory built a series of experimental SHF satellites (LES-1,-2, and -4) and an experimental terminal (LET). LES-1 was launched on 11 February 1965. LES-2 was essentially identical. Both used an all-solid-state X-band (SHF) transponder and an eight-horn electronically-switched antenna. The eight horns gave omnidirectional coverage. While the satellite spun, its sensors determined the direction to earth and chose the appropriate antenna. The orbit was 1500 × 8000nm (32 degrees). Links: 7750 MHz up-link, 8350 MHz down-link (200 mW power). LES-4 was similar to LES-1 and -2, but used a cylindrical rather than a polyhedral shell. It was intended for a geosynchronous orbit, but the booster did not work properly, and after a 21 December 1965 launch it stayed in the transfer orbit (105 × 18,200nm, 26 degrees).

On NRL advice, DSCS internal frequency characteristics (bandpass) were chosen specifically to make shipboard use more practical. Originally DSCS was designed to use a very large ground antenna and a high-powered transmitter.

46. The first seven satellites were launched together on 16 June 1966. For simplicity, they were spin-stabilised, like LES-1 and -2. No attitude-control command system was, therefore, needed. Command system failures had killed Courier and Telstar 1, and problems in command system development had helped kill the Advent programme. Each 100lb satellite carried two 3-watt X-band transponders (one for backup) with 26 MHz bandwidth. Capacity was five commercial-grade voice or eleven tactical-quality voice circuits or 1550 teletype or about 1 Mbps of digital data. Near-geosynchronous altitude was chosen because the satellites had no station-keeping capacity; it was considered better to let them drift. Moreover, drifting limited the effect of individual failures. Telemetry was used only to gather information on satellite performance. The system was very successful, providing near real-time data transmission from Vietnam to Washington. The satellites were launched in groups of eight by Titan IIIB rockets. Design lifetime was 1.5 years, but one survived until 1977.

47. For example, a May 1963 design sketch of a satellite communication antenna (for MISER, the Microwave Space Relay) for the new command ship (which became the current *Mount Whitney*) showed a massive dish requiring a 20ft working circle, ie about 20ft in diameter.

48. SSC-3 deficiencies were cured in a redesigned and more widely used SSC-6.

49. DSCS reportedly carried over 80 per cent of the long-haul communications load during Desert Shield/Desert Storm. DSCS missions include the links to Air Force One and the airborne command posts and links to the Defense Department secure telephone systems as well as to NATO headquarters.

50. According to the official Air Force history, Secretary McNamara authorised preliminary work on a synchronous system in 1964; six contract definition study contracts were awarded in 1965. Compared to IDSCS, DSCS II would provide secure data and command circuits, greater channel capacity, and radiation protection. TRW received the contract in March 1969; six satellites would be launched in pairs by Titan IIIs.

51. Under the same programme the Navy ground antenna was enlarged from 40 to 60ft diameter, and the Army antenna from 9 to 20ft diameter. As of early 1999 only one ship still had the 60ft antenna.

52. This capacity provides SIPRNET (including JDISS [Joint Deployable Intelligence Support System], CTAPS [Contingency Theatre Automated Planning System], and GCCS) at 19.2-128 kbits/sec; JDISS at 16-64 kbits/sec; GCCS at 9.6 kbits/sec; TESS 3 (Tactical Environmental Support System: space-derived weather data) at 2.4 kbits/sec; voice channels at 16-72 kbits/sec; VIXS/VTC (video information exchange/teleconferencing) at 128 kbits/sec; and JWIC at 32-48 kbits/sec.

53. Higher power makes a higher data rate possible in each channel, because with more power per bit the receiver can recognise a shorter signal (a bit) against the background noise. The original DSCS III offered four 60-MHz wide channels, one 85 MHz-wide channel, and one 50 MHz channel, a total of 375 MHz of useable bandwidth. Later satellites (including SLEPs) have 405 MHz of useable bandwidth.

54. The 285lb satellite had two receivers (one standby) and two transmitters (one standby); uplinks were at 7976 to 7978 and 7985 to 8005.1 MHz, downlinks (3.5 Watts) at 7257.3 to 7259.3 and 7266.4 to 7286.4 MHz. Design life was 3 years. Skynet 1A, launched 21 November 1969, operated for 36 months. Skynet 1B was not successfully orbited.

55. The satellite has three 40 W X-band (SHF) transponders feeding four channels: (1) 135 MHz global, (2) 85 MHz European (narrow beam), (3) 60 MHz hemispherical wide beam, (4) 60 MHz narrow (3 degree spot: central Europe). The two UHF channels, using 40 W transponders, are each 25 kHz wide; they use a common helical antenna. Skynet 4D has 50 W X-band transponders; bandwidths for the four channels are 125, 75, 75, and 60 MHz. The 3 degree spot is steerable. The two UHF transponders work at 50 W.

56. Skynet 4A and 4B were scheduled for Shuttle launch in June and December 1986 despite strong French pressure to use Ariane, an important U.S. selling point being a promise to allow a British astronaut on the mission; the British also hoped that the United States would help make Skynet the standard NATO satellite. Once Shuttle launches had been suspended, the British shifted to Ariane for Skynet 4B and 4C, retaining the Shuttle reservation for 4A. As Shuttle problems continued, 4A was shifted to a Titan 3 launcher.

57. The incident raises an interesting net-centric point. The Argentine Super Etendard was cued by a Neptune, which detected *Sheffield* using her APS-20 radar. Because there was no way to link that radar's picture to the Super Etendard, the fighter-bomber had to switch on her own radar to acquire the target in order to feed fire-control data to the missile. Had the Super Etendard been able to use the radar picture generated by the Neptune – had it been fully netted with the Neptune's sensor – it could have fired from below the destroyer's radar horizon, and it would have given the destroyer no ESM warning at all, whether or not the ship had been communicating by satellite. Presumably the Neptune stayed beyond the destroyer's missile range. As it was, the Super Etendard popped up periodically to try to acquire the target, and when it did so the fleet picked up its distinctive radar signature (there is a claim that only bad staff work prevented a Sea Harrier from vectoring out towards the incoming Super Etendard, perhaps in time to save the *Sheffield*). The Soviet 'Backfire' apparently was expected to exploit remote data and to fire without itself acquiring the target. See Chapter 7.

58. The 740lb satellite has both narrow-beam (for Europe) and wide-beam coverage (for the Atlantic). The antennas share the four transponders (three 20 Watt and one 40 Watt). The three communication channels (17, 50, and 85 MHz) can all be used simultaneously. All channels are received through a widebeam antenna. The 50 MHz channel is transmitted widebeam, the others only by narrow-beam (5.4 × 7.7 degrees) horn. Design life was 7 years. NATO IIIB, launched as an orbiting spare, was lent to the United States to fill the east Pacific slot until at least four DSCS II satellites were available; it was returned in January 1979.

59. The SHF footprint covers the Atlantic, as well as Eastern Canada, and parts of Western Europe and North Africa. The UHF footprint covers the Atlantic as well as the Eastern United States, South America, Africa, and most of Greenland. Satellite design lifetime is 7 years.

60. The Hispasat programme was sold to the Spanish Parliament in 1988 on the ground that television capacity was needed to cover the Barcelona Olympics and the Columbus quincentenary. Given Spanish connections with Latin America, the satellite was needed to transmit television programmes there. As approved in April 1989, the two-satellite system (with a ground spare) was intended to provide three to five direct broadcast domestic television channels, two television distribution channels for North and South America, eight to sixteen television/communications channels, and two X-band government/military channels. The Spanish selected a design (Eurostar 2000) similar to the existing Matra Telecom 2 so as to have it in service quickly enough. The system's first military use was in March 1993, when a Spanish unit in Bosnia used a Loral SCT-10 terminal to set up twelve voice/data circuits. There are three (plus a backup) 40-Watt X-band (SHF) transponders with 20-40 MHz bandwidth.

Chapter 6: Finding Targets: Reconnaissance

1. It is not clear when this uncertainty was resolved, nor whether the Soviets lagged behind the Americans much in this respect. To the extent that the Earth was not precisely mapped prior to the advent of satellites, it was impossible to be sure of where even a land-based ballistic missile launcher was, hence of whether it would hit anywhere near its supposed target. This issue is entirely apart from some major later uncertainties, such as whether missiles typically tested in East-West firings would have functioned the same way in flights over the North Pole.

2. Even before satellites entered service, there was apparently evidence that the missile gap was fictitious. However, the key evidence was collected by early photo satellites. The key March 1961 report has now been released by the CIA. McDonald, *Corona*, includes reprints of several satellite photo reports.

3. The May 1955 National Security Council policy papers on the reconnaissance satellite programme advocate the scientific satellite specifically as a means of pressing the principle of a Freedom of Space equivalent to the Freedom of the Seas. They endorse a February 1955 proposal for an intelligence satellite. As it happened, when the IGY was first proposed in 1950 (by James van Allen, later of van Allen belt fame), its advocates were very interested in launching a scientific satellite. They feared that high-altitude research would stall when the supply of V-2 rockets then being used ran out. During the summer of 1954 Army Ordnance and the Office of Naval Research agreed to develop a satellite launcher for the IGY, using a stretched version of the Army's big Redstone as the first stage and bundles of Loki solid-fuel anti-aircraft rockets as upper stages.

ONR would provide the 5lb satellite. On 4 October 1954, on the recommendation of the International Scientific Radio Union and the International Union of Geodesy and Geophysics, the Special Committee for the IGY, meeting in Rome, endorsed the idea of a satellite as part of the international IGY programme. Early in 1955 NRL offered a new rocket, Vanguard, and this proposal went to President Eisenhower in March 1955. On 16 April 1955 the Soviets announced their intention to orbit a satellite. The United States announced its own programme on 29 July 1955.

4. All three services were refused permission to launch satellites before the IGY. For example, Wernher von Braun had developed a boosted version of his Redstone to test nose cones planned for the Army's Jupiter. There was apparently a real fear that he would 'accidentally' launch a satellite, and the Army checked to make sure his upper stages were unboosted.

5. Spinardi, *From Polaris to Trident*, p76.

6. Data for satellites from KH-7 on has not been made public. Data given here are from C P Vick's sketches, as published on the FAS Space Policy Project website. According to Vick, KH-8 used a 45in mirror for its camera, compared to 38in for KH-7; and KH-8 had large solar-power panels (KH-7 had none).

7. KH-10 (Dorian) was the abortive Manned Orbiting Laboratory (MOL). It is sometimes suggested that Hexagon was conceived as a back-up to this programme.

8. The initial SDS-A system comprised two satellites operated by the Air Force as part of its own Air Force Satellite Communication (AFSATCOM) system; each carried a 12-channel UHF transponder for SAC (and probably naval) communications. These satellites also carried nuclear blast detectors (NUDETS) sensors, and they may have been used to control other satellites. Publicly, they were described as links between the Air Force Satellite Control Facility at Sunnyvale, Ca. and seven remote tracking stations. Development began in 1973, and the first satellite was launched on 2 June 1976. The last of six satellites, all in Molniya orbits, was launched on 14 February 1987. It ran out of expected service lifetime in November 1995. These satellites were probably based on the Hughes Intelsat IV or IVA.

The follow-on SDS-B was designed for Shuttle launch. This programme began in FY84 (approved 1983), and SDS-B was probably based on the Hughes Intelsat VI commercial satellite. The first was launched, according to Day, 'Out of the Shadows', on 8 August 1989. The last of four was launched on 2 July 1996. In addition to its data relay equipment, SDS-B carries an IR package (Heritage/Radiant Agate) to detect ballistic missile launches, as an adjunct to the DSP system. The use of a SpaWar code name, Radiant Agate, suggests that this package may have been the Navy-sponsored aircraft detector whose existence was leaked in 1992 under the programme name 'Have Gaze'.

9. This system was developed for communications with NASA satellites and with the Space Shuttle. To relay data from ground-observation satellites such as LANDSAT, it had an unusually high data rate, 300 Mbits/sec. It was the first satellite capable of operating simultaneously in three frequency bands: S (ten transponders), C (twelve transponders), and Ku (ten transponders plus ten spares), the latter presumably for data relay. One satellite can communicate with twenty-six satellites simultaneously. In the 1980s TDRSS was increasingly used for military purposes. Unlike SDS, these satellites are in geosynchronous orbit on the equator, hence cannot link with low-flying satellites in the extreme north. The first Tracking and Data Relay Satellite (TDRS) was launched in 1983. The system's military role was indicated when the launch of TDRS-2 in March 1985 was delayed due to problems in its encryption (for intelligence relays) element. The first satellite in the series was launched on 4 April 1983, and the last of seven (one replacing a satellite lost with the *Challenger*) on 13 July 1995; on 23 February 1995 NASA signed a contract with Hughes for a replacement series of three satellites, to be launched by 2002.

10. According to Peebles, *The Corona Project*, KH-11 grew out of a 1969 interagency study of the value of a real-time reconnaissance satellite. It considered how a real-time satellite might have performed in recent crises (the Cuban Missile Crisis, the October 1967 Mid-East War, and the 1968 invasion of Czechoslovakia). It emerged that in each case it would have provided a President with invaluable early warning and with vital options. Development was ordered on a priority basis. The Air Force offered FROG (Film Read-Out Gambit), which was KH-8 with the film scanner conceived for Samos. The CIA wanted an electro-optical satellite which would store images in electronic form. Unlimited by film supply, it could operate for five years. In 1969 it was estimated that development would take seven to eight years and would cost $10 billion, about half as much as the Moon programme. President Nixon approved the CIA programme instead of FROG. The key technology was an array of charge-coupled devices (CCDs) used to create the electronic equivalent of an image formed by the satellite's optics. The main mirror of the first KH-11 was 92in in diameter; later satellites had slightly larger ones.

11. Harland, *The Space Shuttle*, p181.

12. Burrows, *This New Ocean*, claims that the cost of a Shuttle flight, about $0.5 billion, is roughly that of a new KH-11, but that it was argued that a combination of refuelling and other missions would make the flight cost-effective. It is not clear whether refuelling was ever done.

13. As quoted in the newsletter *Military Space*, Vol 16, No 5 (1 March 1999). Only one imaging satellite remains (Cosmos 2359, launched June 1998), and it is likely to re-enter and burn up soon. Against six EORSATs in orbit in 1990, only one (Cosmos 2347, launched on 15 December 1997) is left. It is not clear how much of the earlier electronic intelligence system is left in service.

14. The account of U.S. ELINT satellites which follows is based entirely on published *unofficial* and necessarily speculative material, including Mark Wade's on-line *Spacepedia* and John Pike's on-line material produced for the Federation of American Scientists. The Russian article cited here is the 1993 Andronov article. It presumably reflects official Soviet thinking based on satellite tracking and, probably, imaging.

15. It had been expected that signals would reflect off surface features of the Moon, creating considerable interference. Instead they returned coherently, presumably because the Moon actually did have an ionosphere. The initial experiment was made at wavelengths down to 1m.
16. The initial appropriation, $60 million, was made in 1959; more money was requested in 1960 and in 1961. Congress capped the project at $135 million, which the Navy estimated was two-thirds of what was needed. It was terminated in 1962.
17. Details of Grab are taken from a booklet, 'Grab: Galactic Radiation and Background: First Reconnaissance Satellite' produced by NRL as part of its 75th anniversary celebration and from Day, 'Listening from Above'.
18. Eisenhower was very sensitive to the possibility that the ELINT mission might be embarrassing. At his behest Grab was designed so that its ELINT system could be turned on or off from the ground; according to Day, 'Listening from Above', he allowed it to be turned on only twenty-three times, and never on successive passes. The successor Kennedy Administration imposed no restrictions whatever.
19. According to Day, 'Listening from Above', during the 6 August 1961 Vostok 2 flight, Soviet cosmonaut Gherman Titov lost contact with the ground. The Soviets turned on all available tracking radars – including their new ABM system.
20. According to Day, op cit, as early as 1953 RAND had pointed out to the Air Force that because Soviet territory was so vast, it would be impossible to intercept inland radar signals using aircraft or ground stations. When RAND proposed the television-type satellite in 1954, it also mentioned the desirability of a Ferret (radar detector), and by 1958 that mission had been folded into the service's WS-117L satellite reconnaissance programme. The Air Force eventually developed three parallel Ferret systems, F-1, F-2, and F-3. By this time NSA had already experimented with an intercept receiver on board a sounding rocket; Day quotes the 1954 RAND study as mentioning a launch which caught the end of the national anthem, as it was broadcast by a U.S. television station.
21. James Wade's on-line *Spacepedia* lists a total of fifty-four Subsatellite Ferrets launched between 18 March 1963 (a failure, launched with the first KH-6) and 8 August 1989. Because Ferret programmes were highly classified, this list probably confuses several different ones. Even the identification of specific sub-satellites with the Ferret mission is controversial. The earliest such satellites weighed 80kg, followed from 1965 on by a 60kg series. Wade mentions that a 27 June 1963 flight ('Hitch Hiker 1') was the first successful flight of a P-11 bus used mainly for radar monitoring. This particular satellite was described as a radiation data gatherer, and it was in an unusual highly elliptical orbit, 323 × 2506km (82.1 degrees). Most of the Ferrets were in much lower orbits, *eg* 156 × 505km (92.9 degrees) for one launched 6 July 1964. Many were in circular orbits at about 500km altitude. Three launched in 1972-5 (and one possibly launched in 1980) were in high circular orbits (about 1400 × 1400km, inclination about 95 degrees); at least one other may have been inserted into such an orbit. The last in the series, launched 8 August 1989, may have been part of a separate Cobra Brass measurement and signature intelligence experiment.
22. Among the first public references to NSA collection of Soviet microwave communication traffic by satellite is Bamford, *The Puzzle Palace*, p197, referring to the quasi-geosynchronous Rhyolite, first launched in 1973.
23. A released CIA report on the build-up specifically mentioned increased bomber activity in the Far East. The reference in the text is from the 1993 Russian article.
24. Many satellites were launched into highly eccentric transfer orbits, then boosted into their quasi-stationary orbits. Published orbital data often refer to the transfer orbits rather than to the final orbits, making it appear that the satellites were in Molniya-type orbits.
25. Zircon was disclosed by a British journalist in January 1987; he claimed that the Skynet 4 communications satellites were covers for ELINT satellites whose product was to be shared with the United States. The argument was that the published location for a Skynet satellite over Asia was ludicrous from a communication point of view, but rational for intelligence-gathering. Total projected cost for a two or three satellite programme was £400-500 million, of which £70 million had been spent by the time the programme had been revealed.
26. According to Richelson, *The U.S. Intelligence Community*, p188, the use of Rhyolite for COMINT was dramatically increased on the specific orders of President Nixon and Dr Henry Kissinger, once they became aware of this capability. According to Richelson, the satellites were used to intercept Soviet and Chinese telephone (*ie* microwave link) and radio communications at VHF, UHF, and microwave frequencies, including tactical radios. Communications were also reportedly intercepted in China, Vietnam, Indonesia, Pakistan, and Lebanon.
27. After the rules had been relaxed, satellite photographs were used to plan the 1980 mission to rescue the U.S. hostages in Iran. Some of them were left behind in the Iranian desert when the U.S. force fled – and were recovered by the Iranians. They were later published, revealing the capability of the KH-11 photo satellite, just as those who had resisted tactical use of satellite data had feared would happen.
28. Much of the early history of the programme is still classified; details are available only of the Block 4 and Block 5 satellites launched from 1965 on. The first of thirteen Block 4A satellites was launched 19 January 1965 (the last was 23 July 1969). These satellites were followed by three Block 5A (launched 11 February 1970 through 17 February 1971); then by five Block 5B (14 October 1971 through 16 March 1974); then by three Block 5C (9 August 1974 through 19 February 1976); then by four Block 5D (11 September 1976 through 15 July 1980); and then by nine Block 5D-2 (21 December 1982 through 4 April 1997).
29. During the Cold War the U.S. Defense Department considered Dushanbe, north of the Afghan border, a laser ASAT test site. The Soviets

claimed that it and a similar facility at Sary Shagan were only for satellite tracking, and the issue remains unresolved.

30. Reportedly the Soviets also feared that the Shuttle would be able to seize their own low-flying satellites in orbit. However, that fear would not explain why their 'space fighter' was abandoned after the Shuttle crash; it was never likely that the Shuttle programme would be abandoned altogether.

31. Korolyev was interested in a scientific satellite, and in 1953-4 he lobbied hard for it. Satellite capability was included in the specification for the R-7, development of which was approved in 1954. However, there was no formal governmental authorisation for a satellite until a decree was issued on 30 January 1956 (Gorin, in McDonald, *Corona*, pp86-7). Two parallel programmes were approved, 'Object D', a scientific satellite (Sputnik 3), and Object OD-1, an imaging satellite. Sputnik 1 (PS) was a spin-off of the Object D programme.

32. Reconnaissance satellites were developed by Branch 3 of Korolyev's OKB-1 (in Samara). It eventually became the Central Specialised Bureau of Space Hardware (TsSKB).

33. At first, these satellites also had Baikal, a television read-out system which scanned film taken by the camera. It proved unsatisfactory. The typical camera arrangement (Ftor-2R) employed four fixed cameras: three with 1m focal length and one with 0.2m focal length, the latter for low-resolution reference photographs with which to locate images taken by the high-resolution cameras. The high resolution cameras were angled so that their frames formed a continuous band across the ground track. Each camera had 1500 frames. Normal photo modes were single (three-frame strip across the ground track), a series of frames along the ground track, and a single picture on either side of the ground track. At an altitude of 200km, swath width was 180km. Ground resolution was reportedly 10-15m. Cameras may have been arranged to produce stereo images. In addition to its cameras, Zenit-2 carried a Kust-12M ('Bush') ELINT system which tape-recorded radio and radar signals for read-out when the satellite communicated with its control centre. From a circular orbit at 200km altitude, Zenit 2 could photograph 8 million square kilometres. The parallel Zenit-4 provided more detailed images, using a pair of cameras with 3m focal length, to cover a strip 46km wide, with a resolution of 3-4m. It also had a low-resolution camera for registration. It lacked the ELINT capability of Zenit-2. At 5500kg Zenit-4 was much heavier than Zenit-2, and therefore required a new launch vehicle, R-7/11A57 (rather than 8A92). Zenit-4 through -6 were sometimes described as second-generation satellites, but in fact they were modified versions of Zenit-2, and they were very nearly contemporary with it. The first Zenit-4 was launched on 16 November 1963, the last on 4 August 1970, a total of seventy-four having been launched. Zenit-2 operations ceased at about the same time (12 May 1970) after eighty-one launches (fifty-eight full successes, eleven partial successes, twelve failures).

Work on the improved Zenit-2M and -4M began in 1968, and they were accepted into service in 1970 and in 1971, respectively. Work on the next update, Zenit-2MK/-4MK, began in 1969. Zenit-4MT was a special mapping (low-resolution) satellite, using three overlapping cameras with 35cm focal length, to produce stereo images. It used special devices to orient its photographs precisely, and it also photographed stars to determine its position. Zenit was so successful that four series of improved versions were developed (in 1972, 1976, 1978, and 1983) with better cameras and film. Improved versions of the Zenit series used solar batteries, which extended their lives to thirty days. They carried three sets of film, two for capsule return, the third to be brought down with the satellite itself. These improved versions of Zenit, tested in 1975-6, used the long-lens camera of Zenit-4 and could look up to 30 degrees to each side.

34. After Khrushchev's demise, Chelomey's main surviving patron was Minister of Defense Grechko. He was opposed by a major engine designer, Glushko, and by Dmitri Ustinov, the Communist Party's main industrialist (and one of Khrushchev's old enemies). Grechko suffered a heart attack early in 1976, after which Ustinov took over the Ministry of Defense – and, according to Mark Wade's *Spacepedia* entry on Almaz, proceeded to strangle Chelomey's projects. Almaz was finally ordered scrapped in 1980, but Chelomey preserved the completed space stations by labelling them 'radioactive matter'. The sudden cancellation of Almaz-K was apparently an attempt to kill off another of Chelomey's programmes; against orders, the satellite was left in place at Baikonur. It was resurrected after both Ustinov and Chelomey had died, the latter being forced to retire in October 1983.

35. Radio amateurs could not detect any transmissions from fifth-generation photo satellites; that suggests that all transmissions were up- rather than downlinked. In addition to Geizer, the Soviets developed the Luch (11F669) satellite (Altair system) for space-to-space communications (comparable to the U.S. TDRSS). Luch was associated with the space station and Buran programs, and is used for some military satellite communications. It also demonstrated SHF retransmission of telephone and telegraph signals. Lifetime was five years. The first of five was launched on 25 October 1985; on 29 March 1986 it ran communications tests with the Mir spacecraft. This satellite (at 95 degrees East) ceased operating in September 1987, but two more were launched (26 November 1987 and 27 December 1989). Together they provided Mir with the ability to communicate with its mission control centre 70 per cent of the time. The three early Luch satellites are sometimes described as transponders on Raduga satellites. Two more Luch were launched later: 16 December 1994 and 11 October 1995, to occupy the system's two stations, at 95 degrees East and 77 degrees East.

36. According to Yutkin and Mozzhorin in Minaev (ed), *Sovetskaya Voennaya Moshch'*, the Tselina series was preceded by a first-generation lightweight (415kg) solar-powered ELINT satellite, developed by the Yangel design bureau (which was also responsible for the entire Tselina series, in more recent times under the name NPO Yuzhnoe, the Southern Scientific Production Association). Presumably this series represented the pre-Tselina experimental stage. These satellites were launched by an Inter-

cosmos rocket (Cosmos-2M, modified R-12/SS-4 missile) into a 530km orbit. It was intended to measure the parameters of surface anti-aircraft and anti-missile radars (frequency, strength, waveform, modulation, etc.). The satellite control system could measure its orbital parameters, but the satellite itself was not precisely oriented, so it could not localise its targets. By way of contrast, Tselina used interferometry to determine the precise location of the sources of the signals it picked up. A list of Soviet space launches shows a series of Cosmos-2M military shots, but stated weights (325kg) are too low, and stated orbits were not circular (typical parameters were 210 × 520km at an inclination, for all the satellites, of 48.4 degrees; the orbits may have been circularised after the initial launch): Cosmos 93 (19 October 1965), Cosmos 95 (4 November 1965), Cosmos 106 (25 January 1966), and Cosmos 116 (26 April 1966). The stated role of these openly military satellites, 'investigation of the upper atmosphere', would have been a typical cover for intelligence satellites. Note, too, that the first attempt to launch a Tselina-O came in 1967, immediately after the end of this series.

37. To the authors of the history of Soviet military space systems, acceptance of Tselina-D in 1976 marked the end of an experimental period. Between 1971 and 1976, fourteen Soviet military space systems had been accepted for service: Tselina-O and -D, US-A and -P, three versions of Zenit (Zenit 2M, Zenit 4M, Zenit 4MT), a geodetic system (Sfera), Tsiklon and Tsiklon-B command/control systems, two versions of Strela, the Korund communications system, and an ASAT system (Lira).

38. When Tselina-2 was encountering launch problems in 1990-1, the Soviet military newspaper *Red Star* stated that it was used to verify START treaty compliance. However, it is not clear how a satellite with very intermittent coverage could pick up telemetry generated in a missile test conducted at an arbitrary time (indeed, presumably timed to avoid exposure to the satellite).

39. Whereas Strela-1 and -1M were quite small (40 to 50kg) and were launched in groups, the approximately contemporary Strela 2 (11F610) was massive (770kg). Its typical orbit was roughly circular, at an altitude of about 700 to 800km (74.1 degrees). Of fifty-seven launched between 28 December 1965 and 20 December 1994, three (16 November 1966, 15 June 1968, and 27 June 1970) failed. The three earliest satellites orbited (1965-8) went into much lower orbits at 56 degrees inclination: 176 × 179, 384 × 399, and 340 × 348km, perhaps indicating that their receivers could not pick up signals from higher altitudes.

Strela-1 was first launched on 18 August 1964, eight satellites being launched together into an elliptical orbit (200 × 760km, later 260 × 1700km; 56 degrees; some also went into 550km or 1500km circular orbits). The last in this series was launched 18 September 1965. The first Strela-1M launch was on 25 April 1970, into a 1400km circular orbit at 74 degrees inclination, which was roughly standard for the series (again, eight satellites per launch). The last in the series was launched on 6 March 1981. Strela 3 (17F13) was presumably a second-generation communications satellite (230 kg), launched in groups of six. The first group was launched 15 January 1985, into a 1400km circular orbit at an inclination of 82.6 degrees; the most recent was 15 June 1998.

40. Raduga is always publicly characterised as a military/government communications satellite. A dummy test vehicle was placed in geosynchronous orbit on 26 March 1974, followed by the first of thirty-two satellites on 22 December 1975. Effective satellite lifetime was one to two years because the 1940kg satellite lacked any means of stationkeeping. Each satellite carries up to six 6/4 GHz transponders (15 W) plus Gals (8/7 GHz; follow-on system which appeared in 1994), Luch P (14/11 GHz), and Volna (1.6/1.5 GHz, compatible with INMARSAT) transponders. It is not clear whether some Radugas carried Luch P (14/11 GHz) transponders to support the Altair space communications system. A new Raduga 1 series appeared in 1989. The launch vehicle is a Proton. Raduga stations are at 12, 35, 45, 49, 70, 85, and 120 degrees East, and at 170 and 25 degrees West. Note that 25 degrees West is over the Eastern Atlantic, in roughly the location ascribed to the geosynchronous ELINT satellite.

A second geosynchronous system, Gorizont, first appeared in December 1978; it was based on experience with Raduga. Gorizont was intended to carry both military and civil traffic. It has six (one 40 W, five 15 W) 6/4 GHz transponders, a Luch transponder (with backup), and a Volna transponder (with backup). Gorizont is apparently primarily for civilian traffic and television (its 40 W transponder is for television distribution via the Moskva system); satellites have been leased to foreign users, and Gorizont is part of the international Intelsat system. However, it also carries the U.S.-Russian hotline previously carried by Molniya. Gorizont stations are at 40, 53, 80, 90, 96.5, 103, and 140 degrees East and at 14 and 11 degrees West. The current Express system is the Gorizont follow-on. There is also Ekran, for television broadcasting to cable users in Siberia and the Russian north and east (first launched 26 October 1976).

41. Meteor was first launched on 28 August 1964. This ten-satellite series (the last of which failed, on 1 February 1969) was followed by thirty-one Meteor-M (launched 26 March 1969 – 10 July 1981), by twenty-one Meteor-2 (launched 11 July 1975 – 31 August 1993), and by six Meteor-3 (launched 27 November 1984 – 25 January 1994).

42. Development (by Martin) began in March 1958, initially to prove the feasibility of an air-launched ballistic missile, but late in the programme the vehicle was tested as an anti-satellite missile. A late two-stage version was fired at Explorer VI as it passed through apogee near Cape Canaveral. Evaluation of tracking data indicated that the missile had passed within 4 miles of the satellite. Martin then proposed a dedicated anti-satellite missile, but by this time the Air Force preferred a surface-based system.

43. A Technical Development Proposal (TDP) for Early Spring, using a Polaris A-2 missile as booster (Advanced Development Objective W17-11X), was issued in November 1962.

44. There was some official interest in non-nuclear ASAT. Work began in 1964, the idea being to add a

homing vehicle to the existing Thor booster. A prime contract was let in June 1967 and test launches were scheduled for 1969. By 1967, however, money was tight due to the Vietnam War, and planned FY68 spending was halved. Then the programme was cancelled and the remaining effort shifted to anti-ballistic missile research. Even so, the Air Force conducted some special tests in 1970. It also prepared a new ASAT plan, but money was not forthcoming. Paper studies prepared at this time were, however, used in later ASAT projects. The Air Force was also interested in using the spaceplane it was trying to develop (Dynasoar, X-20) to inspect satellites.

45. SPASUR employs continuous-wave transmitters and separate receivers, the receivers being about 250 miles from the transmitters. Each transmitter creates a fan of beams. By avoiding pulse operation, the system avoids the usual problems of having a maximum unambiguous range and thus can unambiguously track very distant objects. Compared to the ballistic-missile warning radars, it uses a beam shape extending up rather than out towards likely missile approach paths; thus the missile warning radars cannot effectively detect very high-altitude targets (like satellites).

46. According to Richelson, *The U.S. Intelligence Community*, p242, by 1987 this system reported in four formats. Large Area Vulnerability Reports (LAVR) referred to specific operating areas rather than specific naval formations. Satellite Vulnerability Reports (SVR) were keyed to particular ships or formations transiting or operating outside the areas covered by the LAVRs. Safe Window Intelligence (SWINT) told ships when they were safe from surveillance. CHARLIE data allowed ships to compute their own vulnerability, using an onboard Reconnaissance Satellite Vulnerability Computer (RSVC) programme. Under a CHAMBERED ROUND programme begun in 1988, the Naval Space Command provided deployed units with tailored tactical assessments of hostile space capabilities and with suggested countermeasures. Note that all of these reports entail considerable detailed knowledge of enemy satellites. For example, the effective detection range of a radar satellite will depend on its waveform, on its power, on its beam pattern, on the heading of the target ship, and on the sea state.

47. In 1969, for example, the supplementary radars fell into two categories. In one were big long-range missile trackers with fixed 'fans' of beams: FPS-17s used to detect Soviet missile tests (at Diyarbakir, Turkey, and at Shemya, Alaska), FPS-50 ballistic missile early warning radars (BMEWS) at Thule, Greenland; at Clear, Alaska; and at Fylingdales, England; and the FPS-85 phased array at Eglin, Florida, intended to detect submarine-launched ballistic missiles. In the other category were big mechanically-scanned missile trackers: FPS-49s which supported BMEWS (Thule and Fylingdales); an FPS-92 and an FPS-99 at Clear; an FPS-79 at Diyarbikar; an FPS-80 at Shemya; and a GPS-10 at Ko Kha, Thailand. The Carter Administration (1976-80) upgraded the space surveillance system by creating a Pacific Radar Barrier (PACBAR) comprising three big mechanically-scanned radars: a 60ft dish at San Miguel in the Philippines (Cobra Talon), a 30ft dish in Saipan, and the ARPA Long-Range Tracking and Instrumentation Radar (ALTAIR) at Roi-Namur (Kwajalein). Under this programme the Baker-Nunn cameras were replaced by GEODSS electro-optical ones. The system was to be tied together by a new centre at Cheyenne Mountain. After 1991 the system was gradually cut back. For example, Cobra Talon was shut down in June 1991, and the Saipan radar was cancelled in 1992. ALTAIR is available only part-time, since it is intended mainly to support ballistic missile defence development. The Air Force also operates two global passive systems, the Deep Space Tracking System (DSTS) and the Low Altitude Space Surveillance System (LASS), neither of which is effective against quiet or LPI-transmitting satellites.

48. Until the mid-1980s Baker-Nunn cameras were used. They have now been superseded by a Ground Based Electro-Optical Deep Space Sensor, the difference presumably being that a Baker-Nunn used either film or a human operator, whereas GEODSS is fully electronic. Even satellites in geosynchronous orbits can be imaged (in 1985 a GEODSS photographed a Navy Fleet Satellite).

Chapter 7: A New Kind of Naval Warfare

1. This was a unified system (shore artillery and offshore craft and aeroplanes) co-ordinated by radio. It was the first such system, since the necessary radios had only just become available. Between 1922 and 1933 the Soviets had very close relations with the German military. It is possible that the close control style of German U-boat operations during the Second World War was also inspired by the Flanders experience.

2. The primary roles of these units were attacks on major enemy ships at sea and offensive mining of enemy ports; they were also used to destroy enemy naval shore facilities and to support amphibious and anti-amphibious operations. At the beginning of the 1950s the standard Soviet mine-torpedo aircraft were the Tu-2T, the Il-4T, and the American A-20G obtained under Lend-Lease. Although the Tu-2 bomber had been developed and produced in wartime, Tu-2T was essentially a post-war development, mass-produced in 1946-8. It could carry two torpedoes. A reconnaissance version participated in naval operations against Japan in 1945. Development of the Tu-2T was protracted; the last regiments received it only in 1952, by which time the replacement Tu-14 jet bomber (NATO 'Bosun') was already in production. Meanwhile the A-20G light torpedo bomber was replaced by the Il-28T (NATO 'Beagle'). These new jet aircraft were armed with the new rocket-powered stand-off torpedo (RAT-52). Beginning in 1956 the unsuccessful Tu-14 heavy torpedo bomber was replaced by the large Tu-16 (NATO 'Badger'), of which Tu-16T was the torpedo version. In 1954 there were ten mine-torpedo and twenty fighter divisions, with ten independent reconnaissance regiments. In all, there were 120 regiments and twenty-nine separate squadrons and

groups. At the end of 1958 the U.S. Air Force estimated that Soviet Naval Aviation was 60 per cent fighters (attack units equipped with Tu-2s having been phased out in 1957), 20 per cent mine/torpedo, 10 per cent reconnaissance (including ASW seaplanes), and 10 per cent support aircraft.

3. Wartime attacks were typically made from 5000ft, using a parachuted pattern-running torpedo dropped nearly simultaneously from several directions. Initially the new jet bombers (Tu-14s and Il-28s) dropped singly, but beginning in 1955 they worked in formations of three to five, dropping their torpedoes from altitudes of 13,000 to 19,000ft (about 4000 to 6000m). By the late 1950s torpedoes were often dropped by radar. It is only fair to point out that other navies also tried to increase stand-off range. For example, the Royal Navy fitted its torpedoes with a gyro-controlled air tail, so that it could travel several miles in the air before hitting the water. All navies were aware that co-ordinated torpedo attack tactics (*eg* hammer-and-anvil) gave the greatest chance of success. The Soviets took such ideas much further than anyone else.

4. RAT-52 development began during the Second World War. The weapon weighed only 627kg (warhead weight was 243kg). Dropped like a bomb, it was set to run at a depth of 2 to 8m. After leaving the aeroplane, an airbrake opened to bring the weapon into a vertical attitude, and a parachute opened at a height of 500m. The parachute was jettisoned when the torpedo hit the water. Speed in the water was 58-68kts, and time elapsed from hitting the water to reaching the target was only 35 seconds. However, torpedo range was only 550 to 600m. RAT-52 could be dropped at any height from 1500m up, and at any speed up to 800km/hr. The estimated hit probability, based on tests, was 17 to 38 per cent. RAT-52 was accepted into service on 4 February 1953; it equipped Tu-14s and Il-28s. The weapon passed its operational tests in the Black Sea Fleet in September-October 1953 (fifty-four were fired). According to a contemporary (1957) Soviet manual, it would take eight RAT-52 hits to sink a *Midway* class carrier or an *Iowa* class battleship, or two to sink a *Gearing* class destroyer. However, given the weapon's inaccuracy, it would take numerous shots to make those hits: fifty-eight shots from 7000m, for example, for the eight hits. The judgement at the time was apparently that it would be wise to retain conventional torpedoes, even though they had to be dropped at low altitudes and at low speeds (about 400km/hr). Thus the pre-war 45-36 was modified (as 45-56) with a drag ring. There was also a modernised high-altitude torpedo, 45-54. RAT-52 was exported to China (1954) and a copy was produced as Yu-2. RAT-52 was gyro-guided in the water, and had a contact fuze. Reportedly it tended to porpoise, to break up upon hitting the water at high speed, or to lose its heading. When the mine-torpedo regiments were retired in 1961, some of their Il-28 bombers were transferred: twelve to Indonesia (with fifty RAT-52) and four to Egypt (with 90 RAT-52). In January 1966 the Council of Ministers approved the transfer of six Tu-14T to Egypt (they were delivered by Black Sea Fleet pilots in September 1967).

5. Plans made about 1954 called for a total of 425 Tu-16s, of which the first eighty-five were to join the Northern Fleet in 1956, followed in 1957 by 170 aircraft for the Black Sea and Pacific Fleets, and then in 1958 by 170 more for the Baltic Fleet. In fact in February 1955 the Baltic Fleet air arm began to convert two regiments of Tu-4s to Tu-16Ts. The first four aircraft were accepted on 1 June 1955. The initial aircraft, which were bombers, had to be specially converted to torpedo bombers, armed with four or six RAT-52 stand-off weapons or with six conventional high-altitude torpedoes (45-54). The first operational Tu-16Ts were accepted in May 1957; in all, seventy to eighty aircraft were converted. One problem was that, given the high speed of the Tu-16, a target had to be detected at much greater range (100-120km) in order to be hit. The solution was a special radar-tracking bomb sight (RBP-4). Alternatively, the Tu-16T could carry mines. The Soviet Navy considered obtaining an even more powerful torpedo bomber, in the form of a converted 3M (NATO 'Bison'), but that never happened. Beginning in 1962 some Northern Fleet Tu-16T were modified as ASW aircraft (Tu-16PL), followed the next year by Pacific Fleet Tu-16T bombers. They were equipped with active-passive sonobuoys and with AT-1 homing torpedoes. In this form these aircraft served until 1968. They were replaced by Be-12s and Il-38s. A similar Baltic Fleet conversion of Il-28T torpedo bombers (Il-28PL) was unsuccessful; plans to form two regiments of these aircraft (1966) were rejected, even though the Il-28 could reach a datum more than twice as quickly as a standard ASW aircraft, the Be-6 flying boat. In 1965 other Tu-16T were rebuilt as rescue aircraft (Tu-16S), carrying lifeboats.

6. On 7 May 1947 Chelomey was formally ordered to develop 16Kh for the Tu-2 and Tu-4 bombers; the missile was test-launched by Tu-2s in 1948.

7. 10-KhN weighed 3500kg and carried an 800-1000kg warhead. With a D-3 or D-5 pulse jet, maximum speed was 565-600km/hr. Range was 240km; the missile could be aimed to hit within a 20 × 20km area. 16-Kh weighed 3500kg and carried a similar warhead. With two D-14 or D-16 pulsejets, maximum speed was 858-900km/hr. Range was 100-240km, and the missile could be aimed to hit within a 7.6 × 10.7km area.

8. A special committee on jet engineering was formed, headed by G M Malenkov (who succeeded Stalin in 1953); his assistants were D F Ustinov and I G Zubovits. Ustinov was later Defence Minister under Brezhnev. This committee was responsible for both jet aircraft and jet and rocket-powered missiles. Special responsibility was placed on three ministries: Arms Production (in the past, army artillery), for liquid-fuelled rockets; Agricultural Machinery, for solid-fuelled rockets; and Aircraft Production, for jet-powered (cruise) missiles. The Defence Ministry formed a special research institute and set up the missile range at Kapustin Yar.

9. FX-1400 sank the Italian battleship *Roma* and badly damaged the battleships *Warspite* and *Italia*; it also sank the destroyer *Janus*, and damaged the cruisers *Savannah* and *Uganda*. It also sank several merchant ships. Hs 293s sank the sloop *Egret* (the first warship to be lost to a guided missile, on 27 August 1943), the cruiser *Spartan*, and the destroyers *Inglefield*, *Dulverton*, and *Rockwood*. Effective range

was 4000 to 20,000m (4400 to 22,000yds), and the missile glided towards its target at a speed of about 300 to 400kts.

10. The swept-wing Shtorm was powered by an underslung RD-700 (RD-1) turbojet, to achieve a range of 80km (effective range 40km, about 21nm) at a speed of Mach 0.9-0.95. Three alternative guidance techniques were proposed: active radar, passive infra-red, and television. The missile flew at a maximum altitude of 1500m (about 5000ft) and attacked at 9m altitude. Dimensions were: length 8.3m, fuselage diameter 1.3m, height 3.5m, span 6.9m (sweep angle 35 degrees). Pending development of an appropriate solid-fuel booster, a piloted 'analog' (LM-15) was launched from a Pe-8 bomber (1950-1). There was also an unmanned powered version, LM-15, which in 1952 became the first powered Soviet missile to be launched from a bomber.

11. According to a recent Russian account, the pretext was that Bisnovat was 'too Jewish'; in 1953 Stalin was beginning a major anti-Semitic purge. Russian accounts of the cancellation of Chelomey's OKB-51, which occurred at the same time, and which also eliminated a potential competitor to KS, emphasise that the reason was political rather than technical.

12. Development was formally ordered on 14 April 1948. Both versions were solid-fuelled. The initial Shchuka (FAMT-1400) had straight wings with rather large ailerons, and a butterfly tail with moderate dihedral. It was 6.5 to 6.8m long, with a body diameter of 0.7m and a wingspan of 3.8 to 4.3m. Launch weight was 2 tonnes. The rocket was launched at an altitude of 2km, and it approached the target at 10-12m altitude. The warhead was carried inset under the missile's nose. The follow-on Shchuka-A (FAM-1400A) had a range of 60 rather than 15-20km, using a single-chamber liquid-fuelled rocket motor. Speed was 320m/sec (1250km/hr). Dimensions were similar to those of the original Shchuka (6.7 × 0.7 × 4.0m, weight 2 tonnes), and the contact-fuzed TGAG-5 warhead weighed 620kg (320kg explosive). Development was formally ordered on 27 December 1949. The operator tracked the missile optically, sending it into a terminal dive about 10km from the target. The warhead was designed to separate from the missile about 60m from the target, at a speed of about 200m/sec. Initial trials, using only a pneumatic autopilot, were conducted in 1949, followed by trials using an electric autopilot and captured German radio command guidance (as in the Hs 293) in 1950. A fourth series of rockets using a Soviet-made radio command system (KRU-Shchuka) was tested in August-September 1951. Series production for further tests was ordered on 23 September 1954, and a version carrying a 900kg warhead was ordered for use against ground targets. Shchuka-B (FAM-1400B) was developed under the same order as Shchuka-A. It had prominent wingtip plates. Guidance range was 10-20km, and maximum range was 30km; speed was 1030km/hr. The missile was designed to fly at altitude of 2-10km, and to attack at 60m altitude. Dimensions were 6.8 × 0.7 × 4.55m, and weight was 1.83-1.9 tonnes. The radar seeker passed its tests in 1948-52, but integration proved difficult, and in 1953 the missile was tested without its seeker (but with its radio altimeter; it maintained an altitude of 30m). A new seeker (RG-Shchuka) had to be developed. Five successful tests followed in 1954-5. The programme was cancelled by a Council of Ministers decree dated 3 February 1956.

13. In 1955 Chelomey, having been given a new design bureau (OKB-52) after having been shut down in 1953, looked to GSNII-642 to develop active radar guidance for the big cruise missile he was developing. Presumably he needed an alternative to the system then being developed for the rival MiG bureau and its K-10S missile. On 6 November 1957 GSNII-642 was assigned as a filial organisation to OKB-52, and under an 8 March 1958 Council of Ministers decree it was assigned development of the P-6 and P-35 guidance systems. In this system sense Shchuka was the direct ancestor of the guided version of 'Shaddock'.

14. In 1949 GSNII-642 designed a glide bomb, PAB-750, presumably based on wartime German work. The following year requirements were levied for three guided bombs comparable to FX-1400 (Fritz-X): UB-1600B (armour piercing, like Fritz-X), UB-2000, and UB-6000, the number in each case indicating the weight in kilograms. On 15 October 1951 a decree authorised work on UB-2000F (Chaika) and UB-5000F (Condor) to arm, respectively, the Il-28 and Tu-4/Tu-16 bombers. The first bombs were completed in 1953 and fifteen of the twenty bombs built were tested between November 1954 and February 1955; UB-2000F was accepted into service on 1 December 1955 as UB-2F (4A22). An Il-28 could carry one in its bomb bay; a Tu-16 could carry two underwing. The following year 120 bombs were made and twelve Il-28s (presumably of naval aviation) modified to deliver them. Further versions ordered developed in 1955 were an IR-guided Chaika 2 and a radar-guided Chaika 3, but neither seems to have been developed. The larger Kondor was tested beginning in September 1954, and a television version (comparable, presumably, to the U.S. Razon) was built. By this time true guided missiles were being developed, and by about 1957 the guided bombs had apparently been abandoned as a poor alternative. Chaika weighed 2240kg (warhead 1795kg), was released at an altitude of 5 to 15km, and had a speed after release of 400 to 1200km/hr. These data are from Angelskiy, 'Smertonsnaya Chaika'. Chaika figured in planning sheets prepared for anti-ship attack about 1958, in which it was compared with RAT-52.

15. The Germans who developed Kometa were assembled for the project in October 1947. The system was first tested in 1949. Problems in developing the missile were solved by using a manned version (flown 1951). All but two of the Germans had left by 1952. Kometa was well known to the West because in the early 1950s there was a systematic programme of interviews of returning Germans; some of the reports are in the British Public Record Office.

16. The first Northern Fleet firings were in 1955. During 1956, of forty-two missiles fired, thirty hit. By 3 October 1957 the 124th Regiment had been redesignated a Long-Range Mine-Torpedo Regiment. On the target range, without electronic interference, single Tu-4s managed an overall hitting rate of 81 per cent, at a range of 60-65km and a

launch speed of 340-360km/hr. A standard regiment comprised twelve Tu-16KS, six Tu-16ZShch (tanker), plus Tu-16SPS jammers and an An-2 hack.

17. This problem was overcome: in 1961 eight aircraft fired together, making six hits; their courses were 20 to 30 degrees apart. This type of simultaneous attack was considered very valuable for saturating target defences. It was more difficult to increase launch altitude; the missile's RD-500K engine behaved badly even when started at 4500 or 5000m. It was discovered that if the missile were launched at 6000m or more, it would have to dive sharply away from the carrier, and would not be able to track the target's reflected radiation at sufficient range (at too steep an angle the target would merge with the sea clutter). For example, a missile launched at 10,000m at a target 90km away would not be able to switch to semi-active homing until it was only 24km from the target. Conversely, if the missile were fired at a shallower angle, from 2000m, it would shift to semi-active homing at a range of 43km. The standard anti-ship warhead, which may have been intended to explode beneath the target's waterline (as in the contemporary Shchuka), was FK-1. Warhead weight was 500kg. The surface-launched FKR-1, but probably not the air-launched KS, had a nuclear warhead.

18. U.S. naval intelligence first detected characteristic Kometa electronic emissions in 1955. Most Komets were phased out of Soviet service in 1968, although as late as mid-1973 the missile was still reported in service with the Black Sea Fleet. Beside the Tu-16KS ('Badger B') built for the Soviet fleet, another twenty-five were exported to Indonesia in 1961. Initially the aircraft were manned by Soviet pilots. Indonesia was then involved in a confrontation with the Netherlands over West Irian, which it claimed, and the transfer was intended to deter the Dutch from any naval intervention. A recent Russian account of the Tu-16 describes the force as a 'fleet in being'. After the confrontation, Indonesian pilots took over. Later thirty Tu-16KS were transferred to Egypt. Tu-16s went to Iraq at about the same time; at the outbreak of the 1967 Middle East War, six Iraqi Tu-16s were at Habbaniya. At least sixteen Egyptian aircraft were destroyed on the ground; one which managed to take off was destroyed when an Israeli pilot crashed his burning aircraft into it. The Iraqi aircraft tried to raid Tel Aviv, but three had to turn back because of technical problems, and the only bomber which continued, that of the squadron commander, missed the target and attacked a coastal town, Nataniya, instead. It was hit by an Israeli air-to-air missile (R 530), and then finished off by ground fire.

19. Politics was a major factor in many Soviet programme decisions (note the comments in Chapter 3 about Chelomey and his super-rockets). As another example, a recent Russian account of mine-torpedo units makes it clear that the Tu-14 bomber was unsatisfactory, yet it was bought anyway; by this time Tupolev was Stalin's favourite. Tupolev suffered for his political success (under Stalin) once Khrushchev was in power, to the point that there was a serious attempt to keep him out of the competition for the Tu-22 successor (he regained his clout once Khrushchev was gone).

20. According to the Kuznetzov Academy volume, the first coastal missile units were formed in 1956. By this time it was clear that no matter how well fortified, a fixed coast defence site was vulnerable to air attack. Development of a mobile version of the system, Sopka, was ordered on 1 December 1955. A Sopka battery comprised three radars, four mobile launchers, and eight missiles. Effective radar range was increased to 110km by raising the antennas. This system was credited with a hit probability of 0.7 to 0.8; to destroy a cruiser, three or four would be fired. Missiles could be fired at an interval of 10 seconds to overwhelm shipboard defences. The system passed its production test in 1957 and was accepted into service on 19 December 1958.

21. For details of the projected cruisers, see A S Pavlov, *Warships of the USSR and Russia* (Annapolis and London 1997), ppxxii-xxiii. For drawings of Projects 63 and 64, see I D Snasskogo, *Istoriya Otechestvennogo Sudostroyeniya* Vol V (St. Petersburg 1996), pp162 and 166-167. The profile of Project 64 (p166) shows a small-scale profile of the P-40 missile. This book, the last volume of a history of Russian/Soviet shipbuilding, discusses the various cruiser projects on pp160-168.

22. P-10 was fitted on board a 'Zulu' class submarine (*B-64*) in 1956-7. Between 23 September 1957 and 31 October 1957 the submarine fired four missiles. Range was 600km, flight altitude 200-600m.

23. Work to place a short-range ballistic missile (an early version of 'Scud', R-11FM) on board a submarine began in 1953. In September 1955 R-11FM became the first ballistic missile in the world to be fired from a submarine (in this case, of the 'Zulu' class).

24. P-20 (3 × 30m body) was designed to fly at 30km at Mach 2.5-3 to a range of 3500km. For a time, a force of eighteen Project 653 nuclear P-20 submarines was planned, each to carry two missiles in its 25m hangar. There were also plans to modify the 'November' class (Project 627A) submarine to accommodate this weapon.

25. P-5 was ordered developed for submarine installation. It was first launched (without an engine) on 12 March 1957 at the NII-2 test range; then it was fired from a floating launcher off Balaklava between August 1957 and March 1958. The first complete test, 28 August 1957, failed, as did the second, but the third and fourth succeeded. Trials from the 'Whiskey' class submarine *S-146*, converted at Severodvinsk, were completed on 1 January 1959 (total of twenty-one firings), and the missile was accepted into Soviet naval service on 19 June 1959. P-5 carried either the RDS-4 nuclear warhead (also used in the R-11FM 'Scud', a standard weapon at the time: yield 200 and later 600 kT) or an 800-1000kg HE warhead, and employed autopilot (AP-70A) guidance. Range: 43km at 384m/sec, 574km at 345m/sec, 650 km at 338 m/sec (about 1250km/hr). Flight altitude was 400-800m. P-5 (4K95) could have its initial course corrected to some extent by the launching submarine. P-5D (4K95D) had a Doppler rather than a barometric altimeter, for higher altitude, and could detect drift due to wind, for better accuracy. It armed the 'Echo I' (Project 659) class, had a range of about 600km, and flew at about 250m altitude.

26. KSShch was powered by the same AM-5A turbojet which powered the Yak-25 fighter. Cruise speed was 260-280m/sec, and the missile attained a range of 80km (against a design range of 100km). The missile weighed 2520kg including a 620kg warhead.

27. Design work on a small-boat missile began in 1953. Development of P-15 and of the associated missile boat were authorised by an 18 August 1955 Council of Ministers decree. The tactical-technical requirement (TTZ) for the Project 205 missile boat (to carry 'Styx') was issued on 24 May 1956. The sketch design was approved in July 1956. P-15 was first launched on 28 October 1957, from a moored converted P-6 type torpedo boat (Project 183). Development of an IR-homing (ie thermal) version, P-15T, was ordered in May 1957; it had the first Soviet IR seeker. The Soviets estimated that, given their small radar cross-section, boats armed with P-15 could detect their targets (at a range of 35km) about 15 minutes before they themselves could be detected. Estimated hit probability against a light cruiser was 10 times that of a torpedo (kill probability was estimated as 0.8 or 0.9, versus 0.12-0.16 for two torpedoes). Moreover, the total duration of an attack, conducted from outside the cruiser's gun range, would be 75-110 seconds, compared to 45 minutes for the approach and attack by torpedo boats (beginning at a range of 20-30km). Furthermore, the cost of a missile boat was only 2.5 per cent that of a destroyer and 0.5 per cent that of a cruiser.

28. Harpoon was conceived in 1965 by the U.S. Naval Air Systems Command as a longer-range successor to the Bullpups which patrol aircraft (such as Neptunes and Orions) carried to deal with surfaced submarines; this missile in turn replaced 5in rockets. McDonnell Douglas began its own study of such a missile at about the same time. Late in 1967 the U.S. Navy became interested in a 40nm ship-launched missile, which could be fired from existing surface-to-air missile launchers, carrying a 250lb Bullpup warhead; it is not clear whether this project predated the *Eilat* sinking. After evaluating all existing ship-launched anti-ship missiles, the Navy decided that something new had to be developed (Exocet proved the most attractive of the available options). In 1969 the two projects were tentatively unified as Harpoon, AGM/RGM-84 (this decision was ratified in November 1970 by the Defense System Acquisition Review Council [DSARC]). As of May 1970, according to a paper in the declassified CNO files, first priority would go to patrol aircraft installations, followed by missile destroyers. The high-priority targets were Soviet missile destroyers, followed by surfaced submarines (guiding missiles), then by other ships and then by patrol craft, including those armed with missiles. As CNO, Admiral Elmo Zumwalt, Jr greatly increased the priority of the missile (particularly its surface-launched version) in September 1970. McDonnell Douglas won the airframe contract in June 1971. Note that initially there was no targeting problem. As an air-to-surface missile, Harpoon would attack a target well within the horizon defined by an airborne radar. The 40nm (later 50nm) range initially envisaged for the ship-launched version was within the ESM horizon (Exocet was designed to a similar specification). Targeting became more difficult as specified Harpoon range increased to 65nm.

29. Ships and submarines capable of firing P-6 (4K48) or P-35 could be recognised by the massive radars used to guide the missiles. The radar seeker was tested in 1959 on the P-5RG missile. An anti-radar version of P-5 was tested in 1962. A dual purpose anti-ship/land-attack version, P-7 (4K77), was designed to hit assigned, rather than pre-planned, targets. Developed under a 19 June 1959 decree, it was tested in October and December 1961. It proved insufficiently accurate and never entered service. Ranges: P-6, 500km; P-7, 1000km. Flight altitudes: P-6, 100-7000m; P-7, 100m. Work on the next version, 3M44 Progress, began in 1974. It incorporated a new on-board navigational system and a more automated FCS. This version was accepted for service in 1982.

30. The first successful tests of P-6 were conducted from an 'Echo II' (Project 675) submarine in July-September 1963, using radar data from a converted Tu-16 bomber (Tu-16RTs).

31. There was also a nuclear version, probably with a 6 kT warhead, tested in the autumn of 1962.

32. The YeN target detection/tracking radar was located in the bomber's nose, but the cruise missile guidance antenna was set above the cockpit, and the bomb bay contained both the pressurised cabin for the guidance system operator and the tank for missile fuel. Using YeN, the bomber could detect its target at 220nm range, launch the missile at 100nm, and break away at 90nm. If fired at much higher altitude (33,000ft [10,000m]) and speed (400-430kts) than a KS-1 carrier. Once launched, the missile would climb to about 40,000ft, then at about 40nm from launch it would descend slightly. It would begin a terminal dive against the target at a range of about 70nm from the target. During system development minimum launch altitude was reduced from 5000 to 1500m and flight altitude from 1200 to 600m. Launch range increased from 210 to 325km, requiring an increase in detection range by the YeN radar from 320 to 450km. A version with an improved seeker was designated K-10SN. There was also a jammer version, K-10SP ('Azalea'). It was the first such device in Soviet service. The standard high-explosive warheads were FK-1M (a derivative of the KS warhead, designed to hit below a ship's waterline) and FK-10 (a shaped charge); there was also a nuclear warhead, presumably an RDS-4 derivative. As in KS, the engine was derived from a fighter engine, in this case a constant-rpm derivative of RD-9B (the MiG-19 engine) called M- 9FK.

33. The missile guidance radar (YeS) was tested on board a modified MiG-19SMK fighter, and the system test was conducted in the summer of 1957. The first K-10S missile was delivered in October 1957. Five missiles were launched in 1958, and twelve in 1959, but problems were encountered with the missile and bomber radars and with the engine. A scooped air intake (like that of the contemporary U.S. Regulus II) was proposed for an improved version, which would have armed the new Tu-22. Due to the problems experienced during the 1958 tests, and the new demands made for the missile arming the new bomber (range 300km, speed 2700 to 3000km/hr) the missile had to be redesigned so completely that it became a new weapon, Kh-22. K-10S

would arm only the Tu-16. Given the high failure rate experienced in 1958, its future was in doubt. However, a June 1958 report to the Party Central Committee pointed out the limited effectiveness of the existing ninety Tu-16KS bombers armed with the earlier AS-1 (KS-1) missile. The decision was made to continue making Tu-16s, but to arm them with the new missile. In 1960-2 the K-10 system was in service in seven regiments. A total of seventy-nine launches in 1960 grew to 147 in 1962. The missile itself was improved. Launch range increased to 300-350km in the K-10SD (dalnyy) version, completed in 1966, and it was followed by K-10SDV (by 1971). Minimum launch altitude was reduced from 5000 to 1500 and then to 600m, and the guidance system was modified so that the bomber could order a change in target while the missile was in flight. This was broadly equivalent to the lock-on after launch capability of the later Kh-22M system employed by 'Backfires'. K-10S was also used as a pilotless jamming platform (K-10SP), development of which was completed in 1979 (this weapon was demonstrated in Northern Fleet exercises in May 1981). Beginning in the late 1960s Tu-16K-10 bombers were converted into Tu-16RM reconnaissance aircraft.

34. Given Chelomey's political influence, it also seems reasonable to imagine that he simply killed off P-40 as unwanted competition; his P-500 Bazalt had the sort of high-speed performance P-40 had offered. By 1958 the MiG bureau was in political trouble because it had been too successful under Stalin. For example, Khrushchev revived the Sukhoi design bureau (which Stalin had closed), and it became MiG's competitor. MiG also lost its missile design branch when Raduga was split off.

35. For KSShch, the test bed was the destroyer *Bedoviy*, converted to Project 56-EM beginning in 1955. There was also an abortive design to modify a *Skory* class destroyer with the KSShch launcher, 30BR. The planned platform was Project 57bis, the 'Krupny' class. More 'Kotlins' were converted *after* the 'Krupny' programme had begun. For the short-range 'Styx', the test platform was a modified P-6 class torpedo boat; the planned production platform was the 'Osa' class (Project 205). The productionised test bed was the 'Komar' class (Project 183R, the torpedo boat being Project 183).

36. Of these missiles, Raduga's P-15M (Termit-M) probably arose out of a 1960 or 1961 project for a 50-70km patrol boat missile. A Chelomey alternative, P-25, was rejected (its development, and possibly that of P-15M, was authorised on 26 August 1960). The Chelomey missile was apparently associated with an abortive project, supported by Khrushchev, for a submersible missile boat, to carry four such weapons, and to displace 430 tons; speed would have been 42kts surfaced and 4kts submerged, using a closed-cycle engine. Development of Chelomey's Malakhit (P-120) was authorised on 28 February 1963, under the same directive which authorised P-500 Bazalt. Both were 'universal' missiles, in the sense that each could be launched by both submarines and surface ships. However, Malakhit, like Chelomey's earlier P-70 Ametyst, could be fired by a submerged submarine. Malakhit armed 'Nanuchka' (Project 1234) class missile corvettes and 'Charlie II' (Project 670M) class submarines. It incorporated a new guidance system, probably radar/IR, at least in the submarine version. Probably in 1969 another pair of development requirements, for smaller combatants and for large warships, was issued. Beside Granit (SS-N-19), it probably authorised Chelomey to begin development of Oniks, a missile later offered for export as Yankhont. The 1969 date coincides roughly with the beginning of work on both the 'Tarantul' class missile boat (the 'Nanuchka' successor) and on the *Sovremmennyy* class missile destroyer. In the event, progress on Oniks was apparently slow, and Raduga used the planned powerplant and guidance in the Moskit missile. Moskit development was authorised in 1973, and the missile entered service in 1984. Like Malakhit, it is associated with 'Band Stand' and 'Light Bulb' radomes.

37. Dimensions were: length 8.6m, span 4.6m, weight 3000kg (compared to 2735kg for KS). By way of comparison, K-10S dimensions were: length 9.5-10.0m, span 4.6-4.9m (launch weight 4533kg). The conventional warheads were FK-1M, FK-2, and FK-2N (fragmentation). KSR-2 cruised at 1500 to 10,000m, and reached Mach 1.18 at high altitude. By way of comparison, K-10S cruised at Mach 1.2 and reached Mach 1.8 in a dive; it had slightly longer range (325 vs 180-230km at high altitude). Both missiles were placed in Soviet service roughly simultaneously. Because the performance of KSR-2 was so inferior to that of K-10S, it did not last as long in Soviet service. Only KSR-2 was exported (to Egypt and Iraq), because the Soviets considered its relative simplicity well suited to Third World countries. After the 1967 war, the Egyptians bought about twenty Tu-16K-11-16s to replace their destroyed Tu-16KS missile bombers. The Soviets deployed a squadron of Tu-16R and -16RM reconnaissance bombers to Egypt; they wore Egyptian markings but carried Soviet crews. They left in 1972, when President Sadat ejected Soviet forces from Egypt. When the next Middle East war broke out in 1973, the Egyptians had eighteen missile bombers in service, which launched about twenty-five KSR-2 and KSR-11. Twenty were shot down by Israeli fighters. The other five destroyed two radars and a supply depot. Surviving Egyptian Tu-16s attacked Libyan bases (Tobruk, El-Adem, Alkufra) during the four day confrontation with that country in July 1977; on 24 July they destroyed two radar stations. The Libyans claimed two shot down. No Iraqi aircraft were involved in the 1973 war, but in 1974 they bombed Kurdish rebels (the rebels claimed one shot down). Note that Iraq also obtained Chinese-built Xian-6 (H-6, export designation B-6D) versions carrying C-601 missiles based on the Soviet 'Styx'. They took part in the closing stages of the Iran-Iraq War, attacking Iranian cities with both bombs and missiles. All the Iraqi Tu-16s and H-6s were destroyed during the Gulf War.

38. The adaptation was ordered in August 1959, and the project was cancelled on 5 February 1960.

39. The passive broadband seeker could detect a radar at a range of 180 to 280km at an altitude of 10,000m, or at 150 to 190km at an altitude of 4000m.

40. Development of missiles for the two new heavy bombers ('Bear' and 'Bison') was authorised by an 11 March 1954 Council of Ministers decree, having

been requested the previous year. When the effort to develop a missile for the 'Bison' failed, that aeroplane was taken out of production (the problem was apparently limited ground clearance). For the Tu-95 'Bear' the MiG bureau developed what amounted to an unmanned version of a fighter, Kh-20 (NATO AS-3 'Kangaroo'). The missile designator reflected abortive plans to designate the 'Bear' Tu-20 in service rather than retaining its design bureau designation, Tu-95 (Western writers often referred to 'Bear' as Tu-20). The missile-firing version of the bomber was designated Tu-95K-20; the weapon system was K-20 (Kometa-20). MiG developed the missile airframe and KB-1 its guidance system. The associated surface search radar was Krypton/Rubin-N. Kh-20 was armed with a large thermonuclear warhead (reportedly 1 MT yield). It was designed for use against fixed targets, but trials showed that it could be effective against large moving ships. The missile autopilot was preset at launch time, but course corrections could be fed to the missile by radio link. Launch trials began on 6 June 1957. The system was declared operational on 9 September 1960; in all, forty-nine aircraft (including two prototypes) were built. The Kh-20M version of the missile could fly 650km (about 350nm) at Mach 1.8, carrying a 2500kg nuclear warhead (or 350km at Mach 2). Total weight was 11,000kg, about twice that of Kh-22. In service, Tu-95KDs (the air-refuelled version of Tu-95K-20) often used their big search radars to guide Tu-95MRs to Western warships; the Tu-95MR would overfly the ships and take photographs. In the 1970s and 1980s the Tu-95MRs generally operated alone, which suggests that they were cued by other systems, most likely by satellite ocean reconnaissance.

For a time tactical bombers (Yak-28, the Il-28 successor) survived, probably because the Soviets lacked a suitable stand-off missile (their warheads were probably too large). The only missile developed for the Yak-28 was the Kh-28 defence-suppression (ARM) weapon, intended for an abortive Yak-28N version (1964-5). Yak-28 production may have been curtailed because the Soviet army began to receive medium-range missiles, both ballistic ('Scud') and cruise (FKR-1, a version of KS, and S-6, a land-based equivalent to the naval P-6: NATO SSN-3 'Shaddock').

41. The first large-scale simulated Tu-16 missile exercise known to the U.S. Navy was conducted in September 1958. As of early 1974, U.S. intelligence credited the Soviet Navy with 290 missile attack 'Badgers', of which 190 were armed with AS-2. They were backed by another ninety bomb strike aircraft ('Badgers' and 'Blinders') and by 120 reconnaissance/photo/ELINT aircraft ('Bear D', 'Blinder' and 'Badger') and by eighty 'Badger' tankers. The number of strike reconnaissance aircraft testifies to the vital importance of finding targets for the missile aircraft.

42. As in the United States, which developed the P6M Seamaster, the Soviets experimented with jet-powered seaplane strike aircraft, which could combine the performance of a land-based bomber with the flexibility of a seaplane. The Beriev Be-10 (NATO 'Mallow') was designed to a 1953 naval requirement for a replacement for the existing Be-6 piston-engined seaplane, suitable for reconnaissance, anti-ship strikes, and attacks on shore installations, hence also a replacement for shore-based mine-torpedo aircraft. It first flew on 20 July 1953 and equipped two Black Sea Fleet units. Only sixty were built, and they were withdrawn from service in the early 1970s. To arm a missile version of this aeroplane (Be-10N) and the successor turboprop Be-12, in 1959 a Kh-12 cruise missile (K-12 system) was developed and successfully tested, using a liquid-fuelled rocket motor similar to that of the AS-5 (KSR-2). Its configuration was similar to that of Kh-22, but it was smaller (length 8.36m, span 2.25m, diameter 0.7 to 0.8m, weight 4 to 4.5 tons, warhead 350kg). Range was 40 to 110km at a speed of 2500km/hr. Like K-10S, it used an autopilot plus an active radar seeker. This missile apparently never entered service, and presumably the production run of the Be-10 jet seaplane was accordingly curtailed.

43. According to now-declassified 1962 and 1974 ONI reports, Krug operated at 2 to 20 MHz, the antenna always consisting of 40 dipoles in a circle, in front of a reflecting screen. Estimated accuracy was 0.5 degrees at 5000nm range. It was apparently first deployed in 1952; by 1974 there were thirty sites. Much of the net was built up during a rapid expansion begun in 1958, about when the bombers with their anti-carrier missiles were entering large-scale service. Beginning about 1953 it was supplemented by Thick Eight (2-20 MHz), a 'poor man's Krug' operated by the KGB, and then by a later less expensive fixed DF system, Fix-24. To detect aircraft, a VHF version of Krug (20-63 MHz) had been conceived in 1950, but its status was unclear.

44. The photo reconnaissance version of the 'Badger', Tu-16R, appeared in 1955. In 1961 these aircraft received pairs of trainable ELINT antennas. Presumably they were interim pathfinders for bombers armed with K-10S missiles. Tu-16RM-1 was the K-10 pathfinder version, either converted from a retired Tu-16A (nuclear bomber, not capable of carrying missiles) or built as such. In either case, it was equipped with the same radar (YeN) as a K-10S carrier, and also with ELINT equipment. Tu-16RM-2, also a Tu-16A conversion, had the Rubin-A radar of KSR-2 carriers. This version was first seen, in the West, by fighters from the U.S. carrier *Kitty Hawk*, in 1963.

45. Initially the blip enhancers were separate units (ULQ-5); later they were the omni-directional elements of the ULQ-6 jammer. It was a standing joke in the destroyer force that the blip enhancers were extremely unreliable, destroyermen being unduly aware that they were intended to divert missiles from the carriers.

46. The first electronic support versions of 'Badger' were converted from Tu-16A nuclear bombers and from Tu-16KS missile carriers (of the obsolete AS-1 system). There were two complementary aircraft, Tu-16P Yelka (Fir: 'Badger D') and Tu-16P ('Badger F'). Production of the two versions probably totalled 135 aircraft. Tu-16P Yelka was a bulk chaff layer, to produce a corridor down which aircraft could fly. Tu-16P was the complementary jammer. Both types were updated, the next generation comprising Tu-16PP 'Badger H') and Tu-16P Buket ('Badger J'). Tu-16PP carried up to 9000kg (19,840lb) of chaff,

sufficient for a 60nm-long corridor, in its bomb bay, from which it was cut and then dispensed. The aeroplane monitored the radars it was to jam to determine their wavelengths. Tu-16P Buket was the complementary jammer (for A-I bands). The jamming antennas of the SPS-44 Buket system were carried in a canoe under the former bomb bay, and there were also flat-plate and podded antennas at the wingtips (presumably, to determine an emitter's location and to monitor the effects of jamming). SPS-44 could be used for both escort and stand-off jamming. These aircraft were converted from Tu-16P Yelkas by installing heavy jammers in place of their chaff stowage. A later version of Tu-16P received the NATO reporting name 'Badger L'. There was also an ELINT version of the bomber, Tu-16Ye. Some were fitted with Azalea jammers, becoming Tu-16Ye Azalea. These aircraft may have been intended as pathfinders analogous to the Tu-22R, relying mainly on their ELINT systems to find carrier groups, and providing a degree of stand-off jamming. Reportedly, a typical antiship regiment of the early 1980s consisted of two strike squadrons (each consisting of nine to twelve missile attack bombers), supported by a third squadron consisting of two to four Tu-16PP, one or two Tu-16P Buket, and three to six 'Badger' tankers.

47. Link 11 offered other forms of protection. No single ship transmitted for very long at any one time. Transmitting power was deliberately limited to that needed to reach ships within a net. If ships operated within line of sight, UHF (which could not be intercepted at long ranges) was used instead of HF. If the fleet was so dispersed that HF had to be used, an interceptor would obtain only fleeting indications from each ship, rather than a consistent signal suitable for direction-finding. Presumably the degree of insurance offered by the very short signals waned as intercept technology improved.

48. Tupolev had conducted studies of a variety of supersonic aircraft in 1950-3. His programme was officially authorised by a 10 August 1954 Council of Ministers decree; the earlier Tu-16 had entered production late in 1953. The 1954 programme called for three parallel designs, the Tu-98 tactical strike aircraft, the Tu-103 medium bomber (Project Yu), and the Tu-108 intercontinental missile carrier (presumably a follow-on to the Tu-95K-20 then in development). Tu-103 evolved into Tu-105, with two turbojets aft (a configuration similar to that of Tu-98); it in turn led to the Tu-22 bomber. The prototype was completed in December 1957, and it first flew on 21 June 1958. The second prototype (Tu-105A) was heavily redesigned, with an area-ruled fuselage and new engines; it first flew on 7 September 1959. The first series-production Tu-22 bombers were completed in July-August 1960. The aircraft was first publicly shown in 1961. Plans initially called for two versions, the Tu-22B bomber and the Tu-22R reconnaissance aircraft. Only fifteen Tu-22B were built.

49. Each service received roughly equal numbers of Tu-22Rs. The first two regiments entered service in 1962, two more following in 1965. The Navy Tu-22Rs served with the Baltic and then with the Black Sea Fleets. Peak Navy strength (1969-70) was sixty-two; they were replaced in the 1980s by Su-24MRs. The Navy received a total of eighty of the 311 Tu-22s built.

50. Like K-10S and KSR-2, Kh-22 was developed by Berezhniak's OKB-155, at that time still part of the MiG organisation (later it became the separate Raduga design bureau). Like KSR-2, it used an Isaev liquid-fuelled rocket. The guidance system was developed by KB-1. The Tupolev bureau acted as system integrator. Initial specifications called for a speed of Mach 3 and a range of 300 to 400km (186 to 248nm), the missile being launched at an altitude of 10,000 to 14,000m (32,800 to 45,900ft). Typically the missile cruised at 20,000 to 22,500m (65,600 to 73,800ft) before attacking in a steep dive. The missile had an onboard PG radar cued by the aeroplane's PN ('Down Beat') search/track radar, derived from the earlier Rubin-1A, and housed in a characteristic bulging dome. PN and PG formed a unit, the second letter indicating function: N for noseetel' (carrier, the aircraft), G for golovka (autotracker head). Typically PN acquired a large target at 500km range and a smaller one at 300 to 350km. The unusual nose shape required for the radar disturbed air flow and caused vibration, which tended to derange the electronics inside. The PG radar was typically locked on before launch, the bomber turning away shortly after launch. The missile began a 3-degree terminal dive (at Mach 2.5) when its seeker registered the appropriate depression angle. For its normal flight altitude, that was little short of 80nm; effective minimum range was 70nm. If PN was not working, the missile could acquire a target with its own PG radar before launch. Plans called for a later extension of missile range to 800 to 900km (up to 560nm), using an improved guidance system. Missile aerodynamic range depended on launch speed and altitude: 550km when released at 1720km/hr at 14,000m; 400km when released at 950km/hr at 10,000m. K-22 substituted solid-state circuitry for the vacuum tubes of earlier Soviet systems. There are three versions: radar-homing (naval missiles), autopilot (to attack fixed targets or large naval formations, using a Doppler radar to measure speed over the ground), and anti-radar. Because of its limited accuracy, the land-attack version had only a nuclear warhead (the naval and anti-radar versions could have conventional warheads). Against a ship, the missile's 900kg (1984lb) shaped-charge warhead could tear a $20m^2$ hole and could penetrate 12m (39ft) into a hull.

51. In 1973 replacement of the land-attack version of the obsolete Kh-20 aboard 'Bear' strategic bombers (which became Tu-95K-22) by Kh-22 was finally ordered.

52. The standard Soviet storable oxidant, red inhibited fuming nitric acid, was both unstable and toxic. It was part of a hypergolic oxidant-fuel combination, ie one which would ignite when the two components met. To avoid disasters, Kh-22 missiles were sometimes carried unfuelled. As an illustration of how dangerous such weapons could be (to their users), the Finnish Navy scrapped its P-15M ('Styx') missiles on the explicit ground that if one were accidentally dropped during loading, it might well explode.

53. Note that according to Kuzin and Nikol'skiy the Soviet Navy received a total of about eighty Tu-22K and -22R. This figure may simply reflect the secondary assignment of air force units to naval roles.

According to Gordon and Rigmant, in 1967 the Soviet Navy had thirty Tu-22 in its Baltic Fleet (at Chkalovskaya) and another twenty-seven in its Black Sea Fleet (at Saki).

54. Tu-22R had cameras, a search radar (Rubin-1A), and an SRS-6 Romb-4A or SRS-7 Romb-4B ELINT system (COMINT/SIGINT). The latter systems could detect and record up to fifty-two radars. Normally a Tu-22R carried an SPS-3 Roza jammer. The aircraft could also be fitted with a palletised chaff dispenser (APP-22) in its bomb bay in place of cameras. Presumably Romb could be used to home on a carrier group. The Tu-22P escort jammer (forty-seven in all, including about thirty built as such plus conversions) had a REB-K ELINT station and an SPS-100A Rezeda-A jammer.

55. KSR-5 had about the same speed (Mach 3) and aerodynamic range (250-400km) as Kh-22; it was somewhat lighter (5000 vs 5900kg) but carried a similar warhead (1000kg). The main internal difference between this missile and Kh-22 was that its fuel was 'ampulized' (encapsulated) as in a ballistic missile.

56. Some of the long-range propagation may have been caused by tropospheric scatter of UHF signals, a phenomenon used by the U.S. services to insure communication at a range of about 200nm.

57. For example, ducting was so common in the Indian Ocean that the Indian Navy modified the 'Styx' control radars (NATO 'Square Tie') of their 'Nanuchka' type missile boats to exploit it. The Indians were apparently entirely unaware of Soviet interest in the phenomenon. See Sam Lazaro, Anthony de, *Soviet Electronic Warfare* (Falls Church, Va. 1991). Sam Lazaro was an Indian Navy officer; he commanded an 'Osa' class missile boat in the 1970-1 war with Pakistan, and later he commanded a 'Nanuchka'.

58. Note, however, that by 1950 U.S. submarines regularly used ducting to detect surface ships without themselves being detected (because the destroyer's radar and ESM antennas were above the duct), as recounted in the official Confidential U.S. *Combat Readiness* magazine of the time.

59. According to G S Baskekov, 'Environment and Radio-Electronic Propagation', in Sarkisov, Titanit entered service in 1973, on board 'Nanuchkas' (Project 1234). It combined, for the first time, independent active and passive radar channels, a reception channel for data from an airborne radar (the Success system), means of processing and displaying this information, and a data link (VZOI-VZOR) between ships of a tactical group, as well as a navigational radar. The level of automation made it possible to shrink the system to the point at which it could be accommodated aboard a small combatant. Unfortunately the level of integration made it impossible to install modular parts of Titanit on board other ships, and it also made modernisation very difficult. Hence the development of a next-generation system, Monolit, which was modular in design. Its elements were: the passive target detector/designator, the tracker/data processor, the system to accept and process airborne data, and the data link to transfer target data to other ships. This is the system employed on board 'Tarantul' class (Project 1241) missile boats. Note that early 'Nanuchkas' used a Dubrava system instead of Titanit. The reference to inflexibility in employing Titanit components in other classes probably applies to P-15M, the over-the-horizon version of 'Styx' (P-15), which would have benefited from such a radar. Most P-15M boats were equipped with the same horizon-range radar, Rangout (NATO 'Square Tie') as their predecessors, although later ones (*eg* export 'Tarantuls') had Garpun (NATO 'Plank Shave') which was credited with ducting performance. Presumably the modular design of Monolit made it adaptable to both the Malakhit (P-120) and successor Moskit (P-270) missiles. Note that *Sovremmennyy* class missile destroyers, which also fire Moskit, use the Mineral fire control system. Externally, both Malakhit and Moskit surface ship missiles were associated with a big radome ('Band Stand') and with data link radomes ('Fish Bowl' [probably for Dubrava] or 'Light Bulb' [Titanit/ Monolit/ Mineral]). Presumably they corresponded to the tracking and data link antennas used in the P-35 system ('Scoop Pair' [Binom] and 'Plinth Net').

60. The entire 'Success' system, including the P-6 and P-35 missiles, was developed under a July 1959 Council of Ministers decree. The Tu-95RTs was declared operational on 30 May 1966, production having begun in 1963; in all, fifty-three were built (the first was a converted Tu-95M). They were used to monitor Western naval activity; beginning in the late 1970s they flew from Cuba, Guinea, and Angola. Gordon and Rigmant claim that 'Bear-Ds' monitored the Royal Navy operation in the Falklands in 1982, from the Bay of Biscay south. Note, however, the claims the Soviets made that their ocean reconnaissance satellite system provided this data.

61. Tsiklon-2 was designed by Leonid D Kuchma, who is now President of Ukraine.

62. Work on satellite reactors began in 1960. There were two parallel programmes, the Buk/Topaz-1 thermal reactor using thermionic conversion developed by the Kurchatov Institute of Atomic Energy (IAE) in Moscow; and the Topol/Topaz fast neutron reactor using thermoelectric conversion, developed by the Physical-Power Institute (FEI) of Obninsk. Buk, which powered all the test and operational versions of US-A, weighed 1.35t and produced 5 kW for 344 days. Topol, the multi-cell thermionic reactor (electrons are boiled directly off a hot cathode), did not appear until late in the 1980s. There was also Yenisei/Topaz-2, developed by the Leningrad Central Design Bureau of Machine Building (TsKBM) in conjunction with the Kurchatov Institute: it had a single cell with a uranium crystal fuel element. It weighed 1 ton and produced 6 kW for a year and a half during ground tests. It was intended for a much heavier follow-on satellite, the programme for which was cancelled in the late 1980s. When these reactors were disclosed from 1989 on, they were all given Topaz names (as above), Topaz being the Russian acronym for 'Thermionic Experimental Conversion in the Active Zone'. This confusion caused U.S. experts to think that all RORSATs used thermionic converters, whereas in fact the thermionic design was introduced only in 1987.

63. For a satellite observing a ship 500 miles away, a ship 900ft long is 0.3 milliradians (less than 0.02 degrees) wide. To form a beam that narrow with an

X-band (3cm) radar would require an antenna about 100m across. If the radar spot on the surface is much larger than the ship, reflections from the sea will swamp the reflection from the ship. The required antenna is far too large for satellite installation. The simplest solution is to use a synthetic array technique: the satellite's motion along its orbit extends the effective size of an antenna looking to one side. The radar has a set integration time. During that time, it adds up the reflections it picks up. Its effective antenna size is given by the distance the satellite travels along its orbit during the integration time. Orbital speed is 20 miles/sec, so a 100m (328ft) antenna length equates to about 3 thousandths of a second integration time. The main limit on the technique is the memory capacity of the satellite (so it can add up the received pulses), the curvature of the satellite orbit, and the motion of the targets (synthetic arrays are unlikely to be effective against fast-moving air targets). The United States used this technique (producing much narrower beams) to produce ground images in the Lacrosse system (to provide coverage when clouds blocked optical satellites). Small objects can be seen by space cameras (and by astronauts' eyes) because the wavelength of light is about 30 billionths of that of radar. A small lens will therefore provide the performance of a massive radar antenna.

64. 'Soviet Dependence on Space Systems', an Interagency Intelligence Memorandum, Nov 1975, declassified 1997 under the CIA Historical Review Program.

65. As it was reflected in the annual unclassified (sanitised) versions of the Navy's PEDS (Programme Element Descriptors).

66. Korolyev's proposal called for satellites in 1000km circular orbits, to serve both the Soviet navy and the merchant fleet.

67. According to Minaev, projected navigational accuracy was 500m. Initially it was 150-200m, and ultimately that was reduced to 80-100m.

68. Molniya itself seems to have been inspired by early U.S. communications satellite projects. Work began at Korolyev's OKB-1 late in 1960 or early in 1961, and a 30 October 1961 decree authorised the experimental project. It apparently had only very limited government backing, and was not at first perceived as a military project. Khrushchev had recently demanded that the Soviet space programme be oriented more towards the military use of space. Korolyev had offered a communications satellite system as part of his proposal for the seven-year Space Plan (January 1960). He hoped to orbit an experimental system (using a version of his R-7 rocket) in 1961-63, and then a geosynchronous system, using a much larger rocket (probably the N-1 he planned for Moon shots). According to Hendrickx, the eccentric Molniya orbit was chosen because the most powerful available space booster, the four-stage version of R-7, could place only 100kg in a geosynchronous orbit when launched from Baikonur. They therefore had to consider alternatives. Given the chosen highly eccentric orbit, a 1650kg satellite could be placed in orbit. The satellite could accommodate powerful transmitters, so ground antennas could be quite small. Molniya-1 was conceived as an experimental system, but it was turned into an operational one, and it was the basis for Molniya-2 (first launched October 1971, operational in 1974).

69. The new system used a PN-A radar, sometimes also rendered PNM. Beside the Kh-22M anti-ship missile, there were a Kh-22MP anti-radar version and a Kh-22MA strategic attack (autopilot only) version, the latter with a thermonuclear warhead. Kh-22MP could be carried by some modernised Tu-22Ks. It and Kh-22MA both entered service in 1974, followed by Kh-22M in 1976. Also in 1976 the improved K-22N/Kh-22N system entered service, together with a Kh-22NA strategic version. K-22N employed the PN-B or PNN radar. Kh-22N could be carried by Tu-22M2, -22M3, and by Tu-95K-22.

70. AS-4 flew too high to be shot down by early versions of the U.S. Standard Missile. A hit during its terminal dive would probably strip it of its wings and its fins, but its armoured nuclear warhead might well survive. A version of SM-2 with a new engine was produced specifically to solve this problem, but clearly it was far better to deal with the bomber before it could launch.

71. Projected Tu-22M performance (with NK-144 engines derived from those of the Tu-144 SST) was 2700 km/hr at high altitude (14,500m) and subsonic speed (1100km/hr) at low altitude (50-100m); range was 6000-8000km at subsonic speed (at high altitude) or 4000km at supersonic speed; at low altitude estimated tactical radius was 1500km. At this stage it was assumed that the aeroplane would carry a single Kh-22 missile. In fact Tu-22M2 achieved only 1800km/hr and a subsonic range of only 5100km (combat radius 2200km, with one Kh-22). Zaloga recounts that Soviet pilots called it the 'vsepogodniy defekonosets' (all-weather defect carrier), a play on its official designation of 'vsepogodniy roketnosets' (all-weather missile carrier). A total of 211 were built in 1972-83. Typical regimental strength was eighteen aircraft, and a regiment was generally accompanied by a squadron of Tu-16Ps. In all, there were five divisions of 'Backfire' missile carriers (three air force, two navy: Northern and Black Sea Fleets). U.S. intelligence first spotted 'Backfire-B' in May 1973.

72. This version could be recognised by its steeply-angled 'scoop' air intakes. Improvements included greater wing sweep-back angle (65 degrees, a very secret feature). Tu-22M3 was optimised for low-level flight (50m altitude). Production amounted to 268 aircraft. Tu-22M3 was credited with 2300km/hr at high altitude or 950km/hr at low altitude, and with a range (at high altitude, subsonic) of 5500km. That compared to 4800km for a Tu-16 or 4900km (subsonic) for a Tu-22K.

73. The operational requirement came later, out of a 'Pleyada' study of future heavy-bomber missiles conducted in the early 1970s. It concluded that a future weapon had to climb very high to avoid interception by Phoenix missiles fired by F-14s. A competition for a Tu-22M3 missile attracted proposals by Raduga and also by two other missile design bureaus, Novator and Zvezda. The decision was to retain Kh-22M and to supplement it with a new missile; Raduga won the new-missile contest. The inertially-guided version was Kh-15P. The export version of the anti-ship variant was Kh-15S.

Chapter 8: Dealing With an Emerging Soviet Threat

1. The author remembers attending a table-top wargame at Newport about 1975, in which a 1973 Soviet attack on the Sixth Fleet was simulated. There were so many defensive weapons that it seemed unlikely that the Soviets would get any hits at all; yet those around the table, most of whom had served as destroyer weapons officers, were sure that a Soviet attack would be devastating. First they reduced the assumed performance of the U.S. missiles, but there were so many defensive weapons that this made little difference. Finally one of them found the answer. In a real surprise attack, officers in the fleet would be so stunned that they would take time to react. The Soviets would get the first few shots for free.

2. For example, Grumman conducted a study for the then Naval Electronics Centre (NELC), San Diego.

3. The Rota FOSIF began operations on 18 April 1970. It merged information from all intelligence sources and provided a daily summary of Soviet Mediterranean naval activities, sanitised to the SECRET NOFORN (non-foreign) level, for the Sixth Fleet. Updates were sent out whenever there was any significant change. This data was all in text form, but it was clearly an intermediate step towards computer-to-computer reporting. Note that the FOSIC in London served CINCUSNAVEUR, the C-in-C of U.S. naval forces in Europe. It seems obvious that it drew upon British information and that it and the FOSIF in Norfolk had much the same relation as the wartime London and Washington submarine tracking rooms, which produced a picture of German U-boat activity in a co-operative way. The OSIS Baseline computer system was installed at all the OSIS nodes in 1974. In 1981 the system was ordered upgraded (as OSIS Baseline Upgrade, or OBU) mainly to provide more timely OTH targeting information, but also to support the massive data flow from new sensors.

4. According to Kuzin and Nikol'skiy, *The Soviet Navy*, the first Soviet submarine 'squash' transmitter, Akula, entered service in 1955. Since U.S. warships could not accommodate the necessary receivers, HF/DF antennas generally disappeared from them. The first automated submarine radio system was Molniya (1970, on board the first submarines with automated combat systems: 'Victor II' [Project 671RT] and 'Alfa' [Project 705] class submarines). Later versions were Molniya-L ('Victor III' [Project 671RTM]), Molniya-M, and Molniya-MTS ('Sierra' [Project 945] and 'Akula' [Project 971] classes). An automated submarine communications system, Molniya, appeared in 1970. A captured German VLF transmitter, Goliaf, entered service in 1952, using the Pobeda communications system; it could reach submarines at periscope depth at ranges of several thousand kilometres. Later the Soviets produced their own versions. The first submarine VLF towed antenna system, Paravan, entered service in 1967. It is not clear to what extent the United States tried to (or was able to) intercept VLF signals. Kuzin and Nikol'skiy do comment that, because of the usual Soviet desire for tight operational control, the communications load was considerable, and the number of communications personnel sometimes even exceeded the number of weapons personnel.

5. Durations and locations of 1970s stations are taken from the declassified Cross Report on airborne ASW. Later data are from the FAS web site. According to the NRL's *Awards for Innovation* booklet, the laboratory was responsible for major innovations in U.S. HF/DF realised in Projects Boresight and Bulls Eye. The first of three was *retrospective direction-finding*. Previously all stations in a DF net had to measure their signals while the signals were being received. The shorter the signal, the more difficult it was to get enough stations on the air at the same time to obtain a cross-fix. Recording made it possible to pass data through the net after the signal had vanished. The key aspects were an ability to record significant parts of the HF spectrum (the shorter the signal, the more bandwidth it covered) and a digital technique to overcome recorder instabilities. An earlier (1953) attempt at such recording failed, but in 1960 retrospective DF was achieved in the quick-reaction (presumably, to suddenly shortened Soviet communications) Project Boresight; FLR-7 and FRA-44 equipment was deployed globally. The second innovation was a new type of circularly-disposed wide-aperture DF array significantly improving accuracy and signal collection. It was presumably derived from the German Wullenweber (from which Krug had been developed); NRL produced its first report on the subject in December 1947. A 400ft diameter prototype was built during the 1950s, using an electrically-steered array. The Project Bulls Eye array, FRD-10, was a scaled-up version. Its fast recorders employed FRA-54 and FSH-6 systems. The third innovation was the use of computers (GYK-3) in HF/DF nets. According to the NRL booklet, Bulls Eye innovations 'ma[de] HFDF a principal means of global ocean surveillance, with special capabilities against critical targets'.

6. The Classic Bullseye system was later reportedly renamed Classic Flaghoist. Information from the Air Force HF/DF system is provided via a separate broadcast as Classic Centreboard (associated with Crosshair or Unitary DF, a program to unify all the services' DF operations). Presumably the Navy system is intended primarily to track ships. The Air Force system is probably mainly a means of intercepting messages from command centres, the Wullenweber array's directional capability allowing it to distinguish signals from different transmitters, picking them up simultaneously. Note that the Navy system is FRD-10 (D for DF), whereas the Air Force system is FLR-9 (R for radio receiver, L for countermeasures [ELINT]). The Navy FRD-10s were installed in the early 1970s, replacing earlier installations. At this time the Navy had three HF/DF nets, two in the Pacific and one in the Atlantic. By 1960 the Western Pacific net was detecting about 250-500 'flashes' each day, of which about 40 per cent could be fixed with an accuracy of 100nm. A Wullenweber can measure the elevation angle at which the signals arrive. Since the takeoff angle has to be same, and since the height of the ionosphere is known, this measurement gives the approximate

range to the transmitter, a technique called Single Station Location (SSL) – which was probably the main subject of the modernisation. There was also a project to reduce manning by automating the intercept stations. Nominal range is 3200nm. The intercept operators work in a building at the centre of the array, which is surrounded by two rings of HF antennas. The inner ring (230m diameter, forty folded dipoles) is for longer wavelengths (2-8 MHz); the outer (260m diameter, 120 sleeve monopoles) is for shorter wavelengths. Inside each ring is a wire mesh reflector, to prevent signals from crossing the ring. The entire assembly is called an 'elephant cage'. These antennas are also called Circularly Disposed Dipole Arrays (CDDAs) or Circular Dipole Antenna Arrays (CDAAs). As of the mid-1990s, Bullseye operating locations included Brawdy, Wales; Diego Garcia; Edzell, Scotland; Galeta Island, Panama; Guam; Hanza, Japan; Homestead, Florida; Keflavik, Iceland; Northwest, Virginia; Sugar Grove, Virginia; Rota, Spain; San Diego, California; and Winter Harbor, Maine. There were also associated foreign-operated stations, such as Canadian stations at Gander, Newfoundland, and at Masset, in British Columbia. The first Air Force (Air Intelligence Agency) FLR-9s were installed in 1964; the system was then called Iron Horse. Personnel generally included some from the Navy. Locations reportedly included Augsburg (Germany: closed 1993), RAF Chicksands (UK: ended allied operations 1995), Elmendorf AFB (Alaska), Menwith Hill (UK), Misawa AFB (Japan), and San Vito dei Normanni (Italy: closed 1993); there was formerly one at Clark AFB in the Philippines. There is also an Army HFDF system using FLR-9. Compared to FRD-10, FLR-9 handles three rather than two bands (two lower and one upper). AX-16 ('the Pusher') is a two-band version of FRD-10 (for compactness) used mainly in Britain; it has about half the diameter of the FRD-10 and FLR-9. These data are taken from the FAS web site and from the Cross Report.

7. The EP-3E clearly had functions other than electronic surveillance, but it appeared just as interest in tracking the Soviet fleet took off. EP-3Es worked with carrier-based EA-3s.

8. Clearly there were initial problems. The NRL 75th anniversary awards volume describes the laboratory's contributions to fingerprinting, beginning with a *1977* classified patent disclosure by R L Goodwin, 'System for Classifying Pulsed Radio-Frequency Modulation'. Since the abstract was classified, no patent was pursued. Goodwin received the Navy Superior Civilian Service award on 14 April 1982 for 'performing and directing the research and development efforts that led to the achievement of an important new Navy capability in real-time pulsed-emitter characterization'.

9. Probably after a brand of facial tissue which would wipe away the dirt obstructing the view through a window. It was sometimes suggested that White Cloud was a nickname obscuring a variety of quite different programmes. The account given here is based on unclassified (*ie* unofficial) sources, including a Russian article. All are necessarily somewhat speculative.

10. After the three daughters of Zeus and Themida, presumably indicating the use of a triplet of sub-satellites. According to Andronov (1993), the analogy is quite precise. In Greek mythology, one daughter spins the threat of fate for an individual, a second measures out its length, and the third cuts it as measured. In the Parcae system, one satellite has a wide observation swath (hence cannot get a precise DF 'cut'), a second provides a cross-fix (but with some ambiguity), and the third resolves the ambiguity.

11. According to Richelson, *The U.S. Intelligence Community*, p186, the radar system was developed in response to a 1970 CNO request for a study of ocean surveillance requirements. NRL proposed a radar satellite, which became Program 749.

12. According to Richelson, ibid, the sites, co-located with Navy Regional Reporting Centres, are at Winter Harbor, Maine; Edzell, Scotland; Guam; Diego Garcia; and Adak. The 1993 Russian article adds Blossom Point, Maryland. Adak and Edzell closed during the 1990s.

13. Army participation began in 1985 under a TRUE BLUE program (which was presumably part of Army TENCAP). An Air Intelligence Agency detachment was assigned to the Guam White Cloud site in 1995.

14. The first OBU contract was awarded late in 1982, but a new Decision Co-ordinating Paper was issued in May 1987; presumably it reflected a the decision to add inputs from the new ROTHR. A later OSIS Decision Co-ordinating Paper (January 1990) probably reflected a decision to change to a distributed computer architecture. The OBU system was later sold to the Royal Navy (installation and accreditation were completed in FY90), to the Japanese Maritime Self Defense Force, and to the Royal Australian Navy.

15. LES-3 (launched, with LES-4, on 21 December 1965) was a signal generator operating at 233 MHz in the UHF band. It was intended to test propagation between a satellite and aircraft. At the same time aircraft measured the ambient radio noise in the atmosphere. The last two satellites in the series, LES-8 and -9, were launched together into geosynchronous orbit on 15 March 1976. They tested the concept of a Ka-band cross-link, which later appeared in Milstar (they use UHF up- and down-links).

16. The 230lb LES-5 had a 255.1 MHz up-link and a 228.2 MHz down-link (35 W); it went into an 18,000 × 18,180nm orbit at 7 degrees inclination, drifting (like IDCS satellites) 30 degrees per day. It carried a single 100- or 300-kHz channel. LES-6 (398lbs) had a 302.7 MHz up-link and a 249.1 MHz down-link (130 W maximum) and a single 500 kHz channel. It went into a 3-degree geosynchronous orbit. Although turned off in March 1976 (to avoid frequency conflict with the Marisat launched in February 1976), it was still operable in 1978, 1983, and 1988 tests. LES-7 was a projected X-band satellite with a 100 MHz bandwidth and a multi-beam lens antenna (nineteen feed horns, capable of generating beams as small as 3 degrees, or as large as full-hemisphere coverage). It was never built, although a prototype antenna was built and tested. This was *not* an alternate designation for Tacsat.

17. Tacsat operated at both X-band (7982.5 MHz up-link/7257.5 MHz down-link) and UHF (303.4 and 307.5 up-links/249.6 MHz down-link) with three X-

band transponders (20 W each, two in use at any one time) and sixteen UHF transponders (18.5 W each, maximum 230 W from combiner; typically thirteen on at a time). There were cross-over modes: reception in X-band, transmission in UHF, and vice versa. Capacity: UHF, about forty vocoded voice circuits or several hundred teletype circuits; X-band about forty vocoded voice circuits or 700 teletype circuits to a terminal with a 3ft dish. The satellite carried multiple channels with 50 kHz to 10 MHz bandwidths. The design provided for a varied traffic load, so that it could be used on a demand basis. Estimated life was 2.4 years; the satellite ceased operation when its attitude control failed.

18. NWP-16(C), the 1970 handbook for tactical communications, describes the programme. It is clear that the cross-over capability (SHF up, UHF down) was to have been used to support the Fleet Broadcast. According to the handbook, the new satellite would employ a new technique of 'circuit sets', circuits lumped together into sets defined by the users. There would also be an area broadcast containing three separate circuits: a sixteen-channel Fleet Broadcast using VFCT radio teletypes; a 2400 bit/sec digital Fleet Facsimile channel (which might be analog at first); and a 2400 bits/sec digital data channel for high-speed printers, to be implemented late in the system's lifetime. Proposed lumped circuit sets were: (i) for a major flagship, sixteen teletype, two voice, one wideband data, six data ship to shore; and nine voice, two data, two teletype ship to ship. For a medium flagship, eight teletype, one voice, and two netted data ship to shore; and seven voice, two data, and two teletype ship to ship. For other ships and submarines, one teletype and one voice ship to shore; and four voice, one data, one teletype ship to ship.

19. The Second World War line-of-sight radio, typified by TBS ('talk between ships') operated at VHF frequencies. These frequencies were also used for ship-to-air radio. By the latter part of the war it was clear that VHF offered far too few channels; the only solution was to go to higher frequency, UHF. Experimental radios in this band had been tried as early as 1936, and operational sets were tested in 1944. The fleet switched from VHF to UHF beginning in the late 1940s. The switch was quite expensive, both for the U.S. Navy and for allied navies (such as the Royal Navy) which wanted to maintain communications with U.S. ships.

20. There was another way to reach out using UHF radio: tropo-scatter. UHF signals bend in the troposphere, so that signals from a directional transmitter can be received at a range beyond 250nm. UHF propagation out to 168nm was observed as early as 1932, and in 1940 an early UHF radar on board USS *Yorktown*, CXAM, detected mountains 450nm away. Using the dish of an SK-3 radar, in 1955 the transport USS *Achernar* maintained two-way voice communications with a shore station at a range of 250nm; later another ship received signals out to 630nm. A few years later the national alternative command post ship USS *Northampton* sported a massive tropo-scatter dish on her tower mast, for communication with command centres ashore (USS *Wright* was similarly fitted). Tropo-scatter is probably currently used to detect UHF radars beyond the horizon. For example, many U.S. warships received enlarged Band 1 (UHF) antennas for their SLQ-32 countermeasures sets, presumably to detect Soviet Top Sail and related radars, for missile targeting. UHF (L-band) ducting (probably actually troposcatter) is apparently quite common off the Virginia Capes, where it interferes with U.S. radar tests. Tropo-scatter was and is very widely used for ground radio communications. Tropo-scatter also occurs at SHF (X-band), and at such frequencies it can be supplemented by ducting.

21. The HF Fleet Broadcast survived into the satellite era as a back-up for the Fleet Satellite and its successors. As late as the mid-1990s, long-haul HF radio was used for the fleet primary ship-to-shore, commander's ship-to-shore, as well as for the primary NATO shore-to-ship (fleet broadcast) and HICOM clear voice (non-encrypted) between shore (Fleet Commander) and afloat battle commanders channels. The HF broadcast itself evolved considerably through the postwar period. In 1953 it was a single channel radio telegraph operating at seventeen to twenty-nine words per minute. By 1962 radio telegraphy was secondary, and the broadcast operated mainly by radio teletype, at 100 words/minute. Multiplexing was introduced about 1965. In 1970 the standard was eight parallel radio teletype channels (expandable to sixteen). Typically the channels were: (1) ASW, (2) special purpose, (3) US/allied general traffic, (4) US general traffic, (5) nuclear strike co-ordination, (6) special intelligence (opintel), (7) high command net (could be diverted to opintel if need be), (8) meteorology/environmental data. In 1982 the standard was a full sixteen channels, as in the satellite. The submarine broadcast (LF) could carry eight channels. The VLF submarine broadcast was sent out in either teletype or telegraph (CW) form, the latter at fifty words/min. VERDIN was a multi-channel version (two or four channels, depending on operational requirements).

22. Channels (for HF or satellite) are: (1) destroyer primary (ex-ASW), (2) overload for (1), (3) common channel, (4) primary channel for amphibious/service force/mine force/miscellaneous, (5) overload for (4), (6) opintel, (7) overload for (6), (8) meteorology/environment, (9) destroyer primary for Mediterranean, (10) overload for (9), (11) common channel, (12) primary for amphibious/service force/mine force/ miscellaneous, (13) overload for (12), (14) opintel, (15) meteorology, (16) overload for units with full-period terminations (HF) or timing and synchronising signals (satellite, needed so that the slots appropriate to each channel are stripped out appropriately). Of the sixteen time-division multiplexed channels, eleven carry GENSER (general service, *ie* normally classified, up through Top Secret) data, two carry SI (special intelligence, *ie* 'code word') data, and two carry weather data. Message formats: JINTACCS (Joint Interoperability of Tactical Command and Control System), US Message Text Format (USMTF)(JC-1), Joint Army/Navy/ Air Force Publication (JANAP) 128, and Allied Communications Policies (ACP) 121 and 127.

23. NAVMACS was the shipboard end of the first U.S. Navy computerised message system; the shore end was NAVCOMPARS (Naval Communications

Processing and Routing System). NAVCOMPARS made communication more efficient by screening the messages sent through it, for example for those which have expired (*eg* weather and some intelligence messages). By the mid-1980s originators were required to identify administrative messages; if too many messages built up awaiting transmission, operators could intercept administrative messages, sending them later, when fewer operational messages were being sent. The equipment could also identify those units which had excessive numbers of messages awaiting transmission, then print out the messages to eliminate those which were not urgent.

24. The wideband channel was divided (using FDMA) into five 2400 bits/sec, one 1200 bits/sec, and thirteen 75 bits/sec links, the latter presumably forming a Fleet Broadcast equivalent. Of the two narrowband channels (2400 bits/sec), one was leased to the Army. The Marisats were owned by the U.S. Commercial Satellite Corporation (Comsat). They were turned over to INMARSAT when that organisation was formed in 1979.

25. WSC-3 entered service in 1975. In 1996 its modernisation (to allow access to the 5 kHz channels used by the other services) was cancelled (along with improvements to the USC-38 EHF terminal) due to budget pressures, a step interpreted to mean that surface ship links were not considered as important as those to carriers (which had access to SHF as well as to UHF and EHF). In addition to the United States, it is used by Australia, Canada, Denmark, Egypt, Germany, Indonesia, Japan, Korea, Morocco, Netherlands, New Zealand, Norway, Portugal, Saudi Arabia, Spain, Turkey, and the United Kingdom.

26. SECVOX used a voice encoder (vocoder) to turn a 3 kHz voice channel into a data stream running at 2400 bits/sec. A ship could seek access to the SECVOX channel via a CUDIXS message. If she had only one WSC-3 transmitter, she had to drop out of other nets to talk. The SECVOX channel supported the High Command Net, linking fleet commanders with battle-group commanders. There is also a Fleet Tactical Net.

27. Usually there are two CUDIXS nets within one satellite footprint. Each channel operates at 2400 bps as a half-duplex link. Like an NTDS net, a CUDIXS net operates in master/picket fashion, the controlling terminal transmitting a sequence list to the other subscribers at the beginning of each transmitting cycle. Messages may be of variable length. To fit them in, they are cut into blocks which may be transmitted over several cycles. The sequence list gives the order, net cycle time slice, and total duration of transmission by each subscriber. This type of operation depends on computers: each subscriber's computer cuts its message into blocks fitting the specified time slice. Each block is sent out with a header carrying the precedence level and the number of that block within the message. Net cycle length, typically 2.5 to 3.5 minutes, depends on the number of members in the net, number of blocks authorised for a transmission, and the number of time slices that can be used.

28. N F Direktorov and Vice-Admiral V V Sergeev, 'Scientific Problems of Communication with Submarines,' in Sarkisov. They list alternatives to ELF: seismic waves, blue-green lasers, and hydroacoustic signals. Seismic and hydroacoustic waves penetrate to all depths, but due to their low propagation speeds they are useable only out to several thousand kilometres. Apparently they were at the late experimental stage when the Soviet Union collapsed. Work on blue-green laser communications began in the 1970s, but as of the early 1990s only ELF was a mature technology. ELF transmitters were built in the mid-1980s. The Russian article also makes the interesting observation that ELF signals can be used as geological probes, to find mineral deposits and to forecast earthquakes.

29. These devices all date back to the late 1960s or early 1970s. Active submarine communications projects are listed in a 30 October 1970 CNO memorandum referring to a request for Rapid Development Capability for the SLOT buoy. It had been included as a high priority in Admiral Zumwalt's Project SIXTY, and it was wanted for use by nuclear submarine escorts (for carriers) and to support an anti-SSBN tactic, Squeeze Play. The same memorandum lists other related devices under development: SSIXS, the SLATE buoy, the BIAS buoy, the floating wire (FWA), and the DOLCO (Down-Link Communications System) project. SLOT was expected to be available for service in FY71; it had already proven valuable in Squeeze Play exercises. SLATE, a two-way buoy giving realtime UHF contact with aircraft, was expected to enter service during FY72. Its 6000ft wire limited submarine speed, *eg* to four minutes at 15kts. BIAS was a towed retractable two-way buoy supporting HF and UHF communication. The limit it imposed on a submarine was very depth-dependent: 15kts at 400ft, 6kts at 700ft. The buoy carried an exposed HF whip antenna. It would enter service in FY74. The FWA two-way submarine/air and submarine/surface communication. The Atlantic Submarine Force commander (SubLant) considered it the highest ASW priority among the submarine communications systems, even though it imposed the strictest limits on submarine speed and depth. It was attractive partly because it imposed the least risk of detection. It would probably enter service during FY73. DOLCO was a two-way real time acoustic link using a computer to overcome multipath problems; it was the most feasible near-term acoustic system. Reception would worsen at higher submarine speeds. No money had been reserved for development of the necessary submarine transmitter. SSIXS offered the highest data rate, and could be tested soon; the memorandum asked for more money for planned FY71 tests.

30. The Special Communications System (SCS) uses 5 kHz AFSATCOM channels; Army, Navy, and Air Force Special Forces all have terminals.

31. Since the late 1950s Air Force strategic bombers had been at fifteen-minute standby, so that they could be launched in the event a Soviet ballistic missile attack was detected. Since such warning might be false, policy was for the bombers to orbit waiting for the go-ahead signal. By the early 1980s satellite dishes were compact enough that they could be placed aboard bombers. The further north the bombers could orbit (but still obtain the go-ahead), the more quickly they could reach Soviet targets once released.

32. The Mission-22 (M-22) Tactical Network (MTN) comprises secondary payloads on board classified satellites in highly elliptical orbits, which presumably means SDS satellites. The satellites involved are likely to be deorbited in 2002, and there are no current plans for replacements. This system apparently uses its own unique waveforms (special MTN software is needed to adapt a DSCS terminal) and seems to have been intended primarily to support the army and the Special Forces. It broadcasts a variety of intelligence products. Data rates are 8 to 256 kbits/sec. As part of the JWID-96 exercise it distributed graphics from the Naval Oceanographic Command in Mississippi to Army, Navy, and Air Force processing systems.

33. Examples included the SLQ-33 STADD (surface towed acoustic deception device) and SLQ-34 in an SSQ-74 ICADS (integrated cover and deception) van which could be placed aboard a destroyer.

34. The data which follow are from Mark Wade's Internet *Spacepedia* (as updated to 10 June 1999). According to Wade, the 'cruiser' would have been a conical craft 8.1m long (weight 4500kg) with seventeen rockets at its base, to provide manoeuvring power. The shape would have matched that of Poseidon warheads designed for re-entry. Recovery would have been by parafoil, as demonstrated in classified tests. The MX (Peacekeeper) ICBM would have been an alternative launch platform. There also may have been an air launch option using a Boeing 747 or a C-5 (at about this time the U.S. Air Force fired a Minuteman ICBM after dropping it from a C-5). Wade's illustration is a drawing released in 1983 by DARPA as a 'contemporary study.'

35. Stares, *Space Weapons and US Strategy*, pp207-208, based on JCS studies leaked to the *Washington Post* in 1981. The JCS developed a prioritised list of Soviet satellites in two time frames, 1978-85 and 1985-90. Priority 1 satellites were the EORSATs, which were to be destroyed 'as soon as possible'. Priority 2 satellites were the Salyut space station, 'which can detect missile launches and provide ICBM targeting data'; photo reconnaissance satellites which can be used in tactical situations; navigation satellites which can help direct missiles; and communications satellites. Priority 3 satellites were meteorological and missile warning. Priority 2 and 3 targets were to be destroyed within forty-eight hours.

36. The up-link is at 45.5 GHz (bandwidth 2 GHz), the down-link at 20.7 GHz (bandwidth 1 GHz). The current (1999) Navy share of the low data rate Milstar capacity is 21.6 kbits/sec, amounting to twenty-four channels: eight primary (2.4 kbits/sec each), eight receive only (2.4 kbits/sec each), and eight secondary two-way (300 bits/sec each). Applications include MDUs ([Tomahawk] Mission Data Updates), ATO dissemination, OTCIXS, TADIXS, TACINTEL, and secure voice/data transmission. The Navy share of the Medium Data Rate capacity is 4.8 kbits/sec to T-1: SIPRNET (512 kbits/sec), NIPRNET (128 kbits/sec), JWICS (128 kbits/sec), DSVT (virtual terminal, as used in Desert Storm: 16 kbits/sec), MMT (9.6 kbits/sec), and MDUs (64 kbits/sec).

37. Presumably there were earlier programmes to ensure communications with ships and aircraft in the Arctic. A Cold War Navy programme called Articsat, presumably for a UHF satellite, was cancelled in the early 1990s. The Milstar polar adjunct is the same EHF package that is placed aboard UFOs. Note that the second polar package was originally to have been launched in FY02, but was brought forward as a matter of urgency.

38. This 'Aesopian' approach is typified by Nguyen, *Submarine Detection from Space*. It can be justified in two ways. First, it can be argued that Soviet security precluded any discussion of Soviet systems, but that open discussion was essential; hence the use of supposed U.S. systems as a cover transparent to Soviet readers. The problem with this argument is that there actually were classified Soviet journals (such as those which Colonel Penkovskiy provided to the West in the early 1960s) which would have been better adapted to such delicate discussions. The second possibility is that the Soviets were interpreting real U.S. developments in terms of their own developing technology. Since the interpretations were often dead wrong, due to mirror-imaging, a careful reading could indicate what the Soviets were doing. Moreover, if the Soviets attributed the technology in question to the West, they could hardly avoid discussing it. Nguyen cites a 1979 article by Captains 1st Rank Yu Galich and N Kocheshkov which states (more or less accurately) that U.S. efforts to develop a space-based ocean surveillance system began in 1969. They go on to claim that the first stage was designed to track Soviet surface ships; the second would detect submerged submarines and monitor the launch of their ballistic missiles (later versions of DSP certainly did the latter). A November 1984 Soviet article described a technique in which a carrier battle group detected submarines far beyond its horizon (300km) using a submarine-detecting satellite, a reconnaissance satellite, and a communications satellite; a similar scheme appeared in a 1987 Soviet text on antiship cruise missiles. Both references seem to be mis-readings of the U.S. OTH-T concept. Nguyen reads these statements as 'Aesopian' references to the *Soviet* ocean surveillance system, but in that case the timing would have been about a decade too late. Nguyen cites a 1974 Soviet discussion which mentions the potential of radar to detect a submerged submarine, given detailed knowledge of ocean currents and thermal structure; Nguyen reads this as interest in internal waves. He also cites a May 1976 Soviet Navy article which claims that satellites can detect submarines at moderate (*not* shallow) depths. A standard Soviet text on submarine development carried a similar statement in its 1979 and 1988 editions. Note that at 'great depths' only submarines and fixed acoustic systems could detect submarines.

39. Since the end of the Cold War, U.S. and British submarines have been fitted with oceanographic monitors, generally forward of their sails, to measure salinity, temperature, etc. The U.S. version is TOMS (Total Oceanographic Measurement System); the British equivalent is Type 2081. TOMS can be seen either as a means of detecting a wake (a change in ocean conditions) or as a means of providing the submarine commander with information as to his vulnerability to detection (including nonacoustic) – which is certainly affected by oceanographic conditions.

40. For example, reportedly there was an ASW equivalent to the Kaman Magic Lantern mine detector. For a time a LIDAR, ATD-111, was included in the LAMPS III upgrade. It encountered severe problems, and by the late 1990s it was described as a mine detector in competition with Magic Lantern. Note, too, that unclassified drawings of future U.S. submarines sometimes show topside laser detectors, which would warn the submarine that it was under observation by LIDAR. LIDAR was probably one of four non-acoustic technologies the U.S. Navy was investigating in the early 1990s. According to the heavily censored 1992-3 programme description, the others were Clipper Shale, a wake detector (NA-1/16); NA-4, a high-powered laser; and NA-17, Spotlight (possibly bioluminscence).

41. Quoting the chief of the U.S. SSBN Security Program, Stefanick, *Strategic Antisubmarine Warfare and Naval Strategy*, p19, says that the Bernouilli hump would be no more than a millimetre high (for, say, a submarine running at 5kts at 100m depth). A much shallower submarine at much higher speed (say, 20kts at 30m depth) might generate a hump as much as 15cm (6in) high, which is still quite small (Stefanick, p193). Submarines also generate a characteristic far-field (Kelvin) wake which may persist for tens of kilometres behind it – but which is very weak. It has a characteristic divergent angle of 19.5 degrees on either side of the submarine, which is different from the divergent angle of a surface ship wake. The usual hope of detecting a Kelvin wake is that, because it is systematic, it shows up against the random motion of the sea.

42. Dutton et al, *Military Space*, pp110-111, show Seasat-type SAR images of a surface ship wake. Because the SAR processor assumes the ship is stationary, its image is offset from the wake – by an amount proportional to ship speed. The SAR image on p110 clearly shows both a straight wake pointing aft from the ship and a wider arrow-shaped wake. Its opening angle is apparently constant for all ships, since it is due to the difference between the densities of air and water. Presumably a submarine wake would be quite different, hence (in theory) distinguishable. It would also be far less prominent.

43. According to Stefanick, *Strategic Antisubmarine Warfare*, p19, Edward Y Harper, head of the U.S. SSBN Security Program, testified in Congress that Seasat passed over U.S. missile submarines four times, once over a submarine at 58ft keel depth, at a speed of 5.5kts. In no case was the submarine detected. Stefanick associates the Seasat measurements with possible detection of internal waves rather than of surface wakes.

44. The ROIP association is given in an article in *The Daily Telegraph* (London), 15 May 1999, the conclusion being drawn that British Trident submarines were at risk. The article mentions trials on Loch Linnhe and in the Sound of Sleat (off the Isle of Skye) in 1991. It seems likely that in these relatively confined areas it is easier to model expected wave patterns, which can be tested against observed radar spectra. According to another article in the same issue of the newspaper, trained airborne observers have long been able to identify by eye the surface trace of the 'large pressure wave' a submerged submarine (presumably running at shallow depth) on a calm day. Presumably this is a reference to the Bernouilli hump; a surface wake would be a different proposition. Again, the article reported that 'while [British and American scientists] have made some progress it is only on occasions when the submarine is quite near to the surface and moving quite quickly'. The first report of the Peter Lee affair was published by the *New York Times* on 10 May 1999. According to this account, prosecution was difficult because the U.S. Navy refused to allow discussion of the relevant technology in open court; Lee was convicted only of failing to report the relevant trip to China.

45. Kuzin and Nikol'skiy, *The Soviet Navy 1945-1991*. The authors were Soviet captains, and so might well be authoritative.

46. V N Martynenko and A I Sal'nikov (of the Beriev plant, Taganrog), 'The "Seagull" from Taganrog – Be-12 amphibian story,' *Aviatsia i Vremya* (1997). The 'Gagara' thermal detector used a rotating mirror to scan the sea surface. Company tests began in 1963; the goal was to detect a 0.01 degrees C temperature difference from 500 to 2000m altitude (demonstrated sensitivity, however, was only 0.1 degrees). Experiments showed that at 5kts a nuclear submarine produces a temperature differential of 0.2 degrees, but that it decays rapidly, so that 1km from the source the difference is only 0.01 degrees. This heat wake does rise slowly to the surface. The device was tested over the Mediterranean in 1970 by a Be-12 from the Black Sea Fleet, flying from Mersa Matruh, Egypt. Note that the mechanism the Soviets investigated in 1963-70 was the heat generated by the submarine, rather than a disturbance in the thermal structure of the sea itself.

47. MI-110K, was placed in service in 1963-4 as a result of research which had begun in 1959; presumably it was part of the larger anti-Polaris programme ordered at about that time. The modernised MI-110KM remained in service through the 1990s. There were also electromagnetic and magnetic wake detectors. For example, in 1978 the Soviets placed a static electromagnetic wake detector, Granit, in service; it was distinguished from earlier magnetic loop detectors.

48. In the movie, the satellite system cost the criminal $30 million; at the time the movie appeared, the author well remembers a friend (in Naval Intelligence) saying that such a system would be worthwhile at a hundred times the price.

49. Nguyen, *Submarine Detection from Space*, p12, cites a November 1978 Soviet text referring to 'second-generation SAMOS,' *ie* to DSP. The association is strange because the DSP wavelength was chosen specifically to block out surface emissions, which would be considered noise. The Soviet text refers to naval applications: detection of the stack gases from surface ships, of the inversion track of aircraft, and of the wake track of submarines. It also claimed that DSP satellites could use their IR sensors to detect submarines. A resolution of 30m would suffice. A May 1979 article identified the targets: the 'thermal blob' of a submarine and its turbulent disturbance (presumably its surface wake). Clearly this technique would be useless unless detections could be reported in real time, by radio link; Nguyen associates this requirement with the Soviet fifth-

generation imagery satellites which were first launched in December 1982. He makes the further suggestion that their relatively high perigees were an attempt to widen their search swaths, acceptable because submarine detection did not require very good resolution.

50. Nguyen, p10, cites a 1991 claim by Valentin S Etkin, then head of the applied space physics department at the Space Research Institute, Moscow, that a two-frequency scatterometer (measuring signals reflected from the sea at different angles of incidence) revealed complicated wave interactions at the sea surface. Etkin's statement seems to have indicated an early stage of research, at which basic phenomena would be investigated; Nguyen also reports that Etkin told U.S. personnel that microwaves had been used to detect submerged submarines at significant depths. In 1990 the Soviet air defence journal reported that microwave scatterometers could detect 'naval targets' against a background of sea surface waves, and Nguyen takes this ambiguous phrase to include submarines.

51. Nguyen, p15.

52. Almaz had a checkered history, due in large part to the political problems of its creator, V.N. Chelomey – who was also responsible (at least in part) for the Soviet space ocean surveillance system. The following account is based on one in James Wade's on-line *Spacepedia*, which in turn relies on recent Russian sources. Almaz began in the early 1960s as a 20-ton space station carrying three cosmonauts, competing with Korolev's Soyuz. It would be orbited by Chelomey's huge UR-500K Proton booster. Development was authorised on 12 October 1964 (Soyuz development was authorised on 10 December 1963). Both projects were financed by the Soviet air force (VVS) and by the separate strategic rocket force. In theory both paralleled the U.S. Manned Orbiting Laboratory. When Korolev died in January 1966, Chelomey obtained a decision that Almaz (11F71) would supersede Soyuz, particularly the Soyuz-R reconnaissance version (there was also Soyuz-P, an ASAT interceptor). In final form (about 1970), Almaz was to operate for two to three years at a time, primarily for reconnaissance (it had a recoilless rifle for self-defence). Its three-man crew would be replaced every ninety days. By this time Chelomey's patron, Khrushchev, was long gone. Moreover, Chelomey had made an enemy of D F Ustinov, who eventually became Minister of Defense under Brezhnev. Almaz was finally placed in orbit in April 1973 under the cover name Salyut 2. It failed after thirteen days, but a more successful Salyut 3 was launched in June 1974. Its 1m diameter telescope proved extremely effective; the crew could see numbers on ships and could even identify the types of aircraft aboard carriers. There were also IR and wide-area (topographical) cameras. This craft retained the recoilless rifle for defence against a supposed threat from U.S. Apollo spacecraft, and it carried space-to-space missiles. Crewmen trained for this programme included seamen; for example, Submariner Valeri Rozhdestvenskiy was assigned to develop means of detecting enemy ships and submarines. Almaz was then cancelled in favour of the 'civilian' Salyut and Mir spacecraft, to which its main military experiments were transferred. By this time a big side-looking radar had been developed specifically for Almaz. It was shifted to an unmanned Almaz-K spacecraft, scheduled for launch in 1981. Ustinov apparently finally got his revenge on Chelomey, ordering cancellation of the project ten days before scheduled launch on the ground that it duplicated work on the new Soviet shuttle (Buran). After Chelomey and Ustinov died (two weeks apart), Chelomey's successor as Chief Designer revived Almaz-K, and it was launched on 29 November 1986 – only to fall victim to a launch failure. An Almaz-K was successfully launched (as Cosmos 1870) on 25 July 1987, and a second on 31 March 1991. Nguyen, *Submarine Detection from Space*, p15, interpreted the cancellation of Almaz as a decision in favour of the more compact RORSAT, but current Russian accounts (summarised above) suggest that the two were not competitors at all. Almaz-K was widely compared to the U.S. Lacrosse, and in 1990 the Soviet General Staff journal apparently associated Lacrosse with a submarine-tracking network. Nguyen, p19, reports that in September 1987 Captain 2nd Rank A Makarov and N Chaplygin described a plan, begun in 1976, to deploy up to fourteen side-looking radar (SAR) satellites for submarine detection, which would roughly fit the Almaz programme. However, it would also fit the abortive U.S. Clipper Bow programme, if that were misread as a submarine-detection effort. Nguyen, p20, does say that Etkin claimed that a two-frequency microwave scatterometer was scheduled for deployment aboard an Almaz-2 series of satellites, one of whose purposes was or eventually would be submarine detection. In August 1990 Vice-Admiral V Kalashnikov referred to planned space tests of a 'special detector' to spot surface ships and submarine wakes, and Nguyen takes this to refer to Almaz-1, scheduled for launch late in 1990 (actually launched in March 1991). Almaz-2, scheduled for launch in 1995, was overtaken by the collapse of the Soviet Union.

53. For example, in theory light from a laser tuned to the shifted energy level will be absorbed by atoms affected by the magnetic field, whereas the same laser light will be reflected in the absence of a magnetic field. The differences involved are quite small, and the atoms most affected would be quite close to the submarine, *ie* well below the surface. The intervening water might well absorb so much of the light as to vitiate the technique.

54. As described by Kuzin and Nikol'skiy, op cit.

55. At the time of the Lee revelations in 1999, there were claims in the press that the submarine radar project had begun with Seasat detections of submarine wakes. As for the secrecy of the later systems, the FAS Internet site includes an analysis of U.S. space spending. It notes that overall DoD space meteorology spending did not quite match announced programmes in the late 1980s and early 1990s, but that if two classified Navy programmes, first Link Laurel and then (1989-91) Retract Maple, are added to the total, it does match up. The suggested conclusion is that N-ROSS was simply replaced by one or more black programmes after it was officially cancelled. It is also possible that these black accounts funded oceanographic elements of the Geosat programme, or of the necessary com-

puter analysis. Neither of the two black naval programmes, both apparently associated with ASW, has been declassified.
56. See D T Sandwell and W H F Smith, 'Exploring the Ocean Basins with Satellite Altimeter Data,' a paper on the NOAA National Data Centres website. They mention using data both from Geosat and from the European Space Agency ERS-1 altimeter satellite.

Chapter 9: Enter Tomahawk: OTH Targeting

1. VLF typically operated at 50 bits/sec; SSIXS operated at 4800 bits/sec and had a 64 kbyte (ie 512 kbits) message buffer.
2. Admiral Zumwalt initially included the missile submarine among his Project SIXTY initiatives (a submarine officer pointed out that, given his early statements, simply abandoning it would prove difficult). In March 1971 Zumwalt reported the results of an AdHoc panel on a possible future high-performance submarine. It would be armed with a long-range submerged-launch missile. Zumwalt added that 'I envision that this submarine can ultimately be equipped to utilise satellite reconnaissance for fire control quality detection and localisation'. He therefore proposed submarine tactical missile development as a FY72 initiative. The new submarine reactor was already under development.
3. Typically a ship has a dual-channel receiver with a single processor. The main format is Rainform Gold, the word 'Rainform' indicating 'Rainbow Format', formats described by colour. Phase I accepted data from the largely manual Outlaw Shark Digital Interface Unit (OSDIU). Phase II introduced a new automated link control in place of the OSDIU and introduced the two-way OTCIXS satellite net. It was used for shore-to-ship as well as ship-to-ship communication. Phase III introduced a separate new TADIXS shore-to-ship tactical circuit, leaving OTCIXS for inter- and intra-battle group communication. The current version is TADIXS Phase IV, which is adapted to the new naval communications system. TADIXS is to be integrated into a Global Broadcast Satellite channel, the Integrated Broadcast System (IBS), under a 1996 directive, but that had not yet been done as of 1999.
4. There is one twenty-three-user TACINTEL net in each satellite footprint. As with CUDIXS, one computer in the net controls it by transmitting a sequence order to subscribers. Each subscriber has an ID number, and the onboard processor strips away messages for all others. The DAMA-adapted form of TACINTEL, including indications and warning and OTH-T data, is Phase II (TACINTEL II+). Data can be transmitted at 1.2, 2.4, or 4.8 kbps. A separate OPINTEL broadcast carries information on the location of friendly (Blue) units (BLUFORLOC).
5. OTCIXS is a two-way 2400 bits/sec circuit carrying formatted messages (JINTACCS, USMTF, and OTH-T Gold). Because particular targeting messages may be extremely urgent, OTCIXS has a special FLASH priority level (other messages are coded IMMEDIATE). Unlike other nets, this one does not use a round-robin of transmitters. Its cycle consists of FLASH spots followed by twenty IMMEDIATE slots. When a net member (subscriber) wants to transmit, it sends a request into either the FLASH spot or (randomly) into an IMMEDIATE spot. The first subscriber to gain access transmits, the others waiting for it to finish. Despite the use of a satellite, all messages are repeated three times to insure that they would be received. FLASH messages always supersede IMMEDIATE messages. In addition to the ships of a battle group, the OTCIXS net includes shore sites such as the relevant FOSIF and Submarine Operating Authority (SUBOPAUTH).
6. There are two modes, a group broadcast at a scheduled time, or a reply to a submarine (the system sends the submarine all messages meant for it). The system carries submarine broadcast messages in unformatted text form plus targeting information in formatted computer-readable standard message formats (RAINFORM formats). One 25-kHz wide channel on each FLTSATCOM satellite was allocated to SSIXS (transmission rate was 2400 or 4800 bits/sec). The SSIXS computers (located at the four Submarine Broadcast Controlling Authorities and their back-ups) also assemble the VLF/LF broadcast, which is stored and transmitted by the Integrated Submarine Automated Broadcast Processing System (ISABPS) for the VLF/LF sites. The SSIXS net can span more than one satellite footprint, and it can support up to 120 submarines.
7. TENCAP is now also known as J-TENS (Joint Tactical Exploitation of National Systems).
8. Tactical Receive Equipment (TRE) is the special receiver needed to decode highly classified intelligence broadcasts. TDDS is the TRAP Data Dissemination System, which sets up the broadcasts. Unlike TADIXS-B, TRAP is apparently in the form of high-interest reports plus emitter parameters (so that users can acquire and track the reported contacts). TIBS supports up to ten producers, fifty query nodes, and any number of receivers. It takes its data from systems such as the J-STARS aircraft and the Air Force's Rivet Joint (RC-135), whereas TRAP and TADIXS-B take their data from national systems. Compared to TRAP and TADIXS-B, TIBS offers tracking accuracy rather than simple cueing. The Army's TRIXS is based on its own organic corps-level signals intelligence systems, precise enough for targeting. In addition to the Army's Guardrail (RC-12) and its Airborne Reconnaissance Low (ARL: DASH-7 aircraft), it can use data from Storyteller (EP-3E/E-8 J-STARS), the Contingency Airborne Reconnaissance System (CARS) on board a U-2, and a UAV. The TRAP Data Distribution System (TDDS)/TADIXS B allows multiple data sources to pass data to an unlimited number of users via satellites. The TIBS allows up to twelve providers to pass data to up to 250 addressed terminals via satellites or line-of-sight radio. The TRIXS net allows five data providers to address up to 100 field terminals using airborne relays. BINOCULAR, an NSA broadcast, integrates nine UHF broadcasts to provide electronic order of battle updates; to some extent it can replace TRAP/TADIXS-B. In 1996 the

Assistant Secretary of Defense for C³I ordered all of these broadcasts integrated into an Integrated Broadcast Service (IBS).

9. Neither radar ducting nor the potential (at UHF frequencies) for tropospheric scatter was apparently taken very seriously, on the theory that neither was always available (the Soviets seem to have been willing to use intermittent ducting). Note, however, that a duct radar was included in design possibilities for the 1970 Sea Control Ship, a pre-Tomahawk study.

10. The U.S. Navy initially planned twenty-four Classic Outboards to provide two for each of twelve carrier battle groups; eventually thirty-six were bought. They were installed on board missile cruisers (CGN 36 and CGN 38 classes) and destroyers (at least DD 963-5, 967-70, 972, 976, 983, 985, 988-992). Plans for installation on board large-deck amphibious ships (LHA/LHD) were abandoned in 1993. Outboards were also installed on board some foreign warships: three in Germany, one in Greece, two in the Netherlands (probably on board the *Tromp* class), two in Norway, one in Spain, one in Turkey, twelve in the United Kingdom (three on carriers, later moved to Type 42 destroyers, three on Type 42s, and six on Type 22 Batch 2 frigates). In contrast to the US, British, and Dutch installations, the others were probably on board intelligence-collecting ships (AGIs). A joint US/UK Outboard upgrade programme announced in 1995 called for sixteen new systems for *Spruance* class destroyers (armed with Tomahawk) and twelve new ones for the Royal Navy (for Project Horizon frigates, now cancelled in favour of a new British Type 45 design).

11. The situation was actually worse. Because the necessary computer technology for a programmable set did not yet exist, the PID was hard-wired for a particular radar, 'Head Net C', which in 1975 equipped some of the most important Soviet warships – usually as a secondary set. Some of the most important Soviet warships, such as *Kiev* class carriers, were never fitted with this radar. Apparently the Tomahawk missile was too small to accommodate the lower-frequency receiver needed to use the more important 'Top Sail' as identifier. This problem was never overcome before the anti-ship version of Tomahawk was withdrawn from service.

12. Charger Horse integrated signals intelligence from numerous sources: from all the ships off Vietnam, from aircraft, and from land intercept stations. Thus it included Big Look (BRIGAND) aircraft (probably EP-3s and EC-121s). It was coordinated from the Philippines, using standard teletype links. One effect of coordination was that individual ships could be assigned their own specific sectors and frequency ranges, to improve overall coverage and reduce duplication. In the past, the SIGINT detachment on board a ship had been responsible mainly to NSA, with provision for the ship's self-defence; now the mission expanded to overall tactical intelligence. Signals intelligence, including Big Look, was particularly valuable for tracking aircraft, since it could exploit the North Vietnamese radar system. Charger Horse in turn paralleled Iron Horse, which netted all available radar systems (ships' NTDS/Link 11 nets, the ATDS system, the Marines' radar system, and the Air Force's BUIC system within South Vietnam). Iron Horse became operational in mid-1967.

13. For the origins of BRIGAND, see Price, Alfred J, *The History of U.S. Electronic Warfare* (Washington 1989), Vol 2, pp281-5. In 1999, Racal announced a version of BRIGAND, based on its work in ESM, which could provide a ship or submarine with a covert surface search capability. This concept had apparently been developed entirely independently of BRIGAND.

14. The new capability, introduced in Cruiser-Destroyer Group ONE, was called Battle-Group Data-Base-Management (BGDBM). The launch control group programs the missiles. It is, therefore, used to insert modifications to land-attack missile flight plans.

15. Second Fleet introduced the FOTC about 1987. Initially the FOTC ship carried the OTH picture (including correlation between offboard and organic sensors) and passed tracks to the other shooters. The FOTC ship became, in effect, a bottleneck; the system operated too slowly and delays made shooting inaccurate. The system was soon upgraded to make picture-keeping co-operative, the FOTC ship acting as track manager rather than as sole correlator *and* manager.

16. According to Tuttle, 'A Brief History of JOTS', in DiGirolamo (ed), *Naval Command and Control*, during the Northern Wedding 1981 exercise with the assistance of an analyst who had written many of the existing stand-alone tactical programs and an enthusiastic staff, Carrier Group 8 (USS *Eisenhower*) combined many of the programs into a single package running on a standard non-military computer (HP 9845B). Total cost was about $25,000. Tuttle called it a Tactical Search and Surveillance (TSS) system.

17. Even so, according to Tuttle, the system facilitated submarine interceptions at ranges up to 1000 miles, air interceptions (based on passive data) at 800 miles, and the use of new automated surface surveillance techniques. The software was distributed to all Carrier Group Commanders, and the HP 9020 computer selected as the future platform. This system was called JOTS I. JOTS II, tested on board *Eisenhower* and *Independence* in 1983, was less successful; it was an attempt by the Navy laboratory system to integrate other tactical decision aids into JOTS I. JOTS II+ (*Coral Sea*, 1984) integrated Link 11. JOTS III was networked, so the system could serve several tacticians simultaneously.

18. Tuttle emphasises in his article that, unlike TFCC, JOTS IV had been designed by the staff which would use it in wartime, so it met their needs. It performed well on a six-month deployment, and became standard on board eight carriers as well as the battleships. This system in effect killed the separate TFCC. Tuttle described it as the battle staff's answer to the 'Alfa' class submarine.

19. JOTS systems were renumbered. JOTS I was the version running on a DTC I (Hewlett-Packard HP 9020) computer. JOTS I was USQ-112; JOTS II was USQ-112A (1991), running version 1.15 software. The baseline NTCS-A implemented on DTC II computers is USQ-119. The initial upgrade version (with TAC-3 computers) is USQ-119A. The JMCIS baseline version is USQ-119B. The integrated C⁴I

version is USQ-119C. The TAC-4 version is USQ-119D.

20. Commercial-grade computers were first attractive and then necessary because the technology was changing so fast; according to 'Moore's Law,' computer power doubles every eighteen months. Unfortunately it takes much more than eighteen months to test a computer thoroughly. Bugs do exist, as for example the famous one in the early Pentium chips (which could not divide properly). By the one central military law, Murphy's (or Sod's) Law, such bugs will inevitably affect the most critical functions the computer controls.

21. It will be no surprise, then, that the central elements of the JOTS software are a tactical data base manager (TBDM), which forms correlated tracks) and a geographic data base ('The Chart'). The Chart automatically plots TBDM data for display, and it also provides such navigational graphics as projected tracks.

22. Opcon is a text system. Later it was supplemented by the RAF's ASMA (Air Staff Management Aid) data, a text system providing information on air fields (status, readiness) and opnotes. JOTS uses graphics similar in concept to that used in the U.S. combat direction systems. PFSS is concerned with the surface and air pictures; it does not pass detailed submarine information to the deployed surface force.

23. 'What-if' questions naturally arise as soon as a computer data base becomes available. For example, much of the effectiveness of the computerised P-3C was ascribed to the new ability to compare different possible sonobuoy-planting plans. Standard naval command systems, such as the U.S. NTDS and the British ADAWS, automatically asked whether particular weapons could engage particular targets. However, they lacked the computer capacity to ask the wider planning questions. TDAs or planning aids do not work in the 'real time' of combat direction systems, and thus are natural complements to the time-late data provided by a system such as JOTS.

24. OSS was originally part of the FCC conceived in the 1970s. An upgrade to JOTS II standard (DTC II computers and new large-screen displays) was begun in FY89 (1988; the operational requirement was issued in December 1987). An Increment I upgrade (FY91-92) standardised the centres (which ultimately have about fifteen workstations each) and prepared them for further upgrades by installing a local area network suitable for future computers.

25. According to Borik, 'The Silent Service is on the Air', the BGIXS installation on board a carrier initially involved complex massive equipment. Presumably it was associated with the TFCC. BGIXS II is a lower-capability (2.4 rather than 4.8 kbits/sec) COTS version, using a laptop computer. Both systems are stand-alone; they transfer data to other shipboard systems by moving computer diskettes, rather than by a direct connection.

26. Late versions of SS-N-14 did have a secondary anti-ship capability, the missile being command-guided into a target.

27. According to the NRL 75th anniversary awards booklet, NRL produced a system concept report, 'Electronic Warfare Unintentional-Modulation Processors: System Definition Considerations' in June 1988, and in June 1993 NSA chose its system over those offered by industry and by other government laboratories; it is now standard within the U.S. defence community. The booklet refers to L-MISPE in a test called Musketeer Dixie II. Using the NRL system, particular radar emitters (eg radars on drug-carrying ships) can be catalogued and tracked, and their characteristics transmitted to other sites for continued tracking – eg from an aeroplane to a surface ship. The technique is currently used by all the U.S. services, and as of 1998 one library of over 10,000 distinct signatures had been compiled, and was being shared among tracking sites. The U.S. Coast Guard retained the old WLR-1 ESM set because it could measure radar pulses very precisely, hence make it possible to recognise a specific ship by her radar fingerprints. Note that pulse analysers are being added to the standard U.S. SLQ-32 ESM system.

28. The radar mode was ISAR, inverse synthetic aperture. It uses the motion of the ship target for imaging. The radar sends out a very narrow beam at a precise frequency. Since the ship is moving (by rolling), it applies Doppler to the reflected signal. For a target broadside to the radar, the amount of Doppler shift is proportional to the height of the illuminated part of the ship. As the beam sweeps across the ship, beam width by beam width, the Doppler shift changes, because different parts of the ship are at different heights above the sea. A crude silhouette emerges. Although it offers little to the eye, apparently a computer can use the parameters of this image for quite precise identification. ISAR is attractive because the image is generated very rapidly; the aircraft need not fly the steady course required for a classical synthetic aperture radar (SAR). With the advance of electronics, it is now possible to compensate continuously as the aircraft moves; some projected maritime patrol radars offer 'spot SAR', which ought to give a much better image, particularly of small ships.

Chapter 10: Defending the Fleet: The Outer Air Battle

1. According to the Cross report, which is based on contemporary official documents, at this time the planned peacetime posture had two carriers each forward-deployed to the Mediterranean and the Far East. Four carriers would be on each coast for training and upkeep. The scenario which follows is adapted from one the Navy produced to support its S-3A rationale (19 April 1972, resubmitted 15 February 1973 unchanged). Presumably it reflects Zumwalt's thinking. The scenario depended on the carrier's ability to shift from one role to another (composite, sea control, power projection) by changing its aircraft mix without returning to the United States. At the outset all carriers would have a composite aircraft mix, for maximum flexibility. The Soviet threat as understood at this time included 50 to 100 'Backfires' (plus the earlier missile aircraft), 28 missile destroyers, 75 missile cruisers, and 200 missile and torpedo submarines. Studies already showed that it would take at least three carriers in the Greenland-Iceland-UK (GIUK) Gap to block the Soviet

long-range bombers; they would also deter the Soviets from using their surface fleet, and they would protect the vital ASW base, Iceland. Three carriers further south would be required both for ASW and to stop long-range reconnaissance aircraft (mainly 'Bear Ds') supporting Soviet submarines. In the Mediterranean, the U.S. carriers would provide the main NATO deep-strike force; CINCEUR, the NATO European commander, wanted a total of five carriers, while four would be needed to deal with a determined Soviet naval attack. Ideally two more should be available as reinforcements in the event of battle damage or an unexpectedly strong Soviet attack. At least four were needed in the Western Pacific, where the Soviets might attack over a wider area because local land air defences were weak. Thus ideally the United States needed a total of twenty-one carriers, against a total available force of twelve. The main hope, then as later, was to make up the difference by conducting operations in sequence, first securing sea control.

In the 1973 scenario, on mobilisation (M-day), three of the four Atlantic carriers would be assigned to guard sea lanes in the North Atlantic; the fourth would reinforce the two in the Mediterranean. Three of the four Eastern Pacific carriers would head around South America (because they were too large to pass through the Panama Canal) to reinforce the sea lane force in the North Atlantic. The fourth and the two Western Pacific carriers would cover the North Pacific sea lanes. Carriers on sea lane patrol would change their aircraft mixes for that purpose. S-3s would be flown out to the Western Pacific to provide the two carriers there with the appropriate mix. Once war began, only the three Mediterranean carriers would conduct any strike operations at all. The sea lane protection carriers would try to keep Soviet air and naval forces from penetrating far enough south to endanger shipping (the carriers swung from the Eastern Pacific would provide further defence in depth in the mid-Atlantic. Convoys headed to Europe would be routed as far south as possible to avoid bomber attack. During the second month of the war two of the Atlantic carriers would take on more strike aircraft and would support the land battle from the North Sea and the Bay of Biscay. The submarine threat in the Mediterranean having been brought under control, all the carriers there would also conduct air strikes. Much the same might happen in the Western Pacific. By the third month of the war, the Soviet submarine force would have been largely destroyed, and the carriers would carry out strikes against land targets.

Even so, the twelve-carrier force was clearly insufficient. For example, one or more carriers would almost certainly be in long-term refit at the outbreak of war. The number available in the Pacific would not suffice. Probably all the Pacific Fleet carriers would have had to swing to the Atlantic, leaving the Soviet Pacific Fleet relatively free to act.

To Zumwalt, the obvious shortfall in large-deck carriers justified building a class of far less expensive 'sea control ships' which could deal with submarines but not with bombers. He hoped to build eight for the cost of a single full carrier. They would be used mainly to escort convoys, but it is clear from the scenario that they might also have replaced the three large decks in the Central Atlantic. In that case one of those carriers might have been released to deal with the three-dimensional threat in the Pacific, and the other two might have joined the war in the Mediterranean, helping to meet the otherwise impossible requirements stated above.

The 1972-3 war plan contrasts starkly with the Maritime Strategy. Despite the increased funding available under the Reagan Administration, the carrier force was unlikely to reach the numbers desired. Again, operations would have to be sequential. However, the hope was that the offensives into the North Atlantic and North Pacific would very quickly gain maritime dominance, so that there would be little need to hold carriers back for sea lane protection. That in turn depended on success in the Outer Air Battle.

2. The ASW carriers (CVS) were converted *Essex* class fleet carriers. Originally they had been used to fill the mid-ocean gap in air ASW coverage, due to the limited range of Neptune ASW aircraft based on the Atlantic coasts. When the P-3 Orion entered service, this gap was eliminated.

3. In the late 1960s then-Captain Zumwalt had headed the Navy's studies organisation, Op-965, which was responsible for its long-range plan. In the last report on which he worked (1968), Zumwalt emphasised the likely effect of Soviet submarine silencing, which would drastically reduce the value of the long-range SOSUS sound surveillance system and thus that of the fleet of P-3 maritime patrol aircraft it cued. His proposed solution was a fleet of small ASW carriers. When he became CNO, the small carriers resurfaced as the ultimately-abortive Sea Control Ship (a version of which ultimately appeared as the Spanish *Principe De Asturias*).

4. For example, Nelson's victory at Trafalgar ensured the degree of sea control which made it possible for the Royal Navy to support Wellington's Peninsular War – which in turn badly weakened Napoleon's forces. Conversely, the Royal Navy's failure to secure positive control of the North Sea during the First World War made it impossible to mount major sea-based operations on the flanks of the German Army, and thus, it could be argued, prolonged the agony on the Western Front. As it happened, the one big seaborne operation proposed (in 1917) was designed to support sea control by seizing U-boat bases in Flanders; it would also have outflanked the German army. This history was very meaningful to an analyst, like the present author, working in the 1970s: a NATO *success* on the Central Front would have led to something like the Western Front during the First World War. A maritime-based strategy was a way to avoid that sort of slaughter.

5. The advantage of the passive system was that it would provide the incoming bombers no warning, and thus it could not trigger Soviet deceptive measures. As first conceived about 1978, BGPHES would have been a remote-controlled package on board ES-3s. Then the carrier offensive electronic intelligence aeroplane, the EA-3, was scheduled for retirement. ES-3s were the obvious replacements. The carrier end of the system encountered serious development problems. As a consequence, the system was not fielded until the late 1990s – and it was soon retired.

6. It was usually said that EP-3Es would pick up the

chatter, but they would have been on station only intermittently. In retrospect it seems likely that the reference to these aircraft was intended to conceal the use of the very secret ELINT satellites (via a TENCAP arrangement).

7. For example, in a 1982 study of CAP arrangements for the Outer Air Battle, a squadron could mount a defence in depth: two F-14s in long range CAP positions up to 200nm from the carrier, backed by six in medium range positions, and by two interceptors on deck alert. Once a long-range CAP fighter engaged, medium-range fighters would move up to fill the long-range CAP station. Tactics were predicated on very limited communications, due to the expected intense jamming.

8. Studies of an Integrated Tactical Surveillance System (1982) also showed that an OTH radar was the most cost-effective sensor to be used for wide-area sea surveillance. The CNO Executive Board ratified this decision in December 1982, and production was scheduled for FY88.

9. Three production radars were authorised in FY88-89, but the programme was stopped in 1991 (FY92) due to the end of the Cold War. They became attractive weapons for the drug war. Thus the Amchitka radar, shut down on 15 September 1993, was moved to Virginia (its surveillance area is the Caribbean). Two more were completed about 1995: one for Texas and another for Puerto Rico. The surveillance area for each site is a 64-degree wedge between 500 and 1600nm range (the ionospheric bounce technique rules out shorter ranges).

10. Forward Pass had first been conceived in the late 1950s. During the late 1970s it had been seen as a way of eliminating the big fighters, whose deck requirements kept up the size of U.S. carriers. Earlier in the 1970s there had been interest in a long-range SAM (LRSAM) relying entirely on the Aegis radar. The Mk 26 missile launcher had been designed specifically to allow growth from the solid-fuelled Standard Missile to a future longer-range rocket-ramjet, but the original Standard turned out to have such growth potential that the new missile was never built.

11. The chief proponent of this 'SAM Risk Closure' approach (SM-3) was Anthony Battista, a senior staffer with the House Armed Services Committee. It was an SM-2 Block II with a new tandem booster and a new seeker with an electro-optical adjunct to the usual semi-active seeker and perhaps an active adjunct. Eventually it emerged as the current SM-2 Block IV. SM-3 was advanced as an alternative to a new larger-diameter ASAM/AAAM, which would have used more of the volume available in the new vertical launchers. The Defense Department view was that SM-3 was necessary insurance against the possible failure of the ASAM/AAAM program; Battista reportedly saw SM-3 as a way to kill a wasteful new missile programme, by monopolising the available money (the Navy argued that the SM series was reaching the end of its potential, so SM-3 would only be marginal at the very long ranges desired).

12. JTIDS, the Joint Tactical Information Distribution System (Link 16), offered far more information capacity than its predecessor, Link 4. From the Outer Air Battle point of view, it was probably most significant that using JTIDS a fighter could transfer its tactical picture back to the fleet for use. Earlier links had concentrated on passing orders out to the fighter. The only major drawback was range, which is limited to about 300nm between users. Presumably the assumption was that the fighters in a grid would never be more than 300nm apart, even though the grid might extend over about 500nm from carrier to outer edge. The Navy tried but failed to promote an alternative form of JTIDS (D for distributed), with greater capacity than the basic Air Force-led system. It might be argued that ultimately an Outer Air Battle should use satellite communications, but no such system was (or has been) proposed.

13. LORAINE had a real chance of development because in the late 1980s it was of interest as a way of suddenly reviving U.S. continental air defences. ROTHR radars could, in theory, have been installed very rapidly to cover the approaches to North America, and vertical launchers might easily have been emplaced. The LORAINE vehicle was proposed as an outgrowth of a DARPA manoeuvring re-entry vehicle project of the 1980s, SWERVE.

14. The key to within-horizon fighter control was TACAN, the Tactical Air Navigation beacon, for decades a familiar masthead feature of U.S. warships. It allowed a fighter automatically to find its position relative to a carrier or other ship, and thus to report that position.

15. The 12ft long Schmidt IR camera (with 6000 detectors) onboard a current-generation DSP satellite has its line of sight offset 7.5 degrees to the vertical axis of the satellite, so that it scans continuously as the satellite rotates about its axis at 6 RPM. Phase I satellites used 2048 PbS detectors (2.69-2.95 microns) to detect the hot plume of a missile exhaust against the cold earth (Below the Horizon, or BTH, detection). Detection was simplified because at this particular range of wavelengths the atmosphere absorbs 95 per cent of radiation from the earth, so the earth appears to be dark (some very bright sources are, however, visible). If the sky is cloudy, the missile becomes visible as it breaks through the clouds. A second capability, detection Above the Horizon (ATH), was added to deal with Arctic launches. The cooler second-stage burn was detected against the cold of space. This technique was preferred to the alternative, placing a satellite in a Molniya orbit. Second-generation satellites (DSP-I, for Improved) carried more that twice as many BTH sensors (to locate missile launches more precisely) plus about 750 ATH sensors. One hope was that DSP satellites could identify which missile silos had launched their weapons, so that the others could be destroyed by U.S. weapons fired in retaliation. The second-generation sensor package (SED, Sensor Evolutionary Development) added a second colour (4.3 microns). Using a second colour allows measurement of the temperature of the spot the satellite detects, and helps screen out extraneous data (note that the first two SEDs, on Flights 12 and 13, lacked the second colour). Missiles with less intense exhausts can be detected more effectively. Much the same follows from the increased resolution (due to more detectors) of the telescope. As a result of these and other additions, the spacecraft grew dramatically, from 1950 to 3500lbs. The first second-generation satellite became

operational on 29 May 1985. Ground stations could not handle ATH data until 1986.

Increasing the number of satellites monitoring the earth improved detection of shorter-range missiles with short burn times. Initially three satellites covered the world. However, in 1988 a fourth was launched specifically to cover possible tactical missile firings in Europe, and a fifth was launched in 1991 to cover the Indian Ocean. Combining data from two or more satellites improves launch point accuracy by about a factor of 4. The ground stations are at Alice Springs, Australia, and at Buckley Field, Colorado. There is also a Mobile Ground Station (MGS), conceived during the Cold War to improve system survivability. The Improved version of DSP was to have incorporated a laser satellite cross-link, but due to technical problems it had to be omitted from DSP-I F-14 (June 1989); it was replaced by an ultra-violet sensor proposed by Aerojet/TRW for a strategic defence programme (presumably 'Brilliant Eyes') cancelled when the Soviet Union broke up. This device is to be mounted in other DSP-Is in place of the abortive laser crosslink.

16. 'Backfire' cannot fly at supersonic speed (*ie* on afterburner) with two missiles underwing, but it can fly at such speed carrying a single missile. Hence it can attack at supersonic speed with one missile or it can attack at subsonic speed carrying two missiles underwing, escaping at supersonic speed. The assumption was that a supersonic attack was so much more likely to succeed that the Soviets would generally adopt that option. We now know that attempts to develop an escort jammer version of 'Backfire' succeeded only after the end of the Cold War, and therefore that during the Cold War 'Backfires' would probably have flown out at the subsonic speed imposed by their 'Badger' jammer escorts.

17. Heritage used a staring (mosaic) sensor instead of the scanning sensor of a DSP satellite, to give more observations of a fast-burning rocket. The programme was sponsored by the NRO (the satellite reconnaissance office) rather than by the U.S. Air Force.

18. Under the 1997 programme, as a Phase I upgrade, JTAGS was integrated with the JTIDS (Link 16) communications net, which can connect it with tactical aircraft (which might attack mobile missile launchers) and with the Army's fire support system (which might use missiles such as ATACMS (an Army-developed tactical ballistic missile) to attack the missile launchers). A Phase II upgrade (FY99-FY04) adapts JTAGS to the DSP follow-on system, SBIRS.

19. The current elements of TES are the Air Force Attack and Launch Early Reporting to Theatre (ALERT), JTAGS, and the National Systems Tactical Detection and Reporting (TACDAR, *ie* TENCAP) systems. ALERT is intended specifically to support theatre missile defence.

20. The mid-1992 House of Representatives report on the Defense Authorization Bill apparently inadvertently revealed a black system, Have Gaze, controlled by the previously black Navy-controlled Joint Counter Low Observable Office (JCLO). The satellite used in the experiment was controlled by the Air Force, hence was probably a DSP (the context made it almost certain that detection was by IR). At the time there was widespread talk that the Air Force's claims that its new aircraft were truly stealthy were very excessive, that service largely having ignored the aeroplane's IR signature (which, due to the laws of physics, is very difficult to curb, *eg* due to aerodynamic heating of the airframe).

21. The first Navy study of space-based radar, Albatross, dates from the early 1960s. The Air Force and DARPA spent $500 million developing the Teal Ruby IR system.

22. 'DoD Study Says SBIRS Has Theatre Missile Defense Potential,' *Inside Missile Defense* (7 January 1998).

23. There may be a fear that long-range anti-radar missiles can counter AWACS. At Moscow in 1993 the Russians displayed just such an air-to-air missile.

24. In the mid-1990s the Joint Chiefs announced a new 'national' (*ie* intelligence-gathering) system which would 'consolidate the missions, facilities, and infrastructure of two existing satellites.' According to the 'space programs' web site maintained by the Federation of American Scientists, the new system (for which Martin-Marietta received a contract in 1994) will consolidate the two wide-area ESM surveillance systems (one being White Cloud). The statement that six ground stations would be closed (operations being consolidated at one site) suggests that the new satellite reports via a satellite cross-link such as SDS or TDRSS. The Air Force programme for a space-based radar may be code-named Olympic.

25. The June 1998 list of projected FY01-FY20 launchings was developed to establish requirements for the new Evolved Expendable Launch Vehicle. It includes four classified ones (A, B, C, D), of which B is probably AWACS-B. It is to operate in a low orbit (100 × 100nm), and the first is to be launched in FY04 (which corresponds to the first launch of an AWACS-B in an earlier Air Force document). The table showed Mission B payloads to be launched in FY04, FY06, FY08, FY10, FY12, FY16, and FY18. Of the other classified payloads, A (two in FY02, two in FY04, one in FY05, one in FY07, one in FY09, one in FY10, one in FY11, two in FY12, two in FY14, two in FY17, one in FY18, two in FY19) was in a 'Molniya' orbit; presumably it was replacements for SDS and Jumpseat, which have similar orbits and therefore are equivalent missions for a launch vehicle. C was in a geo-synchronous orbit: presumably it replaces the high-altitude ELINT system (one in FY08, one in FY09, one in FY11, one in FY15, one in FY16, one in FY18). D was in a low polar orbit; presumably it is a photo satellite (one in FY07, one in FY08, one in FY11, two in FY12, one in FY14, two in FY15, two in FY19). The table mentions another classified payload (E), but gives no details. It is probably a lightweight satellite. Note that the table *does not* include any satellite whose orbit seems to correspond to that used for White Cloud. The low-altitude orbit would correspond to a radar warning/imaging satellite.

26. 'Lincoln Labs Study Calls AWACS Sensor Too Heavy for Transition to Space,' *Inside Missile Defense* (18 March 1998).

27. DARPA proposed the lightweight (and presumably low-cost) STARLITE satellite in 1997,

following its own study. STARLITE was then envisaged as a low-cost SAR satellite providing data directly to warfighters. Then a low-cost space-based HRR-GMTI radar concept was added to STARLITE; the Air Force had asked that STARLITE complement its U-2, UAV, and J-STARS. In January 1998 the Defense Science Board Task Force on Satellite Reconnaissance (which had been tasked by the Defense Department, DARPA, and the NRO) recommended that STARLITE be bought as the 'Military Space Radar Surveillance Program'. The demonstration programme was established under a February 1998 Memorandum of Agreement (signed by DARPA, the Air Force, and NRO), and the programme was designated Discoverer II that April. Two System Integration contractor teams are to be selected in January 2000, with a contract award following in May. According to a DARPA data sheet, each satellite would weigh 1000 to 1500kg; two to four could be launched together by a Delta II. Satellites would orbit at 770km altitude (53 degrees inclination), covering the earth up to 65 degrees latitude (North or South). Keys to effective SAR/GMTI operation would include considerable onboard computer memory (over 160 Gbits) and high-capacity downlinks (initially 544 kbits/sec, ultimately 1.096 Gbits/sec). The user would command the satellite by uplink (less than 200 kbits/sec). These figures are lower than those currently planned for UAVs. The satellite radar would be an X-band set with a 600 MHz bandwidth, capable of operating in either SAR (spotlight, strip, scan) or GMTI modes, as commanded from the surface. The beam would be steered electronically, with a pointing accuracy better than 0.02 degrees, and range resolution would be 0.3, 1.0, or 3.0m, as selected by the surface user. The GMTI function would detect targets moving at speeds between 0.4km/hr (about 0.2mph) and 100 km/hr (about 62mph). The surface terminal would be a modified version of the existing UAV downlink terminal.

Chapter 11: Copernicus

1. Data bursts are sent in frames, each containing three data segments consisting of blocks transmitted in parallel. Timing is so critical that the frame includes a special ranging signal, to determine the precise distance from the transmitting ship to the satellite. DAMA has been applied to both the 5 and the 25 kHz channels of the UHF satellites. The 5 kHz DAMA uses an 8.96 sec frame consisting of 1024 building blocks beginning with a forward orderwire and a return orderwire, then communications of various segment lengths. This supports at most one 2400 bits/sec voice channel per frame. Significant round-trip delays make voice communication difficult. The 25 kHz DAMA uses a 1.3866 sec frame with three fixed-size user segments per frame (A: 75-1200 bps; B: 75 bps – 16 kbps; and C: 75-2400 bps; it can use sixteen different fixed formats, stored in the terminal). The delay in voice communications is reduced to two to three seconds (much less than in the 5 kHz DAMA), but this technique is still far better for transmitting data. For voice calls, a user can request (from the DAMA controller via the DAMA link) a dedicated channel (Demand Assigned Single Access, or DASA) either 5 kHz or 25 kHz wide. If the request is approved, terminals automatically change to a clear channel. The channel is maintained for a pre-determined time or until the users hang up.

2. This system replaced an earlier Navy Narrowband Secure Voice system. ANDVT replaces the pseudo-digital Park Hill, which was widely used on HF and some satellite nets. ANDVT could be used to send images. Submarines on reconnaissance duty use a system called Cluster Knave (or Nave) to digitise the photographs they take, which are transmitted back to the United States via the ANDVT channel.

3. OTCIXS Phase II and OTCIXS Phase IV/V are versions suitable for a satellite DAMA channel. OTCIXS II and TADIXS IV became operational (on a limited basis) in FY92. As of 1999, quad DAMAs provide UHF capacity (25 kHz channels only) of 48 kbps (in sixteen to twenty 2.4 kbps nets). This includes the various IXS nets, TDDS, CUDIXS/ NAVMACS, and ANDVT, all at 2.4 kbps each; and the TACINTEL channel, which requires 4.8 kbps.

4. There is a standing operational requirement to fit all P-3Cs with satellite communications equipment. Installations began in FY94. By mid-1997, 171 P-3s had the ARC-187 radio, which provides satellite capability. It also provides a secure radio waveform (Have Quick). Ultimately all 238 P-3Cs and the four Special Projects P-3s (ELINT and EO sensor aircraft externally nearly indistinguishable from conventional P-3s, called Reef Points) are to have the same communications outfit, comprising an ARC-182 VHF radio and two ARC-187 UHF satellite communications radios. The satellite system includes DAMA and secure voice capability (ANDVT). EP-3Es also have satellite capability. The P-3Cs modified for OTH-T (AIP or OASIS III aircraft), the EP-3Es, and the Reef Points aircraft all have, in addition, the special terminals and special receivers necessary for them to participate in the major TENCAP nets (TADIXS B, TRAP, TIBS). The EP-3Es can also originate special messages, as modified under a communications improvement programme called Story Teller. E-2Cs require a new mission computer to accommodate capabilities including satellite communications. This is provided in new Group 2+ aircraft, which have mini-DAMAs and intelligence terminals, so that they can share in the TENCAP nets.

5. These reports would include indications and warning, cueing, tracking, targeting, engagement, target damage assessment (BDA), and retargeting. Presumably this calculation applied to ships without facilities to receive the new computer-oriented links, TADIXS and OTCIXS. HITs were High-Interest Target reports.

6. The COE is the direct descendant of the base software designed for JOTS. It provides communications interfaces, message processing, track data management, track correlation (between sources of data), regional database management (so that a user can filter out information about other areas), and tactical displays. Given the underlying philosophy of Copernicus, it has to be portable, so that it can be installed in newer generations of computers. The Navy COE project began in FY98 as part of the

larger Joint Defense Information Infrastructure (DII) COE.

7. As envisaged about 1993, TADIXS would embrace fourteen channels: OTC Battle Management, ELINT, SEW (Space/Electronic Warfare) Management, ASW Management, AAW (JTIDS), Tactical Intelligence (TACINTEL), Cruise Missile Targeting, High Command, Intelligence Broadcast, the Common High-Bandwidth Data Link (for data from reconnaissance aircraft), an Intelligence Net, a Combined Broadcast, a Single Integrated Satellite Broadcast, and the administrative channel (NAVIXS).

8. The key is IP, the Internet Protocol. The military E-Mail equivalent is the Defense Messaging System (DMS). The new slogan is 'smart push – warrior pull' for information. IT-21 is ultimately to embrace virtually all data exchange, including time-late track data used by command systems.

9. JMCOMS is being deployed in builds. Build 0 established a fleet baseline. It expanded land-based networks (NIPRNET, SIPRNET, JWICS) to sea. However, it provided no media management. Build 1 offers standard Internet services (such as E-Mail) and dynamic bandwidth allocation. Build 2 offers multicast (one broadcaster, multiple receivers, as in the Fleet Broadcast) and standard voice/video services, including the Army's SINCGARS and the Global Broadcast System (GBS). Control of selected radio room equipment is automated. Build 3 offers tactical voice/video services. The prototypes were in the Aegis cruiser *Yorktown* (CG 48) and the amphibious assault ship *Boxer* (LHD 4), with Build 1 first in one battle group, then in two more in FY97, and four more in FY98. Build 2 will be in four battle groups in FY99 and two in FY00, with Build 3 in two in FY00, four in each of FY01 and FY02, and the last in FY03.

10. The integrated terminal is to handle all frequencies above 2 GHz, which means mainly EHF (Milstar) and SHF (DSCS) as well as some satellites operating at similar frequencies.

11. For example, the wideband (2 MHz – 2 GHz) DMR can tune over the 5 kHz-wide UHF channels currently used by the Air Force and by Special Forces. It is also intended to translate between naval messages and those in the formats and waveforms used by the other services.

12. One problem is that this is an average. It is by no means clear how time-urgent communications such as the tactical links (*eg* Link 11) should be handled; their users may prefer slow but reliable data transfer (at 2250 bits/sec) over a dedicated 'stovepipe' to the higher *average* rate of JMCOMS links.

13. The service began with two MARECS satellites (launched 20 December 1981 and 10 November 1982; the initial MARECS B was lost on 10 September 1982) over the Atlantic and Pacific, plus packages on board Intelsats to cover the Indian Ocean. The MARECS shore-to-ship link has thirty-five telephone channels, the return link fifty.

14. The ship terminal uses a 0.85 to 1.2m reflector. This commercial system uses a 6 GHz (C-band) shore-to-satellite uplink and a 1.6 GHz (L-band) ship-to-satellite uplink. Downlinks are 4.2 GHz to shore and 1.5 GHz to a ship. The system currently uses two series of Inmarsats. The contract for Inmarsat 2 (F2) was awarded in March 1985 (bids were requested in August 1983), the first of four being launched on 30 October 1990. Each provides five times the capacity of the original MARECS, 125 voice channels for shore-to-ship and 250 return. This version is used for the 64 kbits/sec U.S. Navy link. The contract for Inmarsat 3 (F3), which provides spot channels specifically to handle mobile telephones (Mini-Ms), was signed on 1 February 1991. These satellites have five spot beams (plus a global beam), among which power can be allocated by the satellite. Power capacity is ten times that of Inmarsat 2. The fifth and final F3 was launched in December 1997; the F2s are now on standby (they have the same frequency plan as the F3s, hence the two cannot be used simultaneously). A fourth-generation Inmarsat P is to be specially adapted to mobile telephones, and thus will compete with systems like Iridium. The shipboard terminals are called Ship Earth Stations (SES). Some of these terminals were used by the U.S. Army in the Gulf War and in Somalia. The standard Inmarsat channel is 2400 bits/sec. The new 64 kbit/sec service uses multiple channels on board an F2 satellite brought back into service specifically for that purpose. To avoid interference with F3s, they are rotated about 40 degrees away from their current positions (near F3s). Probably the two F2s used for the service will be joined by an F3 during the summer of 1999.

15. In March 1997, for example, the IT-21 software standard was Windows NT 4.0, and MS-Exchange was the standard E-Mail system. The standard computers had to have 200 MHz Pentium Pro chips, 64 MB of RAM, a 3.0 GB hard drive, and a 3.5in disk drive, as well as other standard (state-of-the-art) components. Note that the standards were all commercial. The computers are connected by a fast (ATM) local area net on board ship.

16. The system is named after the element Iridium, atomic number 77 (the size of the constellation initially planned). The initial system uses seventy-five satellites, including in-orbit spares; it was completed in 1998. The system employs eleven satellites (plus a spare) in each of six orbital planes, all in 780 km circular near-polar (inclination 86.4 degrees) orbits. Each satellite provides sixteen beams at 1.6/2.5 GHz, with a Ka-band cross-link and a Ka-band link to the gateway. The system is comparable in principle to a cellular telephone system, each satellite handling forty-eight cells, 670km across. Data rates are 2.4 kbits/sec for voice and 4.8 kbits/sec for data. The system was proposed in June 1990 and approved by the Federal Communications Commission in January 1995. The operator is the Iridium Corporation, led by Motorola. Reportedly Iridium was developed after the wife of a Motorola executive complained in 1985 that she could not get through to the United States using her cell-phone. Other, similar, networks, including a space-based Internet, were being financed by late 1998.

17. From the report of the DoD Space Architect, 29 August 1996.

18. The original orbit was 614 × 767km at 89.9 degrees inclination. These store-and-forward satellites each have paired 80C86 microprocessors on board. Each processor has 1.2 Mbytes of RAM including eight 192 kbyte mailboxes. The down-link

operates at 2400 bits/sec, and can transmit 64 kbytes during one pass over a receiver. The Macsats were operated by the Naval Satellite Operations Centre at Pt. Mugu, CA.

Chapter 12: The Global Positioning System

1. Timation provided two-dimensional positioning. At about the same time the U.S. Air Force was making concept studies of the 621B three-dimensional system. The Air Force was selected to consolidate the two into GPS. The first test signals were transmitted from Navigation Test Satellite 2, launched in June 1977.
2. A user automatically chooses the four closest satellites and locks onto their signals, obtaining a pseudorange to each.
3. Worse, three existing satellites (PRN-06, -09, and the newly-launched PRN-23) were experiencing hardware problems. For example, the solar panels of the twelve-year old PRN-06 were down to 60 per cent of their rated power. The Air Force operators therefore turned the GPS payload off when the satellite was not bathed in sunlight (*ie* when its batteries would have been used), so that the batteries could always recharge. The satellite was kept operational throughout the Gulf War. PRN-09 lost its attitude control; it had to be switched to spin-stabilisation, and pointed by ground control during the crucial two-hour period each day when it overflew the Gulf area. PRN-23 accidentally locked its solar panels in place; they had to be slewed on command from the ground.
4. Plans initially called for twenty-one operational and three back-up satellites, cut to eighteen and three in the early 1980s to save money, but then returned to the original figures in 1987; plans then called for the full constellation to be in place by 1993. The first eleven satellites (Block I) were for research and development; they had no SA (selective availability) or A-S (anti-spoofing) capability. They did, however, transmit a useable navigational signal. Design life, set by the atomic clocks and by the attitude control system, was five years (one satellite managed more than double this). The follow-on Block II (twenty-eight satellites) has SA/A-S features. They are radiation-hardened, and they store 180 (vs. 3.5) days worth of 50 Hz navigational message data for transmission. They also automatically detect some kinds of errors, switching to a non-standard signal to protect users from tracking a faulty satellite signal. Design life is 7.5 years. The next twenty-one satellites, built by Lockheed Martin, are Block IIR. They can autonomously navigate (AUTONAV) by ranging on other Block IIR satellites. They generate their own 50 Hz navigational message data. Given this capability, they can operate for up to 180 days without resorting to support from the ground system. The satellite has reprogrammable processors, so that it can be upgraded in flight. It is also better hardened against radiation. Block IIR satellites are about a third less expensive than their Block II predecessors. A full Block IIR system should reduce the net navigational error of the system by 7m. Block IIF is the next series. Satellites are designed for ease of upgrade. The last Block I satellite was launched on 9 October 1985, and at the end of 1993 three of these first-generation satellites were still in service. The first Block II was launched on 14 February 1989 and declared 'healthy,' *ie* useable, on 15 April. The last of the twenty-four Block II/Block IIA satellites was launched on 10 March 1994, and the system became fully operational in 1995. The first Block IIR followed on 22 July 1997. Unit cost is $50.7 million (FY96). The lifetime of these satellites was extended from 5 to 7.5 years. Block IIF should begin replacing the earlier satellites in 2001.
5. The Master Control Station (MCS) is at Falcon AFB, Colorado; there are also five signal monitor stations (to check signal quality) at Falcon, and in Hawaii, on Ascension Island, on Diego Garcia, and on Kwajalein Atoll. There are uplink antennas on Ascension, Diego Garcia, and Kwajalein. A backup control station is at Onizuka AFS, California. The monitor stations send GPS signal data back to the MCS for evaluation.
6. These are the carrier frequencies. Superimposed on the codes is a navigational (NAV) message, transmitted at 50 bits/sec, giving, for example, orbital data for the satellites and ionospheric delays. The coarse service is called the Standard Positioning Service (SPS); the more precise service is the Precise Positioning Service (PPS). In March 1998 the U.S. government announced that a second civilian signal (in the formerly military 1227.6 MHz band) would be added. The military had combined the coarse civilian signal with its own precise signal on the military band; using two bands also made it easier to correct for the distorting effect of the earth's ionosphere, which slows radio signals passing through it (comparison of two signals at different frequencies makes it possible to estimate the amount of the distortion). A third civilian signal, at 1176.45 MHz, was later announced, to be provided beginning with a satellite to be launched in 2005.
7. In this version, the distance between a receiver and a known location is used to calibrate the system.
8. The coarse signal is deliberately randomly distorted as part of a selective availability (SA) feature; the peacetime error thus introduced is held down to less than 100m. In wartime, however, the satellites are instructed to increase the error to about 2000m. Both the coarse and the precise signals are coded; each satellite has its own codes, so a receiver (which contains the codes) can distinguish among them. The precision signal can be further encrypted to make spoofing difficult. The cryptological key is also used to remove the effect of SA distortion of the signal.
9. Ten Block I satellites were launched in 1982-5; design lifetime was a year (fourteen months actual average). They were followed by six Block IIA (1985-1986), by six Block IIB (1987; six of the twelve launched were lost in accidents), and by the current Block IV (thirty-four satellites launched, beginning in 1988). The Phase I system (seven satellites in each of two orbital planes 120 degrees apart) was completed in 1991; it was followed by Phase II (seven active and one spare satellites in each of three planes 120 degrees apart, for completion in 1995). The fol-

low-on is GLONASS-M (1995). Like the U.S. system, GLONASS produces both a coarse signal (at about 1610 MHz) and a precise military signal (at 1250 MHz). Claimed accuracy is 100m for the coarse signal and 10-20m with the military signal. It is not clear whether the satellites have atomic clocks on board; according to one report, they do not, and they rely on timing signals from ground stations to keep them accurate.

10. TERCOM seems to have been invented for an abortive nuclear-powered cruise missile conceived in the late 1950s.

11. The necessary geo-ballistic computer was used as a submarine tactical computer (part of the Mk 113 Mod 9 fire control system) when the submarine was not preparing to launch missiles.

12. The pilots of aircraft dropping laser-guided bombs had to be able to see their targets. To achieve sufficient stand-off range, they generally dropped from medium altitude, so cloud cover could preclude attacks.

13. According to Rip and Lusch, 'The Precision Revolution', the U.S. Air Force had already introduced this type of guidance, integrating a GPS receiver with the pre-existing TERCOM guidance systems of its AGM-86 cruise missiles when these weapons were converted from strategic to conventional weapons (AGM-86Cs). This missile entered service in January 1988. Rip and Lusch claim that the B-52 ALCM strike at the outbreak of the Gulf War, Secret Squirrel, was not officially revealed for a year because the U.S. Air Force did not want to reveal the existence of its GPS-guided missiles (the B-52s were themselves also GPS-guided).

14. In fact modern radio receivers radiate slightly (from their heterodyne circuits), but that radiation (which might leak out via the radio antenna) is far weaker than that from a radar altimeter.

15. The first operational ATWCS was installed on board the destroyer USS *Peterson* in 1996. In this version the original track control group is replaced by a new open-architecture version employing four Tac-series computers. This version is still limited to Tomahawk, because it retains the original LCG, which communicates with the vertical launcher. The next version adds four more Tac-X computers, which replace the separate launch-control group. It still communicates with the VLS and thence with Tomahawks. Finally ATWCS becomes the Advanced *Tactical* Weapons Control System: two more Tac-X workstations are added, and the LAN connects to other naval weapons; for example, in 1996 ATWCS was selected as the foundation of the planned NSFS Warfare Control System, NWCS, controlling the 5in ERGM guided shell.

16. The other new weapon begun under the FY87 programme was SLAM, a land-attack version of the standard anti-ship missile, Harpoon.

17. There was also a requirement for 'quiet,' ie low-signature, launch, which is why JSOW (ex-AIWS) is a long-range glide bomb.

18. According to Rip and Lusch, 'The Precision Revolution', the Israeli Popeye missile, which the U.S. Air Force uses as Have Nap (AGM-142), has a guidance system comparable to that of SLAM, with GPS mid-course guidance (other published accounts claim that the missile uses strapdown inertial mid-course guidance with updates via data link). The Popeye seeker uses imaging IR.

19. As an illustration of the impact of GPS, in 1989 a GPS-equipped RC-135 Rivet Joint ELINT aircraft won the U.S. Strategic Air Command navigation competition, arriving at each check point within three seconds (Rip and Lusch, p173); the command then banned all GPS aircraft from competition until all of its aircraft had been so equipped. According to Rip and Lusch, on the eve of the Gulf War the U.S. Navy had installed GPS receivers on board at least EP-3E, E-2C, A-6E, MH-53E, and SH-60B/F aircraft. The Air Force had fitted the B-52G, C-130, E-3B/C AWACS, E-8A J-STARS, EC-130H Compass Call, EF-111A, F-15E, F-16C, F-111A, MC-130E Combat Talon II, MH-53J Pave Low, RC-135U/V/W, and TR-1/U-2R. Commercial GPS receivers had been installed on board P-3 Orions and aboard Marine helicopters. During the war, the French and the British found that they had to install GPS on board Jaguars and Tornados to meet Allied precision standards.

20. Even existing tactics are affected. Using GPS, Navy minesweeping helicopters in the Gulf War could operate at night and in bad weather, substantially reducing the time required for mine clearance.

21. There is also the problem of OTH communication. The current U.S. Marine Corps solution to both problems is to adapt the Army's PLRS (position and location reporting system), which sets up an HF radio net among all users. The combination of time differences between users defines distances and, ultimately, locations.

22. COBRA is a low-probability of intercept waveform for Collection of Broadcasts from Remote Assets. It is used with the Army's Grenadier Brat remote sensor system.

23. Anyone with a GPS receiver can experience the impact of jamming when the receiver breaks lock in an electrical storm – which is, in effect, a noise jamming signal. GPS can also fail in a city or near hills, because signals will sometimes reflect off buildings or other obstacles, thus lengthening the apparent path to the satellite. A clever opponent might thus be able to shield targets from GPS-guided weapons by properly locating them.

24. However, it might be difficult for a spoofing signal to act sufficiently like one from a moving satellite. A military receiver ought to be able to use the coding of the military signal to distinguish it from a false signal. Also, the coding itself should make it easier to extract a signal from jamming noise.

25. At the same time a GPS receiver is being embedded in HARM to improve its precision, and to prevent it from homing on a target in a protected area.

26. Note that, quite aside from its navigational significance, GPS provides the fleet with a consistent precise timing signal. Modern computer-controlled coding devices switch their keys very frequently. Unless they and their receivers do so together – to a common timing signal – encrypted communication cannot be effective. Without precise timing, keys would have to be retained for much longer, and code-breaking would be far more effective. For that matter, precise timing is the basis for a key tactical

communications circuit, Link 16 (the Joint Tactical Information Distribution System) and, presumably, for all systems which rely on frequency-hopping for their security (such as the link which serves the Cooperative Engagement Capability, CEC). In general, the greater the capacity of the link, the more precise its timing must be – and the more, then, that it relies on the timing signals provided by GPS satellites.

Chapter 13: The Navy and the Battle Ashore

1. The difference between very long-range guns and missiles disappears, because at very long ranges all gun ammunition must be guided in order to be effective. The gun becomes, in effect, the first stage of a multi-stage missile. Because it offers far higher initial acceleration than any conventional booster, the gun makes for much smaller rounds (for a given payload), and a ship can accommodate many more of them. This is the rationale for the projected vertical gun, which may arm the new DD 21 Land-Attack Destroyer (the U.S. Navy's next-generation surface warship). The new Extended Range Guided Munition (ERGM, EX-171), which the existing 5in gun is to fire, is to be the first in a series of gun-launched munitions, some of which may have ramjet powerplants.
2. This is the 'Ship to Objective Manoeuvre' (STOM) which is espoused as part of the Marines' new concept of Operational Manoeuvre From the Sea. Quite possibly the entire movement, at least at the outset, would be by troops aboard V-22 Osprey aircraft, carrying little more than personal weapons and radios. Such tactics demand extremely responsive fire by the fleet – and a very reliable communications link back from the Marines inland to a fleet operating well offshore, out of range of coast-defence missiles, and far enough offshore probably to be free of mines.
3. This programme responded to a 1992 Operational Requirement.
4. As many as eight or nine strike teams are designated before a carrier deploys, to facilitate training and to make for rapid response to an ATO.
5. TAMPS actually does more. For example, it loads the F/A-18 Data Storage Unit (DSU) with flight route data (way points, steering files), air-to-air radar presets, and TACAN and radio channel identification files. It provides missile software files without which weapons such as HARM, SLAM, JSOW, and JDAM cannot be used. It is the primary means of loading key data link (JTIDS) data for the F-14D and the E-2C. Ultimately all naval aircraft planning systems, such as TEAMS (for the EA-6B), are to merge with TAMPS.
6. A current SpaWar project, Radiant Elm, is designed partly to improve TOPSCENE by using a wider range of images (EO, IR, SAR, etc) taken from National sensors.
7. Menner, 'The Navy's Tactical Aircraft Strike Planning Process', pp100-101.
8. The current concept is centralised planning and decentralised execution. In a joint operation, the Joint Force Commanders (JFC) air component commander (JFACC) has his staff prepare the ATO. He uses a Theater Battle Management Core System (TBMCS), which feeds elements of the ATO through the GCCS command/control system to the executing elements (which use CTAPS). Note that the acronym CTAPS is used by the U.S. Air Force to mean Coordination of Tactical Aircraft Planning System. It generates and disseminates the ATO, which is fed to each Wing (via a Wing Command and Control System) and thence ultimately to AFMSS (Air Force Mission Support System), equivalent to the Navy's TAMPS.
9. This sort of operation requires a special wideband receiver on board the F-16, the Improved Data Modem (IDM). Development began in 1990, which dates the beginning of Air Force RTIC efforts. The U.S. Naval Research Laboratory was responsible for the IDM, probably because it had previously been involved in other tactical reception devices. According to the NRL 75th anniversary awards booklet, the IDM was the first high-capacity digital link between fighter aircraft and between fighters and ground stations, providing data in seconds rather than in minutes (by voice). It was used on board F-16s and Navy EA-6Bs in Bosnia and in the Gulf. The Memorandum of Understanding between the Air Force and NRL was signed in December 1990, and the relevant technology handed over in 1993; IDM was operational by 1996. In fact a very limited digital link, a form of Link 4, had previously been established between F-14s on the eve of the Gulf War, but it lacked the capacity to transmit targeting data. Link 16 now offers much the same capability as IDM, but it is only now entering service in quantity.
10. For example, according to an account of Fleet Battle Experiment Bravo (summer of 1997), the Afloat Planning System detachment aboard the carrier *Constellation* managed to locate targets with sufficient accuracy to deliver 100 per cent of JSOW missiles within their allowable target location error (TLE), and to deliver 80 per cent of JDAMS bombs within their TLE (7.2m). Among other things, the experiment demonstrated that new devices used to compensate for distortion in aircraft and UAV imagery were effective.
11. For example, in Exercise Arid Hunter (1994), F/A-18s and F-15s hunted simulated Scud launchers at Fallon, Nevada. In Arid Hunter I they were provided with imagery over RTIC links. Of aircraft which received the imagery, 73 per cent acquired the camouflaged targets. None of those which did not receive imagery acquired the targets. In Arid Hunter II, the targets were not camouflaged. Using coordinates only, aircraft took an average of 9.5 minutes to find targets; if imagery was added, average time fell to 5.33 minutes. In a 1995 exercise, Forward Hunter, a controller used UAV imagery to spot targets, calling in Tomahawk strikes (at one point a missile was flown over Los Angeles). The combination of ground controller and UAV was considered a 'virtual forward air controller' (and used as such in a 1996 exercise). In Project Strike (1995-6) aircraft were re-routed to evade new SAM threats supplied by satellite, initially in coordinate terms but later as imagery, using Radiant Tin compression (Strike II).

Fleet Battle Experiment Bravo (September 1997) used the China Lake Real Time Targeting system, employing the AWW-13 and AXQ-14 pods.

12. Compression is automated simplification of an image, without losing key features. For example, the picture usually has substantial featureless areas. Conventional transmission requires large bandwidth because the picture is sent pixel by pixel, element by element, all elements being treated identically, whether or not they convey much information. Data-compression schemes seek out featureless areas and represent them in simpler ways. This is not too different from the way the human eye works. It, too, has a limited path – the optic nerve – along which it sends its data to the brain. The eye works like a camera, frame by frame. It compresses data by never sending an entire image, only changes from frame to frame. Current techniques are JPEG, Radiant Tin, and Aware. JPEG uses a discrete cosine transform and Hoffman Coding; it handles all parts of the image equivalently, and thus loses some detail. It is widely used in commercial computer systems. Radiant Tin uses wavelets and symbolic transforms. Aware also uses wavelets, but has a different coder. At a given compression ratio, wavelets cause far less distortion. Many compression techniques represent an area of an image, *eg* a black area, with a single symbol rather than repeating the same light level (zero, in this case) for each pixel in the area.

13. Aircraft avionics are currently tied together by 1553B buses, to which are connected their mission computers and various sources of data (such as GPS and data link receivers). On the ground, the TAMPS mission is loaded into the system manually, in the form of a massive memory loader. Similarly, weapons data are manually loaded. Future aircraft, such as the JSF, will have faster, more capacious fibre-optic buses, and TAMPS data may be loaded by electronic message closer to launch time. This architecture also makes it easier to arrange to reload the mission computer once the aeroplane is in the air, via (say) Link 16 or a derivative.

14. Alenia Marconi displayed RTIC at the 1999 Paris Air Show, as an example of 'sensor to shooter integration'. In its scenario, a target was detected while four F/A-18s were already in the air, en route to nearby targets. Marconi already supplies autorouters for the F-117 and the B-2, so it can realistically offer one for the F/A-18, to be incorporated in a TAMPS upgrade. One key to the scenario was that all the F/A-18s were on a common Link 16 net. It could distribute not only situational awareness data (where friendly and enemy aircraft and enemy anti-aircraft weapons were), but it could also directly command aircraft mission computers – and it could transmit compressed images for display on the aircraft's' video screens (it helps that pilots much prefer still images to continuous video, which would take up much more bandwidth). A targeting workstation on a carrier or major amphibious ship commands the airborne fighters via the Link 16 line-of-sight net. No such work station yet exists, but the Joint Targeting Support System Test Bed (JTSST) recently demonstrated for the Air Force's Expeditionary Force C2 Technology Innovation Center showed how it could work. It would be derived from the current DIWS (Digital Imagery Work Station) and the Precision Targeting Work Station used for Tomahawk (the targeting element of ATWCS).

15. The choice of altitude is a compromise between two potential threats: long-range surface-to-air missiles effective at medium to high altitude, and short-range (uncoordinated) surface-to-air missiles and guns operating mainly at very low altitude. In Vietnam, the threat of the long-range missiles drove aircraft down to low altitude, where many of them fell victim to light anti-aircraft weapons. In the Gulf War, the initial attacks disabled the bulk of the Iraqi air defence system, so aircraft could fly well above the effective altitude of the light weapons. In this case light weapons were far more dangerous, because they were uncoordinated (hence could not be disabled by attacks on control centres), and because they were far too numerous to be destroyed one by one. It now seems clear that the larger mobile missiles (like SA-6s) could generally be located and avoided (or neutralised).

16. The carrier was chosen because she had a unique internal communications network (a fibre-optic LAN with a capacity of 100 Mbits/sec, which could support more than 300 workstations, the *George Washington* Information System (GWIS). The carrier used a 1.5m receiving dish (the shore terminal has a 3.6m antenna). Challenge Athena (on *Mount Whitney*) used a 3m dish.

17. 'Challenge' indicates a short-term CNO initiative paid for out of a special fund.

18. The system uses leased transponders on one Comsat (Comstar) satellites for zonal coverage around the Continental United States and on three Intelsats for the Atlantic, Mediterranean/Persian Gulf, and Pacific. Throughput is about 3 Mbits/sec per satellite area (using a 36 MHz C-band transponder). Current naval applications of this wideband capacity are: primary imagery (772 kbits/sec), telemedicine (56 kbits/sec), STU-III secure telephones (352 kbits/sec for twenty-two phones), VTC/VTT (teleconferencing) at 128 kbits/sec, QOL (quality of life, *ie* to reach home) phones at 128 kbits/sec, JWICS at 64 kbits/sec, SIPRNET at 128 kbits/sec, NIPRNET at 128 kbits/sec, and a capacity to accept surge DSCS traffic (or to replace DSCS in an emergency) at 128-384 kbits/sec.

19. The original receiver was retroactively designated WSC-8. The contract for the new WSC-8(V)X was let in the summer of 1998. There are to be two variants, single- and double-dish. The first is on the carrier *Constellation*, the second on the amphibious carrier *Belleau Wood*. All should be in place by 2002, on board all the carriers and the large amphibious ships and flagships.

20. According to Boyd, Commander Austin, 'GBS ... The Satellite that Junior Officers Built', *Space Tracks* (May/June 1998), the GBS concept originated in 1994 with Commander Dave Baciocco, who was then action officer for SHF satellite communication on the Navy Space Systems Division staff. Baciocco wanted to overcome bandwidth scarcity; he envisaged a kind of satellite-cable television for ships. He incorporated his idea in a TENCAP project, Radiant Storm, demonstrated at the Joint Warrior Interoperability Demonstration in California, and

in a subsequent exercise, Roving Sands. Without flag-level approval, he attended an International Radio Advisory Committee, where he demanded a Department of Defense reservation for part of the Ka-band spectrum, even though the necessary satellite had not yet been approved, let alone funded. Baciocco suggested that the tenth UFO satellite be modified to carry the Ka-band broadcast. Support quickly built, so that the programme had been funded by 1995. That year the new Director for Space and Electronic Warfare, Vice-Admiral Walt Davis, visited the Hughes UFO plant; the company wanted to promote installation of its Direct TV system on UFO-10. Davis agreed that this was a great opportunity, and the GBS/UFO programme was formally begun.

21. The GBS Mission Need Statement (MNS) was approved in August 1995, and that year the full digital broadcasting system waveform and the system's multi-level access technique (to protect compartmented information) were demonstrated (in the SP-95 trial). As in Copernicus, GPS uses asynchronous transmission (ATM) techniques, which in its case was demonstrated in two exercises, AFC4A and JWID 95. As of 1999, projected naval uses for GBS included transmission of UAV video (6 Mbits/sec), television (including training videos, at 6 Mbits/sec), ATOs (0.5 Mbits/sec), MDUs (0.5 Mbits/sec), imagery (BDA [battle damage assessment]/MSI[multi-spectral imagery]/DMA [Defense Mapping Agency] at 3 Mbits/sec), DS TPFDD (TPFDD time-phased force deployment database [as in Desert Storm] at 6 Mbits/sec, JSIPS-N (intelligence imagery) at 1 Mbits/sec, and METOC (meteorology data) at 1 Mbit/sec.

22. Both the cruiser *Princeton* and the amphibious carrier *Tripoli* triggered mines in water thought not to have been mined. Reports that the areas had already been declared cleared were incorrect.

23. See Bouchard, Captain Joseph F, 'Guarding The Cold War Ramparts: The U.S. Navy's Role in Continental Air Defense, *Naval War College Review* (Summer 1999). Bouchard points out that the old Continental Air Defense mission was much more affordable, from a Navy point of view, because the hulls involved were pulled from the reserve ('mothball') fleet. They did, however, soak up new radars, which were in short supply in the mid-1950s. After the mid-ocean barriers were abandoned in 1965, the pickets were used for blockade duty (Operation Market Time) in Vietnam; the long endurance which made them effective pickets also made them useful blockaders. Note, incidentally, that in the late 1960s (as part of the abortive Safeguard system) there was a serious proposal to build specialised missile defence cruisers for boost-phase interception of Soviet or Chinese ICBMs. The idea was rejected on the ground that the ships were vulnerable to underwater attack.

Chapter 14: A New Kind of War?

1. According to the classic definition, it takes a capital ship to be sure of sinking another capital ship. Carriers superseded battleships because their aircraft could sink battleships at long range, whereas their aircraft could detect the battleship in time for the carrier to avoid contact. Thus, at least in theory, a battleship could not sink a carrier by gunfire. The great Second World War exception was HMS *Glorious*, sunk by gunfire in 1940. Her Commanding Officer apparently forbade flying while the ship was en route home from Norway, thus dooming her when the two German battlecruisers *Scharnhorst* and *Gneisenau* accidentally encountered her. The only other carriers sunk by capital ship gunfire, USS *St Lo* and *Gambier Bay*, were sunk off Samar while en route to support the Leyte Gulf landings in 1944; neither they nor their consorts had searched for enemy ships (they had reason to think the Japanese surface fleet, found by U.S. aircraft, had fled).

2. Aircraft can still deliver far heavier tonnages of ordnance on a much more sustained basis. Surface warships cannot currently take on Tomahawks at sea, which is why, during the 1998 Iraqi strikes, a U.S. naval formation attacked, then withdrew to be replaced by a second formation carrying fresh missiles. Aircraft are also far more flexible than missiles, and they can be recovered if they are mistargeted. Perhaps most important of all, by flying near or over a country's territory, aircraft can impose naval presence to a degree a surface ship cannot. That is probably the one role no surface ship, no matter how well connected to offboard sensors, can fill as well as a carrier can.

3. As a measure of just how difficult it would be to use imaging satellites for sea surveillance, in 1963 the British government seriously considered developing its own space imaging system to Air Staff Target 9003. It envisaged satellites orbiting at 200 miles in 97 degree (sun-synchronous) orbits (much like Corona) carrying two cameras with television read-out plus an ELINT antenna covering a limited bandwidth. Resolution would be 25 and 5 yards for swath widths of 45 and 7 miles, respectively. Such a system would obtain 80 per cent coverage of any given area of the Soviet Bloc within two to three months. The 25-yard camera would operate continuously, transmitting on command to two to six ground stations. Among the system requirements was the need for accurate measurement of the satellites' orbits, which could be by radar (as in the U.S. BMEWS, one set of antennas for which was being installed in Britain) or by Doppler (as in the U.S. Navy SPASUR system). The narrow-field camera would function on command. ELINT would be used to intercept airborne communications, VHF radio relays (*ie* microwave links), and would search for ABM radars and for UHF transmissions. ELINT data would be transmitted via the imagery link, and might be correlated with images if the ELINT package used a directional antenna. According to the British analysis, 20 per cent of satellite weight could be saved if the TV link was eliminated in favour of film recovery, but in that case the satellite would last only about 10 days before it ran out of film. According to a Royal Navy analysis (ADM 1/28880, September 1963), ships could barely be identified at 25-yard resolution (a suitably-fuzzed photograph was prepared), the 5-yard strip being far too narrow for surveillance. However, to obtain sufficient coverage would require a total of thirty satellites, all in precise polar orbits. If satellites could not be

maintained in precise orbits, even more would be needed. Data reduction would be a heavy burden, because most of the sea would always be empty. Although resolution is now better than the British imagined, swath widths of high-resolution satellites are, if anything, narrower than the 45 miles envisaged in 1963.

4. Note that the French planned a geosynchronous satellite, Zenon, as a combined ELINT *and* tactical communications system.

5. A distinction is made between hyperspectral (HSI) and multi-spectral (MSI) imagery, the former involving different colours within a band. A current TENCAP (Spawar) project, Radiant Beryllium, is intended to develop and exploit strategic HSI sensors (*ie* satellite sensors) for 'aided target identification'.

6. The Chinese orbited a series of military surveillance satellites (the 1100kg Ji Shu Shiyan Weixing series). Of six attempts between September 1973 and November 1976, three succeeded. The first two were placed in 190 × 400km orbits, which might be considered typical of imagery satellites; the third was in a 190 × 2100km orbit, which might be better adapted to ELINT. This series was launched by the FB (Feng Bao, Storm)-1 booster developed by the by the Shanghai Academy of Space Technology, which was directly backed by Mao Zedong. FB-1 was an alternative to the CZ-2 booster (both were derived from the DF-5 ballistic missile) developed by the more successful China Academy of Launch Vehicle Technology. When Mao died, so did the Shanghai programme. Because no re-entry vehicles were observed associated with the first two launches, there has been speculation that they were radio-link imagery satellites. However, radio links entail technology probably much more sophisticated than what would have been available in China in the early 1970s.

Then the Chinese launched three generations of recoverable satellites (Fanhui Shi Weixing [FSW] series). Four test launches (1974-8) were followed by six operational FSW-0 flights (1982-7, launched by CZ-2C); then there were five FSW-1 flights (1987-93, launched by CZ-2C); and then there were three FSW-2 flights (1992-6, launched by CZ-2D). Lifetimes are, respectively, three, five, eight, and fifteen days. Officially these are all remote-sensing satellites. The FSW programme was designed apparently to meet both military and civil requirements. Work began in 1966, and a feasibility study was completed (and the programme approved) the following year. Like the Soviet recoverable-body satellites, these craft drop their film pods on land, the recovery zone being at Liuzhi, Guizhou Province, in southwest China. FSW-1 reportedly carried both a CCD camera (50m resolution, radio down-link) and a film camera (10-15m resolution); a typical orbit was 210 × 310km at 57-63 degrees inclination. Maximum recoverable payload was 150kg. FSW-2 carries a heavier payload (350kg recoverable) and can manoeuvre in orbit. It operates at lower perigee (presumably its onboard motor can protect it from orbital decay until it is ordered to re-enter). An FSW-3 series was mentioned in a 1989 scientific paper, but no such satellite had emerged by 1999. None of the earlier series offers very much sustained capability, and certainly none seems well adapted to any operational military use. There has been speculation that some later Chinese satellites, nominally for communications satellite research, were ELINT craft: in particular the three Shi Jian (Practice) satellites (masses 257, 28, and 483kg) launched together on 20 September 1981 into a 200 × 1600km, 59.4 degree orbit. Externally these satellites are very similar to Telstars, and thus could serve as Grab-type ELINT relays. Presumably the high apogee gives the satellite a longer effective intercept range. The effectiveness of these satellites in any such role would depend on the availability of down-link sites. Certainly no dedicated Chinese ELINT satellites had been identified in the open literature as of mid-1999, although Chinese communications satellites may have some ELINT capability.

The French Helios, launched 7 July 1995, uses the same bus as the commercial (Spot-4) imaging satellite. It follows a low (450km) sun-synchronous orbit and has a laser cross-link to the Syracuse communications satellite. Project partners were Italy and Spain (France paid 79 per cent of the cost and took an equivalent share in the project). The satellite had a three-year design lifetime. A second satellite is planned for launch in November 1999. The satellite uses 4096 pixel and 2048 pixel linear CCDs to achieve a resolution of 1-5 m. During the Gulf War, the French government announced plans for Helios-2, with additional IR sensors. It would be complemented by a digital radar imaging satellite, Osiris, and by an ELINT satellite, Zenon. There were also proposals for early warning satellites. Up to four Helios-2 were envisaged. The French saw Helios-2 as a national programme but sought international funding for Osiris. The Osiris project collapsed in 1994 when Germany, which had been expected to provide the bulk of the money, refused to participate. A reduced version of the project, Horus, died in April 1998. That left a reduced version of Helios-2 (two satellites), planned (as of Spring 1999) for launch beginning in 2002. They will be paid for mainly by France, with some participation by Spain, Italy, and possibly Belgium.

7. The Navy's high-powered laser programme (MIRACL) was always directed at short-range ship defence. Navy laser ranges are inherently limited because the beam operates in the dense lower atmosphere. The beam path used by the Air Force laser is in the thin upper atmosphere, hence it can enjoy very long range. Range is greatest when the laser is pointed up into space, towards a satellite.

8. Similarly, an Air Force General and an Admiral rotate in command of the U.S. Strategic Command, which combines Air Force and Navy strategic forces.

Bibliography

In addition to the books and articles listed below, I have made extensive use of the material on two Internet web sites, that of the Federation of American Scientists (FAS) and of James Wade's *Spacepedia*.

Amato, Ivan, *Pushing the Horizon: Seventy-Five Years of High Stakes Science and Technology at the Naval Research Laboratory* (Washington 1997).

Andronov, Major A, 'The U.S. Navy's "White Cloud" Spaceborne ELINT System', *Foreign Military Review* (in Russian) No. 7 (1993), pp57-60.

-----, 'American Geosynchronous SIGINT Satellites', *Foreign Military Review* (in Russian) No. 12 (1993), pp37-43.

-----, (Lt Col) and Garbuk, Captain S, 'U.S. IMEWS [missile warning] Space Systems and Creation of an Advanced Ballistic Missile Launch Detection System' in *Foreign Military Review* (in Russian) No. 12 (1994), pp34-40.

Angelskiiy, Rostislav D, 'Smertsonsnaya Chaika' [The Deadly Gull], *Tekhnika i Oruzhia* 1-96.

-----, 'Perviye Otechestvenniye Protivokorabel'niye Samoletiye–Snaryadiy', *Nevskiy Bastion* 1-1997 (the KS story).

-----, 'The Supersonic Kipper – K-10 Anti-Shipping Missile', *Kryla Rodini*, 12-97.

-----, 'Kel'ti Rvutsya K Palestine: Raketa KSR-2 Dlya Tu- 16', *Kryla Rodini* 9-98.

Artemev, Anatolii, 'Morskaya Raketonosnaya', *Aviatsiya i Kosmonavtika* No. 32 (Nov-Dec 1997).

Baker, David, *The Rocket: The History and Development of Rocket & Missile Technology* (London 1978).

Bamford, James, *The Puzzle Palace: A Report on NSA, America's Most Secret Agency* (Boston 1982).

Barry, William P, 'The Missile Designers and Soviet Piloted Spaceflight Policy' a paper presented at the 28th National Convention of the American Association for the Advancement of Slavic Studies, 14-17 November 1996.

Borik, Cdr Frank C, 'The Silent Service Is On The Air: How Advanced Communications Will Revolutionize Submarine Warfare', *Submarine Review* (July 1999).

Brannigan, John, *Space Radio Handbook* (Bath: Radio Society of Great Britain, 1991).

Breitler, Alan L, and Nguyen, Hung Q, 'U.S. Navy Mass Communications Options', *Proc. IEEE* (1995) (0-7803-2489-7/95).

Burrows, William E, *This New Ocean: The Story of the First Space Age* (New York 1998).

Butrica, Andrew J, *Beyond the Ionosphere: Fifty Years of Satellite Communication* (Washington 1997; NASA publication SP-4217).
Clark, Phillip, 'The Decline of Russian Orbital Reconnaissance', *Launchspace* (March/April 1999).
R F Cross Associates, *Sea-Based Airborne Antisubmarine Warfare 1940-1977* (a formerly classified U.S. Navy report, dated 28 April 1978 and declassified in 1990; 3 volumes). This document is referred to below as the Cross Report.
Day, Dwayne A, 'Out of the Shadows: The Shuttle's Secret Payloads', *Spaceflight* (February 1999).
-----, 'Listening from Above: The First Signals Intelligence Satellite', *Spaceflight* (August 1999).
Deyneka, V G, *Aviatsiya Rossiskogo Flota* [Aviation of the Russian Navy] (St Petersburg 1996).
DiGirolamo, Lt-Cdr Vinny (ed), *Naval Command and Control: Policy, Programs, People & Issues* (with a foreword by Vice-Admiral Jerry O Tuttle) (Fairfax, Virginia 1991).
Dmitriev, Gennady (ed), *Tupolev Tu-16* (Kiev 1997).
Dutton, Sqdn Ldr Lyn, de Garis, Wg Cdr David, Winterton, Sqdn Ldr Richard, and Harding, Dr Richard, *Military Space* (Brassey's Air Power: Aircraft, Weapons Systems, and Technology Series, Vol 10; London 1990).
Gebhard, Louis A, *Evolution of Naval Radio-Electronics and Contributions of the Naval Research Laboratory* (NRL Report 7600) (Washington 1976).
Geckle, William lecture notes on 'RTIC/RTOC,' for a lecture given for the Technology Training Corp., Washington, D.C., 3-4 December 1998.
Goeller, Dr Lawrence (Com Dev Ltd.), lecture notes on 'Military Satellite Communications', from a lecture presented at Cambridge, Ontario, 28-30 October 1998 (courtesy Technology Training Corporation).
Gordon, Yefim and Rigmant, Vladimir, *Tupolev Tu-95/Tu-142 'Bear'* (Earl Shilton: Midland Publishing, 1997; an Aerofax book).
-----, *Tupolev Tu-22 'Blinder' and Tu-22M 'Backfire': Russia's long-range supersonic bombers* (Earl Shilton: Midland Publishing, 1998: an Aerofax book).
Gunston, Bill and Gordon, Yefim, *MiG Aircraft Since 1937* (London 1998).
Harland, David M, *The Space Shuttle: Roles, Missions, and Accomplishments* (Chichester 1998).
Hendrickx, Bart, 'The Early Years of the Molniya Program', *Quest* 6:3
Hezlet, Vice-Admiral Sir Arthur, *The Electron and Sea Power* (London 1975).
Jenkins, Dennis R, *Space Shuttle: The History of Developing the National Space Transportation System: The Beginning Through STS-75* (Marceline, Mo. 1997).
Karpenko, A V (ed), *Rossiyskoye Raketnoye Oruzhiye 1943-1993* [Russian Military Missiles 1943-1993] (St. Petersburg 1993).
-----, 'I Istokov Sozdaniya Reketn'ikh Korableiye', *Gangut* No. 16 (1998).
-----, Ganin, S M, and Kolnogorov, V V, *Aviatsionniye Raketyi Bol'shoi Dal'nostiy* [Missiles of Long-Range Large Bombers: Vol 1 of a series on aviation weapons], (St Petersburg 1998).
Kim, John C, and Mueldorf, Eugen I, *Naval Shipboard Communication Systems* (Englewood Cliffs 1995).

Kotelnikov, Vladimir, Arten'yev, Anatolii, and Yurtensona, Cherteshi Andreya, 'Minno-Torpednaya: Aviatsiya Osobogo Roda' [Mine-Torpedo Aviation: Unknown Aviation] *Aviatsiya i Kosmonavtika* No.15 (April 1996).
Kuzin, V P, and Nikol'skiy, V I, *The Soviet Navy 1945-1991* (in Russian) (The Historical Naval Society; St Petersburg 1996)
Kuznetzov, Lt-Col A N, 'Technical Capabilities of U.S. Space Assets', *Military Thought* No. 8-9 (Aug-Sept 1993).
Lardier, C, 'La Russie Devoile Certains Satellites Secrets', *Air & Cosmos/Aviation International* No. 1638 (3 Oct 1997).
LePage, Andrew J, 'NOTSNIK: The Navy's Secret Satellite Program', *Spaceviews*, an online magazine (www.spaceviews.com) of space exploration (July 1998 issue).
Martin, Donald H, *Communication Satellites 1958-1995* (El Segundo 1996).
McDonald, Robert A (ed), *Corona: Between the Sun and the Earth, The First NRO Reconnaissance Eye in Space* (Bethesda 1997).
Menner, William A, 'The Navy's Tactical Aircraft Strike Planning Process', *Johns Hopkins APL Technical Digest*, Vol 18, No. 1 (1997).
Minaev, A V (ed), *Sovetskaya Voennaya Moshch'* [Soviet Military Industrial Complex] (Moscow 1999).
(U.S.) Naval Research Laboratory, *75th Anniversary Awards for Innovation* (Washington 1998).
Nguyen, Hung P, *Submarine Detection from Space: A Study of Russian Capabilities* (Annapolis 1993).
Peebles, Curtis, *The Corona Project: America's First Spy Satellite* (Annapolis 1997).
Richelson, Jeffrey T, *The U.S. Intelligence Community* (3rd, Boulder: 1999).
-----, *America's Space Sentinels: DSP Satellites and National Security* (Lawrence 1999).
Rip, Michael Russell, and Lusch, David P, 'The Precision Revolution: The Navstar Global Positioning System in the Second Gulf War', *Intelligence and National Security* Vol. 9, No. 2 (April 1994).
Sarkisov, A A (ed), *Rossiyskaya Nauka-Voenno-Morskomy Floty* [Russian Naval Science] (Moscow 1997). Collection of papers on Soviet naval technology, prepared for the 300th anniversary of the Russian navy.
Shirokorad, A V, 'Raketnyi Nad Morem' [Rockets Over the Sea], *Tekhnika i Oruzhiye* [Technology and Weaponry] 2/96, pp2-44, and 3/96, pp2-28.
-----, *Sovetskiye Podvodniye Lodkiy Poslevoennoiy Postroikiy* [Soviet Submarines] (Moscow 1997).
Siddiqi, Asif, 'Staring at the Sea: The Soviet RORSAT and EORSAT Programmes', *Journal of the British Interplanetary Society* Vol. 52, No. 11/12 (November/December 1999).
Spinardi, Graham, *From Polaris to Trident: The Development of US Fleet Ballistic Missile Technology* (Cambridge 1994).
Spires, David N, *Beyond Horizons: A Half Century of Air Force Space Leadership* (Peterson Air Force Base: Air Force Space Command, 1997).
Stares, Paul B, *Space Weapons and U.S. Strategy* (London 1985).
Stefanick, Tom, *Strategic Antisubmarine Warfare and Naval Strategy* (Lexington, Mass. 1987).

Strategic Rocket Forces (Russian), *Voenno-Kosmicheskiye Sil'iy (Voenno–Istoricheskiy Trud)*, Vol 1: *Kosmonavtika i Vooruzhenniye Sil'iy* (Moscow, 1997). Commemoration of 40 years in space, 1957-1997.

U.S. Navy, NWP 4, *Basic Operational Communications Doctrine* (July 1982) (formerly Confidential, now declassified).

-----, NWP 16-1, Basic Operational Communications Doctrine (1953) (formerly Confidential, now declassified).

-----, NWP 16 (A), Basic Operational Communications Doctrine (31 December 1962) (formerly Confidential, now declassified).

-----, NWP 16(B), Basic Operational Communications Doctrine (21 October 1965, with revisions of 26 May 1966 and October 1967) (formerly Confidential, now declassified).

-----, NWP 16(C), *Operational Communications Doctrine* (8 May 1970) (formerly Confidential, now declassified).

Voenno-Morskaya Akademiya Im. Admiral N. G. Kuznetzova [Naval Academy Named After Admiral N.G. Kuznetzov], *Perspektiviya i Puti Sovershenstvovaniya System Vooryzheniya s Krilatimi Raketami Morskogo Bazirovaniya: Posvyashchena 40-letiyo Lervikh Puskov Krilatikh Raket P-5 i P-15* (St Petersburg1 999). Papers marking the 40th anniversary of the P-5 and P-15 missile).

Werrell, Kenneth P, *The Evolution of the Cruise Missile* (Air University Press 1985).

Winter, Frank H, *Rockets Into Space* (Harvard 1990).

Yerokhin, Yevgeniiy ' 'Shtorm' Zakonchilsya 'Shtilem'.' [Storm Becomes Calm], *Kryla Rodini* 10-97.

Zaloga, Steven J, 'Tupolev Tu-22 'Blinder' and Tu-22M 'Backfire" ', *World Air Power Journal* Vol. 33 (1998).

Glossary

ABM	Anti-Ballistic Missile
AEW	Airborne Early Warning
AFB	Air Force Base (U.S.)
AM	Amplitude Modulation
ASAT	Anti-Satellite
ASW	Anti-Submarine Warfare
ATACMS	Army Tactical Missile System (U.S.)
ATH	Above the Horizon
BTH	Below the Horizon
bit	measure of information content, 1 or 0; minimum piece of information; information rate is measured in bits per second,
bps	bits per second, measure of information rate
byte	8 bits, each bit being a 0 or a 1; measure of information; often indicated by upper-case B in contrast to b for bit
CAP	Combat Air Patrol
CEC	Co-operative Engagement Capability
CIA	Central Intelligence Agency (U.S.)
CIC	Combat Information Centre (U.S., equivalent to Royal Navy Action Information Centre operated by Action Information Organization [AIO])
CNO	Chief of Naval Operations (U.S.)
DF	Direction-Finding
DSP	Defense Support Program (IR Warning System)
ECM	Electronic Countermeasures
EHF	Extremely High Frequency (30-300 GHz)
ELF	Extremely Low Frequency (below 3 kHz)
ELINT	Electronic Intelligence
EMP	Electromagnetic Pulse
ESM	Electronic Surveillance Measures
EW	Electronic Warfare
FLIR	Forward-Looking Infra-Red (*ie* thermal imaging)
FM	Frequency Modulation
FY	Fiscal (financial) year (*eg* FY98) (in U.S., begins 1 November; before 1977, began 1 July; other countries use different financial years)
Gb	Gigabits, billions of bits
GCCS	Global Command and Control System
GHz	GigaHertz, billions of cycles per second
GPS	Global Positioning System
GRU	Soviet all-service military intelligence organization, parallel to KGB (see below)

HF	High Frequency (3-30 MHz)
Hz	Hertz, cycles per second (formerly c/s) (measure of signal frequency)
ICBM	Intercontinental Ballistic Missile
IR	Infra-Red
IRBM	Intermediate-Range Ballistic Missile
JCS	Joint Chiefs of Staff (U.S.)
JMCIS	Joint Maritime Command Information System
JOTS	Joint Operational Tactical System ('Jerry O. Tuttle System')
JSTARS	Joint Standoff Airborne Radar System (U.S.: British equivalent is ASTOR)
JTIDS	Joint Tactical Information Distribution System (Link 16)
kb	thousand bits (actually 1024 bits)
KB	Design Bureau (Soviet; prefixes 'O' and 'Ts')
kHz	kiloHertz, thousands of cycles per second (formerly kcs)
KGB	Soviet civilian intelligence and internal security agency (letters in Russian mean committee for state security), parallel to GRU for foreign intelligence, but included border control and internal security roles
LF	Low Frequency (30-300 kHz)
LPI	Low Probability of Intercept (*ie* stealthy emission)
Mb	million bits, actually 1,048,576 bits)
MF	Medium Frequency (300 kHz – 3 MHz)
MHz	MegaHertz, million cycles per second (formerly mcs)
MIRV	Multiple Independently-targeted Re-entry Vehicles (missile warhead)
MRBM	Medium-Range Ballistic Missile
MT	megaton, million tons of TNT equivalent
NASA	National Aeronautical and Space Administration
NATO	North Atlantic Treaty Organisation
NSA	National Security Agency (U.S.)
NTW	Navy Theater-Wide (ballistic missile defence)
OAB	Outer Air Battle
OTH(-T)	Over the Horizon (-Targeting)
SAC	Strategic Air Command (U.S.)
SALT	Strategic Arms Limitation Treaty (1970s, U.S.- Soviet)
SAM	Surface-to-Air Missile
SAR	Synthetic Apeture Radar
SHF	Super High Frequency (3-30 GHz)
SIGINT	Signals Intelligence
SSM	Surface-to-Surface Missile
START	Strategic Arms Reduction Treaty (1980s-1990s U.S.-Soviet), successor to SALT
THAAD	Theater High Altitude Area Defence (missile)
UAV	Unmanned (sometimes, Uninhabited) Air Vehicle
UHF	Ultra-High Frequency (300-3000 MHz, *ie* 300 MHz – 3 GHz); in radar, typically metric frequency, such as 225-400 MHz (used in Second World War air search sets)
VHF	Very High Frequency (30-300 MHz)
VLF	Very Low Frequency (3-30 kHz)

Index

Ship and spacecraft names in *italic*.

Abbreviations
AFB = Air Force Base (U.S.); Fr = France; HMS = Her Majesty's Ship; Rus = Russia/Soviet Union; U.S. = United States; USS = United States Ship.

A-6 Intruder bomber 302
AAAM (Advanced Air-to-Air Missile) 240-1
AADCs (Anti-Air Defense Centers) 296
ABL (Airborne Laser) system 298, 312
Able boosters 25
Abraham Lincoln, USS 81, 257
ACM (Advanced Cruise Missile) 210
Aden 82
Admiral Nakhimov (Rus) 139
Admiral Senyavin (Rus) 163
ADCSP (Advanced Defense Communication Satellite System) 75
Advent project 75
Aegis system 81, 195, 233, 240-1, 244, 256, 296
Aerobee booster 25, 91
Aerojet-General company 23, 25, 243
Aerospace Corporation 123, 242-3
AFATDS (Army Field Artillery Tactical Direction System) 295
Afghanistan 7, 57, 110, 304
AFRTS (Armed Forces Radio and Television Service) 264-5
AFSCN (Air Force Satellite Control Network) 79-80
Agena stages 22, 24-6, 91-2, 94
AIWS (Advanced Interdiction Weapon System) 274
Albany, USS 292
ALI (Aegis LEAP Intercept) 297
Almaz satellites 40-1, 117-18, 160-1
Aloha net 254
Altair stage 25, 33
Amelko, Admiral N N 159, 161
America, USS 219-20, 270
Annapolis, USS 182
Apache cruise missile 281, 316
Apache helicopter 276
APAR radar 69
Apollo spacecraft 118
APS (Afloat Planning System) 149-50, 153-4, 272-3
Aquacade programme 108
Area Defense programme 296-7
Argon shipboard command system 167
Argus satellites 108-9
Ariane booster 45, 86
Ark Royal, HMS 189
Arkon reconnaissance satellite 40, 119-20
Arlington, USS 182
ARPA (Advanced Research Projects Agency) 75, 126-7, 250, 254
ASAM (Advanced Surface-to-Air Missile) 240
ASAT (Anti-Satellite Systems):
 Russian 10, 19, 37-9, 41, 97-8, 112-15, 117, 155
 U.S. 19, 104, 122-6, 160, 163, 195-7, 311-12
Ascension Island 59
ASCIET (All-Services Combat Identification Evaluation Team) exercises 279-80
Asmara (Ethiopia) 59
ATCU (Air Transportable Communications Unit) 59-60
Atlantique aircraft 85
Atlas booster 21-4, 28, 75, 92
 Atlas Agena 75, 94, 123
 Atlas Centaur 20, 24, 75
 Atlas I 24
 Atlas II 24, 79
 Atlas III 32
ATOs (Air Tasking Orders) 282-7, 286-7, 290, 292
AT&T corporation 76
ATWCS (Advanced Tomahawk Weapons Control System) 273
Australia 51, 86, 109, 176, 224, 228, 243, 261

B-2 bomber 303
B-47 bomber 33, 122
B-58 bomber 22
B-70 bomber 168-9

373

'Backfire' aircraft see Tu-22M
Bad Aibling 107, 109
'Badger' aircraft see Tu-16
Baikonur 114
Baker-Nunn cameras 126
Bazalt missile 166-7, 170
BCIXS (Battle Cube Information Exchange System) 256
Be-12 seaplane 203
'Bear' aircraft see Tu-95
Bedoviy (Rus) 141
Belknap class (U.S.) 219
Bell Aircraft Company 28, 123
Berezhniak, Aleksandr Ya 134, 136, 142, 145
Beria, Lavrenti 135, 138
Beria, Sergei L 135
Beriev bureau 140
Berkut missile 136
BGIXS (Battle Group Information Exchange System) 224
BGPHES (Battle Group Passive Horizon Extension System) 236
BI rocket fighter (Russia) 134
bin Laden, Osama 304
Bisnovat, M R 134, 138
'Blackjack' aircraft see Tu-160
'Blinder' aircraft see Tu-22
Blossom Point 127
Blue Ridge, USS 80
Boeing corporation 22-5, 32, 34, 43
Bold Orion missiles 122
Boorda, Admiral 265
Bottom-Up Review 199-200
Boyd, Colonel John 131, 305
Brezhnev, Leonid I 118, 169, 231
BRIGAND (Bistatic Radar Intelligence Generation and Analysis) 215-17
Brilliant Pebbles satellite 35
Britain 11, 34, 52, 59, 63, 76, 82-4, 83, 85, 107, 131, 133, 163, 176, 185, 189-90, 221, 224, 235, 239, 264, 316, 316-17
British Aerospace 83, 108
Buckley Air National Guard Base 108-10
Bull, Dr Gerald 34-5
Bullseye global sensor 175-6, 179
Buran space shuttle 41-2, 44
Burner II stage 25

C-601 missile 145
Cable and Wireless 86
Caleb booster 125
Cambodia 57
Camp Roberts 76, 79
Canada 26, 159, 224, 228
Canberra, USS 77
Canyon satellites 106-8
Cape Canaveral 13-14, 22-3, 29-31
Carl Vinson, USS 286
Carter, President Jimmy 96, 121, 213, 233
Castor boosters:

Castor I 20, 25, 33
Castor II 26
Castor IV 26
Castor IVA 24, 27
Cauthen, Captain Hal 215
CBC (Common Booster Core) 32
CCCs (Combined Command Centers) 256, 291
CDMA (Code Digital Multiple Access) 69, 72-3, 264
CEC (Cooperative Engagement Capability) system 246, 298-9
Centaur stages 21-4, 32
Centaur-G 22-3
CESs (Coastal Earth Stations) 258
Chaika glide bomb 135
Chalet programme 107-8, 110
Challenge Athena system 291-3
Challenger (U.S. space shuttle) 23, 28, 30-1, 42, 78, 84, 96, 98, 115, 267
Chelomey, Vladimir N, and bureau 37-9, 41-2, 113, 117-19, 133, 136, 138-40, 142, 144, 155-6, 160, 166-7, 204 *see also* US-A *under* US satellites
Cheyenne Mountain 128
China 14, 20, 31, 43-4, 45, 145, 202, 246, 297, 303-4, 311, 315
CIA (Central Intelligence Agency) 87, 91-2, 96, 111, 157, 159, 196, 313
CIC (Command Information Center) 51-2, 131, 218
Clarke, Arthur C 65, 75
Clinton, President Bill 263
Clipper Bow satellites 177-8, 215
clustering 36, 43-4
CMSAs (Cruise Missile Support Activities) 270
codes and code-breaking 52, 56, 63, 104, 108, 132, 155, 213-14, 267
COE (Common Operating Environment) 253
Columbia (U.S. space shuttle) 96
commercial satellites 10-11, 19, 24, 44-5, 75-6, 82, 86, 103, 164, 190, 307-8, 310-12
communications 8, 10, 40, 54-86, 95, 131-3, 163-5, 182-91, 197-200, 237-8, 249-65, 289-94, 298-300, 311-12, 316-17
Comsat corporation 75
Condor bomb 274
Constellation, USS 57, 219, 257
Copernicus system 253-6, 258
Coriolis/Windsat system 208
Corona satellites 91, 93-4, 104
Coronado, USS 80
Cosmos satellites 37, 120
Cosmos-4 115
Cosmos-158 163
Cosmos-185 114
Cosmos-336 164
Cosmos-367 159
Cosmos-514 163-4

Cosmos-516 159
Cosmos-574 164
Cosmos-626 159
Cosmos-651 159
Cosmos-654 159
Cosmos-699 161
Cosmos-700 164
Cosmos-777 161
Cosmos-785 159
Cosmos-905 116
Cosmos-954 161
Cosmos-1027 164
Cosmos-1176 162
Cosmos-1402 159
Cosmos-1426 119
Cosmos-1567 160
Cosmos-1579 160
Cosmos-1735 161-2
Cosmos-1737 161-2
Cosmos-1818 160
Cosmos-1867 160
Cosmos-1870 160
Cosmos-1932 160
Cosmos-2031 119
Cosmos-2122 162
Cosmos-2242 119
Cosmos-2266 164
Cosmos-2267 119
Cosmos-2290 119
Cosmos-2344 119
Cosmos-2359 119
Cosmos-2367 162
Cospas satellite 164
Courier satellite 103
Crystal satellite *see* KH-11
CSCI (Commercial Communications Satellite Initiative) 292
CSS-5 missile (China) 297
CTAPS (Contingency Theatre Automatic Planning System) 285
Cuba 103, 121;
    missile crisis 36, 55-6, 182, 255
CUDIXS 186, 220, 251
cueing 141-2, 147-50, 289, 298
CZ satellites 43-4

DAMAs (Demand-Assigned Multiple Access Devices) 250-2, 254-6, 259, 263, 284-5
DD 21 warship (U.S.) 314, 316
DDN (Defense Data Network) 255
de Gaulle, General Charles 93
Decree satellite system 75
Delta boosters 21-2, 24-7, 28, 37
    Delta II stages 32
    Delta III 32
    Delta IV 32
    Delta IV Heavy 32
    Delta 104 25
    Delta 900 26

Delta 1000 series 26
Delta 2914 27
Delta 3000 series 26-7
Delta 3914 27
Delta 4920 27
Delta 6000 27
Delta 6925 27
Delta 7000 (Delta II) 27
Delta 7325 27
Delta 7425 27
Delta 7925 27
Delta A stage 25
Delta B upper stage 25
Delta D upper stage 26
Delta E 26
Delta G 26
Delta II 24
Delta J 26
Delta L 26
Delta M 26
Delta M-6 26
Delta N 26
Delta N-6 26
Diego Garcia 59
DISA (Defense Information Supply Agency) 292
Discoverer satellites 25, 91-3
    Discoverer-13 100
    Discoverer-17 25
    Discoverer-37 104
    Discoverer I 25
    Discoverer II 246-8
    Discoverer V 123
    Discoverer XIV 93
DISN (Defense Information Support Network) 79-80, 263
DMA (Defense Mapping Agency) 271, 304
DMRs (Digital Modular Radios) 256
DMSP (Defense Meteorological Support Program) 112, 226
DOPLOC radar system 126
Douglas corporation 24-6
DSCS (Defense Satellite Communications System) 19, 69, 73, 76-82, 83, 85, 183, 185, 197, 226, 252, 255, 262, 285, 293
    DSCS I 71, 77-8, 83
    DSCS II 71, 78-9, 81, 83
    DSCS III 24, 71, 78-81, 81, 84
    DSCS IV 82
    SLEP 81-2
DSMAC (Digital Scene-Matching Correlation) 269-70
DSP (Defense Support Programme) 23, 93, 110, 203, 227, 242-4, 284, 293, 298
    DSP-A 243
DTDs (Data Transfer Devices) 270, 272
DTS (Diplomatic Telecommunications System) 79-80

E-2 aircraft 184
E-2C aircraft 236-7, 242, 252
EA-6B aircraft 288
Early Bird satellites 76
Early Spring missile 122
early warning satellites 10, 40, 117
Earth Coverage horns 78, 80
Echo-1 satellites 25, 64
Edwards AFB 29
EELV (Evolved Expendable Launch Vehicle) programme 23-4, 32, 43, 79
Eilat (Israel) 142, 174
Eisenhower, President Dwight D 55-6, 75, 88, 92-3, 100, 102
Ekran satellites 40
ELF systems 187-8, 224
ELV (Expendable Launch Vehicle) programme 31
E-Mail 260-1
EMP (Electromagnetic Pulse) 19, 125, 190, 265
Energiya booster 42
Enterprise, USS 230, 257, 259-60
EORSAT 122, 157, 196, 204
EP-3E aircraft 177, 252
ESM receivers 84, 101, 155-6, 214, 226-7, 236
E-Systems company 108
European Geostationary Overlay Service 267
European Space Agency 267-8
European Union 267, 315-16
EWCM (Electronic Warfare Coordination Module) 218

F4D-1 aircraft 33
F-14 aircraft 152, 233-4, 236-42
F-16 aircraft 288-9
F-86 Sabre aircraft 131
F-111 aircraft 276
Falklands War 83-4, 163, 258
FCCs (Fleet Command Centers) 181-2, 188-90, 211, 222-3, 249, 256, 270
FDMA (Frequency-Domain Multiple Access) 72-3, 81
Feniks programme 116
Ferret programme 89, 100, 101-2, 104-5, 105-6, 108
    Ferret 2 satellite 104
FKR-1 coast defence system 138
Fleet Battle Experiment Echo 206
Fleet Broadcast 83, 250-1, 254
Fleet Satellite system 19, 78, 182-8, 190-1, 212-13, 217, 225-6, 243, 249-50, 252, 270, 292
FOBS combat re-entry vehicle 38-9, 125
Ford Aerospace 85
Forrestal, USS 220
FOSIF 210, 217, 221
France 14, 20, 43, 76, 84-5, 93, 98, 131, 190, 224, 281, 311, 315-16
Frants, Captain K K 155

French Guiana (Kourou) 14
FT2000 missile (China) 246
fuels 20-1, 23, 26, 32, 96
    cryogenic 21, 24, 29, 32, 37, 41-3
    liquid 20-1, 23, 31, 33, 35-7, 43
    solid 20-7, 31-3, 38, 41
FX-1400 glide bomb 134-5

Gage signals 103
Galileo system 267-8, 316
Gambit satellite 94
Gapfiller transponders 82, 185, 190, 258, 293
Garpun radar 153
GBS (Global Broadcasting System) 66, 251, 257, 264, 293-4
GCCS (Global Command and Control System) 254-5, 255-8
Geizer satellites 119, 121
Gemini-4 91, 117
General Dynamics 23
GEO-IK system 164
George Washington, USS 291-2
Geosat (Geodetic Satellite) 207-8
Germany 35, 47, 57, 63, 69, 79, 84-5, 87-8, 100, 103, 107, 109, 129, 131-4, 136, 147, 155, 175, 201
GFO (Geosat Follow-On) system 208
Globalstar 262
GLOBIXS 254-6, 263
GLOMR (Global Low Orbit Message Relay) 265
GLONASS (Rus GPS) 267
Glushko, Valentin P, and bureau 35, 41
GMFs (Ground Mobile Forces) 79-80
GMTI (Ground Moving Target Indication) 247
Goonhilly Down 76
Gorizont satellites 40, 109, 166
Gorkiy 99
Gorshkov, Admiral Sergei 155
GPS (Global Positioning System) 208, 228-9, 242, 266-82, 288, 303-4, 314-16.
    see also Navstar satellites
GRAB (Galactic Radiation and Background) 101-4
Granit missile *see P-500*
GREB (Galactic Radiation Energy Background) 102
GRU 122
GSC-49 ground terminal 80
GSR launch aircraft 41
GTE corporation 291
Guam 105
guidance 24, 38, 143-4, 168-71, 268-77, 286
Gulf War 7-8, 68, 80-1, 81, 98-9, 107, 109, 221, 225, 227-9, 232, 243-4, 253, 266-9, 271-2, 275-7, 282-4, 286, 291-2, 295-6, 316
guns:
    satellite launcher *see* HARP
    in space 117-18

Index

Haiti 316
HARM missiles 281, 288
HARP (High Altitude Research Project) 34-5
Harpoon missile 142, 180, 211, 213, 225, 275
Harry S Truman, USS 257
Hawaii 56, 77, 124, 127, 261-2, 270
Hayward, Admiral Thomas 234-5
helicopters 141-2, 168, 276, 278, 316
Helios II programme (France) 315
Helios photo-reconnaissance satellite 85
Hermes, HMS 84
Hexagon ('Big Bird') (KH-9) 94-5
Hi-Hoe system 125
Hispasat satellites 85
Honeywell corporation 208
Horn, Kenneth 242-3
Horus programme (Fr) 315
HOTOL project 34
Hs 293 missile 134
HS-601 shuttle 86
Hubble space telescope 96
Hughes Space and Communications corporation 31, 44, 86, 183, 190, 200, 309
HULTEC 177-8

IBM corporation 243
ICBMs 21, 23-4, 36, 38-40, 46, 92, 110, 122, 193
IDCSP (Initial Defense Communications Satellite Programme) 75, 82, 84
IFF (Identify Friend or Foe) 279, 286
Ikar booster 120
Il-4 bomber 133
Illustrious, HMS 52
Ilyushin bureau 140
Independence, USS 221
Indigo programme 98
INMARSAT (International Maritime Satellite Organisation) 257-9, 262, 264
Inner Air Battle 237
INTELCAST 251
intelligence-gathering 67, 95-7, 100-11, 120-2, 161-6, 188, 291, 303-5, 307, 309-10
Intelsat series 76, 264
internet 250, 259-60
Intruder system 110
Iowa class (U.S.) 219
IR (infra-red) sensors 33, 93, 110, 113-14, 117-18, 123, 203, 207, 242-3, 245-6, 276, 297, 311
Iran 108
Iraq 7-8, 34, 98-9, 221, 225, 227-9, 266-7, 271-2, 276-7, 283, 290, 295-6
  see also Gulf War
IRBMs 21, 36, 38
Ireland 50
Iridium satellites 11, 35, 40, 261-2
IS (Istrebitel Sputnikov) 113-14, 156
Isaev, Aleksei M 134
Israel 34, 142, 244

IT-21 system 206, 256, 259-61, 292, 317
Italy 86
ITP (Integrated Terminal Programme) 257
ITW (Integrated Warning/Attack) system 246
IUS (Initial Upper Stage) 22-3, 30

jamming 68-9, 72-3, 79-81, 85, 115, 170, 188, 237-8, 280-1, 281
Japan 20, 26, 51-2, 76, 79, 86, 175
JCTN (Joint Composite Tracking Network) 246, 299
JDAMS (Joint Direct Attack Munition System) 275, 303-4
JMCIS (Joint Command Information System) 222
JMCOMS (Joint Maritime Communications System) 256-8, 317
JMPS (Joint Mission Planning System) 287
'Joe' *see* PAMOR
John F Kennedy, USS 217, 219, 259
Johns Hopkins University 35, 48
Johnson Island 124
Johnson, President Lyndon B 56, 75, 77, 92
Joint Vision 2010 7, 301-2, 305-6, 311, 315
Josephus Daniels, USS 219
JOTS ('Jerry O Tuttle System') 220-3, 227-8, 253-4, 317
  JOTS II 270
Jouett, USS 221
JRSC (Jam Resistant Secure Communications) 80
JSAT 86
J-STARS (Joint Surveillance Targeting Attack Warning System) 245, 247, 253, 276
JTAGS (Joint Tactical Ground Station System) 244-5, 284, 297-8, 313
JTIDS 240
Jules Verne Launcher 35
Jumpseat satellites 104, 108;
  Advanced Jumpseat 109-10
Juno series rockets 21
Jupiter IRBM 21
JWICS (Joint Worldwide Intelligence Communications System) 256

K-10 missile 143-4, 147, 149, 152
  K-10S 143-9, 150-2, 234
K-14 system 14
K-16 system 146
K-22M system 169
K-26 missile152
K-56 (Rus) 133
Kaliningrad 118
Kamchatka 105
KB Arsenal bureau 162
KB-1 bureau 113, 135-6, 154-6, 163
KB-51 bureau 136
Keldysh, M V 39

Kennan satellite *see* KH-11
Kennedy, President John F 55-6, 255
Kg-22M 170-1
KGB, the 111, 122
KH ('Key Hole') satellites 30-1, 93-7, 104, 108, 116
    KH-11 ('Big Bird') 22, 95-8
Kh-15 missile 171, 236
Kh-20 missile 145, 149
Kh-22 missile 150-2, 169-71, 234, 243
    Kh-22M 236
Kh-23 missile 234
Kh-35 missile 168
Kh-45 missile 168
Kh-2000 missile 171
Khrushchev, Nikita 36, 38-40, 56, 93, 117, 122-3, 132, 139-40, 144, 146, 150, 155, 168-9, 306
Khrushchev, Sergei 38, 139
Kiev 110, 154, 162
Kiev class (Rus) 166-7, 172, 226
Killiam Commission 88-9
Kingpost, USS 77
Kingsport, USS 76
Kirov class (Rus) 99, 167, 172, 226
Kissinger, Dr Henry 192
Kitty Hawk, USS 211, 217, 259
Klakring, USS 260
KM space station 41
Kobalt radar 119, 149
    Kobalt-N 137, 149
Kometa system 136-7, 145-6, 203
Kometa-10 missiles *see* K-10
Koptev, Yuri 97
Korea 267
Korean War 131, 137, 216
Korolyev, Sergei P, and bureau 35-41, 43, 113, 115, 117
Korund command system 165
Kourou launch site 14
Kozlov, Dmitri I 118
Krasnyi Kavkaz (Rus) 137
'Kresta I' class (Rus) 225-6
'Kresta II' class (Rus) 225-6
Kristal communication system 164
'Krivak I' class (Rus) 225-6
KRM 146
Krug global sensor 147, 149, 175, 192, 194
'Krupny' class (Rus) 139, 141-2
KS missile 135-8, 142-5, 149; KS-1 145-6, 149
KSR missiles 145-6;
    KSR-2 145-6, 149, 152, 234
    KSR-5 152
    KSR-11 146, 152
KSS coast defence system 139
KSShch missile 139, 141
Kuban television system 119
Kudriavtsev, I V 154
Kust radio receiver 120
Kuznetsov class (Rus) 10, 167, 226

Kwajalein 123-4
'Kynda' class (Rus) 225-6

La Salle, USS 80
Lacrosse satellites 95, 98
Lake Erie, USS 297-8
LANDSAT 98
Landstuhl 79
LANTIRN IR sensors 276
Lanyard satellite 94
lasers 19, 41, 114-15, 187, 202, 275, 298, 312
LASM (Land Attack Standard Missile) 275
'Lastochka' missile 133
LCAC landing craft 278, 294
LEAP missile 296-7
Leasat-5 satellites 86, 190
Lee, Peter 202-3
Legenda system 162-3, 168
Lehman, John 230, 234-6, 239-40
Leningrad 99-100, 132
LES satellites 183, 185
Libya 8, 291
Lightsats 265
Link 11 system 52, 60-1, 74, 149, 185, 188, 194, 211-12, 217-18, 223, 228-9, 238, 257, 298, 316
Link 14 system 61
Link 16 system 257-8, 289-90
littoral warfare 282-300, 310-11
LK-1 manned lunar orbiter 40
LKS spaceplane 42
Lockheed Martin Missiles and Space corporation 22-3, 32-3, 91, 188, 200, 217, 223, 242, *see also* Titan
Long Beach, USS 219
LORAINE ballistic missile 241
Loral corporation 31, 44, 86
Loran system 47, 49-51, 276, 278
    Loran A 47
    Loran C 47, 50-1, 280
Lorenz system 47
Los Angeles class (U.S.) 62, 209-11
Lourdes (Cuba) 103, 121
LPI (Stealth) 66, 204
Luch satellites 40
Luna 37

MACSATS (Multiple Access Communication Satellites) 265
MAD (Magnetic Anomaly Detection) 201
Magnum programme (Argus) 109
Makeev bureau 38
Malakhit missile 145
Malinovsky, Marshal 117
manned spacecraft 27-8, 37, 40, 89-91, 115, 117-18, 204
Marconi Space and Defence Systems 83
Marisats (Maritime Commercial Satellites) 185
Maritime Strategy 162, 191, 230, 233, 234-7

Martin-Marietta company 241
Martlet projectile 34
Mashinostroyeniy bureau 160
Maverick missile 274
Mayaguez (U.S.) 57
Mayo, Reid D 101
McDonnel-Douglas corporation 32
McNamara, Robert S 75, 77, 103, 123-4, 232
Mentor programme 109
Menwith Hill 107
Menzel, Donald 101
Mercury programme 23, 27, 107
Meteor weather satellite 122
MI-110 sensor 203
MIDAS (Missile Detection Alarm System) 25, 92-3, 242
Midway, USS 77, 219
MiG bureau 41-2, 113, 136, 136-8, 143-5
Mikoyan, Anastas 138
Mikoyan, Artom 138
Milstar satellites 23, 65-7, 69, 82, 197-200, 262-3
    Milstar 1 199
    Milstar 2 199
    Milstar LDR 73
    Milstar MDR 73
mine countermeasures 277-8, 295-6
Minitrack receiver system 126-7
Minuteman missile 38
Mir space station 42, 118
    Mir-2 41
MIRV capabilities 44
Mishin, Vasiliy P 41, 118
Mission 22 191
Mitchell, Brigadier-General Howard 313-14
Mitsubishi 86
MKS system 42
MMBA (Multifunctional Multi-Beam Broadband Antenna) 257
MOL (Manned Orbiting Laboratory) 23, 91
Molniya orbits 17, 65, 95, 104, 108-10, 165, 199, 243
Molniya satellites 17, 36-7, 42, 153, 162-6, 168-9, 196
    Molniya-1 163, 165
    Molniya-2 162, 165
    Molniya-3 165
    Molniya-M 36-7
MOMS (Map Operator and Maintenance System) 287
Monino 41
Moon, the 16, 20, 27, 40-1, 64, 101, 117
    Moon landing 26, 44, 76, 118
Moskit missile 145
Motorola company 262
Mount Whitney, USS 80, 291-2
MRBMs 36, 38
MUOS (Mobile User Objectives System) 263
MX system 193, 200-1

N-1 booster 40-1
Nadezhda navigational system 164
NASA (National Aeronautics and Space Agency) 21-2, 28, 30-1, 34, 95, 112, 206-7
National Security Space Systems Architect 313
NATO 53, 61, 74, 82-3, 84-6, 85-6, 150, 154, 162, 174, 224, 225, 230, 232-3, 235, 289, 305-6, 315-16
NATO satellites:
    NATO-I 84
    NATO-II 84-5
    NATO-III 85
    NATO-IV 27, 85
Naval Air Weapons Center, China Lake 289
NAVDACs 51
navigation 8-11, 46-53, 163, 266-91, 302
NAVIXS 254-5
NAVSPASUR (Naval Space Surveillance System) 195
Navstar satellites 27, 266-7
NAVWEPS (Naval Weather and Environmental Prediction System) 112
Navy Wide Internet 259
NCTAM communication system 251, 259
NEMO (Naval Earth Map Observer) 208
NESP (Navy EHF Satellite Communications Programme) 197
NEST 82
NETCAP (National Exploitation of Tactical Capabilities) 213
net-centric warfare 7-9, 129-73, 205-6, 311, 317
Netherlands 52, 69, 189, 228
New Zealand 33
NFCS (Naval Fire Control System) 295
NICS (NATO Integrated Command System) 85
NII bureaus 35, 134, 139, 154, 157, 162-3
Nike Zeus missile 123-4
Nike-X missile 124-5
Nikolaev 10, 99
Nimitz, USS 219, 272
NIPRNET (Non-classified Internet Protocol Router Network) 256
Nixon, President Richard M 28, 192-3
Noginsk 162
NORAD (North American Air Defense) Command 299, 312-13
North Korea 297
Norway 63, 170, 232, 235
NOSC 254
NOTS (Naval Ordnance Test Station) 32-3, 125
Notsnik project 32-3, 125
NPO Energiya bureau 41
NPO Lavochkin bureau 119
NRL (Naval Research Laboratory) 64, 74-5, 77, 101-3, 126-7, 254, 266
NRO (National Reconnaissance Office) 96, 103, 265, 273, 313-14
N-ROSS (Navy Remote Ocean Sensing

Satellite) 207-8
NSA (National Security Agency) 96, 104-6, 110-11, 215-16, 237
NSC (Naval Space Command) 227
NSCP (National Space Communications Programme) 264
NSWPC (Naval Strike Warfare Planning Center) 287
'NTM' ('National Technical Means') 95, 114
NTW (Navy Theater Wide) programme 296-7
nuclear power and propulsion 18, 117, 144, 157, 160
nuclear weapons 9-10, 19, 36, 53, 113, 122-5, 132-3, 138, 140-1, 192-3, 230, 236, 264, 273, 305, 308. *see also* Tomahawk *and individual weapons systems*
Nudelmann cannon 118
Nurrungar (Australia) 243
NUS (No Upper Stage) satellites 23
NWC (Naval Weapons Centre) 269

OAB (Outer Air Battle) 8, 230, 237-9, 240-4, 249, 252, 266, 294
OASIS (OTH-T Airborne Sensor Information System) 229
Ocean Surveillance System 230
Ogarkov, Marshal Nikolai 305-6
O'Grady, Captain Scott 286-7
OK space shuttle project 41
OKB bureaus 35-6, 113, 115, 118, 133-6, 138-9, 155-6
Okean naval exercises (Rus) 159, 171-2, 181, 192
Okean spacecraft 161
OK-M spaceplane 42
Oktan satellite 116
Omaha, USS 228
Oman 176
Omega navigation system 51
OODA cycles 131, 305-7
Optus satellites 86
orbits 15, 17-18, 20, 25, 31, 41, 75, 122-3, 125, 157, 160
    geostationary 15-16, 86, 106-7, 242
    geosynchronous 15-16, 22-4, 26, 29, 40, 43-4, 65, 74-6, 78, 85, 95, 104, 109, 113, 121-2, 127, 166, 183, 245, 309-10
    inclined 16-17, 75-6, 102, 104, 120-1, 127, 162, 207
    low 22-4, 26, 33, 43-4, 65, 75, 89, 95-6, 102-3, 110, 116, 127
    transfer orbits 22, 24, 26-7, 32-3, 43-4, 127
    see also Molniya orbits *under* Molniya
Orion programme 109
Orlets satellites 119
Orlov, M V 141
OS spaceplane 41
OSIS (Ocean Surveillance Information System) 175, 177-82, 188, 211-13, 217-18, 223
OTCIXS (Officer-in-Tactical-Command Information Exchange System) 212, 219-20, 223-4, 228-9, 254, 270
OTH-T (Over-the-Horizon Targeting) 8, 142, 145, 180-2, 188-9, 211-12, 214, 218-21, 223-7, 231, 237, 239-40, 242-3, 249, 252, 254-5, 266, 278, 288, 294-5, 310, 316
Outboard HF/DF system 176
Outer Space Treaty 125

P-3 aircraft 184, 228, 252, 288
P-5 missile 140, 142, 144-5
P-6 ('Shaddock') missile 135, 142-3, 153, 166-7, 173, 180, 192, 210, 225
P6M Seamaster bomber 154
P-10 missile 140
P-15 ('Styx') missile 142, 145, 153, 214, 225
P-20 missile 140
P-35 missile 135, 142-4, 153, 167, 173, 180, 192, 210, 225
P-40 missile 144
P-500 Granit missile 166-7
Packard, David 195
PAMOR (Passive Moon Relay) 101
Parus system 164-5
Patriot missile 244, 300
Pe-8 bomber 133, 136
Pegasus rocket 33, 265
Pelton, Roger 105
Perry class (U.S.) 257
Perry, W 265
Philco-Ford 83
Phoenix missile 152, 233-4
PID (Passive Identification Devices) 214
Pine Gap 108-9
Planet One telephone service 258
planetary exploration 39-40, 44, 157
Plesetsk 162, 164
Pleumont-Boudou 76
Polaris submarines 21, 46-51, 54-5, 62, 88, 94, 106, 122, 269
Polyot system 37, 113
'Poppy' programme *see* Grab
Port Royal, USS 297-8
Poseidon missile 51, 195, 269-70
Potok satellites 40, 119
Program 505 *see* Project Mudflap
Project Babylon 34
Project Compass Link 77
Project Corona *see* Corona
Project Mudflap 123
Project Pilot 33
Project Shepherd 126
Project Tattletale 101-2
Proton boosters 30, 39-40, 42-3, 117, 121
    Proton-K 119
    Proton-KM 40
Providence, USS 77

Index 381

PSK (Phase-Shift Keying) 70
PTPS system 287
Pueblo, USS 57

Quicksat programme 80-1

R-1 missile 35
R-4 missile 36
R-5 missile 36
R-6 missile 24
R-7 missile 21, 36-7, 40, 92, 113, 115, 158
    R-7A 37
R-12 missile 36, 38
R-14 missile 36, 38, 120
R-16 missile 38
R-26 missile 38
R-36 missile 38-9, 114, 156
R-46 missile 39
R-56 missile 41
R-790 missile 164
Radiant Clear programme 296
Radiant Ivory project 243
Radiant Mercury 288
radio:
    EHF bands 64, 66, 82-3, 187, 191, 197-9, 199-200, 257-9, 262-4, 298, 310
    HF bands 54-5, 57-64, 64, 71, 82-3, 104-5, 149, 166, 172, 175-6, 179, 182, 184, 194, 214, 224-5, 249, 252, 257, 299
    HF/DF bands 57, 63, 132, 147-9, 168, 175-6, 201, 214, 254
    jamming 68-9, 72-3, 79-80
    LF bands 57, 61-2, 152, 257
    MF bands 57
    SHF bands 64, 66, 69, 76-7, 81-3, 85, 122, 183, 185, 187, 191, 258-9, 262-4, 284-5, 310
    UHF bands 11, 60-1, 64, 66, 69, 77, 80, 83, 85, 106-7, 152, 164-5, 179, 183-8, 191, 195, 200, 206, 212, 213, 224, 251, 255-8, 261-2, 264-5, 278-9, 284, 293, 298
    VHF bands 60, 64, 69, 164
    VLF bands 62, 184, 210
Raduga bureau and satellites 40, 109, 122, 136, 168, 171
    Raduga/Potok system 166
RAFOS navigational system 48
Rainfall equipment 108
Raketoplan 39-41
Ramrod system 109
RAND research institute 91
Ranger, USS 219
Raspletin, A A 155
RAT-22 torpedo 130
RAT-52 torpedo 135
RB rocket 41
Reagan, President Ronald 57, 205, 219, 230, 234-5, 267, 305
reconnaissance 27-30, 33, 37, 44, 88-112, 115-20, 126-8, 155-66, 171, 173-88, 194-7, 200-8, 230, 285-94, 301, 305-7, 311
    space surveillance 126-8
REDUX navigational system 47-8
Regulus missile 47, 50, 140
Rhyolite satellites 107-8, 110
Rickover, Admiral H G 210
Rockwell corporation 266
RORSAT 157, 161, 196, 239
ROTHR (Relocatable Over-the-Horizon Radar) 239-42, 268
RTIC (Rapid Targeting in the Cockpit) 287-90, 290
Rubikon radar 146
Rubin 149
Rucheiy command system 165
Ruffer ground processing system 108, 110
Runway system 107
Russia 8, 10, 14, 18, 20-1, 30, 32-44, 47-8, 51, 53
    communications 40, 131-3, 163-6, 188
    guidance 38, 143-4, 168-71
    intelligence-gathering 120-2, 161-6
    navigation 163
    reconnaissance 37, 97-8, 115-20, 129-72, 155-66, 171, 194-7, 200-6, 230, 305-6;
    targeting 39, 147-55, 157, 166
    tracking 166, 192

S-5 missiles 140
S-35 missiles 142
SABER (Situational Awareness Beacon and Response) 278-80
SAC (Strategic Air Command) 87-8
Saenger, Eugen 28
Safeguard system 125
SAINT (Satellite Inspection and Neutralisation) programme 104, 123
Salt Lake City, USS 228
SALT (Strategic Arms Limitation) treaties 95, 108, 114, 193
Salyut space stations 118, 204
SAMOS (Satellite and Missile Observation System) 92-3
Sampson radar 69
SAR (Synthetic Aperture Radar) 177-8, 206-7, 248
SATRAN (Satellite Reconnaissance Advanced Notice) 128
Saturn rockets 20, 26, 28, 40
Saudi Arabia 244, 284, 296
Savage aircraft 148-9
SBIRS (Space-Based IR System) 245-6, 298
SCORE (Signal Communication by Orbiting Relay Equipment) 75
SCOT receivers 84-6
Scotop receivers 100
Scout rockets 122
Scud missile 244, 267
SDI (Strategic Defense Initiative) 35, 104, 205
SDS (Satellite Data System) 95, 191, 264, 308

Sea Eagle missile 189
Sea Sparrow missile 174
Seaplan 2000 233-4
Seasat 202-3, 204, 206-7
SECOMSAT (Sistema Espanol de Communicaciones Militares por Satelite) 86
SEI (Specific Emitter Identifier) 226-7
'self-consistent gridlock' 53
Sentry programme 93
Serbia 283, 286-7, 289, 303, 306, 315
Severodvinsk 99-100
SGEMP (System Generated Electromagnetic Pulse) interference 19
'Shaddock' *see* P-6 missiles
Shchuka missile 135, 138, 141
Sheffield, HMS 84
Shiloh, USS 298
Shrike missile 195-6
Shtorm missile 135, 138
Sicral (Sistema Italiana de Communicazione Rizervante Allarmi) system 86
Sidewinder missile125
Signaal SCOUT radar 310
'Singleton' programme 104
SINS (Ships Inertial Navigation System) 50-1, 268, 276
SIOP (Single Integrated Operational Plan) 88, 102
SIPRNET (Secret Internet Protocol Router Network) 206, 256, 261
'Siss Zulu' programme *see* Grab
Six-Day War 173, 183
Skoriy (Rus) 135
Skybolt missile 168-9
Skylab space station 18, 118
Skynet 82-6, 183
Skyray aircraft 33
Skywarrior aircraft 149
SLAM missile 274-5, 286, 289
Slava class (Rus) 166-7, 226
SM (Standard Missile) 240-1, 296-7
Socotra Island 59
SOFAR navigational system 48
solar batteries 117-18
solar panels 18, 118, 157, 161, 165
SOLRAD satellites 102-3
SOSUS (Sound Surveillance System) 175, 179, 201
Souda Bay 60
Sovremennyy class (Rus) 153, 226
Soyuz spacecraft 36-7, 40-1, 43, 119
    Soyuz 7K-LOK 117
    Soyuz 7K-VI (Vesta) 37, 117
    Soyuz K7-S 118
    Soyuz VI 117-18
    Soyuz-R 117
Space Command (U.S.) 11, 299, 312-14
Space Communications Corp. of Japan 86
Space Shuttle, the 14, 16, 21-2, 27, 27-32, 29-31, 34, 41-2, 44, 78-9, 79, 96, 98, 109, 115, 206, 265, 267
Space Systems corporation 86
SPACE TRACK data centre 126
Space Transportation Systems 27-8
spaceplanes 27-9, 41-2, 44
Spain 85-6, 175
Sparrow missile 122
Spartan missile 195
SPASUR (Space Surveillance System) 126-8, 157, 227
SpaWar Command (NSEC) 249-50, 254, 280, 291
Special Forces 79-80, 276-7, 280
SPINS (Special Instructions) 283, 285
Spiral programme 34, 41-2
'Spook Bird' satellite 106
Spruance class (U.S.) 219, 257
SPS-49 radar 195
Sputnik satellites 3, 32, 36, 37, 48-9, 92-3, 126, 158;
'spy satellites' *see* strategic reconnaissance
SPY-1 radar 244, 297-8
SSIXS link 210-13, 228
SSN (Space Surveillance Network) 128
Stalin, Josef 35-6, 105, 132, 134, 140
STAM rocket 209-10
Starlite satellites 247-8
Starsem consortium 43
STOM (Ship to Objective Manoeuver) 278
Straight-8 rockets 24, 26
'Stray Cat' 128
Strela system 138-9, 164, 164-5
    Strela 2 121,165
    Strela 3 164-5
Stump Neck 101
'Styx' *see* P-15
SUBROC rocket 209
Success system 154
Sudan 7, 303
Sukhoi bureau 168
Sun, the 16, 30, 42-4, 96
Superbird satellites 86
'Surf' missile 133
'Surgut' system 165
Sverdlov class (Rus) 133, 135, 139, 172
Svinets missile launch detector 117
SWRS (Slow Walker Reporting System) 93, 242-5, 254
Syncom satellites 25-6, 76-7
Syracuse satellites 85, 190

T-4 bomber 168-9
TACAMAO aircraft 62
Tacan beacons 279
TACINTEL (Tactical Intelligence Circuit) 212, 251
Tacsat programme 183-5, 197
TAD (Thrust Augmented Delta) 26
TADIXS (Tactical Data Information Exchange System) 182, 189, 211-13, 249,

254-6, 270, 292-3, 314
TALC (Tactical Airborne Laser Communication) system 187-8
TAMPS (Tactical Aircraft Mission Planning System) 285-7, 290
targeting 8, 10, 315
  Russian 39, 147-55, 157, 166; U.S. 7-8, 81, 87-8, 91-2, 99-100, 209-29, 269-73, 283-95, 302-7. *see also* OTH-T
TARPS 291
Tascat satellites 183
TAT (Thrust-Augmented Thor) 25-6
Taurus 33
TDMA (Time-Domain Multiple Access) 72-3
TDRSS (Tracking and Data Relay Satellite System) 95
TEAMS (Tactical EA-6B Mission Support System) 286-7
Telecom satellites 85
telephone systems 110, 121, 165, 258, 261-2, 309
television systems 76, 89-91, 95-6, 119-20, 123, 166, 264, 292; T1 channels 252, 291-3
TELs (Transporter-Erector-Launchers) 284
Telstar I 76
TENCAP (Tactical Exploitation of National Capabilities) programme 111, 182, 213-16, 217, 225, 227, 244, 271, 284, 314
TERCOM (Terrain Comparison) system 268-72
Termit-M missile 145
TERS (Tactical Event Reporting System) 243
TES (Tactical Event System) 244-5, 284
TFCCs (Task Force Command Centers) 188-90, 210-12, 217-23, 225, 256, 272
TGR television system 119
THAAD (Theater High Altitude Defense) missile 297-300
Thailand 176
Thor rocket 21, 24-6, 91-2, 94, 123-5, 195
  Long Tank Thor 24, 26;
  LTTAT (Long Tank Thor Augmented Thrust) 26
  Thor Able 25
  Thor Able Star 25, 102;
  Thor Agena 25-6, 104
  Thor ASAT 124-6
  Thorad 26
TIBS (Tactical Information Broadcast System) 213-14, 293
Timation 266
TIROS (Television and Infra-Red Observing Satellite) 112
Titan boosters 21-2, 24, 28-9, 43, 96, 107, 109
  Titan-2 208
  Titan-34D 22, 79
  Titan-401 22-3
  Titan-402 22-3
  Titan-403 23
  Titan-404 23

Titan-405 23
Titan-II 21-3, 104
Titan-III 22-3, 75, 94
Titan-IV 22, 22-3, 32, 109
TMPCs (Theater Mission Planning Centers) 270, 272
Tobol (Rus) 163
Token signals 103
Tokyo 76
Tomahawk missile 7, 7-8, 8, 10, 81, 98, 182, 193, 197, 210-12, 213-15, 217-20, 222-3, 225, 228, 236, 252-3, 264, 268-73, 281-2, 285-6, 290-1, 293, 302-4, 311, 314, 316
  Tactical Tomahawk 307
TOPSCENE system 286
tracking 8, 126, 166, 175, 181, 192, 200, 221, 228, 239-48, 309
Transit navigational system 18, 25, 46, 49-51, 88, 94, 102, 163, 226, 266, 268
  Transit 1A satellite 49
  Transit 1B satellite 49
  Transit 5A-1 satellite 49
TRAP (Tactical Receive Equipment Applications) 182, 213, 293, 314
TREE (Transient Radiation Electronic Effect) 19
Trexler, James 101
Trident missile 51, 121, 193
Trimilsatcom system 85
Tripoli (Libya) 8, 59
TRIXS (Tactical Reconnaissance Exchange System) 213
Trumpet system 108-10
TRW corporation 108
TSCMs (Tactical Strike Coordination Modules) 287
Tselina satellites 120-1
Tsikada navigational system 164
TsNII bureau 161-2
TsUKOS command system 165
Tsunami navigational system 163-4
Tsyklon boosters 38-9, 114, 116, 156, 162-5
  Tsyklon-2 39, 156, 158, 196
  Tsyklon-3 39, 120
  Tsyklon-M series 164
Tupolev bureau 168-70
Tu-2 aircraft 133-4
Tu-4 aircraft 135, 137
Tu-16 ('Badger') aircraft 131, 137-8, 143-50, 152, 154, 168, 170, 230, 237
Tu-22 aircraft ('Blinder') 143, 150, 150-2, 168-9, 171
Tu-22M aircraft ('Backfire') 169-71, 233, 235-7, 239-40, 242-3
Tu-22P aircraft 151
Tu-22R 151-2, 171
Tu-95 ('Bear') aircraft 143, 145, 149, 154, 166-7, 174
Tu-105 aircraft 143;
Tu-142 aircraft 167-8

Tu-160 ('Blackjack') aircraft 10
Tuttle, Admiral Jerry O 220, 221-2, 225, 249-50, 252-4, 271, 273
TV-DTS (Television Direct to Sailors) 264
TWCS (Tomahawk Weapon Control System) 219-20
'Typhoon' class (Rus) 100, 206
Tyuratam site 14, 42-3, 162, 196

U-2 spyplane 87, 89, 91-3
UAVs (Unmanned Airborne Vehicles) 281, 289-90, 294, 301, 307
UFO system 19, 86, 262
Ukraine 45, 173
Universal Modem 69, 73, 264
UNT (Unified Networking Technology) 254
UR boosters:
  UR-100 38
  UR-200 37-9, 113-14, 156
  UR-500 (Proton) 38-40, 117
  UR-700 38-41
Uragan spaceplane 42, 115
U.S. Army 11, 21, 64, 75-6, 103, 123-4, 126, 191, 95, 213, 233, 244, 255, 263, 274, 276, 295, 310, 312-13
U.S. Marine Corps 11, 80, 227, 278-9, 282-4, 294, 296, 310-11, 313
US satellite series (Rus) 38, 113, 156
US-A 157-62, 196;
US-M 162; US-P 157, 160-2, 195-6;
US-PM 162;
US-PU 162
United States of America. 7-14, 18-35, 41-5
  communications 54-64, 66-7, 70-83, 95, 182-91, 197-200, 237-8, 249-65, 289-94, 298-300
  guidance 24, 268-77, 286
  intelligence-gathering 67, 95-7, 100-11, 291
  navigation 46-53, 266-91
  reconnaissance 27-36, 33, 88-112, 126-8, 173-88, 200-3, 205-6, 285-94, 301
  targeting 7-8, 81, 87-8, 91-2, 99-100, 209-29, 269-33, 283-95, 302-7
  tracking 126, 175, 181, 200, 221, 228, 239-48, 309
USAF 7, 10-11, 21-3, 25, 27-30, 32, 54, 75, 78, 80, 87-9, 91-3, 100, 102, 104, 106, 122-6, 128, 131, 148-9, 182-3, 185, 191, 193, 195-6, 200-1, 204, 212-13, 227, 233, 233-4, 241, 243-7, 255, 263, 274-6, 281, 286-7, 298, 312-15;
Uspekh radar link system 153-6

V-1 missile 133
V-2 missile 35, 88, 102
V-3 gun 34
Vandenberg AFB 13, 23, 30-1, 42, 109
Vanguard boosters 21, 25, 33, 91-2, 101-2, 126-7

Vector system 139
Vega system 110
Venus 40
Vesta 117
Vietnam 86
Vietnam War 56-7, 76-7, 95, 110-11, 174, 182-4, 215, 216, 230-2, 274
Vigilante aircraft 149
Vinson, USS 260
Vladivostok 105, 291
'Volna' missile 133, 135-6, 138-9
von Braun, Werner 21, 36
Vortex programme 110;
  Advanced Vortex 107-8
  see also Chalet programme
Voskhod boosters 36-7, 117
Vostok programme 36-7, 115, 117
Vulkan missile 167

Walker, John 56, 213
Walleye bomb 274-5
Warsaw Pact 98
weather 111-12, 122, 155, 185, 208, 228, 315
WHCA (White House Communications Agency) 79
'Whiskey' class (Rus)144-5, 163
White Cloud programme 104-5, 122, 176-80, 200, 214-18, 225-6, 245
Willow programme 108
WS-117L system 89, 91
WSC receivers 81, 85, 184, 186, 256-7, 285, 292-3
WWMMCCS (World-Wide Military Command and Control System) 255

X-20 Dynosaur 28
X-33 booster 34
X-37 booster 34
X-40 booster 34
X-band radars 153, 157-8, 293

Yangel, Mikhail K, and bureau 36-9, 41-3
Yantar satellites 116
  Yantar-2K 116, 118
  Yantar-4K 119
YeN-D radar 146, 149, 152
Yield programme 108
Yugoslavia 290, 304, 316
see also Serbia

Zelenodolsk 99
Zenit boosters 32, 37, 41-3
  11K77 (SL-16) 42-3
  Zenit-2 37, 115, 119, 121
  Zenit-4 37, 115-16
  Zenit-6 116
Zhukov, Marshal 140
Zircon programme 108
Zumwalt, Admiral Elmo 209-11, 230-3, 235